Cognitive Technologies

For further volumes:
http://www.springer.com/series/5216

Oleg Anshakov · Tamás Gergely

Cognitive Reasoning

A Formal Approach

With 6 Figures and 13 Tables

Springer

Authors
Prof. D.Sc. Oleg M. Anshakov
Institute for Linguistics
Intelligent Systems Department
Russian State University
for the Humanities
6 Miusskaya Square
Moscow, 125993
Russia
oansh@yandex.ru

Prof. D.Sc. Tamás Gergely
Applied Logic Laboratory
H-1022 Budapest
Hankoczy u. 7
Hungary
gergely@all.hu

Managing Editors
Prof. Dov M. Gabbay
Augustus De Morgan Professor of Logic
Department of Computer Science
King's College London
Strand, London WC2R 2LS, UK

Prof. Dr. Jörg Siekmann
Forschungsbereich Deduktions- und
Multiagentensysteme, DFKI
Stuhlsatzenweg 3, Geb. 43
66123 Saarbrücken, Germany

Cognitive Technologies ISSN 1611-2482
ISBN 978-3-540-43058-2 e-ISBN 978-3-540-68875-4
DOI 10.1007/978-3-540-68875-4
Springer Heidelberg Dordrecht London New York

Library of Congress Control Number: 2009940796

ACM Computing Classification (1998): I.2, F.4, F.1, F.2, H.1, I.1, J.4

Cover design: KünkelLopka GmbH, Heidelberg

Printed on acid-free paper

Springer is part of Springer Science+Business Media (www.springer.com)

Preface

Understanding cognition and modelling, designing and building artificial cognitive systems are challenging and long-term research problems. In this book authors wish to take a step towards a better understanding of cognition and its modelling by providing a well-founded integrated theory. Though this book is primarily not on cognitive system development, the need for an account of a genuine, logically well-founded formal framework seems to us especially important for system designers too. The authors' belief is that the practice of even those who most explicitly reject the need for theoretical foundation of intelligent systems on a constructive cognitive reasoning framework inevitably involves tacit appeal to insights and modes of reasoning that can be understood, designed and implemented.

This book had a long prehistory intertwined with the development of an important new method of plausible reasoning. The authors both participated in the development of the plausible reasoning technique which was called by its initiator V. K. Finn the JSM method of automatic generation of hypotheses in honour of the British thinker John Stuart Mill. This development has a long history of several decades, having already started in the former Soviet Union at the All-Russia Institute of Scientific and Technical Information (VINITI), Russian Academy of Sciences, Moscow. Some intensive periods of this development took place in cooperation with VINITI and the Applied Logic Laboratory, Budapest, Hungary.

The authors not only participated in the development of this method but also played key roles in the elaboration of the main results of this method.

Earlier, around the millennium, a book was planned to be written on the results obtained in the area of plausible reasoning provided by this method. Intensive and productive joint work was started with the participation of V. K. Finn, T. Gergely, S. O. Kuznetsov and D. P. Skvortsov on the preparation of the manuscript. It proved difficult at first, because coherence among the different viewpoints and techniques was not easy to establish. Soon after O. M. Anshakov joined the author team, replacing D. P. Skvortsov. The authors of this present book felt a pressing demand for the development of a new integrated approach that could provide a logic foundation for the entire area of cognitive reasoning. While working on the manuscript this feeling became stronger and stronger. Finally, having almost completed the manuscript, work on it was suspended and the two authors started to realise their general goals. This took place around 2006. From then on the majority of the work was done in Budapest at the Applied Logic Laboratory. Authors wished to contribute to a better understanding of cognition and its modelling by developing a formal approach. This could provide a platform on which various existing approaches could be presented and handled.

Acknowledgements

We appreciate the long-standing cooperation and fruitful discussions we have had with Professor V.K. Finn and his research team.

We also thank the Applied Logic Laboratory for providing all the facilities and conditions for writing this book.

Special thanks go to Springer for its endless patience in every sense, as well as to the team that supported us to shape the final version.

Budapest–Moscow,
October 2009

Tamás Gergely
Oleg Anshakov

Contents

Part IV Handling Complex Structures

Part V JSM Theories

Chapter 1
Introduction

Recently the creation of artificial cognitive systems has been one of the main challenges in the field of information and communication technology. Despite the many alternative approaches a complete and convincing artificial cognitive system has not yet been developed and, even further, it has not been known how to design and build artificial cognitive systems. This is why understanding cognition and modelling, designing and building artificial cognitive systems are challenging but long-term research problems. In this book we wish to take a step towards a better understanding of cognition and its modelling by providing a well-founded integrated theory together with methods and tools for designing and building artificial cognitive systems.

In order to understand the nature and scope of the difficulties of building artificial cognitive systems let us first consider what is this cognition that we expect artificial cognitive systems to realise.

1.1 What Is Cognition?

This term originates etymologically from the Latin root *cognoscere* meaning to become acquainted with. *Cognoscere* itself consists of *co-* + *gnoscere* = to come to know. In other words, cognition basically is gaining new information and knowledge by providing the missing information necessary to reduce uncertainty.

Cognition is the subject of various scientific disciplines such as psychology, cognitive science, philosophy and artificial intelligence (AI). However, the term "cognition" is used in various loosely related ways in these disciplines.

In **psychology**, it refers to an *information processing* view of an individual's psychological functions. According to the definition given in en.wikipedia.org, "In psychology the term *cognition* is used to refer to the *mental processes* of an individual, with particular relation to a view that argues that the mind has internal mental states (such as *beliefs*, desires and intentions) and can be understood in terms of *information processing*, especially when a lot of abstraction or concretisation is involved,

O. Anshakov, T. Gergely, *Cognitive Reasoning*, Cognitive Technologies,
DOI 10.1007/978-3-540-68875-4_1, © Springer-Verlag Berlin Heidelberg 2010

or processes such as involving knowledge, expertise or learning for example are at work". Many psychologists argue that cognition is merely the manifestation and/or the result of thinking. Moreover, cognition directly connects thinking and problem solving. Cognition — as a thinking process oriented towards problem solving — is involved in any human activity, be it practical, scientific or of any other type. Problem situations appear in any type of activities: cognitive, educational and in research and development. An important class of problem situations is characterised by incomplete and uncertain information. Therefore the solution process requires finding and generating the required information and processing it in order to find some solutions, even if only hypothetical ones. Some psychologists define thinking as problem solving, and Polya suggested that problem solving is based on cognitive processing that results in "finding a way out of a difficulty, a way around an obstacle, attaining an aim that was not immediately attainable" [Polya, 1954].

Cognitive science considers the term "cognition" in a more general sense, namely it aims to determine the main characteristics of the cognitive processes independently of the subject (human, animal or machine) that will realise cognition. Here cognition is related to information and regularity extraction, knowledge acquisition and improvement as well as environment modelling and model improvement. Thus, at its most general, cognition is the ability to reduce uncertainty (or entropy from the information-theoretic point of view), supporting the adaptation of the cognizing subjects to their environment by improving the models about their environment, by extracting new information and generating and improving their knowledge, and also reflecting their own activity and abilities in order to become ready to improve them. The improvement of the model of the environment permits understanding of how things might possibly be, not only now but at some future time too, and take this into consideration. Thus cognition also permits the use of the past to predict the future and then assimilation of what does actually happen to adapt and improve the anticipatory ability of the cognizing subjects.

Note that, in contemporary cognitive science, cognition is studied in a multidisciplinary way by using logic, psychology, philosophy and artificial intelligence.

In **philosophy** *epistemology* is the branch which is concerned with the origins, structure, methods and validity of knowledge and thus it also studies cognition as the ability to generate new information and improve knowledge. Here the central subject is how information discovery influences knowledge development in a general setting, i.e. cognition is interpreted in a social or cultural sense to describe the *emergent* development of knowledge and concepts within a group. Here epistemology without the cognizing subject is the central subject of study (see, e.g., [Popper, 1974]).

Scientific discovery is an important subject of the *philosophy of science*, which is closely related to *epistemology*. It seeks to explain, among other things, the nature of scientific statements and the way in which they are produced. Discovery is a way to produce such statements. Discovery in general means bringing to light something previously unknown. The following three types of activities belong to it: (i) exploration of an object or event which previously existed but was not known, (ii) finding new aspects (properties, way of use, etc.) of something already known, and (iii) introduction and creation of new ideal objects, or creation of new material objects. All these types are related to cognition. The first two types are connected with experimentation and data processing. The second type may be connected with theoretical investigations. The third type is related to theoretical investigations and constructions, e.g. by introducing new notions, while the last type is connected mainly with engineering activities. In the philosophy of science, discovery and evaluation are the subjects of investigation. To understand how scientists discover and evaluate hypotheses requires the consideration of cognizing subjects. Therefore here epistemology with cognizing subjects is the central subject of study [Thagard, 1993; Pollock, 2002].

In **artificial intelligence** research the development of intelligent systems is one of the main goals, in relation to which cognition is an important objective to study. Intelligent systems with cognitive ability are so-called cognitive systems. There have been various attempts to develop cognitive systems, which mainly differ in their approaches that reflect different understanding of cognitive processes. The three main ways of understanding are the cognitivist, the emergent and the hybrid approaches. The cognitivist approach attempts to create unified theories of cognition, i.e. theories that cover a broad range of cognitive issues, such as learning, problem solving and decision making from several aspects including psychology and cognitive science. The emergent approach practises a very different view of cognition. Namely, cognition is a process of self-organisation whereby the cognitive system is continually re-constituting itself in real time to maintain its operational identity through moderation of mutual system–environment interactions. The hybrid approach combines aspects of the emergent systems and cognitivist systems. These approaches give different proposals about the cognitive processes to be imitated. The main tool to study these processes is computational modelling, which explores the essence of cognition and various cognitive functionalities. The computational models provide detailed, process-based understanding of representation, mechanisms and processes. These models represent cognition by computer algorithms and programs; that is, they produce executable computational models.

The computational approach introduced new perspectives in the study of cognition not only in artificial intelligence but in all the other above-mentioned disciplines; e.g. we can develop computation models in order to understand how cognizing agents model their environment or scientists discover and evaluate hypotheses. These models employ data structures and algorithms intended to be analogous to generic cognizing processes in the case of cognizing agents, or to human mental representations and procedures in the case of scientific discovery. Computationally supported philosophy of science views topics such as discovery and evaluation as open to investigation using the same techniques employed in cognitive science. This means that study of cognition became multidisciplinary, and the computational models are often constructed by integrating sources from various disciplines, possibly including conceptual models from psychology and cognitive science, algorithmic solutions from artificial intelligence, philosophical ideas from epistemology and ontology, and so on. That is, the study of cognition requires the diversity of methodologies that these fields have developed. Regarding computational modelling see, e.g. [Sun, 2007b].

We can conclude that various ways of understanding cognition range from the extraction of new information and regularities to all mental processes, which of course includes even those related to information extraction. In this book we will use cognition in the following sense: (i) cognition extracts new information and relationships; (ii) the possible sources of information to be extracted are experimentation, systematic observation and theorising the experimental and observational results and knowledge; (iii) the extraction is carried out by the analysis of the initial data, generation of hypothesis, hypothesis checking and acceptance or rejection of hypothesis.

We are aware that the term "cognition" has a dual character. It stands for both intellectual processes and for the content of those processes. In this book we will consider both of these characteristics of this term and also the interplay between static contents and dynamic processes.

1.2 Cognizing Agents

Cognition is a very complex, multilevel and multidimensional ability, the realisation of which presupposes a system of high complexity. A system with this ability should perceive, understand, reason, learn, develop and handle its own knowledge through interaction with its environment. The main functional components of this ability are perception, reasoning, learning, remembering, using experience and communication. These functions need to be integrated into a coherent whole, since only in their wholeness can we get a truly cognitive system.

Thus it is expected that cognitive systems are able to understand their environments while operating either fully autonomously or in cooperation with other cognitive systems in complex, dynamic environments. Moreover, these systems should be able to respond intelligently to gaps in their own knowledge and to give sensible responses in unforeseen situations.

Understanding goes through perception, while responses are realised as actions. Therefore, interaction with the environment is realised through perception–action cycles, which can be continuously improved by cognition. This improvement is connected with the ability to learn, e.g. new procedures to accomplish goals, new plans of actions and methods to improve control and coordination of perception–action cycles. Thus cognitive systems have the capability to extract the necessary but missing information for their adaptation by learning about the physical and information environment through experimentation, systematic observation and theorising their results and pre-existing knowledge.

There have been various attempts to develop cognitive systems, which mainly differ in their approaches which reflect different understanding of cognitive processes. Vernon et al [2007] give an overview of the autonomous development of mental capabilities in cognitive systems, presenting a broad survey of the various paradigms of cognition, addressing cognitivist (physical symbol systems) approaches, emergent systems approaches, encompassing connectionist, dynamical, and enactive systems, and also efforts to combine the two in hybrid systems.

Various approaches differ also in the structure that is proposed for cognition, which is the so-called *cognitive architecture.* Cognitive architectures specify the underlying infrastructure for cognitive systems, which include a variety of capabilities, modules and subsystems. A cognitive architecture is also connected with an underlying theory of cognition. A cognitive architecture includes the generic functions that are constant over time and across different application domains and it explains how its capabilities and functioning arise out of the capabilities, functioning, relationships and interactions of the constituents. Thus this architecture is a relatively complete proposal about the structure of a cognitive system, particularly of human cognition. Cognitive architectures are connected with computational modelling. A cognitive architecture is a broadly scoped, domain-generic computational model of cognition which captures the essential structure and processes of cognition (see, e.g. [Sun, 2004, 2007b]).

There are various cognitive architectures that the reader can meet in the literature. Also there are good review papers that help the reader take one's bearings; see, e.g. [Vernon et al, 2007; Langley et al, 2008; Sun, 2007a]. An important issue that the reader confronts in a cognitive architecture is how it accesses different sources of information and knowledge and how it handles information. For example, information about the environment comes through perception; information about implications of the current situation comes through planning, reasoning and prediction; information and knowledge from other cognitive systems comes via communication; and knowledge from the past comes through remembering and learning. The more such capabilities an architecture supports, the more sources of information and knowledge it can access to improve its behaviour.

Further on we will use the term *cognizing agent* for cognitive systems, thus emphasising that it has the cognition ability. Thus cognizing agents will be the main actors in our approach. *cognizing agent* means either a living entity (particularly a human being), or a group of them or a technical system, which can *adapt* to the changing conditions of the external environment. In the adaptation process a cognizing agent perceives information (data) about the environment and generates reactions. How a cognizing agent will react in a given situation depends also on the cognition ability it implements, i.e. it depends on the peculiarities that cognition can provide linking perception to action. These peculiarities are connected with the information processing the cognition provides. Therefore the perception–action cycles of a cognitive agent are determined by the underlying information processing methods that are allowed by the cognition.

Learning and reasoning are important constituents of information processing that provide the ability to reduce uncertainty and to move from ignorance to knowledge. This is connected with information extraction from the initially available data and facts. For this information and knowledge extraction appropriate initial data are needed, which can be collected from different sources: (i) from observation and experimentation, (ii) from processing the available knowledge, (iii) from external information and knowledge bases, and (iv) from model experimentation based on the available knowledge. The extraction may result in new information which can then be used to augment the knowledge either about the environment by improving the model about it or about a problem domain. In the latter case the uncertainty connected with a problem situation will be reduced. In the first case a cognizing agent can use the improved model for a better understanding of how things might possibly be, not only now but at some future time, and to take this into consideration when determining how to act in the environment. Therefore cognitive reasoning provides the possibility of predictive activity for a cognizing agent beyond the usual reactive activity. Therefore cognition breaks free of the present in a way that allows a cognizing agent to act effectively, to adapt and to improve. For this, information processing is strongly intertwined with knowledge. However, a cognizing agent would not only need to reason with knowledge, but would also be required to remember its experiences and recall them exactly when needed as well as learning from them.

Therefore, a cognizing agent is capable of (i) recognizing the world in which it exists, (ii) acquiring, storing, maintaining and enhancing its knowledge in order to learn about its environment and/or solve problems and (iii) reflecting on its own knowledge and processes. Thus the basic capacities of cognitive systems are *learning, reasoning*, *knowledge management* and *self-reflection*. From these components the system synthesises its cognitive processes, such as *perception*, *cognitive learning* and *problem solving*.

1.3 Cognitive Reasoning

Cognitive reasoning is a common form of reasoning that takes place in the case of insufficient and incomplete information in situations with uncertainties.

Cognitive reasoning is the driving force of cognitive processes of cognitive agents. As reasoning in general it is responsible for the organisation of the necessary information processing. Therefore reasoning forms, e.g. the problem-solving and the decision-making processes. Thus reasoning may be characterised as the organisation of the processes that connect some given information-premises (e.g. results of observations and experiments, facts and data) to information-conclusions by the use of judgments, estimations or inferences. Thus the possible sources of information to be processed by reasoning are experimentation, systematic observation and theorising concerning the experimental and observational results and preliminary knowledge.

Reasoning may be classified from various points of view. Depending on the orientation of the reasoning processes one distinguishes in the literature practical and theoretical reasoning. Practical reasoning leads to (or modifies) intentions, plans, and decisions, i.e. it is directed towards action. Theoretical reasoning leads to (or modifies) beliefs, models and knowledge and thus expectations, i.e. it is directed towards belief and knowledge. Of course it is also possible that reasoning of either type leaves things unchanged. Moreover, changes may lead to contradiction. Therefore one important aspect of both practical and theoretical reasoning is resolving contradictions. Any given instance process of reasoning may combine both theoretical and practical reasoning.

Now let us consider the types of reasoning depending on the reliability of the information used and generated during the reasoning process. We distinguish between reliable and plausible (non-reliable) reasoning. A reasoning process is reliable if it takes us from truths to further truths, or more precisely if the degree of reliability of the consequences will not be less than the minimum of the reliability degrees of the premises. In the case of plausible reasoning the degree of reliability of the consequences may be less than that of the premises.

A reasoning process consists of elementary information processing steps, which are generated by so-called inference or reasoning operators. These operators form a reasoning method or a reasoning process. These operators may be either reliable or plausible. During a reliable reasoning process each step preserves at least the initial level of reliability. Deduction is an example of reliable reasoning. In the deduction process, reasoning is a sequence of formulas (statements) that are either axioms of a given axiomatic theory or formulas obtained from the previous ones of the sequence by the use of reliable inference rules.

Plausible reasoning is a series of statements which are either given as initial ones or obtained by the use of non-reliable operators, which may lead to the change of the degree of reliability of the consequences in comparison with the reliability of the initial statements. The truth value of the consequence may be less or more or without any changes with respect to the minimum of the truth value of the premises. An important example of this type of reasoning is induction.

Another important view for classification is from the information aspect. It considers the information content of reasoning processes and whether a reasoning process can provide new information. Note that reliable reasoning, in which conclusion follows given true premises, teaches nothing new about a given environment, i.e. about the corresponding subject domain. In plausible reasoning, on the other hand, if the premises are true, then the conclusion follows with some uncertainty and may contain information new to the problem domain and to the given information-premises. The novelty of information is especially important in the case of cognitive reasoning. Since the generation of new information is connected with plausible reasoning processes, cognitive reasoning processes are therefore strongly connected with plausible reasoning.

In cognitive psychology, reasoning was traditionally considered in a deductive logic focus and some researchers used the term "thinking" in order to indicate that they go beyond this tradition by considering other ways of arriving at new conclusions, including induction and analogy (cf. [Markman and Gentner, 2001]). An important theme is the relation between abstract reasoning and concrete domain-specific reasoning, or, in other words, abstract and knowledge-based approaches to reasoning. Note that within psychology the study of reasoning has focussed largely on the use of content-independent logical rules (e.g. [Johnson-Laird and Byrne, 1991; Rips, 1994]). However, other research shows that the content being reasoned about influences human reasoning ability, even for tasks to which logical rules are applicable (e.g. [Cheng and Holyoak, 1985; Cosmides, 1989; Wason and Johnson-Laird, 1972]).

Some authors have taken the extreme opposite of the logicist view, arguing that there is no utility to a general notion of representation or process. One such view is the situated cognition approach that assumes all thinking is fundamentally context governed (see, e.g. [Suchman, 1987; Clancey, 1997]).

In cognitive science deductive reasoning is considered as the processing of premises by using specifiable operators — similar but not identical to formal logical operators; that is, deduction is viewed as a series of specific processes, or operations performed on information. The main focus of this approach is the syllogism, which consists of two premises and a conclusion. Aristotle claimed that syllogistic reasoning represented the highest achievement in human rational thought. Thus many psychologists have focussed on the investigation of syllogistic problem solving with the hope of providing information about the basis of human rationality.

Knowledge plays a significant role in cognitive reasoning. Therefore epistemology, which is the branch of philosophy concerned with the origins, structure, methods and validity of knowledge, has a foundational role. This epistemology should be an epistemology of a computer-aided cognizing agent, which is the opposite to K. Popper's epistemology [Popper, 1974]. The latter rejects both the cognizing agent and dealing with heuristic procedures, induction in particular. The required epistemology will be related to a special type of experimental computational epistemology that is related to the evolutionary epistemology too. Evolutionary epistemology studies the process of changing knowledge. This process can be considered as a transition from ignorance to knowledge and from approximate solution of problems to posing new problems.

The organisation of cognitive reasoning processes should take into account the characteristics of the environments (worlds) containing the events and the objects to be studied. According to the characteristics the following three world types can be distinguished:

(i) *Stochastic world* that consists of random events and entities studied by methods of the theory of probability and/or statistics;
(ii) *Deterministic world* that consists of events and entities studied by deterministic formal methods;
(iii) *Mixed world* that unifies the characteristics of the above two types of worlds; i.e. there are both random changes of events and deterministic relations among events affecting the final states of this world.

Note that the relationships among the events or the characterisation of the objects and events may often be of cause–effect type.

In order to cognize events and objects that belong to these various worlds adequate methodology of information processing and regularity extraction is required. As we have already mentioned new information can be obtained by the use of plausible reasoning. Therefore we require various methods of organising plausible reasoning processes. According to the method to be used, plausible reasoning can be divided into the following classes: (i) probabilistic reasoning (for example, using

Bayesian rules, see e.g. [Pearl, 1997; Jaynes, 2003; Chater et al, 2006]), (ii) approximate reasoning (for example, using the apparatus of fuzzy sets or possibility theory, see e.g. [Bouchon-Meunier et al, 1999]), and, finally (iii) heuristical plausible reasoning (for example, explanation reasoning, see e.g. [Alisheda, 1997], or JSM-reasoning, see e.g. [Finn, 1999, 1991]). Various methods support the realisation of these classes of plausible reasoning, providing appropriate reasoning operators and rules.

Considering the cognitive reasoning processes we emphasise the following three characteristics:

(i) The reasoning processes are of a dynamic nature. Their dynamic nature has two aspects: informational and representational. The informational aspect means the ability to produce substantively new information, while the representational aspect means the possibility to modify the history of a reasoning process.
(ii) The reasoning processes are referential, i.e. they presuppose content (semantic) character connected with meaning, the internal representation of objects and their structures.
(iii) The reasoning processes are not strictly determined in the sense that they are reasoning processes even though they seem incorrect.

The investigation of cognitive reasoning should consider both the representation of the related information and the information processing methods. The following three main approaches of investigation can be distinguished:

- Philosophical–methodological
- Logical–mathematical
- Engineering–computational

The philosophical–methodological approach provides the conceptual background for cognitive reasoning theories. The key concept in most of these theories is that of heuristics, the guide in extracting new information. Heuristics plays a significant role in cognitive processes, especially in the development of knowledge. Note that the development of heuristics is one of the focuses in methodological investigations. Heuristics can be obtained, e.g. as a result of historical analysis of scientific practice in the case of philosophy of science or that of psychological experiments in the case of cognitive psychology.

The formalisation of the conceptual cognitive reasoning theories needs the formalisation of the heuristics used in the cognitive processes. In most cases formalisation is done by the use of the engineering–computational approach, which however does not change the attitude of the researchers in their investigation to a more formal one, because the proposed computational systems which realise the heuristics in question lack any logic foundation. The other way of formalisation is based on the use of a logical–mathematical approach; e.g., J.S. Mill tried to create a logical apparatus for formalising reasoning about cause–effect dependencies that define the nature of *the deterministic world* (see [Mill, 1843]). Mill's methods of agreement, difference and concomitant variation — which are some heuristics — were a development of the idea of the tables of F. Bacon (see [Bacon, 2000]). As a foundation for the acceptance of conclusions deriving from the premises of these methods of reasoning, J.S. Mill formulated the law of the uniformity of nature, from which was derived the principle that in similar conditions similar events lead to the same things. Note that this very principle was denied by D. Hume and K. Popper, who shared the sceptical view of Hume on induction (cf. [Hume, 1739]). But their sceptical view throws out heuristics from the sphere of knowledge studied by exact methods. This last circumstance is an obstacle to investigating the synthesis of cognitive procedures that forms the body of cognitive reasoning.

Another example of logical formalisation is connected with C.S. Peirce, who presented a very productive idea of abduction [Burks, 1946; Peirce, 1995]. Abduction, according to Peirce, is reasoning that leads to the acceptance of hypotheses that explain facts or initial data, and Peirce called the testing of emergent hypotheses "retroduction". In fact, a cognitive activity, according to Peirce, is a synthesis of abduction, induction (retroduction) and deduction.

Now let us turn to the review of the logical–mathematical approach to the formal investigation of cognitive reasoning.

1.4 Logic and Cognitive Reasoning

Logic is connected with the systematic study of human reasoning by understanding it as a managed formalised system of cognitive procedures. Consequently, logic is a discipline that studies reasoning and knowledge, their structure and representation. However, the study of formal aspects of cognitive activity of human beings yields the possibility of developing models of cognitive processes and of supporting the creation of artificial cognizing agents, synthesising cognitive procedures. This generates a demand for unifying logicism and psychologism in the broad sense of these terms in order to adequately model the cognitive reasoning processes.

In the last century reliable reasoning was in the focus of the systematic studies by logicians. Note that this preferred role of reliable reasoning in logical study was connected with the foundational research in mathematics followed by mathematical logicians at that time. Similarly, in cognitive science, most of the approaches that deal with reasoning are also based on traditional deductive logic. Namely, reasoning reflects in these settings the conditions due to which a statement, a proof or a syllogism will be true and false. However, in order to formalise cognitive reasoning it is important to go beyond this traditional reliable focus of the term "reasoning" and include other ways of arriving at new conclusions, including, e.g., induction, analogy and abduction. Earlier in the 19th century reasoning was treated in a broader sense. At that time logicians and philosophers considered both the reliable and non-reliable types together. They studied deduction, induction, confirmation and various further forms of reasoning. Authors such as Mill and Peirce included various non-deductive modes of reasoning (induction, abduction). Note that, according to Peirce: "Abduction is the process of forming an explanatory hypothesis. It is the only logical operation which introduces any new idea" [Peirce, 1903]. These modes remain central to a logical understanding of human cognitive processes. Recently, this older, broader agenda is coming back to life, but now pursued by more sophisticated techniques — made available, incidentally, by advances in mathematical logic. The study of induction and abduction and also of combinations of deduction, induction and abduction has been very extensive in recent years. There are various solutions in the literature that use a logical approach for handling some aspects of plausible reasoning. However, it is impossible to give an exhaustive overview of this exploding field here. Therefore we limit ourselves to giving only some characteristic references from the literature: see, e.g., [Flach and Kakas, 2000; Alisheda, 2004; Gabbay and Woods, 2005; Flach et al, 2006].

An important step was done in the 1980s with non-monotonic logics for default reasoning coming from AI, which model more closely how humans would approach problem-solving or planning tasks (see, e.g., [Antoniou, 1997]). Many further broad ideas in the literature reflect human practice, such as linear logic that models the processes and resource use (see, e.g., [Girard, 1987; Troestra, 1992]) or the substructural logics that also model the role of resources in establishing inferences (see, e.g., [Paoli, 2002; Restall, 2000]).

Moreover, logic should support the consideration not only of the products of cognitive reasoning such as inferences and propositions but also the study of the information flow and processing and their mechanisms themselves; i.e. logic should study both the product and process aspects of cognitive reasoning, i.e. its static and dynamic aspects (see, e.g., [van Benthem, 1996]). Therefore here, beyond, for example, the static notion of correctness, its dynamic counterpart correction should also be studied (see, e.g. [van Benthem, 2008]). Appropriate logical methods are required to adequately represent the reasoning processes which fundamentally differ from the logical proofs (cf. [Joinet, 2001]).

1.5 Requirements for a Formal Cognitive Reasoning Theory

Understanding cognition and modelling, designing and building artificial cognitive systems are connected with challenging long-term research problems. An important set of problems concerns the understanding of cognitive reasoning in such a way that it would reflect the entire range of cognitive information-processing demands. The investigation of cognitive reasoning should consider both the representation of the related information and the information-processing methods. Thus this investigation is also directly connected with the understanding of representation, acquisition and management of knowledge.

An adequate understanding of cognition and the related cognitive information processing and their investigation and modelling will require all three, the philosophical–methodological, the logical–mathematical and the engineering–computational, approaches. Moreover it is challenging to develop a theory the result of which will provide a scientific foundation for artificial cognitive systems at the same time. It is desirable that this foundation be constructive in order to support directly the design and development of artificial cognitive systems.

In order to develop the logic foundation of cognitive reasoning, i.e., the logic that reflects the main characteristics of the process of learning new information and discovering new regularities in the subject domain in question, the development of an appropriate logic-based framework is required that follows the traditional rich view of logical investigation for the representation and modelling of human reasoning processes but which will use the latest advanced techniques of mathematical logic. It is expected that this formal framework will permit the integration of reliable and plausible reasoning operators into reasoning processes, and moreover that various mathematically expressed plausible reasoning methods (fuzzy, probabilistic, statistical, etc.) could be also involved in these processes as constituents.

Moreover, the required framework should provide formal logic tools to handle: (i) the dynamic nature of cognitive reasoning, in contrast to the static proofs considered by the traditional approach; (ii) the "semantic" or "content" aspects of reasoning, whereas proofs articulate statements inferentially, according only to their shape, without regard to reference; and (iii) indeterminacy and temporal contradictions of the reasoning processes opposite to correctness of proofs.

Beyond the logic foundation explicitly or implicitly the philosophical–methodological approach should also provide a model cognitive reasoning. Namely epistemology, which is the branch of philosophy concerned with knowledge and its changes and transformations, has a foundational role for the theory of cognitive systems. One of the main requirements for epistemology is that it should also deal with the structure and growth of knowledge. The problem of creating a precise epistemology for artificial cognitive systems and its systematic study is still unsolved, however many steps have been made in this direction. The required epistemology should be an epistemology of a computer-aided cognizing agent as mentioned earlier. Thus it is important to develop a formal epistemology for cognitive systems that particularly supports the construction of descriptive and normative theories of cognitive systems and generally provides a constructive view for the philosophy of knowledge.

The epistemology that should be developed will be interrelated with (i) the formal methodology of dealing with information and knowledge, (ii) the methods of knowledge discovery, (iii) the control of the knowledge generated by special methods of falsification and justification and (iv) the methods of growth of the knowledge base. This epistemology could be related to both a special type of experimental computational epistemology and the evolutionary epistemology. It could deal with the processes of knowledge discovery realised by cognitive procedures and with the growth of the knowledge base resulting from them.

Thus the foundation should be appropriate for the representation of the process from ignorance to knowledge, that is, the learning and knowledge acquisition process.

1.6 Objectives

In this book we wish to develop a scientifically well-founded general approach which allows us (i) to model the information-processing processes related to cognitive reasoning and (ii) to construct cognitive reasoning processes of cognizing agents. We require for this approach that it should provide appropriate methods and tools for handling cognitive reasoning processes at the three levels of abstraction: conceptual, formal and realisational. The coherent collection of appropriate methods and tools is to form a framework, which will be called a *cognitive reasoning framework*, or *CR framework* for short.

(i) At the conceptual level this approach should provide a *CR framework* which is appropriate to give a structural and functional model of cognizing agents in the form of cognitive architecture. This framework should also support the description of the cognitive reasoning processes of a cognizing agent.

(ii) At the formal level this approach should provide a formal logical *CR framework* that will permit the modelling of the cognitive activity and provide a constructive foundation of artificial cognizing agents. It should provide tools for representing cognition in a dual way, covering the processes and the product. The formal *CR framework* should have a strong logic foundation which can support two alternative approaches: one for the simulation of cognitive processes *by means of the standard deductive technics* and another for directly constructing, representing and realising the cognitive reasoning processes *by means of original formal tools*.

Moreover, for this CR framework we require to have adequate formal methods and tools that can support:

- The description of the observed events and objects (descriptive language)
- The description and representation of the cognizing agent's knowledge (subjective knowledge)
- The description and representation of the knowledge about a given environment and about the corresponding subject domain (objective knowledge)
- The representation of cognitive processes from observation to subjective knowledge and from subjective knowledge to objective knowledge
- The construction, representation and implementation of the cognitive reasoning strategies of the cognitive reasoning processes

(iii) At the realisational level the approach should provide an object-oriented architecture which will adequately represent the main characteristics of the cognitive processes and which satisfies the formal logic foundation and permits the realisation of efficient computation tools for each constituent of the cognitive architecture. Thus, at this level, the framework should provide appropriate tools for:

- Representing the description of the perceived data and information (results of experiments and observations)
- Representing and handling knowledge
- Formalising and representing cognitive reasoning processes

Therefore, the required realisational tools and methods should be adequate for the design and implementation of a "cognitive engine" which is the kernel of the cognitive architecture and which represents the "brain" of the cognizing agent.

We suppose that the formal approach that will meet the above requirements may bring us to a new understanding of the semantic counterpart of reasoning processes and will provide a better understanding of cognitive reasoning.

1.7 The Formal Approach to Be Developed

In this book we propose a formal approach which will satisfy the main requirement described above and will permit the above-formulated objectives to be achieved.

The proposed approach will provide a CR framework at the conceptual level, which will support the description of the (conceptual) structure, tools, techniques and methods. The conceptual structure together with all the constituents will be represented as a "cognitive architecture". When we come to develop a cognitive architecture for modelling cognition we will, on the one hand, generalise from domain-specific characteristics based on results from the study of cognition and, on the other hand, hypothesise more detailed computational mechanisms for the implementation of general cognitive abilities.

At the formal level, the approach will provide a formal CR framework which contains methods and tools to represent and process the information considered in the cognitive architecture at the conceptual level. At this level the approach will be broadly logical. We will define logic as the disciplined description of the information processes which can provide a model of how thinking agents reason and argue. The approach will provide the logic that reflects the main characteristics of the process of learning new information and of extracting new regularities in a given environment, or more precisely, in the corresponding subject domain by the use of cognitive reasoning processes. Thus the proposed formal CR framework will be logically well based.

The proposed approach will concern both the rationalist and the empiricist theories of knowledge. The rationalist tradition focusses on the reasoning processes on the basis of the existing information and knowledge, while the empiricist theory concentrates on the processing of the data and facts obtained from experimentation. Intuitively, cognition is oriented towards the analysis of the initial data, generation of hypothesis, hypothesis checking and acceptance or rejection of hypothesis.

The formal CR framework supports the formulation of the cognitive reasoning processes and the representation of knowledge. The latter is supported by an original logical theory, the so-called open cognitive (quasi-axiomatic) theories. The knowledge of a cognizing agent about its environment will be formulated in the form of an open cognitive theory.

The formal framework will also permit the use of models for experimentation and accepts the data so obtained as being about the *external world* (environment). The results of experiments will be represented as observational sentences, i.e., as facts.

Thus a cognizing agent may generate its own hypothetical inputs, as in the case of modification rules, e.g. in the case of abduction, induction or theory formation.

The formal CR framework will provide formal tools to handle (i) the dynamic nature of cognitive reasoning; (ii) the "semantic" or "content" aspects of reasoning; and (iii) indeterminacy and temporal contradictions of the reasoning processes. Namely, these formal tools will provide:

(i) A model of dynamics of cognitive reasoning processes, based on the representation of a cognitive reasoning process as a motion from ignorance to knowledge. The proposed formalism will deal with a specific syntactic structure, the so-called modification inference, which contains applications of non-deductive rules.
(ii) A dual (semantic–syntactic) approach, which has the so-called modification calculus and a collection of axioms which entirely and uniquely describes the semantic structure. The latter is considered as a model of our initial knowledge about the subject domain.
(iii) A formal approach to deal with logical contradictions. Contradiction will have temporal character; it may appear at one of the internal stages and then it may disappear.

The proposed logical CR framework will support two alternative approaches. The first one supports the simulation of cognitive processes by using usual reliable (deductive) derivation techniques. The second one will provide original calculi that support the representation and realisation of the cognitive reasoning processes. Moreover this framework will provide a logical "platform", which permits the development of very complex reasoning processes by using reasoning operators of various types; e.g. this platform permits the combination of reliable and plausible reasoning operators in the development of the same reasoning process. Moreover various plausible reasoning operators such as probabilistic and heuristic ones can be applied together.

The formal CR framework will be developed by the use of highly sophisticated techniques — made available, incidentally, by advances in mathematical logic.

Beyond permitting the modelling of the cognitive reasoning processes, the formal logical framework will provide a constructive foundation of artificial cognitive systems on the basis of which the formal approach provides the possibility of realisation at the computational realisation level. At this level an important subject is the relation between abstract logical reasoning and concrete domain-specific reasoning or, in other words, between abstract and knowledge-based approaches to reasoning.

1.8 Overview

Conceptual theory of cognitive reasoning. Part I of the book is devoted to the development of the conceptual CR framework that provides a conceptual model of cognitive reasoning. In order to characterise and model the cognitive reasoning processes the main actor of our approach, the so-called cognizing agent, is defined. The structure and functioning of cognizing agents are represented in the form of cognitive architecture. The cognitive reasoning will be modelled with respect to this architecture, able to interact with its environment in order to obtain data, facts and information, and to process these in order to extract new information and regularities. These will be used to augment the system's knowledge about the environment and/or about the corresponding subject domain.

The model shows how data, knowledge and reasoning should be combined to ensure the required reasoning process that will realise plausible and reliable (deductive) reasoning. The goal of these processes is the extraction of relationships over the data that can be interpreted as the generation of new information.

The model contains appropriate tools for (i) the description of the observed events and objects, (ii) the description and representation of knowledge and (iii) the representation of cognitive reasoning processes from observation to extracting information and knowledge.

For the representation of the processes that are taking place in the "brain" of the cognizing agent the CR framework provides two classes of theories: (i) the open cognitive or quasi-axiomatic theories and (ii) the modification theories.

A theory is called open because it is open for the representation of the interaction between the cognizing agent and its environment. It should be open for (i) obtaining new facts for analysis, (ii) adopting and internalising knowledge from other similar theories and (iii) modification of the reasoning rules. An *open cognitive theory* is used to represent the activity of a cognizing agent as a history. Thus an open cognitive theory is an essential component of *self-reflection* of an agent.

Theories are called modification theories because, during the construction of a process in theories that will model the reasoning process, not only can a new statement be added, but the process itself can be modified. Modification theories use special techniques of inference for modelling the reasoning processes. Modification theories permit the representation of dynamics of cognitive processes, understood as movement from ignorance (uncertainty) to knowledge (definiteness). Moreover, it is explained how modification theories can handle contradictions.

The information processing activity of a cognizing agent is represented in the form of several embedded cycles. The internal cycle consists of iterative reasoning algorithms repeated until the completion condition is met. This cycle is modelled by the use of a modification theory. The external cycle represents repetition of perception and reasoning phases and proceeds from the beginning to the end of the activity of the cognizing agent. The external cycle is modelled by the use of an open cognitive theory.

Part I ends with an example on modelling of a cognitive reasoning process in a sample case.

Logic foundation. Part II of the book is devoted to the development of the logic foundation of the formal approach that should be able to provide a logically well-founded CR framework. Such a foundation is proposed that supports two alternative approaches: (i) the simulative one that permits the simulation of the cognitive processes by using standard deductive derivations and (ii) the direct one which provides original calculi appropriate to represent and realise cognitive reasoning processes.

The logic foundation will be based on a special family of many-valued logics which we call pure J logics (PJ logics). The PJ logics are logics with two sorts of truth values: external and internal. The internal (empirical) level corresponds to observations and experiments of the cognizing agent that interprets the results of these activities. Therefore the corresponding internal knowledge represents the individual or subjective knowledge. The external knowledge can be considered as the conventional or objective knowledge. The external truth values are the traditional "true" and "false". The internal truth values are selected for handling results of observation, experimentation and plausible reasoning. One of the internal truth values is interpreted as "uncertain". Any other internal truth value represents various kinds of certainty.

The discussed logics are said to be J logics because they contain the so-called J-operators that are used for encapsulation of internal knowledge into external knowledge. By using these operators we can apply empirical knowledge to theoretical constructions.

A special class of PJ logics, the class of *finite* PJ logics (FPJ logics), is introduced, having finitely many internal truth values and such a set of J-operators that contains the characteristic functions of all one-element subsets of the set of internal truth values and only these. Any FPJ logic will be appropriate for describing and representing a *knowledge state* at a certain moment of time. However, this will be a *static* description. Standard deductive techniques do not permit representation of the dynamics of the cognitive processes in these logics. For the latter the so-called iterative logics are defined, which are PJ logics with an infinite (countable) set of internal truth values.

For any FPJ logic there exists a unique *iterative version* up to selection of an internal truth value representing uncertainty. Each *definite* truth value of a source FPJ logic (i.e. internal truth value different from "uncertain") corresponds to an infinite sequence of definite truth values of the iterative version of this logic. The internal truth values of the iterative version of an FPJ logic are used to evaluate the results of a cognitive process, including applications of formal analogues of plausible reasoning rules. We assume that the application of plausible rules leads to a decrease of certainty of obtained statements.

The iterative versions of FPJ logics will permit the description of the history of cognitive processes that include several repetitions (iterations) of certain cognitive procedures. A cognitive process includes reasoning with the application of plausible inference rules (induction, analogy, abduction, etc.). In the iterative versions of FPJ logics, the so-called cognitive axioms correspond to these rules.

Cognitive axioms and iterative logics allow us to represent cognitive reasoning dynamics by the use of ordinary logical inferences. In this case we will have a *deductive simulation* of plausible reasoning. These iterative versions can also be interpreted as tools for representation of the history of a cognitive process. If we restrict ourselves to finite structures, then a cognitive process, considered as motion from ignorance to knowledge, will be completed necessarily. The sign of completion will be so-called stabilisation or saturation, which is the situation when we cannot achieve further narrowing of the uncertainty domain. In a finite structure the uncertainty domain is also finite and cannot be narrowed infinitely. The stabilisation situation will also get a formal representation in the language of iterative logic in the form of stabilisation axiom schema.

In Part II we will consider both the propositional and the first-order many-sorted FPJ logics and their iterative versions; namely, the syntax and semantics and the calculi of these logics will be described. The discussion will follow the structure of the presentation of the syntax and semantics of the classical first-order logic. The proofs of the statements will be similar to those of the corresponding statements of the classical first-order logic.

The basic results of this part are the Soundness Theorem and Completeness Theorem, which are analogues to classical cases. We prove these theorems for the FPJ logics and their iterative versions. To prove the Completeness Theorem, first we define the relations that connect derivability and validity in classical logic and in PJ logics. Then the Main Lemma will be proved that provides sufficient condition for PJ calculi to be complete. Further we prove that this sufficient condition holds for FPJ logics and their iterative versions. From this it follows that the corresponding calculi are complete.

The formal CR framework. Part III is devoted to the development of a formal cognitive reasoning framework that will be able to describe and represent information processing of cognitive reasoning, i.e. the processes that allow cognizing subjects to gain new knowledge.

The formal CR framework will use an original non-monotone derivation technique obtained by adding specific rules to the standard deductive technique. This new technique will play the role of the plausible constituent of cognitive reasoning, which supports data analysis and the discovery of principally new relationships and regularities by the use of the non-deductive technique of reasoning. This discovery has its price such as the incorrectness of non-deductive reasoning, the possibility of obtaining incorrect conclusions even from valid premises and the possibility of obtaining contradictions.

In the proposed formal CR framework the cognitive reasoning processes are modelled as a *motion from ignorance to knowledge*. In the course of this motion the uncertainty domain is permanently narrowing. The proposed formalism will deal with a specific syntactic structure, the so-called *modification inference*. In this formalism, adding formulas to an inference by non-deductive rules may cause modifications in those constituents already in the inference. This is why non-deductive rules are called modification rules in our formalism.

A modification inference has a complicated structure that reflects the discrete structure of cognitive reasoning in accordance with our view about this matter. First of all this inference splits into *stages*. It is supposed that there is a cognitive routine that takes place at each stage of reasoning. Every stage of a modification inference has to reflect the structure of this cognitive routine. Following this structure each stage has to be split into the same number of *modules*. A completed module corresponds to some *knowledge state*, which can be expressed by its formal *description*.

In any module, a simple assertion of the same type must be estimated by internal truth values. These simple assertions are written as atomic formulas with the same predicate symbol. The set of all such atomic formulas represents an *internal predicate*, i.e. a predicate that takes values from the set of internal truth values of an FPJ logic (instead of the usual "true" and "false").

A motion from ignorance to knowledge is represented by modification inference as a process of transforming internal predicates. In this process a formula estimated as "uncertain" can be modified by replacing this estimation with one of the definite internal truth values. So this process results in the narrowing of the *uncertainty domains* of internal predicates.

Calculi of a special type are defined that allow us to build modification inferences. In the present book these calculi are called *modification calculi*. To define a modification calculus it is necessary to describe:

- A system of *modification rules* (any such rule is a formal analogue of a plausible reasoning rule)
- An *initial state* of knowledge represented by its formal description

An interesting effect of the proposed formalism is the temporal character of the obtained contradictions. A modification inference may be free of contradiction at the initial stage. Contradiction may appear at one of the internal stages and then may disappear. In the subsequent stages of this inference a contradiction may also appear but for a different reason.

We obtain a *modification theory* if we add a set of non-logical axioms to a modification calculus. These axioms postulate some requirements for the initial knowledge state represented in a modification calculus. For example, these axioms can require some algebraic structure for a model of the subject domain (to be a lattice, a Boolean algebra, a group etc.).

We define a model of a modification theory as a sequence of structures, where each structure corresponds to some knowledge state. For such a *notion of model* we prove analogues of the Soundness Theorem and the Completeness Theorem.

The other important results of Part III are the assertions that describe the relation between modification theories and iterative theories (i.e. theories based on iterative versions of FPJ logics). In these assertions we propose a way of constructing a corresponding iterative theory for each modification theory. These theories will have coherent sets of theorems.

Handling complex structures. Part IV is devoted to the extension of the techniques introduced in the previous parts. A cognizing agent often meets situations where the events or objects to be observed and investigated are very complex and have many interacting constituents in their structure. Such situations are typical, for example, in biology, medicine and pharmacology as well as geosciences and sociology. In order to handle the cognitive processes in such situations the proposed, logically well-based CR framework should be augmented:

- With a more powerful language that can handle complex structures with interacting components. (This requirement concerns both syntactic and semantic as-

pects. In this language the objects should be interpreted as formal structures such as sets, strings, graphs, etc.)
- With a calculus that can handle the complex language units in order to find hidden relationships concerning the features of complex objects and their structures.

The required extension is realised with the tools that permit the handling of so-called set-sorts, i.e. sorts, the objects of which can be interpreted as sets. This permits the consideration of formal structures of complex objects in the form of sets. The set-sort objects should behave like sets and be arguments of internal predicates. The extension concerns the languages, their syntax and semantics, the inference technique as well as the modification calculi.

For the sake of simplicity first a special class of set-sorts is investigated, the so-called atomic sorts. Objects of an atomic sort considered as arguments of internal predicates behave as isolated and independent from each other. This independence can be formally expressed in the form of a locality requirement, which informally means that the value of a predicate for an object of atomic sort does not depend on the value of the same predicate for any other object of the same sort.

Special tools are introduced to represent the sets of properties in the modification calculi. First of all the definition method for modification rules is defined. Then the transformation of modification rules into generator rules is introduced. Different classes of modification calculi are investigated. Among them the so-called *regular* and *perfect* classes are introduced and investigated. For the perfect calculi an appropriate class of models can be formulated for which the analogue of the theorems on correctness and completeness holds.

JSM theories. The abbreviation JSM originates from the initials of John Stuart Mill. The JSM method uses some formalised modification of Mill's *method of agreement* that solves an important task of extracting knowledge from data. Peirce's abduction and Popper's falsification are the two other philosophical sources for the ideology of the JSM method. V. K. Finn, the initiator of this method, defines it as a synthesis of three cognitive procedures: *induction, analogy* and *abduction*, beyond which deduction can play some auxiliary role. Part V is devoted to the JSM method, which will be considered as a concrete example of application of our formal CR framework. The JSM method solves the following two tasks:

- Acquisition of a characteristic relation between structured objects and sets of target properties
- Prediction of the presence or absence of the target properties for objects with known structure and unknown presence or absence of the target properties

The JSM method uses the method of similarity based on the ideas of J.S. Mill. Mill's method of agreement considers similarity between the objects, phenomena or events under investigation. We show that *similarity* in the JSM method is understood operationally. In order to extract causal regularities the JSM method uses plausible reasoning based on the use of induction and analogy. Many mathematical results from the previous parts will be oriented towards the JSM method. These will be developed with the aim of including the JSM method into our paradigm.

The logical–mathematical basis of the JSM method is investigated in our interpretation. We give an interpretation of this basis from a more general point of view in order to (i) insert it into the context of cognitive theories considered in the previous chapters and (ii) provide a uniform description for the rules of various versions of the JSM method. The mentioned basis for the JSM method is developed by using the techniques of the modification calculi.

Looking back and ahead. Part VI is devoted to the discussion of the results obtained in the book. It summarizes the results concerning the three main investigation methods — the philosophical–methodological, the logical–mathematical and the engineering–computational ones. First the algorithmisation of the formal components is discussed and an object-oriented model is developed for the realisation of the cognitive reasoning processes. Then the main results are summed up and some important and very interesting open problems are formulated. Finally, the philosophical–methodological aspect of the proposed approach is discussed.

Part I
Conceptual Theory of Cognitive Reasoning

Chapter 2
Introductory Explanation

In this chapter we develop the first constituent of the proposed formal approach, namely the conceptual theory of cognitive reasoning. In the course of this discussion, the key topics included in this book emerge at the conceptual level. We will describe the main processes and their cognitive reasoning constituents which permit going beyond the readily available information. We will develop a model which represents the processes occurring as common forms of reasoning in the case of insufficient and incomplete information. In order to characterise and model cognitive reasoning processes we will describe the systems where these processes take place. Thus we will model cognitive reasoning with respect to a general cognitive system able to interact with its environment to obtain data, facts and information, and to process these in order to extract new information and relationships. These will be used to augment the system's knowledge about the environment or about a problem domain. The model of cognitive reasoning will be developed with respect to a model of the cognizing agent.

The proposed approach will concern both the rationalist and the empiricist theories of knowledge. The rationalist tradition focusses on the reasoning processes on the basis of the existing information and knowledge, while the empiricist theory concentrates on the processing of the data and facts obtained from experimentation. That is, the possible sources of information to be extracted are various internal or external bases of knowledge, facts and data and are also results of specially planned experiments. Cognitive processes are strongly intertwined with knowledge, which is not passively received, e.g. by way of communication, but is actively built up by using cognitive reasoning. Intuitively, cognitive reasoning processes are oriented to the analysis of the initial data, generation of hypothesis, hypothesis checking and acceptance or rejection of hypothesis.

Developing the conceptual level of our approach we intend to achieve the following goals:

(i) The theory should provide appropriate methods and tools for:

- The description of the observed events and objects
- The description and representation of knowledge

O. Anshakov, T. Gergely, *Cognitive Reasoning*, Cognitive Technologies,
DOI 10.1007/978-3-540-68875-4_2, © Springer-Verlag Berlin Heidelberg 2010

- The cognitive reasoning processes from observation to extracting information and knowledge

(ii) The theory should provide a special framework, which permits application of different types of reasoning operators in the development of one and the same cognitive reasoning process. Thus different types of information-processing operators can be combined in one process.
(iii) The methods and tools to be introduced should be appropriate for further formalisation.

While developing the conceptual theory of our approach the following postulates will be taken into account:

(a) Cognitive processes aim to extract new information and knowledge from the data and facts obtained from the environment.
(b) Cognitive reasoning is the skeleton of these processes which is realised by the appropriate information-processing processes according to the cognitivist approach.
(c) The main actor that can realise cognitive reasoning is the cognizing agent that, we suppose, has no other motivations than the formation of an adequate model of the environment.
(d) A cognitive process is a discrete process in which perception and reasoning alternate. Within reasoning a cognizing agent deals with data and facts that contain uncertainties and are obtained as results of perception. During reasoning the agent extracts regularities and formulates hypothese which reduce the domain of uncertainty.
(e) The decrease of the domain of uncertainty results in a better model of the environment w.r.t. the actual situation. This model will support more adequate decision making and problem solving related to the environment.
(f) The dynamics of cognitive reasoning processes can be represented as a motion from ignorance to knowledge.
(g) In a cognitive reasoning process contradictions may appear with a temporal character. This means that a contradiction may appear at one of the stages of a reasoning process and then it may disappear.
(h) A cognizing agent possesses knowledge and experience about itself and about the situations of the environment it met. A collection of statements can be used to represent the knowledge of a cognizing agent.

In the next chapter the main notions on which our approach will be based will be introduced and discussed. The main aim of this chapter is not to define these notions but to explain how they will be understood and used in our approach in the forthcoming sections.

In Chap. 4 we will describe the structure, processes, techniques and methods at the conceptual level that, according to our approach, will be necessary (i) to understand cognition and (ii) to support its modelling. This conceptual structure together with all the constituents is what we call a "cognitive architecture". This architecture will be developed w.r.t. a cognizing agent, which is the main actor in our approach.

In Chap. 5 we will describe the cognitive reasoning framework (CR framework), which permits the realisation of a cognitive reasoning process by the use of different types of information-processing operators. The cognitive reasoning processes will be discussed in detail in this framework.

Chapter 3
Basic System of Concepts

The objective of the present chapter is the informal introduction of the main notions that will serve as the basis of our approach for modelling cognitive reasoning. Most of these notions should be well known to readers. However we aim not to give them a formal definition but to provide an explanation of how these notions will be understood and applied by our approach in the forthcoming sections.

The basic aim of the cognitive processes is to enlargen knowledge by discovering new regularities in the results of experimentation and observations. These results are described by the use of a language as facts. The processing of the letters will provide in our approach new hypothetical causal regularities which will be turned into knowledge after justification. This is why facts and knowledge will be the first notions to be discussed (see Chap. 2).

A model of cognitive reasoning should permit not only the evaluation of information used and generated during processing the facts but also the dynamics of this process with the development of knowledge as well. For this purpose a special theory of truth will be discussed in Sect. 3.2.

In the development of the cognitive reasoning theory of course, the notion of reasoning will have a crucial role. A working explanation of this notion will be given in Sect. 3.3 by attempting to impose the least restrictions onto this notion. During the discussion another important question will also be discussed: the possible monotonicity of reasoning. Moreover, we will also consider the situations where the premises are doubted: if they should be rejected and replaced by others. I.e. the reasoning process should permit not only to obtain consequences but also to falsify and modify the premises. The rules that generate the reasoning processes will also be discussed.

3.1 Facts and Knowledge

Cognition is done by processing information obtained from observation and experimentation while the existing knowledge is considered to different extent and

O. Anshakov, T. Gergely, *Cognitive Reasoning*, Cognitive Technologies,
DOI 10.1007/978-3-540-68875-4_3,

cognition results in new information that will extend the existing knowledge. In this process two main notions are considered, namely facts and knowledge. Therefore these are the first notions that we intend to discuss; both notions can be defined in various ways from which we take two: (i) the origin and the content aspects and (ii) the representation aspects.

Thus a fact is:

(i) A datum obtained as a result of observations and experiments, according to the first aspect
(ii) A concrete elementary statement without parameters (that has the same format as experimental data), which may be represented in the form of atomic formulas without variables, according to the second aspect

A fact in sense (i) is a reliable information (at least, it is not less reliable than the experiment itself is). A fact in sense (ii), i.e. in a pure formal sense, may also be obtained as a result of any formal transformation (processing) of the corresponding atomic formulas without variables. Thus it may be obtained not only as a result of an experiment but, e.g. as a result of a logical inference which uses heuristics (as happens in expert systems). In this case, of course, it will be only a hypothesis.

When knowledge is defined it is usually opposed to facts. We emphasise the following features of knowledge:

(i) It is impossible to obtain knowledge directly from the results of experiments because formation of knowledge occurs on the basis of experimental data analysis. In the analysis we reason, and our reasoning may be not only deductive but plausible also, e.g. we can use inductive generalisation. The process of knowledge formation may require auxiliary experiments for checking the hypotheses obtained by way of using reasoning.
(ii) Knowledge expresses general regularities that can be represented in the form of formulas with universal quantifiers or in the form of (production) rules that contain variables.

For both facts and knowledge in definition (i) the characteristics connected with the origin are listed, and in definition (ii) the peculiarities of the representation format are given. For computer implementation, including AI, the definition via the form of representation appears to be more convenient. However, it is worth noting that the representation form may not be unique.

Since in our approach causal regularities are important elements in developing knowledge, let us consider what a statement about cause–effect relations represents. Information on the presence (or absence) of cause–effect relations is not included directly in the experimental data. It can be obtained only as a result of analysis of these data. Sometimes such an analysis will be very simple, but it should necessarily be present. Thus from the point of view of content the information about cause–effect relations is knowledge.

A statement about the cause–effect relationship between two concrete events from the point of view of the representation form is a concrete elementary statement without parameters, i.e. a fact. For example:

Electric discharge in the air (lightning) is a cause of a *loud sound* (thunder).

On the other hand it can be interpreted as a universal statement (statement with universal quantifiers) that describes regularity:

In each case after an electric discharge in the air (lightning) a loud sound (thunder) is audible.

Then a statement on the cause–effect relations should be considered as *an element of knowledge*.

How are we going to interpret the notions *fact* and *knowledge* in the present work?

First, we do not exclude any of the possible interpretations of these notions. As to which interpretation is to be selected, it should be clear from the context or will be specified in each case when these terms are used

Second, we prefer a pragmatical point of view. Namely, it is convenient to consider a statement as a fact or knowledge (element of knowledge) depending on the actual situation under scrutiny. Moreover, if it is convenient to consider one and the same statement as a fact at one stage and as knowledge at another one, then we will do so.

All in all, these are the ways in which we are going to use the terms "fact" and "knowledge" in most cases in the present work:

- A **fact** is a statement *represented* (in the context of functioning of a system or application of certain methodology) *in the form of a concrete elementary statement without parameters*,
- **Knowledge** is a statement *the meaning of which is a reflection of certain general regularity* regardless of the form of representation of this statement (in the context of functioning of a considered system or application of a certain methodology).

It is evident that, according to the above definitions, we can consider *one and the same statement as a fact as well as knowledge*. We are really going to do so.

We intend to introduce some general schema that permits to analyse *facts*, to obtain *regularities* as a result of this analysis and to form *hypotheses*. However, from the formal point of view, facts, regularities and hypotheses—which we are going to consider in the frame of the proposed approach—are all facts, i.e. concrete elementary statements which can be described by atomic formulas without parameters. On the other hand, we cannot consider the observed (in the process of applying our methodology) regularities as something different from *knowledge*. We consider our problem as the problem of knowledge discovery.

3.2 Truth Values: Informal Discussion

Our approach to model and investigate cognitive reasoning will permit the evaluation and handling not only of the constituents of information processing but also the dynamics of this process with the development of knowledge as well. For this purpose a special theory of truth is required, together with appropriate sets of truth values.

What do we mean by calling a sentence true? A sentence may be called true from four different aspects. The first one is the correspondence aspect, where the notion of truth will reflect the correspondence of facts to events and objects of the observed and experienced environment. During the information processing the logical consistency (or inconsistency) of the obtained sentences with the knowledge that is held to be true will be reflected by the notion of truth. That is, this notion of truth requires the obtained new statements (i.e. information) to fit properly within a whole system of statements (i.e. knowledge). This evaluation will reflect the coherence aspect of the notion of truth. Moreover, the new information obtained as a result of the information process (i.e. cognition) will also be evaluated from the point of view of how it can explain certain facts, i.e. a statement will be true if it allows the explanation of certain facts or, in other words, it allows a cognizing subject to interact more effectively and efficiently with the environment. Thus a statement that expresses a hypothesis will be true, i.e. it will be acceptable, if it is found to be pragmatically useful. This is the pragmatic aspect of the notion of truth. Moreover, the theory of truth will also support the evaluation of the degree of certainty which permits the handling of the dynamics of the growth of knowledge. In this case the proposed theory of truth provides a process of generation of truth values that is constructive and iterative. This will be the constructive aspect of the notion of truth.

The theory of truth will be developed in the framework of a non-classical version for many-valued logic. At the informal level this means that there may be truth values differing from the classical "true" and "false".

There are various intuitive interpretations for these differing, i.e. auxiliary, truth values; e.g. the frequently used interpretation of truth values in three-valued logic is: "true", "false" and "undefined". If the set of truth values of many-valued logic is ordered such that "false" is the least and "true" is the greatest element of this set then we can speak about the truth values in terms of the "degree of truth".

We divide all truth values of the many-valued logic into two classes: the class of external truth values and the class of internal truth values. There are two external truth values: the classical "true" and "false". There are at least two internal truth values, of which one is called "uncertain", which means the complete absence of information about the question of interest. The others may express the degree and nature of certainty and being informed. Note that this division into two classes is similar to the classical philosophical tradition of dividing truths into analytical and factual ones. The factual truth values will be subdivided into the values of facts that are given at the start of the cognitive process and the values of hypotheses that will be generated during this process.

The degree and the nature of certainty may be connected with the quantity and significance of arguments which will be either for or against the statement characterised and evaluated by internal truth values. We give an example in which there are four internal truth values, among which "uncertain" is also present. These may be explained by the presence or absence of justifying and rejecting examples (arguments for or against). Let φ be a statement which can assign internal truth values. Then:

- A statement φ takes the truth value "empirically true", if
  - There are arguments for the statement φ;
  - There are no arguments against the statement φ.
- A statement φ takes the truth value "empirically false", if
  - There are arguments against the statement φ;
  - There are no arguments for the statement φ.
- A statement φ takes the truth value "empirically contradictory", if
  - There are arguments for the statement φ;
  - There are arguments against the statement φ.
- A statement φ takes the truth value "uncertain", if
  - There are no arguments for the statement φ;
  - There are no arguments against the statement φ.

We emphasise that the value "empirically contradictory" expresses the nature of certainty, it is an example of a definite internal truth value. According to our assignment uncertainty means the complete absence of information while empirical contradiction means the existence of information, although it is contradictory.

We give one more example: let p be the quantity of arguments "for", and c be the quantity of arguments "against". Then the fraction

$$\frac{p}{p+c}$$

expresses the "proximity to the empirical truth". This fraction is equal to 0 when there are no arguments "for" and there are arguments "against"; this situation corresponds to the truth value "empirically false". This fraction is equal to 1 if there are arguments "for" and there are no arguments "against"; this corresponds to the truth value "empirically true". The uncertainty of the form $\frac{0}{0}$ corresponds to the value "uncertain".

Another expression which contains the quantity of arguments "for" and arguments "against" can also be suggested. For example the fraction

$$\frac{p-c}{p+c}$$

can also be used for the qualitative characterisation of the nature of certainty. This fraction is equal to:

- $+1$, if there are arguments "for" and there are no arguments "against" (empirically true)
- 0, if there are arguments "for" and "against" and their quantity is equal (empirically contradictory)
- -1, if there are arguments " against" and there are no arguments "for" (empirically false)

Similarly to the previous example the uncertainty of the form corresponds to the value "uncertain". In the remaining cases the fraction characterises the certainty from the point of view of proximity to the values "empirically true" and "empirically false". The positive numbers correspond to the values which can be described by the terms "rather true than false", and the negative numbers can be described as "rather false than true".

To each argument (both for and against) its weight coefficient may be assigned. Then p and c should be interpreted as the sum of the weights of the arguments "for" and arguments "against", respectively. In such an interpretation of p and c we can use the previous expressions for the calculation of the nature of certainty.

In all the above examples we expressed the nature of certainty by the use of internal truth values. In order to express the certainty degree simultaneously with its nature the truth value may be represented by an ordered pair. For example in the pair

$$\left\langle \frac{p-c}{p+c},\ p+c \right\rangle$$

the first component expresses the nature (quality) of certainty and the second one expresses the degree (quantity) of certainty. Evidently, the higher certainty is, the more information there is, i.e. the more both justifying and rejecting examples there are.

We will consider internal truth values represented as ordered pairs. The first component of this pair is the nature of certainty, which can be computed by one of the above methods. The second component is the number of stages of iterative process that the calculation of the certainty nature has completed. The proposed iterative process is organised so that at each stage we obtain results less certain than at the previous ones. In this case the second component should be considered not as the degree of certainty but as the *degree of uncertainty*.

Among the logics with two classes of truth values we distinguish the logic which has exactly two internal truth values: "*uncertain*" (unknown) and "*certain*" (known). In this minimal case we accept only arguments "for" and we are not interested at all in the arguments "against". Then

- A statement gets the truth value "*uncertain*" if there are no arguments "for" this statement;
- A statement gets the truth value "*certain*" if there are arguments "for" this statement.

Now let us see the intuitive motivation for dividing truth values into two classes: the class of internal and the class of external truth values. We suppose that there are two levels of knowledge:

- External, theoretical or conventional (objective)
- Internal, empirical or individual (subjective)

At the external level there are general statements from a generally accepted theory. We suppose that knowledge at the external level, i.e. the objective knowledge, is subordinated to the principles of the classical logic.

At the internal level there are facts (results of experiments and observations) and hypothetical empirical regularities which can be obtained by direct analysis of the experimental results (e.g. by the use statistical methods) without developing an advanced theory. At the internal level we can use many-valued logic and formal analogues of the plausible rules, as will be seen later. The truth values of the generated hypothetical empirical regularities will be defined by these plausible inference rules recursively cutting down the set of cases with the evaluation "uncertain" in the initial set of facts.

3.3 Reasoning

The term "reasoning" has various meanings as we have seen in Sect. 1.3. Therefore we are compelled to explain here what will be understood as "reasoning" in the present book.

We consider reasoning as a version of a cognitive activity including two components: argumentation of some statement or set of statements on the one hand and generation of the new information, which may also include new knowledge. However not any such activity can be called reasoning.

We will understand as reasoning both the process of application of reasoning rules, and the result (history) of this process, represented in the form that permits one to check if the rules are applied correctly.

Thus an important feature of reasoning is that it is submitted to rules. A reasoner should in advance accept a set of rules as applicable. Reasoning rules contain premises and the conclusion. If the reasoner considers all premises of a rule credible it should consider the conclusion credible as well.

Now we discuss the rules that generate the reasoning processes. Reasoning rules can be divided into two classes, reliable and plausible:

- A rule is *reliable* (deductive), if the truth of the premises *necessarily* implies the truth of the consequence.
- A rule is *plausible*, if the truth of the premises of the rule does not necessarily imply the truth of the consequence. However, we will suppose that the truth of the premises of plausible rules implies the truth of the consequence in a *significant quantity of cases*.

We emphasise that the requirement of plausibility of reasoning is connected with the possibility of using the results in practice. It is impossible to give some qualitative restrictions in advance, and it is even more impossible to state that the conclusion of a plausible rule is valid in most of the cases when the premises are valid. It can only be said that, if the premises of a plausible rule are valid, then the conclusion may sometimes also be valid. However, this is sufficient for us if it helps in practice.

Although reliable rules do not allow a false conclusion to be reached from true premises, they also do not make it possible to receive essentially new information, to make a discovery. Plausible reasoning on the contrary gives a chance to generate essentially new information, though does not guarantee the truth of the received conclusions. New information (including knowledge) received by means of plausible reasoning will have the status of a hypothesis and will require verification.

The rule *modus ponens* may serve as a good example of a reliable rule:

$$\frac{\Phi, \quad \Phi \to \Psi}{\Psi}.$$

From the validity of premises of this rule the validity of the conclusion follows.

Abduction, on the other hand, is an example of plausible rules:

$$\frac{\Phi \to \Psi, \quad \Psi}{\Phi}.$$

The validity of premises of this rule can lead to false conclusions.

Despite the unreliability of plausible rules, they are widely applied both in everyday life and in scientific research, in particular in natural science. The conclusion of a plausible rule is not necessarily true (even if the premises are valid). Therefore this conclusion is usually accepted as a hypothesis which needs to be checked later. Plausible rules are powerful tools for discovering laws and regularities, for searching possible explanations of a studied phenomenon, for prediction etc. Together, they are a rather effective mechanism used by a reasoning agent in order to adapt to environmental conditions.

Although usual logic language is traditionally used for the formulation of plausible rules, it does not always adequately reflect the intuitive sense of these rules. Let us consider, for example, abduction reasoning. The rule scheme given above can be redefined by using the *quantifiers* as follows:

$$\frac{\forall x\ (\Phi(x) \to \Psi(x)), \quad \Psi(a)}{\Phi(a)}.$$

Formal logical interpretation of abduction may result in rather funny examples of reasoning:

All hens are birds.
An eagle is a bird.
Hence, an eagle is a hen.

This reasoning perfectly corresponds to the following abduction scheme:

$$\frac{\forall x \ \ (\mathrm{Hen}\,(x) \rightarrow \mathrm{Bird}\,(x))\,, \quad \mathrm{Bird}\,(\mathrm{Eagle})}{\mathrm{Hen}\,(\mathrm{Eagle})}$$

In this case we interpreted the symbol $\rightarrow$ as implication, that is as a logical entailment. Such interpretation of this symbol in the abduction scheme can lead to rather strange (and obviously incorrect) results. However, it does not mean that the abduction rule is bad. It is really not reliable, but in many cases it can still be useful.

Let us consider for example the following reasoning that could have been performed by a prehistoric person:

> If there is a lion nearby a loud roar may be heard.
> A loud roar is heard.
> Hence, there is a lion nearby.

This reasoning is also carried out by one of the variants of the abductive scheme. The practical use and high degree of plausibility of such reasoning is obvious. However, the relation between the simple statements in the first premise of the rule does not yield a logical implication. In fact, there is a cause–effect relation between these statements. The scheme of abductive reasoning

$$\frac{\Phi \rightarrow \Psi, \quad \Psi}{\Phi}$$

is possible to interpret in this case as follows:

> The event Φ is the cause of the event Ψ.
> The event Ψ occurred.
> Hence the event Φ should occur.

In fact the event Ψ cannot occur spontaneously; it should have a cause and it is Φ. Therefore, Φ has also occurred (it had occurred earlier than Ψ, as the cause precedes the consequence in time).

However, the event Ψ can have other reasons, which are different from Φ. The credibility of abductive reasoning increases if the most plausible and probable cause is used to explain the fact that the event Ψ has occurred.

In practical reasoning, the choice of a cause may be aimed at minimisation of risks, i.e. the most risky cause is chosen, because considering this and therefore taking necessary precautions permits the reduction of the possible negative consequences.

So, if in the abduction scheme we interpret the symbol $\rightarrow$ not as implication but as a cause–effect relationship, we will not receive absurd examples of application of this rule. Besides that, the degree of credibility of this rule will increase, if the most probable cause is considered as the conclusion.

Below, we will show that, for other rules of plausible reasoning, the symbol $\rightarrow$ should also be interpreted not as an implication (logical entailment) but as an indication of the presence of a cause–effect relationship.

We will consider, for example, inductive reasoning. One of the variants of induction schemes can be written as follows.

$$\frac{\begin{array}{c}\Phi(a_1) \rightarrow \Psi(a_1) \\ \ldots \\ \Phi(a_n) \rightarrow \Psi(a_n)\end{array}}{\forall x\ (\Phi(x) \rightarrow \Psi(x))}.$$

This scheme of inductive generalisation is well interpreted in a language of cause–effect relations. If we know that, in the cases $a_1, \ldots, a_n$, after the event Φ there is an event Ψ we make a conclusion that, in any case, the event Φ implies the event Ψ, i.e. Φ is the cause of Ψ.

I.e. in this case in the premises of the rule the symbol $\rightarrow$ may be interpreted as an indication that the event denoted on the left of the symbol $\rightarrow$ occurs not later than the event denoted on the right of the symbol $\rightarrow$. In the conclusion of the rule the symbol $\rightarrow$ points to the presence of a cause–effect relation.

The number of examples sufficient to draw a conclusion on the cause–effect relation depends on a concrete situation, namely on the importance of a consequence for the reasoner and on an estimation of the risk of the consequence. It is clear that even one example will be sufficient for a child to draw the conclusion that *touching a flame leads to a burn*.

Speaking about the cause–effect relations in the context of plausible reasoning it is necessary to note that naive inductive reasoning does not guarantee detection of the *true* cause of the phenomenon. It only permits establishing the fact that there is a connection between the considered phenomena and that one phenomenon accompanies another. Bright light (e.g. lightning) is not necessarily the cause of a loud sound (e.g. thunder), but both of these phenomena are consequences of the electric charge, which in turn is a consequence of the accumulated potential difference between a thundercloud and the surface of the Earth.

For detection of the true (real) cause, plausible reasoning is usually not sufficient; theory development is required as well. By means of plausible (particularly inductive) reasoning empirical knowledge is formed: the knowledge of laws, the knowledge about combining some phenomena, properties of objects or events or the impossibility of combining them. This empirical (pre-theoretical) knowledge is also important as it is the basis of the theory development. Besides, this empirical knowledge permits predictions to be used in practice.

In the examples considered in this book we will often speak about cause and effect relations, about searching for such relations, about the use of knowledge of cause–effect relations etc. In most cases our wording about cause–effect relations will be simply a convenient metaphor. What we will call cause will in fact only seem to be a cause. We suggest that such illusionary cause should be called a *phenomenological cause* as it is perceived as the cause by the consideration of clearly observable *phenomena*. At the same time detection of illusionary causes allows for finding real laws, which on the one hand can serve as a basis for theory construction and on the other hand can be used for predictions.

However this reasoning is not deductive since the cause–effect relation (like the relations of compatibility, regular support etc. which *appear* to be of cause–effect) is not identical to logical entailment. In this case we deal with plausible reasoning where the truth of the premises does not guarantee the truth of the conclusion.

For example, consider the following the plausible rule:

$$\frac{\text{Bright-Coloured}\,(\text{Milabris})\,, \forall x\,(\text{Bright-Coloured}\,(x) \rightarrow \text{Venomous}\,(x))}{\text{Venomous}\,(\text{Milabris})}$$

Really, there are brightly coloured insects that are not poisonous. They use bright colouring like poisonous insects for scaring away birds.

In the case of logical entailment, the truth of a statement Φ and the truth of the statement $\Phi \rightarrow \Psi$ imply the truth of the statement Ψ. In the case of the cause–effect relation that there was an event Φ which can serve as the cause of the event Ψ, this does not imply that the event Ψ will necessarily occur. Firstly, there can be circumstances which prevent the occurrence of the event Ψ even in the presence of its possible cause Φ. These circumstances (inhibitors) do not allow event Φ to occur as the cause of the event Φ. Secondly, cause–effect relations can have a statistical property, and the consequence is present not always, but only in a certain percentage of cases.

However, predictive reasoning similar to that discussed above is often called deductive because of formal similarity of the rules. We would like to apply the term "deductive" only to reliable reasoning that is reasoning preserving truth. To plausible predictive reasoning, in our opinion, another term should be applied, at least because this reasoning explicitly or implicitly uses the concept of *similarity* of objects or events, and similarity presence is the main condition in reasoning by analogy (cf. [Finn, 1991]).

Note that in this case similarity is understood as a general characteristic (a set of general characteristics) of some objects or events. I.e. similarity is understood as a separate entity, and not as a relation between objects or events. In our example the similarity is the bright (red or orange with black markings) colouring of an insect.

However, in our opinion, reasoning of the above-described type is already quite different from reasoning by analogy. Since we assume that it is not correct to call such a rule of reasoning deductive, another term should be offered, differing from both deduction and analogy; for example, it is possible to call this rule *causal prediction*.

The considered examples of widely known plausible rules show that these rules should be formulated not in the language of logical entailment (implication), but in the language of cause–effect relations. This will prevent possible logical errors and lead to more adequate understanding of the character and degree of plausibility of the examples of application of this rule.

Examples in this chapter show which rules of plausible reasoning can be used for the generation of new information and new knowledge. Such rules can serve as samples when defining their formal analogues, which we will call modification rules. Certainly there are a lot of possible variants of rules of induction, abduction and analogy. For example, in the premise of the induction rule it is possible to use various statistical criteria. Moreover, there is a possibility of definition of essentially new rules or the rules which combine the well-known ones.

However the concrete examples considered above are only to illustrate the possibilities of our approach. The book does not aim at detailed consideration of concrete rules and methods of plausible reasoning. Its main task is the description of a general scheme of knowledge extraction from data using formalised plausible reasoning. The rules of such reasoning, generally speaking, can be arbitrary. The only restriction is that they should allow formalisation through the methods and techniques offered in the subsequent parts of the book.

Chapter 4
Constructing a Model of a Cognizing Agent

In this chapter we describe the structure, processes, techniques and methods at the conceptual level, which according to our approach will be necessary (i) to understand (e.g. human) cognition and (ii) to support its modelling. This structure, together with all the constituents forms, we call a "cognitive architecture". This architecture will be developed w.r.t. a cognizing agent, which is the main actor in our approach. A *cognizing agent* means either a living entity (particularly a human being), or a group of them or a technical system, which can *adapt* to the changing conditions of the external environment.

A cognitive architecture specifies the underlying infrastructure for a cognizing agent. A cognitive architecture includes the generic functions that are constant over time and across different application domains, and it explains how its capabilities and functioning arise out of the capabilities, functioning, relationships and interactions of the constituents.

The architecture of almost all existing models of cognition include three subsystems:

- A sensing subsystem, which is responsible for receiving information from the agent's environment
- An information-processing subsystem, which forms a "world model", which the agent uses as a guide in a decision-making process
- An effecting subsystem, which realises decisions of the agent, thus influencing the environment

These constiuents support the achievement of the main objective of a cognizing agent's functioning, which is adaptation to the constantly changing conditions of the environment. Along the adaptation process a cognizing agent perceives information (data) about the environment and generates reactions. In this process the model of the environment, i.e. the word model of the cognizing subject, plays a significant role. A given world model is a set of statements regarding what exists and what does not exist (either in actuality or in principle) in the environment, what the experiences are and of what type they are, etc. That is, a world model defines what is known about the environment, what has been done with it and how this can be known or

O. Anshakov, T. Gergely, *Cognitive Reasoning*, Cognitive Technologies,
DOI 10.1007/978-3-540-68875-4_4, © Springer-Verlag Berlin Heidelberg 2010

done. The world model is used for the formation of the reaction. The more adequate and flexible the model of the world of the intelligent agent, the more correct his response to the influence of the environment and the more well adapted it will be to the environment.

As for the world model itself, it is obvious that the most important role in it belongs to knowledge represented in one form or another. For example, machine learning and soft computation can to be used for the formation of the world model.

The structure of a cognizing agent described in the present chapter also includes the three basic subsystems mentioned above. It also forms a world model based on knowledge. The degree of adequacy of this model is of course connected with the adaptation abilities of the cognizing agent.

However, in the proposed model we mostly focus on the *cognitive abilities* of the cognizing agent which are part of the adaptation mechanism. We do not study here questions connected with the maintenance of the activity of the cognizing agent which is distinct from knowledge acquisition and world model formation.

We can assume, for example, that this agent is a component of a multiagent system, where different agents have different functions. *The main task of the cognizing agent is to acquire knowledge* which it can share with other agents, providing the adaptation of the system as a whole.

A cognizing agent can be of any nature: a living creature (in particular a human), a group of people, an institution, a technical device, or it can be a hybrid object, such as a group of people using certain hardware and software.

The conceptual model of the cognizing agent which we are going to develop in this chapter results from *abstraction*, *idealisation* and *generalisation* of those characteristic features of different types of cognizing agents. This inevitably leads to omitting some details, which may be important for concrete realisations of the cognizing agent, but are insignificant for the description of its general scheme. Moreover, certain restrictions will be imposed upon the model of the cognizing agent, and these restrictions, which make it less general, will in return allow for the formalisation and algoritmisation of this model.

The basic restrictions to our model of the cognizing agent are as follows:

- For knowledge extraction a cognizing agent uses *reasoning in a broad sense*, which may include deductive inference, plausible argumentation and computing procedures;
- Activity of a cognizing agent is *discrete*; it consists of alternating phases of perception of the environment and reasoning directed at formation of new knowledge;
- In the course of functioning of the cognizing agent a *history* of its cognizing activity is formed.

The requirement of discreteness is connected, first of all, with the necessity to define such a model that permits formalisation and algorithmization. However, in our opinion, the process of analysis of the available information and the process of acceptance of this information for consideration do not occur simultaneously even in natural cognizing agents.

Generally speaking, we represent a cognizing agent as a *researcher*, not as a hunter or a soldier. It is not the speed of reaction, but the completeness, validity and depth of the knowledge that will be the criteria of success of such an agent. Therefore it is quite justified for such a cognizing agent to act thoroughly, systematically and even slowly. The discreteness of the activity of this agent manifests itself in the fact that *first of all* it accumulates information and only *after* that analyses it and forms new knowledge using all data available to the analysis.

Below, we will consider the history of activity of a cognizing agent as a theory of a special type. This theory will be called *open cognitive* or *quasiaxiomatic theory*. The history of activity of the cognizing agent, presented as an open cognitive theory, is an essential component of self-reflection of such an agent.

4.1 The Structure and Functioning of the Cognizing Agent

The structure of a cognizing agent includes the following subsystems:

- A sensing subsystem
- An information-processing subsystem
- An effecting subsystem
- The long-term memory which stores history of the activity of the cognizing agent

It is the information-processing subsystem that we will study in detail. The description of the formalisation and the ways of algorithmic and program realisation for this subsystem are given in this book. The other subsystems will be characterised only in a very general way. The structure of the cognizing agent is represented in Fig. 4.1.

The *sensing subsystem* serves for receiving signals from the environment, in which the physical and information parts can be distinguished. The reception of signals from a physical environment causes formation of data, for example data of observation and experiments. The information environment is understood as a set of other cognizing agents or their components which have the possibility to transfer information to the agent in question. For example, information can be transferred from a user to an artificial cognizing agent realised as a program. Signals from an information environment can be interpreted as additional information, advice or instructions. These will be transformed by the sensing (accepting) subsystem into data or knowledge.

The *effecting subsystem* is used for carrying out some actions which can be interpreted as a response to environmental influence. Since we have agreed that the main task of our agent is cognition (knowledge acquisition), its actions should also promote cognition. For example, carrying out additional observation and experiments can be such actions.

Long-term memory stores history of the discrete activity of the cognizing agent, which we consider here as a theory of a special kind. This history can be interpreted

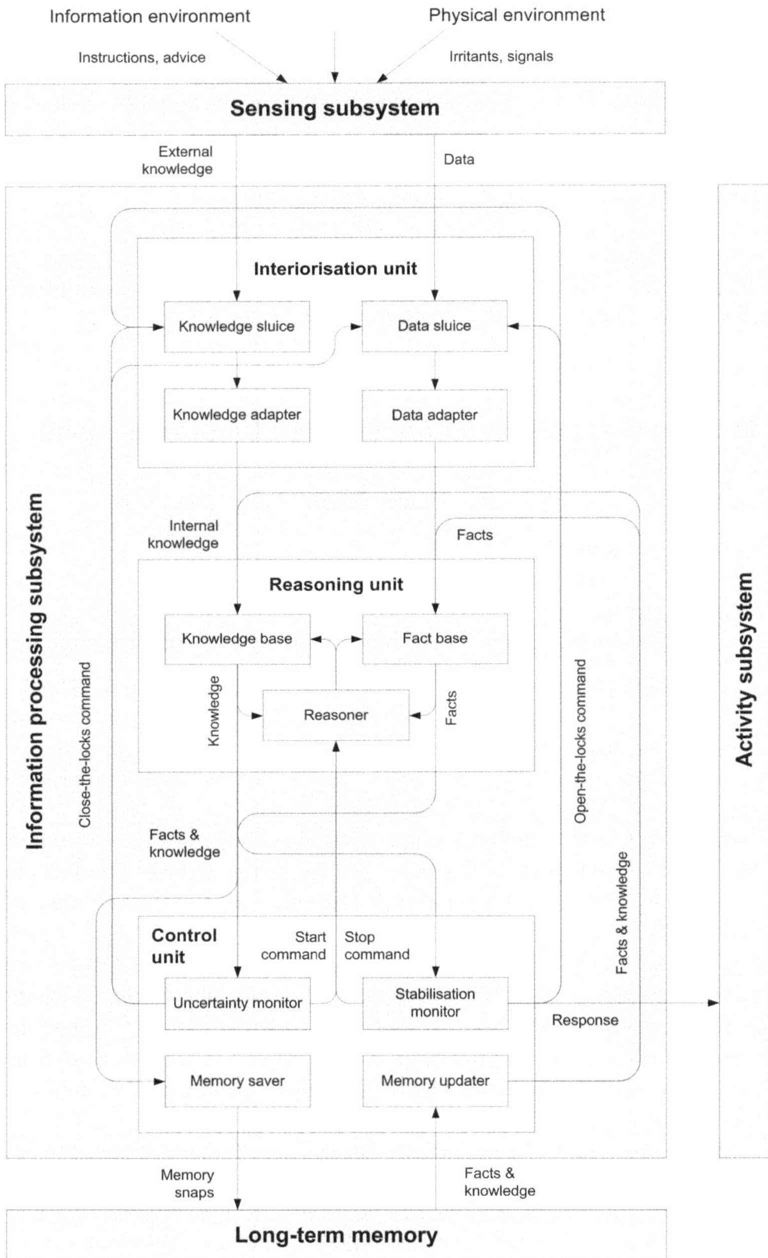

Fig. 4.1 General scheme of the cognizing agent

as a sequence of *knowledge states*. The functioning of the long-term memory will be described in detail in the next chapter.

Now let us study the information-processing subsystem, which is presented in Fig. 4.1 essentially in much more detail than the other subsystems. *The information-processing subsystem* contains the following units:

- The interiorisation unit
- The reasoning unit
- The control unit

The *interiorisation unit* is used for the accumulation of the information arriving from the environment and for its transformation into an internal representation (an internal format) of the cognizing agent. The sensing subsystem transfers to the interiorisation unit the data and knowledge in external formats. They are considered *external* data and knowledge. The data and knowledge are collected in the corresponding buffers (the *data buffer* and the *knowledge buffer*). This accumulation can occur in parallel with other actions of the cognizing agent, for example cognitive reasoning.

When it becomes necessary, the data and knowledge from the buffers will be transformed by corresponding converters (the *data converter* and the *knowledge converter*) into an internal format of the cognizing agent. The data in the internal format of the cognizing agent are called *facts*. The transformation of the data and external knowledge into facts and internal knowledge is called *interiorisation* of data and knowledge.

The *reasoning unit* is the kernel of the considered subsystem. It contains the *fact base*, the *knowledge base* and the *reasoner* (the mechanism of reasoning). The fact base and the knowledge base together form the short-term (working) memory. The reasoner introduces changes in the working memory during its operation (both in the fact base and in the knowledge base). Changes in the knowledge base can be interpreted as acquisition of new knowledge, and changes in the fact base can be interpreted as generation of predictive hypotheses.

The formalisation of the operation of the reasoning mechanisms proposed in this book is such that the facts and knowledge in this formalisation will have similar formats of representation. Roughly speaking, knowledge as well as facts will be represented in the form of simple concrete statements. Despite a similar syntax, the meaning of facts and the meaning of knowledge strongly differ:

- *Facts represent statements which can be checked directly*. Facts are usually representations of results of observation or experiments, or are statements which could describe the results of observations (experiments) if these observations (experiments) were to be performed.
- Knowledge represents natural regularities in a subject domain which cannot be found out directly, but can be discovered by means of reasoning.

For example, the simple concrete statement "After the event the event occured" is the representation of an *individual, concrete fact*, but the statement "The event is the cause of the event" represents a *general law*, i.e. knowledge. (This statement can be read as "After the event always the event occurs".) The general law cannot be seen directly, but it can be revealed, for example, by means of inductive generalisation.

In accordance with the way of formalisation of reasoning developed in this book, simple concrete (atomic) statements representing facts and knowledge should be supplied with *truth values*. In the course of operation of the reasoner the truth values of facts and knowledge will be changed. This change is directed at an increase of definiteness.

In the elementary case (which is considered in our formalisation) the value "undefined" is replaced with one of the defined truth values. Thus, in the course of operation of the reasoner a narrowing of the area of uncertainty takes place in both the fact base and the knowledge base. We interpret such narrowing of the uncertainty area as movement from ignorance to knowledge.[1]

Operation of the reasoner is carried out by a certain algorithm by means of a sequence of sets of rules. This algorithm also belongs to knowledge. From our point of view, it is an example of *metaknowledge*. The operation algorithm of the reasoner generates the implementation of iterative procedures of the same type. The algorithm itself cannot be changed as a result of such iteration, but it can be changed after a cycle of the reasoner operation is completed when the reasoning unit can be retuned.

Thus, the structure of the knowledge base is not homogeneous. On the one hand, there is a part of the knowledge base that can be changed as a result of reasoning. This part includes subject knowledge presented in the form of atomic statements about natural regularities between certain events, phenomena, structural and functional characteristics etc.

On the other hand, there is a part of the knowledge base which does not change within one cycle of the reasoner's operation. This part contains some a priori knowledge which can be derived from the accepted theory of the given subject domain and which is beyond doubt. Besides that, this part contains metaknowledge necessary for the operation of the reasoner, for example, plausible reasoning rules which the reasoner uses in the operation. The operation algorithm of the reasoner is also in the same part, and it cannot be changed inside one cycle either.

The fact base is also heterogeneous. It contains a part which is not subject to change within one cycle of operation of the reasoner. For example, in such a part there may be some obvious facts connected with the description of the structure of objects in a certain subject domain.

In the framework of the conceptual model we will not specify the internal structure of the reasoner. We only note that the reasoner should possess its own *con-*

[1] Of course such a representation of the reasoning process is an idealisation and simplification of the real situation. The reasoning process could be reduced to the change of truth values if the cognizing agent had at its disposal all possible atomic statements representing facts and knowledge. Although the cognizing agent deals with the final fragment of a subject domain, this final fragment can be so big in real problems that its exhaustive search would be impossible. Therefore in a reasoning process the atomic statements representing facts and knowledge can be generated. Some additional mechanisms built into the reasoner will be responsible for the generation. These mechanisms remain outside the formalism which is introduced in this book, but will be mentioned in connection with realisation.

trol mechanism which manages only the reasoning process. (Do not confuse this mechanism with the *control unit*, which is intended to control the operation of the cognizing agent as a whole.) The control mechanism of the reasoner includes an interpreter of the reasoning algorithm. This mechanism implements the rules in the order demanded by the reasoning algorithm.

Note that the reasoner can include some mechanisms which perform those actions which are not reduced to application of rules. Among such actions can be calculations, statistical estimations, generation of structures of some objects etc.

The control unit is intended for managing the activities of the cognizing agent. Actually this unit controls the switching over of the operating modes of the cognizing agent. Our model distinquishes two modes (or phases) of operation of the cognizing agent:

- The mode of interiorisation of the perception
- The mode of cognitive reasoning.

The main task of the interiorisation phase of perception is updating (retuning) of the reasoning unit. This updating can affect all the components of this unit. The basic action of the interiorisation phase of perception is conversion of data and knowledge into an internal representation (format) of the cognizing agent.

During a perception phase data and knowledge taken from the long-term memory can be added to the data and knowledge which the cognizing agent receives from the environment. This addition, generally speaking, is a rather complex action. It can be accompanied by analysing whether the added information is compatible with the already available information, searching for the information critical for the further operation of the agent etc. However, in studying the conceptual model of the agent we will not concretise the description of the technique used for information extraction from the long-term memory.

The operation algorithm of the reasoner also belongs to knowledge. It can also be replaced in the interiorisation phase of perception. For example, a new algorithm can be accepted from information constituents of the environments (as advice or instruction) or it can be taken from the long-term memory (in this case the agent returns to one of its previous reasoning algorithms).

The activities taking place in the *reasoning phase* are as follows:

- Proper reasoning, i.e. operation of the reasoning mechanism described above in detail
- Accumulation of data and knowledge in the buffers of data and knowledge
- Actions (operation of the effecting subsystem) aiming at reception of new data and knowledge from the environment, for example carrying out observations

These three activities can occur in parallel.

There are also actions which take place between the phases, namely:

- After the completion of the interiorisation phase and before the beginning of the reasoning phase the content of the short-term (working) memory gets saved in the long-term memory;

- The saving of the content of the working memory in the long-term memory similarly takes place after the completion of the reasoning phase and before the beginning of the interiorisation phase.

Thus, for each cycle of the reasoner's operation, in the long-term memory these remain two snapshots of the working memory, fixing two knowledge states: prior to the beginning of reasoning and after the completion of the reasoning.

Two checkers are used to manage the switch over of the operating mode of the information-processing subsystem.

The *starting condition checker* monitors the condition which has to be satisfied in order to start the reasoner. This monitoring takes place within the interiorisation phase. The starting condition can be defined in various ways. Generally the fulfilment of this condition means that the cognizing agent has enough facts to start to extract knowledge from them. This sufficiency can be expressed in terms of quantitative (for example, statistical) criteria.

If the starting condition is satisfied, then the following actions are carried out:

(i) The starting condition checker sends the command "lock" to the data and knowledge buffers ;
(ii) The data and knowledge buffers are locked;
(iii) The interiorisation process of perception, i.e. the data and knowledge conversion into an internal format of the cognizing agent, stops;
(iv) The process of filling the fact and knowledge bases stops;
(v) The snapshot of the working memory is saved in the long-term memory (the "Memory saver" subblock is used);
(vi) The starting condition checker sends the command "start" to the reasoner;
(vii) The reasoner starts working.

The *stop condition checker* monitors the condition which is to be satisfied in order to stop the reasoner's operation. This monitoring takes place within the reasoning phase. The stop condition can be defined in various ways. For example, the reasoner can suspend operation after a given period of time, or upon the completion of a given number of iterations of its operation cycle. The stop condition can also depend on quantitative criteria, for example on the relative size of the uncertainty area. The most rigid stop condition is the *stabilisation or saturation condition*, i.e. the occurrence of a situation in which it is already impossible to receive new knowledge and to narrow the uncertainty area as a result of reasoning.

If the stop condition is satisfied, then the following actions are carried out:

(i) The stop condition checker sends the command "stop" to the reasoner;
(ii) The reasoner stops the operation;
(iii) Operations with various kinds of memory and effecting subsystem are carried out (these operations will be considered in more detail below);
(iv) The interiorisation process of perception, i.e. the conversion of the data and knowledge into an internal format of the cognizing agent, starts again;
(v) The process of updating the fact and knowledge bases begins, which permits data and knowledge to be received not only from the environment, but also, with the use of the "Memory updater" subblock, from the long-term memory.

Now we will describe two action sequences connected with various kinds of memory, and with the effecting subsystem in detail. These action sequences, mentioned above in (iii), can be carried out in parallel.

The actions related to the effecting subsystem are as follows:

(i) The results of the operation of the reasoner are evaluated and actions for obtaining new data and knowledge are planned;
(ii) A message is sent to the effecting subsystem;
(iii) The effecting subsystem begins its operation.

The actions related to the different kinds of memory are the following:

(i) The snapshot of the working memory is saved in the long-term memory (the "Memory saver" subblock is used);
(ii) The working memory is prepared for updating (in the simplest case it is cleared);
(iii) The stop condition checker sends the command "unlock" to the data and knowledge buffers;
(iv) The data and knowledge buffers are unlocked.

So, we have completed the sketch description of the processes occurring in the cognizing agent.

Chapter 5
Cognitive Reasoning Framework

In this chapter we aim to provide a conceptual theory of cognitive reasoning processes, i.e. processes that permit a cognizing agent to gain new knowledge.

According to our approach the conceptual model should provide the preparatory stage for the development of the formal theory of cognitive reasoning. Therefore the main constituents of the formal theory should be prepared on the conceptual level. Our conceptual theory should meet the following main requirements.

First, this theory should allow the adequate representation of and the dynamics of the cognitive reasoning processes. This dynamics in particular includes the possibility to modify the truth value of statements that constitute the reasoning processes. In an elementary case, the truth value "uncertain" will be changed to "true" or "false". However, even in this (elementary) case this will result in a contradiction because one and the same statement cannot have several truth values; it cannot be simultaneously *uncertain* and *true*. Therefore the theory should support the representation of non-monotonic processes. In this theory during the construction of a reasoning process as an inference we should not only be able to introduce new statements as the conclusions of inference rules but also to revise those statements which have already been deduced.

Second, this theory should consider plausible reasoning. Plausible reasoning, for example reasoning by induction or by analogy, allows the incorporation of essentially new statements (to integrate new knowledge). However, it also allows contradictions since the validity of the premises of plausible statements does not predetermine the validity of the conclusion. Therefore the proposed theory should be able to handle contradictions arising from non-reliable reasoning. Note that these contradictions can arise despite the ability of a theory to revise previous statements.

Third, this theory should be closely connected with reasoning processes of the cognizing agent. Hence, it should reflect the structure and functions of the architecture units of the cognizing agent, responsible for reasoning. In particular, it should include components which correspond, e.g. to the fact base, knowledge base or operation algorithms of the reasoning unit.

According to the last requirement the theory should provide appropriate tools for:

O. Anshakov, T. Gergely, *Cognitive Reasoning*, Cognitive Technologies,
DOI 10.1007/978-3-540-68875-4_5, © Springer-Verlag Berlin Heidelberg 2010

- The description of the observed events and objects
- The description and representation of knowledge
- The cognitive reasoning processes from observation to extracting information and knowledge

In Sect. 5.1 we develop all the theoretical constituents important to define a coherent theory satisfying the above requirements. These constituents will describe and represent the internal and external information processing related to cognitive reasoning. Two classes of theories, the open cognitive and the modification theories will be the main constituents that play a significant role in the definition of the cognitive reasoning framework, i.e. of the CR framework. This latter will provide a coherent theory to model the cognitive reasoning processes of a cognizing agent at the conceptual level. It will be shown in Sect. 5.2 how this modelling can be realised by using the proposed CR framework.

5.1 Theories of the CR Framework

The information-processing activity of a cognizing agent can be represented in the form of several embedded cycles, from which we emphasise two embedded cycles. The internal cycle consists of iterative reasoning algorithms until the completion condition is met. The external cycle represents repetition of perception and reasoning phases and proceeds from the beginning to the end of the activity of the cognizing agent. Note that the completion condition of the activity of a cognizing agent will not be defined in the theory. Therefore this cycle can be considered potentially infinite. In practice it can be carried out, for example, until exhaustion of resources (physical or informational) of the cognizing agent.

It is desirable that the conceptual theory would represent both the internal and the external cycles. These cycles will determine the cognitive reasoning processes of a cognizing agent. The model that is able to handle the embedded cycles of information processing related to cognitive reasoning will be called a *cognitive reasoning framework* — a *CR framework*.

For the representation of the processes that are taking place in the "brain" of the cognizing agent we suggest the use of a CR framework which will support two classes of theories: (i) the open cognitive or quasi-axiomatic theories and (ii) the modification theories.

In our view these theories should first of all provide convenience and freedom in the process of building models of the environment and then these theories should permit fairly radical changes of the paradigm in the cognitive process.

A theory of the open cognitive or quasi-axiomatic theories will be called open, as it is open to the interaction between the cognizing agent and its environment. It should be open for:

- Obtaining new facts for analysis
- Adopting and internalising knowledge from other similar theories

- Modification of reasoning rules
- Modification of the language
- Modification of the internal form of the representation of data and knowledge

The open cognitive theories will model the external cycle of the operation of a cognizing agent. The detailed consideration of the open theories will be linked to the second class, i.e. the modification theories.

We will now consider the modification theories in more detail, which will represent the internal cycle of the operation of the cognizing agent. This cycle is connected with reasoning. At the conceptual level we are not going needlessly to concretise the reasoning algorithm implementation which is repeated in this cycle. We only note that we mean here a simple enough procedure of *traversal* of some distinguished part of the working memory including a part of the fact base and a part of the knowledge base. This traversal is accompanied by the analysis of the facts, extraction of knowledge and transformation (modification) of the fact and knowledge bases.

These theories are called modification theories because, during the construction of a process in such theories that will model the reasoning process, not only can a new statement be added, but the process itself can be modified. Modification theories use a special technics of inference for modelling the processes. An inference in the modification theory will be called a *modification inference*. For the construction of a modification inference, *modification rules* will be used. Modification rules will modify: (a) the truth values of the statements entering into an inference and (b) the modification inference itself. These rules correspond to the plausible reasoning rules.

The modification rules change not only the inference, but also the truth value of some statements; they replace the truth value "undefined" with one of the given truth values, for example with "true" or "false". However, such values can be as many as wished: they will express the degree of trust to an estimated statement, i.e. the degree of its truth or falsity.

Thus, modification theories permit the representation of dynamics of cognitive processes, understood as movement from ignorance (uncertainty) to knowledge (definiteness).

Now let us see how a modification theory is related to the architecture of the cognizing agent. This theory includes: (i) axioms (of the fact base and of a part of the knowledge base) and (ii) inference building (modification) rules (the residual part of the knowledge base). That is, the axioms of the modification theories are analogues of the facts and knowledge, hence the content of the components (such as the fact base and knowledge base) of the reasoning unit of the cognizing agent is represented in the modification theories

Modification rules in the theory are organised into a system analogous to the reasoning algorithm of the cognizing agent. This system controls the construction of a modification inference that is divided into stages, each of which is an analogue of one iteration of the implementation of the traversal of the fact base and knowledge base.

Non-monotonity of a modification inference is connected with the deactivation of some formulas entering an inference. That is, non-monotonity is connected with the modification of inferences, whereas those statements whose truth value has been changed are declared to be inactive. Those statements which depend on the deactivated ones are declared inactive too. The inactive statements cannot be used as premises in further inference extension any more; i.e. they cannot influence further construction of the inference. Actually, it is possible to consider these formulas as removed (marked as removed) from the inference. However, these features of an inference in the modification theories (of a modification inference) do not eliminate the possibility of contradictions in the use of modification rules. However, a modification inference possesses a rather unusual property of *temporal contradictions*: contradictions which have appeared in an inference can later disappear because of a modification of the inference, and contradictions can appear again, etc.

Thus, modification theories can handle contradictions.

The inference in a modification calculus can be considered as a history image of the reasoning of the cognizing agent. Namely, they are such reasonings that are not interrupted by a perception phase and therefore are carried out entirely in a reasoning phase. We say in this case that reasonings do not fall outside the limits of one operation cycle of the reasoner. We will notice that this cycle, generally speaking, consists of several iterations of a traversal belonging to the fact and knowledge bases. This traversal is accompanied by the analysis of facts and extraction of knowledge, as already mentioned.

As a model of a modification theory a sequence of structures is considered. Each structure reflects on the one hand a knowledge state about a considered fragment of a subject domain, whereas on the other hand it is an image of the working memory (of the integration of the fact base and the knowledge base) of the cognizing agent at some stage of reasoning. Application of modification rules leads to modification of the structures, i.e. it appears as *modification of the semantics*.

The above can be considered as a set of arguments that modification theories represent the structure and functions of the reasoning unit.

A modification theory is an adequate representation of the situations and the processes corresponding to one cycle of the reasoner's operation. However, we would like to describe the theory which would correspond to the activity of the cognizing agent as a whole. For this we need open cognitive theories. Note that a theory of this class will also be *open* in the following sense: it will not be reduced to generating consequences from a set of postulates (and the consequences received by the means of both reliable, and plausible, reasoning), but it will be directed to receiving information from an external world (it will be open for external data and knowledge).

Such an open theory should reflect the various processes connected with the activity of the cognizing agent. In particular it should possess all possibilities of modification theories, i.e. it should have means for representing the operation of the reasoner in a reasoning phase. Moreover, it should reflect the possibility of radical reorganisation of initial postulates due to the receipt of new data and knowledge from both the physical part and the information part of the environment.

In the description of such an open theory there are points considered informally. For example, we are not going to include any restrictions on the methods of reception of the information from the external world. We note that it would be an interesting problem to define a reasonable set of methods of information search in an external world and by that to describe a subclass of open theories discussed here. Besides, we do not impose restrictions on the format of data and knowledge representation (except for a quite simple and natural requirement of representability in a language of a many-sorted first-oder logic).

Nevertheless, some definition of the open theory is necessary, not as a process or a potential possibility, but as a result, as systems of the artefacts that represent a fragment of the world model. From a naive point of view any theory can be represented in the form of a set of texts. This set can be extended and can be specified, but during each moment of time we deal with a fixed state of this theory and besides we can also know the history of the development of this theory. The knowledge of history of the theory can help further develop this theory.

Therefore, we define an *open cognitive theory* as a history of the activity of the cognizing agent, beginning with an initial state and coming to the end at a current time moment. To achieve the effective formal description of such a history we will fix only some key knowledge states.

For the explanation of what these key knowledge states represent, we will recall that in the activity of the cognizing agent two phases alternate: a perception phase and a reasoning phase. We will consider those states as key ones which surround a reasoning phase, i.e. a knowledge state of the cognizing agent before the beginning of a reasoning phase (it arises after the termination of a perception phase) and after the termination of a reasoning phase.

For each knowledge state it is possible to construct a corresponding modification theory. Therefore we define an open cognitive theory as a final sequence of pairs of modification theories representing the history of activity of a cognizing agent from the beginning of its activity to a current moment of time. It is obvious that such a theory can always be continued.

Now let us consider how an open cognitive theory as a description of the history of cognitive processes of a cognizing agent corresponds to the constituents of the cognitive architecture. Compact description of a history exists in the long-term memory of the cognizing agent. The activity schema of the cognizing agent is such that in the long-term memory the snapshots of the contents of fact base and knowledge base are stored in the following two cases:

- After the renewal of the short-term memory and before the start of the reasoning process
- After the occurence of the stabilisation, i.e. after the termination of the reasoning process

We will interpret the content of the fact base and of a part of the knowledge base as a collection of statements about the subject domain. These statements may be represented in a language of certain formal theory. The content of the remaining part of the knowledge base can be interpreted as a *system of formal rules* that correspond to the rules of reliable and plausible reasoning. Thus *the content of the short-term memory of the cognizing agent can be interpreted as a collection of axioms and rules of building inference for a certain theory.*

It could be supposed that the suggested schema for the activity of the cognizing agent is appropriate for a wide class of theories. Apparently this might be so, nevertheless we aim to restrict the class of theories which, according to our model, should be included in the history of the cognitive process.

We agree that the content of the fact base and of the knowledge base:

- Directly before the start of the reasoning process
- Directly after its completion

should be interpreted as the description of a modification theory, which includes:

- Axioms (of the fact base and of a part of the knowledge base)
- Inference building rules (the residual part of the knowledge base)

According to the above proposed model of the cognizing agent the cognitive process has a discrete character. In this process two phases alternate continuously:

- Reasoning phase: in this phase the cognizing agent is connected with the environment, but it does not take into account the signals coming from the environment in its reasoning;
- Perception phase: in this phase the cognizing agent consecutively solves two tasks:
  - Conversion of facts and knowledge into the internal form (adapters of data and knowledge are responsible for the solution of this task)
  - Formation of new content of the short-term memory from new and old facts and knowledge (the most intellectual component of the cognizing agent, the updater, which is a part of the memory controller, is responsible for the solution of this task)

Renewal of the content of the short-term memory is a task difficult enough to model. Its solution includes the following actions:

- Comparison of the newly received facts and knowledge with those which are already in the base in order to detect identity, similarity, difference and contradiction
- Selection of data and knowledge from the union of new and old ones by certain criteria (some of the possible criteria will be discussed below)
- Formation of new fact and knowledge bases; it is preferable (but not obligatory) that the union of these bases could be represented as a *consistent* set of formulas of some language, but in any case, *the fact base should be consistent*

Note that interiorisation of data and external knowledge will be completed only if a new fact base and knowledge base are formed. Only in this case will *new data and external knowledge be built into the internal system of facts and knowledge* of the cognizing agent.

A cognizing agent has two ways of acquisition of new facts and knowledge.

The *first* one is rather conservative. It provides a use of reasoning for generation of hypotheses about relationships between the entities from a subject domain. Some such relationships will have a general, universal character; these will be considered as knowledge. Others will have a partial character. The cognizing agent considers such hypotheses as facts. In this way new facts and knowledge acquisition are applied in a *reasoning phase* of the cognizing agent.

The *second* way of new facts and knowledge acquisition is more radical. It consists of receiving information from the environment. This environment has physical and information constituents. From the information constituent the cognizing agent receives knowledge which include declarative and procedural components. The operation algorithm of the reasoner for example belongs to procedural knowledge. The algorithms for the transformation of external data and knowledge into the internal representation of the cognizing agent also belong to procedural knowledge. Reception of the information from the environment occurs in a *perception phase* of the cognizing agent.

Note that in this phase there can be a radical updating of the fact and knowledge bases, for example:

- The reasoning algorithms can be changed,
- The algorithms for transformation of formats of the data and knowledge can be changed,
- The formats of internal data and knowledge representation, i.e. actually, the descriptive language for the facts and knowledge, can be changed.

A cognizing agent starts its cognitive activity with the perception phase. However, the initial perception phase is reduced, because it does not contain the comparison of new and old facts and knowledge, intelligent selection and formation of the new system of facts and knowledge. The first perception can be practically reduced to the conversion of data and knowledge into the internal format. Thus we can say that the initial phase is the phase of *non-critical perception*.

According to our view the initial phase should significantly differ from all the other phases of the activity of the cognizing subject. At this phase the initial model of the world will be formed, though not entirely independently. We think that at least a part of the initial knowledge base, which contains reasoning rules, should be formed under the effect of external cognizing subjects.

After the completion of the initial perception phase the *initial modification theory of a cognizing agent* will be formed. This theory will immediately be saved in the long-term memory. After this the reasoning phase will follow.

During the reasoning process the *modification theory is under development*. During this development the set of axioms of the modification theory can be modified but not the set of rules of building inferences. The development of the modification theory starts when the reasoner is initiated. This development goes through many intermediate stages and terminates when stabilisation is achieved. The modification theory obtained as the result of reasoning (stable stage of the modification theory) will be saved in the long-term memory.

After the completion of the reasoning phase a new (now complete) perception phase starts. At the end of this phase a new modification theory will be formed. Further on a reasoning phase starts again, etc. until the termination of the activity of the cognizing subject. Potentially, cognitive activity can be continued infinitely and termination may be connected with external causes.

Above an agreement was formulated that allows us to interpret the content of the fact base and knowledge base as a modification theory. Thus the content of the long-term memory of a cognizing subject is a finite sequence of modification theories. We will call this finite sequence an *open cognitive theory*.

An open cognitive theory is a finite sequence of modification theories each of which is a snapshot of the short-term memory of a cognizing subject obtained

- Either directly before starting,
- Or directly after the completion

of the reasoning phase.

Thus we identify an open cognitive theory with the content of the memory of the cognizing agent. This content can be considered as a reduced description of the history of cognitive process. The long-term memory contains only the key moments of this history.

If we are limited to the reception of new information only through reasoning, then the modification theories will be quite sufficient for us. If we wish also to reflect the result of reception of information from an external world we should use the open cognitive theories discussed above. A visual representation of the structure of an open cognitive theory is shown in Fig. 5.1.

We remind that an open cognitive theory is defined as a sequence of pairs of modification theories $\langle T_i, T_i' \rangle$, where the theory T_i corresponds to a knowledge state of the cognizing agent prior to the beginning of some reasoning phase, and T_i' corresponds to knowledge state after the end of this reasoning phase.

Note that the theories T_i and T_i' have some general set of knowledge which cannot be changed in the course of reasoning, because, for example, in the course of reasoning the reasoning algorithm does not change, and some axioms describing the model of an external world will also not change.

In the structure of a cognizing agent the content of the short-term (working) memory corresponds to modification theories, and the content of the long-term memory corresponds to open cognitive theories. In the long-term memory the history of activity of the cognizing agent is kept.

A perception phase of the cognizing agent corresponds to the transition from one pair of modification theories to another in an open cognitive theory. In this phase

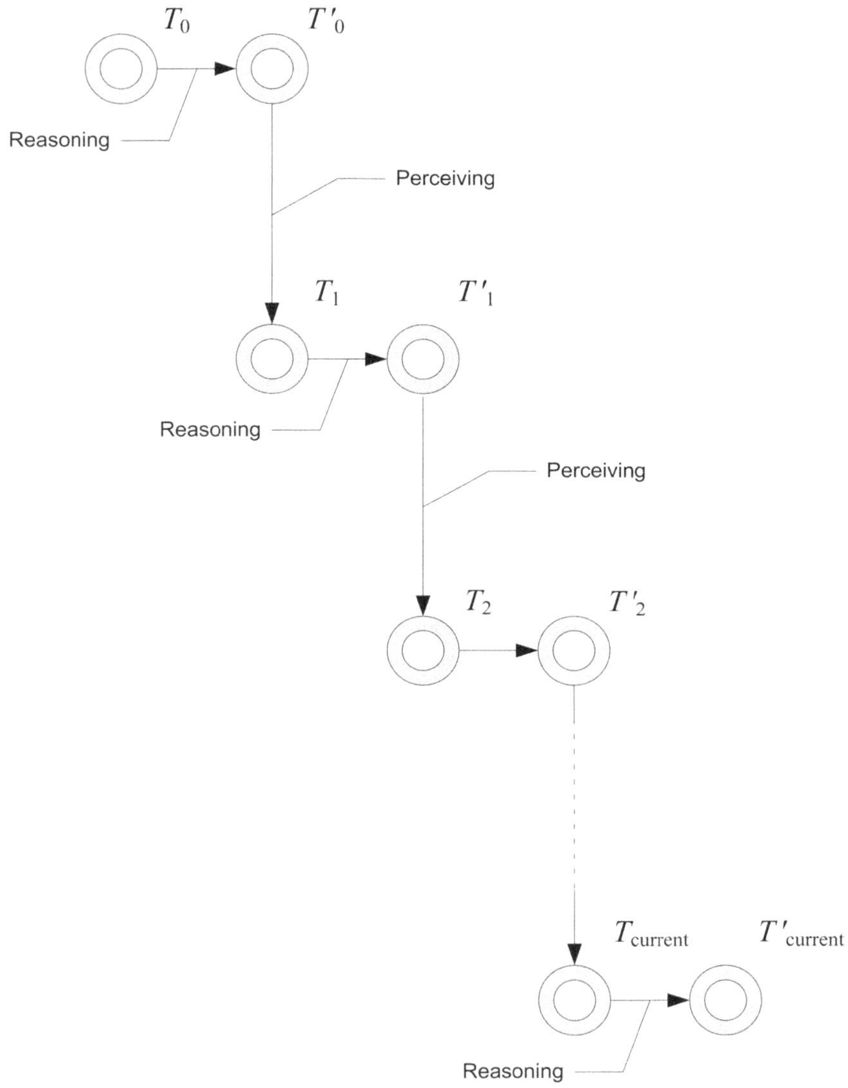

Fig. 5.1 The structure of an open cognitive theory

there is (generally speaking) more radical updating of the fact base and the knowledge base. However, management of the cognizing agent on a perception phase is outside the scope of the present book.

5.2 Modelling Cognitive Reasoning in the CR Framework

The CR framework supports the representation of a cognitive reasoning process as a *motion from ignorance to knowledge*. In the course of this motion the uncertainty domain is permanently narrowing. For this representation a specific syntactic tool, the so-called *modification inference*, is provided. For the construction of a modification inference, modification rules will be used, which can modify: (a) the truth values of the statements entering into an inference and (b) the modification inference itself. As we know, these rules correspond to the plausible reasoning rules.

Note for comparison that the application of traditional inference rules can only add statements to an inference but they cannot do anything with the statements which are already in the inference. Thus the traditional notion of inference cannot be modified by the application of traditional inference rules.

Of course, any approach imposes certain restrictions and puts the cognitive reasoning processes into a certain frame. Our view of the dynamics of cognitive reasoning processes is that they are *discrete* and rather rigidly regulated. Let us now consider the structure of the representation of a reasoning process. A cognitive reasoning process is split into stages (the number of which is unlimited). Each stage is further split into modules (the number of which is finite and is the same for all stages). Each module is divided into two phases: the *rapid phase*, where plausible reasoning appears, and the *torpid phase*, where there is only reliable (deductive) reasoning. Thus the structure representing dynamics has three levels of embedding: stage, module and phase.

In order to go into details of the structure of an inference a digression is necessary.

We know that a modification inference is built in a modification theory, which apart from the modification rules, contains axioms. Such axioms are analogues of facts and knowledge, for the representation of which appropriate language is needed. With respect to this language we need to introduce a notion important to represent statements, namely a frame that we call *predicate*.

E.g.: "an event A has occurred yesterday" is a statement represented by the use of the predicate "... has occurred ...".

With respect to these languages we suppose that they use two sorts of truth values, which can be called internal and external. This is so because the degrees of validity of empirical statements and statements received by means of plausible reasoning should differ from the degrees of validity of the statements received by means of strict and reliable proofs. The reason for this difference is not that some statements are considered better justified than others, but simply that the nature of justification of statements of different types is different. Therefore it is quite natural to define and use two sorts of truth values:

- The *internal* truth value is for the evaluation of empirical statements and statements received by means of plausible reasonings,
- The *external* truth value is for the evaluation of statements which are supposed to be proved by means of strict proofs.

From the point of view of the oppositions empirical–theoretical and plausible–reliable, the internal sort corresponds to the empirical (plausible) components of cognition and the external sort to the theoretical (reliable) ones. We can also say that the internal sort of truth values is intended to work directly with empirical data and allows for using plausible rules and makes it possible to estimate the degree of validity of the results of such reasoning. As for the external sort, it is for working with already completed evaluations of the results and allows for regarding empirical and plausibly justified statements from the outside, as if being at a metalevel.

We distinguish two external truth values: the classical *True* and *False*, which will be denoted by **t** and **f**, respectively. The language for describing and representing the facts and knowledge will use the classical logical connectives: conjunction $\wedge$, disjunction $\vee$, implication $\rightarrow$, negation $\neg$ and equivalence $\leftrightarrow$, which are defined on the set of external truth values.

As for the internal truth values, we suppose that there are at least two of them. There is one, which is called *Uncertain* or *Unknown* and denoted here by '?'. This truth value means the complete absence of information about the question of interest. All the other internal truth values are considered as defined and express the character and degree of certainty and definiteness.

Note that this division of truth values into two classes is similar to the classical philosophical tradition of dividing truths into analytical and factual ones. The factual truth values will be subdivided into the values of facts that are given at the start of the cognitive process and the values of hypotheses that will be generated during this process.

The external truth values are intended for use at the theoretical level of knowledge. The internal truth values are especially for estimation of the results of observations, experiments and plausible reasoning. That is, it is possible to say that the internal truth values are intended for use at the empirical level of knowledge.

In the example below we will use elementary language suitable for representing a modification theory. This language will contain '?' and one additional truth value: *Certain* or *Known*, which will be denoted by '!'.

Now we continue to discuss the structure of an inference in more detail. Each module processes an internal predicate, i.e. a predicate which is used by the language for describing internal statements. A stage includes the processing of all internal predicates of a structure. The order of processing the predicates is defined by the modification calculus. This order will be the same for all inferences in a given modification calculus and it cannot be changed within an inference.

When the processing of the internal predicate next in order starts, the next module of a stage begins. When a new module starts, the inference is in the modification phase. In this phase we can apply modification rules as well as the usual deductive technique. We have to apply modification rules to each statement of the truth value *uncertain*. A modification rule can be applied to such a statement just once within a module. Repeated application of the modification rules is prohibited even when the application of a rule does not change the truth value. Note that from the start of a module it is not required to apply one of the modification rules to each uncertain statement. In order to have a possibility to apply a rule some preparatory work may be necessary, e.g. the derivation of the major premise of the rule.

However, a well-designed system of modification rules should always allow for completion of the modification phase. The modification phase of a module is completed if the modification rule is applicable to the last uncertain statement with the current predicate symbol. Then there is a switch to the deductive phase, which may be of any duration. In the deductive phase only deductive techniques are allowed to be applied. Switching to the deductive phase means certain stabilisation, or even stagnation. This is why an inference in the modification phase is called *rapid* and in the deductive phase it is called *torpid*.

The modification inferences combine reliable and plausible reasonings. However the combination of application of plausible and reliable rules in the same inference is not a trivial problem. The solution of this problem requires several ideas, two of which we are going to consider in this chapter. The first idea is the use of two sorts of truth values. The second one provides a tool which permits to carry out, in an explicit form, self-estimation of the degree of validity of those statements which are the subject of reasoning. The simplest way to introduce the required tool for this self-estimation is to provide special operators, such as modal operators. Since we are interested not only in whether a statement was justified but also in the degree of its validity then we define the operators ? and !, which correspond to the internal truth values. The operators ? and ! are mappings of the set of internal truth values into the set of external truth values. By a formal definition,

$$?\alpha = \begin{cases} \mathbf{t},\ \alpha = ?, \\ \mathbf{f},\ \alpha \neq ?, \end{cases} \qquad !\alpha = \begin{cases} \mathbf{t},\ \alpha = !, \\ \mathbf{f},\ \alpha \neq !, \end{cases}$$

where α belongs to the set of internal truth values.

The intuitive meaning of the expressions $?\varphi$ and $!\varphi$, where φ is an elementary empirical statement, are "*it is unknown that* φ" and "*it is known that* φ", respectively. Some more exact interpretations of the last statement will be "φ *is plausible*" or "*there are serious arguments for* φ". We will also say in this case that "φ *is empirically true*".

Now we have the minimum set of tools for the construction of a modification theory.

Example 5.2.1. To explain the structure of the modification theory we will give an obvious, though fictitious, example. We will assume that a robot (let us call it Mick) has been sent to a planet belonging to a remote star system for exploration, observation and experimental investigation. Moreover it should simply try to survive in a hostile environment.

Mick observes natural phenomena and tries to discover their regularities. Besides other items, Mick's fact base contains a log-book for registering observations of natural phenomena. There are tables in this log-book which contain information about phenomena that occur following other phenomena within a time interval of 10–30 min. The time interval is important for Mick because of the conditions of its operation. Mick lives at a restless ocean. On the shore it has some shelters (caves) in which it can hide. Besides that, if need be, it can climb high rocks. It takes Mick 10–30 min to reach the nearest shelter or to climb a rock. Table 5.1 can be considered

Table 5.1 A fragment of Mick's fact base

...occurred		
Date	**Phenomenon**	**Truth value**
08.01	Purple clouds	!
08.01	Storm	!
08.02	Large moon	!
08.02	Rising tide	!
08.04	Purple clouds	!
08.04	Storm	!
08.05	Large moon	!
08.05	Rising tide	!
08.07	Purple clouds	!
08.07	Storm	?

...occurred following ...within a short time interval			
Date	**Phenomenon**	**Phenomenon**	**Truth value**
08.01	Purple clouds	Storm	!
08.02	Large moon	Rising tide	!
08.04	Purple clouds	Storm	!
08.05	Large moon	Rising tide	!
08.07	Purple clouds	Storm	?

as a representation of the *internal predicates*, which are mappings from a set (of sequences) of objects in a given subject domain into the set of internal truth values. Mick's fact base contains a binary predicate "...has occurred ...", which depends on the date and event, and a ternary predicate "...after ...within the interval of 10 to 30 minutes occurs ...", which depends on the date and two events. A fragment of Mick's fact base is presented in Table 5.1.

In this fragment four phenomena are presented: "Purple clouds over the ocean", "Storm", "Large moon over the horizon" – occurrence of the planet's biggest sattelite over the horizon, and "Rising tide". The truth value ! means confidence in justice of the statement, and ? means uncertainty, the unknown. In general the last rows of the tables contain the descriptions of events which have not occured yet, but can occur on the day specified in these rows.

Mick's knowledge base contains formal analogues of the plausible reasoning rules: – the modification rules. In order to formulate these rules we will introduce the following notations. $\mathrm{A}(d,e)$ will denote the sentence "On day d there occurred an event e". $\mathrm{F}(d,e_1,e_2)$ will denote the sentence "On day d within a time interval of 10–30 min after the event e_1 there occurred an event e_2". $\mathrm{C}(e_1,e_2)$ will denote the sentence "e_1 is the phenomenological cause of e_2". Then modification rules will have the following form:

$$\frac{?\mathrm{C}(e_1,e_2),\ \exists d_1 \exists d_2 (d_1 \neq d_2 \wedge !\mathrm{F}(d_1,e_1,e_2) \wedge !\mathrm{F}(d_2,e_1,e_2))}{!\mathrm{C}(e_1,e_2)} \tag{5.1}$$

$$\frac{?\mathrm{F}(d,e_1,e_2),\ !\mathrm{A}(d,e_1) \wedge !\mathrm{C}(e_1,e_2)}{!\mathrm{F}(d,e_1,e_2)} \tag{5.2}$$

$$\frac{?\mathrm{A}(d,e_2),\ !\mathrm{A}(d,e_1) \wedge !\mathrm{C}(e_1,e_2)}{!\mathrm{A}(d,e_2)} \tag{5.3}$$

The rule (5.1) will be called the induction rule. Informally it can be explained as follows:

Let:

- It be *not known* whether the event e_1 is the phenomenological cause of the event e_2
- Exist not fewer than two various days when it is *known* that the event e_2 follows the event e_1 within a time interval of 10–30 min.

Then:

- The statement that the event e_1 is the phenomenological cause of event e_2 is *plausible*.

Note that it is impossible to consider the given rule as a usual inference rule. It changes the status (and consequently the truth value) of the statement "the event e_1 is the phenomenological cause of the event e_2". The status "uncertain" is modified to "plausible", and accordingly, the truth value ? is replaced by !.

The rules (5.2) and (5.3) are called rules of causal prediction. The rule (5.3) can be described informally as follows.

Let:

- It be *not known*, whether an event e_2 should occur on day d;
- It be *known* that on day d an event e_1 occurred;
- The statement that e_1 is the phenomenological cause of the event e_2 be *plausible*.

Then:

- The statement that on day d the event e_2 should occur is *plausible*.

The rule (5.2) can be described informally in a similar way.

Note that the expressions "it is known" and "it is plausible" correspond to the same internal truth value. Certainly, in this case the situation is less precisely described, but it enables us to use very simple logic and simple modification rules.

Mick's knowledge base contains a table describing the internal predicate "... is the phenomenological cause of ... ". The fragment of this predicate can be represented, for example, by the Table 5.2.

Table 5.2 A fragment of Mick's knowledge base

... is the phenomenological cause of ...		
Phenomenon	**Phenomenon**	**Truth value**
Purple clouds	Storm	?
Large moon	Rising tide	?

Mick's knowledge base also contains a reasoning algorithm which prescribes:

- First, to traverse the table corresponding to the predicate C (*... is the phenomenological cause of ...*), trying to replace the value ? with the value ! everywhere where it is possible; the rule to be applied is the rule of induction (5.1);
- Second, to traverse the table corresponding to the predicate F (*on day ... following an event ... the event ... occurs within an interval of 10–30 min*), also trying to replace the value ? with the value ! everywhere where it is possible; the rule to be applied is the rule of causal prediction (5.2);
- Third, to traverse the table corresponding to the predicate A (*on day ... there occures an event ...*) trying to replace the value ? with the value ! everywhere where it is possible; the rule to be applied is the rule of causal prediction (5.3).

Implementing the reasoning algorithm, the robot Mick will find out that the statements "The occurrence of purple clouds over the ocean is the phenomenological cause of the storm " and "The occurrence of the large moon over the horizon is the phenomenological cause of the rising tide" are plausible and it will replace the value with the value in the corresponding rows of the last table.

Then the robot will apply a rule of causal prediction to the last rows of the tables corresponding to predicates "on day ... following an event ... the event ... occurs within an interval of 10–30 min" and "on day ... there occures an event ... ". It can replace the value with the value in these rows.

The robot knows that after an occurrence of purple clouds over the ocean with an interval of 10–30 min a storm will begin. But the clouds have already appeared, about which the robot made a record in the log-book of observations. Hence, the robot should go to the nearest shelter (cave) and wait there until the storm is over.

We assume that Mick's knowledge base contains a set of a priori knowledge which improves the robot's adaptation to the conditions of the environment. In particular, this knowledge base can contain a set of patterns according to which new knowledge will be organised into a system. It is even more important that among a priori knowledge a set of behavioural algorithms can be found.

These algorithms prescribe the robot to carry out certain actions if some events happen (or if they are predicted). For example, if a storm is predicted then the robot should go to the nearest cave and, if a rising tide is predicted, then it should climb up the nearest rock.

The cognitive activity of the robot can be reflected in the modification theory by means of a modification inference in which the reasoning algorithm is taken into account. This algorithm is determined by a certain sequence of internal predicates. A module of the inference corresponds to the *traversal* of one internal predicate, and a stage of the inference corresponds to the *traversal* of all the above finite sequence of internal predicates. In the example above the entire inference fits into one stage containing three modules.

In the present chapter we are not going to give a detailed definition of modification theories and modification inference. They will be strictly defined and studied in Part IV. Here we consider an example of modification inference almost without any explanation. We hope that everything will be clear from the context.

In modification calculation it is possible to present an inference in the form of the table containing the following columns:

- Serial number of a row of the inference
- Formula representing statement
- The basis, i.e. the reason why the formula can be included into the inference
- Stage number
- Module number
- Row number, which deactivated (blocked) the given row of the inference

The last item on the list requires a more detailed explanation.

Earlier we considered the processes which occur in the fact and knowledge bases and found that, when modification rules are applied, the uncertain truth value (?) is replaced with the certain truth value (!). Such is the semantic interpretation of the modification rules. But syntactically, when a modification inference is built, such a replacement of the truth values corresponds to adding the formula $!\varphi$ to the inference that already contains the formula $?\varphi$.

If the formula $!\varphi$ is simply added to the inference, then we will immediately receive a logical contradiction, since $!\varphi$ and $?\varphi$ cannot be simultaneously true. If this were possible, the formula would accept two different truth values ! and ? simultaneously. Hence, when adding $!\varphi$ we have to block $?\varphi$. Moreover, in order to avoid the emergence of contradictions, we also have to block all the formulas deductively dependent on $?\varphi$, i.e. all such formulas which have been received by means of usual deductive (reliable) rules with the use of the formula $?\varphi$.

Now we consider an example of an inference in modification calculus in a shortened notation. The reduction is reached at the expense of the following inaccuracies:

- We will write several formulas in one row of the inference,
- As the reason for including a formula into the inference, we will simply write "deduction", meaning by this a consecutive application of various authentic rules,
- We will not repeat those formulas which have already received certain truth values (strict definition of a modification inference on the subsequent stages demands such repetition).

The example of the modification inference corresponding to Mick's reasoning is presented in Table 5.3. In this table, in addition to the above-introduced notations, we will use the following abbreviations: PC (purple clouds), S (storm), LM (large moon), RT (rising tide).

Table 5.3 Modification inference in shortened notation

#	Formula representing statement	Reason	Stage	Module	DN
1	$08.01 \neq 08.04$	Axiom	0	0	
2	$08.02 \neq 08.05$	Axiom	0	0	
3	$?\mathrm{C}(\mathrm{PC},\mathrm{S})$	Axiom	0	1	12
4	$?\mathrm{C}(\mathrm{LM},\mathrm{RT})$	Axiom	0	1	13
5	$!\mathrm{F}(08.01,\mathrm{PC},\mathrm{S}),\ !\mathrm{F}(08.04,\mathrm{PC},\mathrm{S})$	Axiom	0	2	
6	$!\mathrm{F}(08.02,\mathrm{LM},\mathrm{RT}),\ !\mathrm{F}(08.05,\mathrm{LM},\mathrm{RT})$	Axiom	0	2	
7	$?\mathrm{F}(08.07,\mathrm{PC},\mathrm{S})$	Axiom	0	2	15
8	$!\mathrm{A}(08.07,\mathrm{PC})$	Axiom	0	3	
9	$?\mathrm{A}(08.07,\mathrm{S})$	Axiom	0	3	16
10	$\exists d_1 \exists d_2\, (d_1 \neq d_2 \wedge !\mathrm{F}(d_1,\mathrm{PC},\mathrm{S}) \wedge !\mathrm{F}(d_2,\mathrm{PC},\mathrm{S}))$	1, 5, Deduction	1	0	
11	$\exists d_1 \exists d_2\, (d_1 \neq d_2 \wedge !\mathrm{F}(d_1,\mathrm{LM},\mathrm{RT}) \wedge !\mathrm{F}(d_2,\mathrm{LM},\mathrm{RT}))$	2, 6, Deduction	1	0	
12	$!\mathrm{C}(\mathrm{PC},\mathrm{S})$	3, 10, Induction	1	1	
13	$!\mathrm{C}(\mathrm{LM},\mathrm{RT})$	4, 11, Induction	1	1	
14	$!\mathrm{A}(08.07,\mathrm{PC}) \wedge !\mathrm{C}(\mathrm{PC},\mathrm{S})$	8, 12, Deduction	1	1	
15	$!\mathrm{F}(08.07,\mathrm{PC},\mathrm{S})$	7, 14, Prediction	1	2	
16	$!\mathrm{A}(08.07,\mathrm{S})$	9, 14, Prediction	1	3	

Now we will give a short explanation of the modification inference represented in the table.

The first nine rows of the inference correspond to the zeroth stage. The zeroth stage describes a situation prior to the beginning of plausible reasoning. The description of an initial state of the fact base and the knowledge base (i.e. the state which was prior to the beginning of reasoning) is represented by axioms of the modification theory.

The modules of any stage correspond to members of a finite sequence of internal predicates, which are determines the reasoning algorithm. In our case the first predicate of the sequence is C, the second is F and the third is A. The zeroth module corresponds to the situation prior to the beginning of the traversal of the above sequence. In the zeroth module can be written either the axioms not connected with representation of internal predicates or the results of application of deductive rules. However, we can apply deductive rules in any place of a modification inference.

Let us continue commenting on the modification inference. Row 10 marks the beginning of the first stage of the reasoning. Rows 10 and 11 contain premises of the induction rule, received by means of deductive rules from formulas located above.

Rows 12 and 13 contain conclusions of the induction rule. They belong to module 1, which is connected with the predicate C. Simultaneously to the addition of formula 12 we block formula 3, which contradicts formula 12. By the way, formula 3 is one of the premises of the rule according to which formula 12 was received.

Simultaneously to the addition of formula 13, formula 4 (which is one of the premises of the rule used to obtain formula 13) is blocked.

Row 14 contains a premise of the predictive rule received by means of deductive rules from the formulas located above in the inference.

Rows 15 and 16 contain results of application of a predictive rule. Simultaneously to the addition of row 15, row 7 is blocked; simultaneously to the addition of row 16, row 9 is blocked. Rows 7 and 9 contain premises of the rules used for obtaining formulas from rows 15 and 16, respectively.

So, we have discussed a simple example of an inference in the modification theory and have shown the correspondence of this inference to the reasoning of the cognizing agent. It is obvious that the robot which we used for demonstration of the possibilities of the cognitive reasoning and of its formal analogues was given in this chapter as a hypothetical example of the realisation of the architecture of the cognizing agent considered above.

The example of a modification inference given in this chapter is too simple to demonstrate rather interesting features of such inferences, for example the possibility to generate new knowledge by means of consecutive iterations of the above traversal of the sequence of internal predicates or the effect of temporal contradiction. Suitable examples will be considered later in the detailed investigation of modification theories.

Part II
Logic Foundation

Chapter 6
Introductory Explanation

In this part we are going to develop the logic foundation of our approach which is to provide an appropriate formalism to represent, simulate and generate the cognitive reasoning processes of cognizing agents. Thus, in our view, the foundation should be appropriate for the representation of the process from ignorance to knowledge, that is, the learning and knowledge acquisition processes.

The logic foundation of cognitive reasoning should be able to reflect the main characteristics of the process of learning new information and discovering new regularities in the subject domain in question. Moreover, this foundation should be appropriate to support the development of logical methods and tools to support:

- The description of the observed events and objects (descriptive language)
- The description and representation of the cognizing agent's knowledge (subjective knowledge)
- The description and representation of the knowledge about a given domain (objective knowledge)
- The representation of cognitive processes from observation to subjective knowledge and from subjective knowledge to objective knowledge
- The representation and construction of the cognitive reasoning processes permitting reasoning operators of various types to be combined into one process

Moreover, we also require that the logic foundation provides different methods for supporting different kinds of representation of cognitive reasoning processes that follow different goals of modelling. We will provide a logic foundation which supports two alternative approaches to the formalisation of cognitive reasoning processes:

(i) The simulative approach, that is, construction of a logic frame in which a special theory can be defined such that the following conditions hold:

   - It can handle both deductive and plausible constituents of cognitive reasoning;
   - Its provable statements can be derived by the use of usual means of logical inference (usual deductive technique);

O. Anshakov, T. Gergely, *Cognitive Reasoning*, Cognitive Technologies,
DOI 10.1007/978-3-540-68875-4_6,

- It should include specific axioms that correspond to the rules of plausible reasoning. In this theory both constituents of cognitive reasoning will be simulated by the use of deduction. Thus this approach can be called deductive simulation of cognitive reasoning.

(ii) Direct approach: development of an original non-monotone derivation technique by adding specific rules to the standard deductive technique; this new technique will play the role of the plausible constituent of cognitive reasoning.

The logic foundation will be based on a special family of many-valued logics which we will call PJ logics. The state logic will be the PJ logic that can represent the current states of knowledge a cognizing agent operates with wich. It will be the basis for the development of the cognitive derivative technique required for the realisation of the cognitive reasoning processes. The iterative logic will be the PJ logic that will serve as the framework within which the deductive simulation of the cognitive processes will take place, which realise the simulative approach. This logic will also permit the handling of the dynamic aspects of cognitive reasoning processes.

Chapter 7
Propositional Logic

The main purpose of this chapter is to introduce some classes of many-valued logics, namely the class of *pure J logic* (PJ logics) and two its subclasses: the class of *finite pure J logics* (FPJ logics) and the class of *iterative versions of finite PJ logics*. In these subclasses we select important examples that can be used in existing and possible application of the technique developed in this book: they are the so-called *state logic* and *iterative logic*, denoted by **St** and **It**, respectively. **St** is an example of FPJ logics and **It** is an iterative version of **St**.

First we introduce the propositional case. Here we abstract from the functional properties of the considered calculi and concentrate on their logical features. The singularity of our approach is the definition of many-valued logics by their algebras. Therefore, and to understand the section, one must be acquainted with the basic algebraical notions such as: *signature*, *algebra*, *homomorphism* and *free algebra*. For the same reason one must be familiar with the basic concepts of naive set theory and be skilled at working with mathematical texts.

Let L be a logic. Then there are two algebras connected with L: the algebra of logic L and the algebra of formulas of logic L.

The algebra of logic L is denoted by $\mathfrak{A}(L)$.

By $\mathfrak{F}(L)$ we denote *the algebra of formulas of the propositional logic L*.

The algebras $\mathfrak{A}(L)$ and $\mathfrak{F}(L)$ share the same signature; $\mathfrak{F}(L)$ is a free algebra in the class of all algebras of this signature.

In algebraic works, when considering algebras of the same signature, one usually denotes *corresponding operations* by *the same character*. E.g., by + one can denote addition in a linear space and in the ring of integers. It is assumed that it is always clear from the context which operation is meant. Here, we follow this convention and, for instance, use identical notation for corresponding operations in the algebras $\mathfrak{A}(L)$ and $\mathfrak{F}(L)$.

O. Anshakov, T. Gergely, *Cognitive Reasoning*, Cognitive Technologies,
DOI 10.1007/978-3-540-68875-4_7, © Springer-Verlag Berlin Heidelberg 2010

7.1 Notation

The set of logical operations of logic L (i.e. the set of propositional connectives) is denoted by $\mathscr{O}(L)$.

In this book, we discuss logics that have one-sorted or two-sorted algebras. If logic L has a one-sorted algebra, then it has one *set of truth values*, denoted by $\mathscr{V}(L)$.

If logic L has a two-sorted algebra, then it has two disjoint sets of truth values then:

- The set of *internal truth values* is denoted by $\mathscr{V}(L, \mathsf{I})$,
- The set of *external truth values* is denoted by $\mathscr{V}(L, \mathsf{E})$,
- The set $\mathscr{V}(L, \mathsf{I}) \cup \mathscr{V}(L, \mathsf{E})$ is denoted by $\mathscr{V}(L)$.

If the algebra of logic L is one-sorted then we denote:

- The set of *propositional variables* of logic L by $\mathscr{P}(L)$
- The set of *formulas* of propositional logic L by $\mathscr{F}(L)$

For a two-sorted algebra, we define two disjoint sets of propositional variables and two disjoint sets of formulas:

- The set of *internal propositional variables* of logic L is denoted by $\mathscr{P}(L, \mathsf{I})$,
- The set of *external propositional variables* of logic L is denoted by $\mathscr{P}(L, \mathsf{E})$,
- The set $\mathscr{P}(L, \mathsf{I}) \cup \mathscr{P}(L, \mathsf{E})$ is denoted by $\mathscr{P}(L)$,
- The set of *internal formulas of logic L* is denoted by $\mathscr{F}(L, \mathsf{I})$,
- The set of *external formulas of logic L* is denoted by $\mathscr{F}(L, \mathsf{E})$,
- The set $\mathscr{F}(L, \mathsf{I}) \cup \mathscr{F}(L, \mathsf{E})$ is denoted by $\mathscr{F}(L)$.

Then

$$\mathfrak{A}(L) = \langle \mathscr{V}(L);\ \mathscr{O}(L) \rangle, \quad \mathfrak{F}(L) = \langle \mathscr{F}(L);\ \mathscr{O}(L) \rangle$$

for logic L with one-sorted algebra and

$$\mathfrak{A}(L) = \langle \mathscr{V}(L, \mathsf{I}), \mathscr{V}(L, \mathsf{E});\ \mathscr{O}(L) \rangle,$$

$$\mathfrak{F}(L) = \langle \mathscr{F}(L, \mathsf{I}), \mathscr{F}(L, \mathsf{E});\ \mathscr{O}(L) \rangle$$

if the algebra of logic L is two-sorted.

It should be noted that $\mathscr{P}(L)$ is the set of free generators in the free algebra $\mathfrak{F}(L)$. For any logic L considered further in the book, $\mathscr{P}(L)$ is a *countable* set. We do not care about the nature of propositional variables. However, we make use of the convenient assumption that the sets of propositional variables are disjoint for different logics.

7.2 Classical Propositional Logic (Syntax and Semantics)

In this section, we define some notions and discuss some standard and well-known results concerning the classical two-valued logic. We must do this for the following purposes:

- To unify terms and notation,
- To be able to refer to definitions and statements.

From now on, **C** denotes the *classical two-valued logic*.

Definition 7.2.1. *The algebra of logic* **C** is the ordered pair

$$\mathfrak{A}(\mathbf{C}) = \langle \mathscr{V}(\mathbf{C}); \mathscr{O}(\mathbf{C}) \rangle,$$

where

$$\mathscr{V}(\mathbf{C}) = \{\mathbf{t}, \mathbf{f}\}$$

(**t** designates the truth value TRUE; **f** designates the truth value FALSE),

$$\mathscr{O}(\mathbf{C}) = \{\rightarrow, \neg\},$$

where $\rightarrow$ is a binary operation and $\neg$ is a unary operation on the set $\{\mathbf{t}, \mathbf{f}\}$ such that

$$\neg\ (a) = \begin{cases} \mathbf{t}, \ a = \mathbf{f}, \\ \mathbf{f}, \ a = \mathbf{t}, \end{cases}$$

$$a \rightarrow b = \begin{cases} \mathbf{t}, \ a = \mathbf{f} \quad \text{or} \quad b = \mathbf{t}, \\ \mathbf{f}, \ a = \mathbf{t} \quad \text{and} \quad b = \mathbf{f}. \end{cases}$$

We have chosen *implication* $\rightarrow$ and *negation* $\neg$ as the main operations of logic **C**, since this makes it easier to define the propositional calculus.

Definition 7.2.2. Operations of the classical logic different from $\rightarrow$ and $\neg$ are defined by the following equalities:

$$\begin{aligned} a \vee b &= \neg(a) \rightarrow b && (\textit{disjunction}), \\ a \wedge b &= \neg(a \rightarrow \neg(b)) && (\textit{conjunction}), \\ a \leftrightarrow b &= (a \rightarrow b) \wedge (b \rightarrow a) && (\textit{equivalence}). \end{aligned}$$

Agreement 7.2.3. To make less complicated expressions with the operation characters $\rightarrow$, $\neg$, $\vee$, $\wedge$ and $\leftrightarrow$, *we define the operation priority*:

- Negation $\neg$ has the highest priority,
- Conjunction $\wedge$ and disjunction $\vee$ are of lower priority (they have the same priority),

- Implication $\rightarrow$ and equivalence $\leftrightarrow$ have the lowest priority (and they also have the same priority).

Definition 7.2.4. *Formulas of propositional logic* **C** (**C**-*formulas*) are recursively defined as follows:

(i) Every propositional variable of the logic **C** (i.e. every element of the set $\mathscr{P}(\mathbf{C})$) is a **C**-formula.
(ii) Let φ and ψ be **C**-formulas. Then $(\varphi \rightarrow \psi)$ and $(\neg\varphi)$ are also **C**-formulas.
(iii) There are no other **C**-formulas besides those defined by (i) and (ii).

Definition 7.2.5. Following the notation of the previous section, denote the set of **C**-formulas by $\mathscr{F}(\mathbf{C})$. Then *the algebra of formulas of logic* **C** (*the algebra of* **C**-*formulas*) is the algebra

$$\mathfrak{F}(\mathbf{C}) = \langle \mathscr{F}(\mathbf{C});\ \mathscr{O}(\mathbf{C}) \rangle ,$$

where the operations $\neg$ and $\rightarrow$ in $\mathscr{O}(\mathbf{C})$ are defined as follows:

- The operation $\neg$ applied to a formula φ results in the formula $(\neg\varphi)$;
- The operation $\rightarrow$ applied to formulas φ and ψ results in the formula $(\varphi \rightarrow \psi)$.

Definition 7.2.6. We introduce propositional connectives $\vee$, $\wedge$, and $\leftrightarrow$ as the following abbreviations:

- $(\varphi \vee \psi)$ is an abbreviation of $((\neg\varphi) \rightarrow \psi)$;
- $(\varphi \wedge \psi)$ is an abbreviation of $(\neg(\varphi \rightarrow (\neg\psi)))$;
- $(\varphi \leftrightarrow \psi)$ is an abbreviation of $((\varphi \rightarrow \psi) \wedge (\psi \rightarrow \varphi))$.

Agreement 7.2.7. To simplify notation, we use the following *agreement on living out parentheses*:

- We drop the outmost parentheses of formulas;
- We drop parentheses when this is allowed by the agreement of operation precedence defined above for the operations $\rightarrow$, $\neg$, $\vee$, $\wedge$ and $\leftrightarrow$.

Definition 7.2.8. *A classical valuation* (**C**-*valuation*) is a homomorphism of algebra $\mathfrak{F}(\mathbf{C})$ to algebra $\mathfrak{A}(\mathbf{C})$.

An arbitrary mapping $u\colon \mathscr{P}(\mathbf{C}) \rightarrow \mathscr{V}(\mathbf{C})$, where $\mathscr{P}(\mathbf{C})$ is the set of propositional variables of the classical propositional logic and

$$\mathscr{V}(\mathbf{C}) = \{\mathbf{t}, \mathbf{f}\},$$

can be uniquely extended to a **C**-valuation by induction. The induction step is as follows:

$$u(\neg\varphi) = \neg u(\varphi), \quad u(\varphi \rightarrow \psi) = u(\varphi) \rightarrow u(\psi),$$

where φ and ψ are formulas of propositional logic **C**.

Definition 7.2.9. Let Γ be a set of formulas of logic **C** and u be a **C**-valuation. We say that u is *a model of* Γ if $u(\varphi) = \mathbf{t}$ holds for any $\varphi \in \Gamma$.

C-valuation u is called a **C**-*model of a* **C**-*formula* φ if u is a model of the set $\{\varphi\}$, i.e. if $u(\varphi) = \mathbf{t}$.

Definition 7.2.10. Let Γ be a set of **C**-formulas and φ be a **C**-formula. We say that φ is a *logical* (*semantic*) **C**-*consequence* of the set Γ (and denote this by $\Gamma \models_{\mathbf{C}} \varphi$) if any model of Γ is a model of φ.

Definition 7.2.11. Formula φ of logic **C** is called a **C**-*tautology* (or simply a tautology) if it is a logical **C**-consequence of the empty set of formulas, i.e. if every **C**-valuation is its model.

7.3 Classical Propositional Logic (Calculus)

In this short section we discuss some well-known facts about classical propositional logic. The basic results introduced below are the Soundness Theorem and Completeness Theorem for the classical propositional logic. Later we will prove the analogues of these theorems for two many-valued two-sorted logics, the state logic (**St**) and the iterative logic (**It**).

Now we introduce *the axiom schemata of the classical propositional calculus*. This calculus is also called *zero-order classical calculus* and is denoted by $\mathbf{C}^0$.

Axiom Schemata 7.3.1 (Axiom schemata of classical propositional calculus). Let φ, ψ and θ be **C**-formulas. Then, the following **C**-formulas are axioms of $\mathbf{C}^0$:

(C1) $\varphi \to (\psi \to \varphi)$,
(C2) $(\varphi \to (\psi \to \theta)) \to ((\varphi \to \psi) \to (\varphi \to \theta))$,
(C3) $(\neg\varphi \to \psi) \to ((\neg\varphi \to \neg\psi) \to \varphi)$.

Inference Rules 7.3.2 (Inference rules for $\mathbf{C}^0$). The only inference rule in $\mathbf{C}^0$ is *modus ponens*:

$$\frac{\varphi, \quad \varphi \to \psi}{\psi},$$

where φ and ψ are **C**-formulas.

Definition 7.3.3. Let Γ be a set of formulas of **C**. An *inference from* Γ *in* $\mathbf{C}^0$ ($\mathbf{C}^0$-*inference*) is a finite sequence of **C**-formulas such that everyone of its element:

- Either is an axiom of $\mathbf{C}^0$,
- Or belongs to Γ,
- Or follows from preceding formulas by *modus ponens*.

A $\mathbf{C}^0$-inference from the empty set of **C**-formulas is called a *proof in* $\mathbf{C}^0$ ($\mathbf{C}^0$-*proof*).

Definition 7.3.4. Let Γ be a set of **C**-formulas and φ be a **C**-formula. We say that φ is *derivable from* Γ *in* $\mathbf{C}^0$ (written: $\Gamma \vdash_{\mathbf{C}} \varphi$) if there exists a $\mathbf{C}^0$-inference containing φ from the set Γ. A **C**-formula φ is called *provable in* $\mathbf{C}^0$ or *a theorem of* $\mathbf{C}^0$ (written: $\vdash_{\mathbf{C}} \varphi$) if φ is derivable in $\mathbf{C}^0$ from the empty set.

Below, we present several well-known theorems for classical propositional logic. Proofs of these theorems can be found in any major textbooks on mathematical logic, e.g. Mendelson [1997].

Theorem 7.3.5 (Deduction Theorem for classical propositional calculus). Let Γ be a set of **C**-formulas and φ and ψ be **C**-formulas. Then

$$\Gamma \cup \{\varphi\} \vdash_{\mathbf{C}} \psi \quad \text{implies} \quad \Gamma \vdash_{\mathbf{C}} \varphi \to \psi.$$

Proof. See, e.g. Mendelson [1997]. □

Theorem 7.3.6 (Soundness Theorem for classical propositional calculus). Let Γ be a set of **C**-formulas and φ be a **C**-formula. Then

$$\Gamma \vdash_{\mathbf{C}} \varphi \quad \text{implies} \quad \Gamma \models_{\mathbf{C}} \varphi.$$

Theorem 7.3.7 (Completeness Theorem for classical propositional calculus). Let Γ be a set of **C**-formulas and φ be a **C**-formula. Then

$$\Gamma \models_{\mathbf{C}} \varphi \quad \text{implies} \quad \Gamma \vdash_{\mathbf{C}} \varphi.$$

Proof. See, e.g. [Mendelson, 1997]. □

7.4 Propositional PJ Logics (Syntax and Semantics)

In this section, a class of logics will be developed, to which two many-sorted logics belong, that further will be used to formalise cognitive (plausible) reasoning. One of them is the state logic that permits one to describe "the state of knowledge". The other is the iterative logic that permits one to describe the history of cognitive processes, which includes several repetitions (iterations) of certain standard cognitive procedures. The state logic is a logic of finitely many values, while the iterative logic is that of countably infinite many values. The algebras of both logics are two-sorted. Each of these logics has two sets of truth values: internal and external. The set of internal truth values of the state logic consists of four elements, while the iterative logic has a countable set of internal truth values.

However, these two logics are very similar. They both belong to the wide class of J-definable J-compact logics [Anshakov et al, 1989]. Proofs of many statements will be identical for both logics, therefore it is not reasonable to investigate these logics separately. For their investigation we will use a simpler technique by considering a special class of logics, the so-called pure J logics, instead of using the complex techniques of J-definable J-compact logics. The state logic and the iterative logic belong to this class. The concepts will be defined for the class of PJ logics and, if necessary, we will specify them separately for the state logic and for the iterative logic. Statements will also be formulated for all PJ logics and, wherever possible, the schemes of the proofs will also be general. When necessary, we will prove the statements separately for these logics.

Definition 7.4.1. Logic L is called *pure J logic (PJ logic)*, if

$$\mathfrak{A}(L) = \langle \mathscr{V}(L,\mathsf{I}), \mathscr{V}(L,\mathsf{E}); \mathscr{O}(L)\rangle,$$

where

(i) $\mathscr{V}(L,\mathsf{I})$ contains at least two elements,
(ii) $\mathscr{V}(L,\mathsf{E}) = \mathscr{V}(\mathbf{C}) = \{\mathbf{t},\mathbf{f}\}$,
(iii) $\mathscr{O}(L) = \mathscr{J}(L) \cup \mathscr{O}(\mathbf{C})$,
(iv) $\mathscr{O}(\mathbf{C}) = \{\rightarrow, \neg\}$ (operations of $\mathscr{O}(\mathbf{C})$ can be defined as in Definition 7.2.1),
(v) There is a set of indices $\mathrm{JI}(L) \neq \emptyset$ (the elements of the set $\mathrm{JI}(L)$ will be called *J-indices of the logic L*) such that:

  (i) $\mathscr{J}(L) = \{\mathrm{J}_\alpha \mid \alpha \in \mathrm{JI}(L)\}$ (operations of $\mathscr{J}(L)$ are indexed by the set $\mathrm{JI}(L)$),
  (ii) For any $\alpha \in \mathrm{JI}(L)$, J_α is a surjection of $\mathscr{V}(L,\mathsf{I})$ onto $\mathscr{V}(L,\mathsf{E})$,
  (iii) For any $\alpha, \beta \in \mathrm{JI}(L)$, if $\alpha \neq \beta$ then $\mathrm{J}_\alpha \neq \mathrm{J}_\beta$ (i.e. J_α and J_β represent different mappings of $\mathscr{V}(L,\mathsf{I})$ onto $\mathscr{V}(L,\mathsf{E})$).

Operations of $\mathscr{J}(L)$ are called *J-operators of the logic L*. Such operations were initially studied in [Rosser and Turquette, 1951]. Obviously, J-operators are characteristic functions of non-empty proper subsets of the set of internal truth values.

Definition 7.4.2. Let L be a PJ logic, $\alpha \in \mathrm{JI}(L)$. Then the set

$$\{\delta \in \mathscr{V}(L,\mathsf{I}) \mid \mathrm{J}_\alpha(\delta) = \mathbf{t}\}$$

is called the *truth-value set that is characterised by the J-operator* J_α and is denoted by $\mathrm{Ch}(\alpha)$.

The operator J_α is the characteristic function of the set $\mathrm{Ch}(\alpha)$, i.e.

$$\mathrm{J}_\alpha(\delta) = \begin{cases} \mathbf{t}, \ \delta \in \mathrm{Ch}(\alpha), \\ \mathbf{f}, \ \delta \notin \mathrm{Ch}(\alpha). \end{cases}$$

The set $W \subseteq \mathscr{V}(L,\mathsf{I})$ is called *J-characterised* if there exists an $\alpha \in \mathrm{JI}(L)$ such that $W = \mathrm{Ch}(\alpha)$.

The collection of all J-characterised subsets of the set $\mathscr{V}(L,\mathsf{I})$, i.e.

$$\{\mathrm{Ch}(\alpha) \mid \alpha \in \mathrm{JI}(L)\},$$

is denoted by $\mathrm{JCh}(L)$.

Remark 7.4.3. Let L be a PJ logic. Obviously a one-to-one correspondence may be established between $\mathcal{J}(L)$ and $\mathrm{JCh}(L)$ (see 7.4.1 (v)). Moreover, it is sufficient to define the set $\mathrm{Ch}(\alpha)$ $(\alpha \in \mathrm{JI}(L))$ in order to determine the operator J_α. Further on these properties will be used in defining concrete PJ logics.

Definition 7.4.4. Let L be a PJ logic. We say that L is a *finite PJ logic* (an FPJ logic) if the following condition hold:

- The set $\mathcal{V}(L,\mathsf{I})$ is finite;
- $\mathrm{JI}(L) = \mathcal{V}(L,\mathsf{I})$;
- $\mathrm{Ch}(\alpha) = \{\alpha\}$ for any $\alpha \in \mathcal{V}(L,\mathsf{I})$.

Thus the J-operators from $\mathcal{J}(L)$ are characteristic functions of the one-element subsets of the set $\mathcal{V}(L,\mathsf{I})$), i.e.

$$\mathrm{J}_\alpha(\beta) = \begin{cases} \mathbf{t}, \; \beta = \alpha, \\ \mathbf{f}, \; \beta \neq \alpha \end{cases}$$

for arbitrary $\alpha \in \mathcal{V}(L,\mathsf{I})$

Definition 7.4.5. Let L be an FPJ logic. Select a $\tau \in \mathcal{V}(L,\mathsf{I})$ and give it a special role. Let τ be called *uncertain*. The set $\mathcal{V}(L,\mathsf{I})\setminus\{\tau\}$ is called the set of *definite truth values* of logic L and is denoted by $\mathcal{D}^\tau(L)$. We say that I is an *iterative version of L* w.r.t. τ if the following conditions hold:

(i) $\mathcal{V}(I,\mathsf{I}) = \{\langle\varepsilon,n\rangle \mid \varepsilon \in \mathcal{D}^\tau(L), n \in \omega\} \cup \{\tau\}$ (here and below ω denotes the set of natural numbers $\{0,1,2,\ldots\}$),

(ii) $\mathrm{JI}(I) = \{(\varepsilon,n) \mid n \in \omega, \varepsilon \in \mathcal{D}^\tau(L)\}$ (in fact (ε,n) is the same as $\langle\varepsilon,n\rangle$, but to avoid confusion we introduce different notations for the ordered pairs considered as indices and the ordered pairs considered as truth values),

(iii) $\mathrm{Ch}(\varepsilon,n) = \{\langle\varepsilon,i\rangle \mid i \leq n\}$ $(n \in \omega, \varepsilon \in \mathcal{D}^\tau(L))$,

(iv) $\mathrm{Ch}(\tau,n) = \{\tau\} \cup \{\langle\varepsilon,i\rangle \mid i > n, \varepsilon \in \mathcal{D}^\tau(L)\}$ $\quad (n \in \omega)$.

We write $\mathrm{Ch}(\varepsilon,n)$ instead of $\mathrm{Ch}((\varepsilon,n))$.

Remark 7.4.6. For any FPJ logic L and any $\tau \in \mathcal{V}(L,\mathsf{I})$ there exists a unique iterative version w.r.t. τ.

We denote the iterative version of FPJ logic L w.r.t. $\tau \in \mathcal{V}(L,\mathsf{I})$ by $\mathrm{I}^\tau(L)$.

For the logic $I = \mathrm{I}^\tau(L)$, the elements of $\mathcal{V}(L,\mathsf{I})$ are treated as *truth value types*.

- The value τ (uncertain) is used as both a *truth value* and a *truth value type*, and it is the only *truth value* of this type.
- Elements of $\mathcal{D}^\tau(L)$ are *types of definite truth values*.

We suppose there exists a procedure of hypothesis generation; the hypotheses generated by this procedure include their own evaluation by truth value types. Below this procedure is called *main*.

The intuitive meaning of internal truth values of logic I is as follows:

- τ is the *uncertain* that cannot be revealed by means of plausible reasoning (τ is the only truth value of type τ);
- If the truth value of internal formula φ is $\langle \varepsilon, n \rangle$ (where $\varepsilon \in \mathscr{D}^{\tau}(L)$, $n \in \omega$), then we say that φ has a value of definite type ε and the corresponding hypothesis was generated at step n of the main procedure.

If the formula $\mathrm{J}_{(\varepsilon,n)}\,\varphi$ is true, where $\varepsilon \in \mathscr{D}^{\tau}(L)$, then the truth value of φ belongs to $\mathrm{Ch}(\varepsilon, n)$. According to 7.4.5 (iii) this means that the formula φ has a value of definite type ε and a hypothesis about this fact can be established *not later* than at the *n-th step* of the basic procedure.

If the formula $\mathrm{J}_{(\tau,n)}\,\varphi$ is true then the truth value of φ belongs to $\mathrm{Ch}(\tau, n)$. According to 7.4.5 (iv) this means that *until the n-th step* we could not establish a definite truth value type of φ. It could happen that this could never be established (if the truth value of φ is equal to τ).

Let us denote important examples of FPJ logics and their iterative versions.

Definition 7.4.7. **Si** (*the simplest PJ logic*) is an FPJ logic such that

$$\mathrm{JI}(\mathbf{Si}) = \mathscr{V}(\mathbf{Si}, \mathsf{I}) = \{?, !\}.$$

The informal interpretations of ? and ! are "uncertain" and "certain" respectively. Note that ! can be also understood as "plausible".

For brevity, the J-operators $\mathrm{J}_?$ and $\mathrm{J}_!$ are denoted by ? and ! respectively.

We had used **Si** in Sect. 5.2 when we considered the example of a modification inferene.

Definition 7.4.8. Symbols $+1$, -1, 0, τ are called *internal truth symbols*. The set $\{+1, -1, 0, \tau\}$ is denoted by ITS. Symbols $+1$, -1, 0 are called *definite truth symbols*. DTS denotes the set $\{+1, -1, 0\}$. The internal truth values $+1$, -1, 0, τ are called "empirically true", "empirically false", "empirically contradictory" and "uncertain", respectively.

Definition 7.4.9. **St** (*state logic*) is an FPJ logic such that

$$\mathrm{JI}(\mathbf{St}) = \mathscr{V}(\mathbf{St}, \mathsf{I}) = \mathrm{ITS}.$$

Definition 7.4.10. **It** (*iterative logic*) is an iterative version of FPJ logic **St** w.r.t τ.

Lemma 7.4.11. Let L be an FPJ logic, $I = \mathrm{I}^{\tau}(L)$, $n \in \omega$. Then

(i) $\bigcup_{\varepsilon \in \mathscr{V}(L)} \mathrm{Ch}(\varepsilon, n) = \mathscr{V}(I, \mathsf{I})$,

(ii) $\mathrm{Ch}(\varepsilon, n) \cap \mathrm{Ch}(\delta, n) = \emptyset$ for $\varepsilon, \delta \in \mathscr{V}(L)$, $\varepsilon \neq \delta$.

Proof. See 7.4.5. □

Lemma 7.4.12. Let L be an FPJ logic, $I = \mathrm{I}^{\tau}(L)$, $n, k \in \omega$, $k > n$. Then:

(i) $\mathrm{Ch}(\varepsilon, n) \subseteq \mathrm{Ch}(\varepsilon, k) \quad (\varepsilon \in \mathscr{D}^{\tau}(L))$,

(ii) $\mathrm{Ch}(\tau,k) \subseteq \mathrm{Ch}(\tau,n)$.

Proof. See 7.4.5 (iii) and (iv), respectively. □

Agreement 7.4.13. Further on we will suppose that L is a PJ logic.

Definition 7.4.14. Operations on the set $\mathscr{V}(L,\mathsf{E})$ (the set of external truth values of logic L) are called *external operations of logic L*. Obviously, operations of $\mathscr{O}(\mathbf{C})$ are external.

Agreement 7.4.15. We define external operations $\wedge$, $\vee$ and $\leftrightarrow$ similar by to how this was done in 7.2.2 and define the priority of external operations as in 7.2.3. J-operators have the same priority as the negation $\neg$.

Definition 7.4.16. Internal propositional variables of logic L (elements of the set $\mathscr{P}(L,\mathsf{I})$) are called *internal formulas of propositional logic L (internal L-formulas)*. They are also called *atomic formulas of propositional logic L (atomic L-formulas)*.

Definition 7.4.17. Expressions of the form $(\mathrm{J}_\alpha\, p)$, where $p \in \mathscr{P}(L)$, $\mathrm{J}_\alpha \in \mathscr{J}(L)$, are called *J-atomic formulas of propositional logic L (J-atomic L-formulas)*.

Definition 7.4.18. *External formulas of propositional logic L (external L-formulas)* are defined recursively as follows:

(i) Every external propositional variable is an external L-formula;
(ii) Every J-atomic L-formula is an external L-formula;
(iii) If Φ and Ψ are external L-formulas, then $(\Phi \to \Psi)$ and $(\neg\Phi)$ are also external L-formulas.

Definition 7.4.19. Following the notation introduced in 7.1, we denote the set of internal formulas of propositional logic L by $\mathscr{F}(L,\mathsf{I})$, and the set of external formulas of L by $\mathscr{F}(L,\mathsf{E})$. Then the *algebra of formulas of propositional logic L* is the algebra

$$\mathfrak{F}(L) = \langle \mathscr{F}(L,\mathsf{I}), \mathscr{F}(L,\mathsf{E}); \mathscr{O}(L)\rangle,$$

where

$$\mathscr{O}(L) = \mathscr{J}(L) \cup \mathscr{O}(\mathbf{C}),$$

operations of $\mathscr{O}(\mathbf{C})$ are defined as in 7.2.5, and operations of $\mathscr{J}(L)$ are defined as follows: operation J_α applied to propositional variable p yields a J-atomic formula $(\mathrm{J}_\alpha\, p)$.

Definition 7.4.20. An *L-valuation* is a homomorphism of algebra $\mathfrak{F}(L)$ to algebra $\mathfrak{A}(L)$.

The algebra $\mathfrak{F}(L)$ is free in the class of all algebras of the same signature with the generating set $\mathscr{P}(L)$ (in terms of the notation from 7.1, $\mathscr{P}(L)$ denotes the set of propositional variables of logic L). Then, any mapping

$$v\colon\ \mathscr{P}(L) \to \mathscr{V}(L),$$

where $v(\mathscr{P}(L,\mathsf{I})) \subseteq \mathscr{V}(L,\mathsf{I})$, $v(\mathscr{P}(L,\mathsf{E})) \subseteq \mathscr{V}(L,\mathsf{E})$, can be uniquely extended to an L-valuation by induction. The induction step is as follows:

(i) $v(\mathrm{J}_\alpha p) = \mathrm{J}_\alpha(v(p))$ $(p \in \mathscr{P}(L), \mathrm{J}_\alpha \in \mathscr{J}(L))$;
(ii) $v(\neg\Phi) = \neg v(\Phi)$ (Φ is an external formula of logic L);
(iii) $v(\Phi \to \Psi) = v(\Phi) \to v(\Psi)$ (Φ and Ψ are external formulas of logic L).

Since $\mathscr{F}(L,\mathsf{I}) \cap \mathscr{F}(L,\mathsf{E}) = \emptyset$ and $\mathscr{V}(L,\mathsf{I}) \cap \mathscr{V}(L,\mathsf{E}) = \emptyset$, we can define a homomorphism v of the two-sorted algebra

$$\mathfrak{F}(L) = \langle \mathscr{F}(L,\mathsf{I}), \mathscr{F}(L,\mathsf{E}); \mathscr{O}(L)\rangle$$

into the two-sorted algebra

$$\mathfrak{A}(L) = \langle \mathscr{V}(L,\mathsf{I}), \mathscr{V}(L,\mathsf{E}); \mathscr{O}(L)\rangle$$

as a mapping $v\colon \mathscr{F}(L) \to \mathscr{V}(L)$ such that

$$v(\mathscr{F}(L,\mathsf{I})) \subseteq \mathscr{V}(L,\mathsf{I}), \qquad v(\mathscr{F}(L,\mathsf{E})) \subseteq \mathscr{V}(L,\mathsf{E})$$

and (i), (ii) and (iii) hold (recall that $\mathscr{P}(L,\mathsf{I}) = \mathscr{F}(L,\mathsf{I})$).

Definition 7.4.21. Let Γ denote a set of external formulas of logic L and v be an L-valuation. We say that v is a *model of the set Γ of external formulas* if $v(\Phi) = \mathbf{t}$ for all $\Phi \in \Gamma$. A valuation v is called *model of external formula Φ* if v is a model of $\{\Phi\}$, i.e. if $v(\Phi) = \mathbf{t}$.

Note that the models have been defined only for (sets of) *external* formulas of logic L.

Definition 7.4.22. Let Γ be a set of external formulas of logic L and Φ be an external L-formula. We say that Φ is an *L-consequence* of Γ (written: $\Gamma \models_L \Phi$) if every model of Γ is a model of Φ. An external L-formula Φ is called an *L-tautology* if it is an L-consequence of the empty set of formulas, i.e. if any L-valuation is a model of Φ (written: $\models_L \Phi$).

7.5 PJ Logics (Calculus)

The basic results of this section are the Soundness Theorem and Completeness Theorem, which are analogues to 7.3.6 and 7.3.7. We prove these theorems for two two-sorted PJ logics defined above as state logic and iterative logic. To prove the completeness theorem we use the technique introduced by D. P. Skvortsov (cf. [Anshakov et al, 1989]). First, we define the relations that connect derivability and validity in classical logic and in PJ logics. We also prove the series of intermediate statements. Then we prove the Main Lemma that provides sufficient condition for propositional PJ calculi to be complete. Further we prove that this sufficient condition holds for both PJ logics **St** and **It**. From this it follows that the corresponding calculi are complete.

Let L be a PJ logic. The propositional calculus for L is also called the zero-order calculus for L and is denoted by L^0.

Axiom Schemata 7.5.1 (Axiom schemata of the propositional calculus for L (L^0)). Let Φ, Ψ and Θ be external L-formulas. Then, the following formulas are *axioms of* L^0:

(A1) $\Phi \to (\Psi \to \Phi)$,
(A2) $(\Phi \to (\Psi \to \Theta)) \to ((\Phi \to \Psi) \to (\Phi \to \Theta))$,
(A3) $(\neg\Phi \to \Psi) \to ((\neg\Phi \to \neg\Psi) \to \Phi)$.

The above axiom schemata form the group (A). The calculus L^0 should also include the group (B) of axiom schemata by which the J-operators are described. Different PJ logics possess different sets of J-operators. Therefore, the corresponding calculi include different axiom schemata of the group (B^L). Instead of giving a general method of axiomatisation of PJ logics we only suppose that L^0 contains a group of axiom schemata in which the interrelation between the J-operators is described completely and adequately. This is equivalent to the description of the relation between the J-characterised subsets of the set $\mathscr{V}(L,\mathsf{I})$.

Notation 7.5.2. Let $\top$ denote an always-true statement (for example $\Phi \vee \neg\Phi$ for an external formula Φ).

Let $\bot$ denote an always-false statement (for example $\Phi \wedge \neg\Phi$ for an external formula Φ).

The notation is introduced by induction following:

BASIS.

$$\bigwedge_{\Phi\in\emptyset} \Phi = \top, \quad \bigvee_{\Phi\in\emptyset} \Phi = \bot.$$

INDUCTION STEP.

$$\bigwedge_{\Phi\in\Gamma\cup\{\Psi\}} \Phi = \bigwedge_{\Phi\in\Gamma} \Phi \wedge \Psi \qquad \bigvee_{\Phi\in\Gamma\cup\{\Psi\}} \Phi = \bigvee_{\Phi\in\Gamma} \Phi \vee \Psi.$$

Axiom Schemata 7.5.3 (Axiom schemata of group (B) for an FPJ logic L). Let L be an FPJ logic and p be an internal propositional variable of L ($p \in \mathscr{P}(L,\mathsf{I})$). Then the following formulas are axioms of the calculus L^0:

(B1) $\bigvee_{\varepsilon\in\mathscr{V}(L,\mathsf{I})} \mathrm{J}_\varepsilon\, p$,
(B2) $\mathrm{J}_\varepsilon\, p \to \neg \mathrm{J}_\delta\, p \qquad (\varepsilon,\delta \in \mathscr{V}(L,\mathsf{I}),\ \varepsilon \neq \delta)$.

Axiom Schemata 7.5.4 (Axiom schemata of group (B) for the iterative version of an FPJ logic L). Let L be an FPJ logic, $\tau \in \mathscr{V}(L,\mathsf{I})$, $I = \mathrm{I}^\tau(L)$, $p \in \mathscr{P}(I,\mathsf{I})$. Then the following formulas are axioms of the calculus I^0:

(B1) $\bigvee_{\varepsilon\in\mathscr{V}(L,\mathsf{I})} \mathrm{J}_{(\varepsilon,n)}\, p \qquad (n \in \omega)$,
(B2) $\mathrm{J}_{(\varepsilon,n)}\, p \to \neg \mathrm{J}_{(\delta,n)}\, p \qquad (\varepsilon,\delta \in \mathscr{V}(L,\mathsf{I}),\ \varepsilon \neq \delta,\ n \in \omega)$,
(B3) $\mathrm{J}_{(\varepsilon,n)}\, p \to \mathrm{J}_{(\varepsilon,k)}\, p \qquad (\varepsilon \in \mathscr{D}^\tau(L),\ n,k \in \omega,\ k > n)$,
(B4) $\mathrm{J}_{(\tau,k)}\, p \to \mathrm{J}_{(\tau,n)}\, p \qquad (n,k \in \omega,\ k > n)$.

Below, L is meant as a PJ logic.

Inference Rules 7.5.5 (Inference rules for L^0). The only inference rule of L^0 is *modus ponens* defined as follows:

$$\frac{\Phi, \quad \Phi \to \Psi}{\Psi},$$

where Φ and Ψ are external L-formulas.

Definition 7.5.6. Let Γ be a set of external formulas of L. An *inference from Γ in L^0 (an L^0-inference from Γ)* is a finite sequence of external L-formulas such that every its element is:

- Either an axiom of L^0
- Or a member of Γ
- Or follows from preceding formulas by *modus ponens*

An L^0-inference from the empty set of L-formulas is called a *proof in L^0 (an L^0-proof)*.

Now we show that the analogues of some well-known theorems hold either for all PJ logics or for our "selected" PJ logics **St** and **It**. We especially mean the Deduction Theorem and Soundness Theorem. In a general way, their proofs are similar to the corresponding ones for classical logic.

Definition 7.5.7. Let Γ be a set of external L-formulas and Φ be an external L-formula. We say that Φ is *derivable from Γ in L^0* (written: $\Gamma \vdash_L \Phi$) if there exists an L^0-inference containing Φ from the set Γ.

An external L-formula Φ is called *provable in L^0* (a theorem of L^0) if Φ is derivable in L^0 *from the empty set*. The fact that φ is provable in L^0 is denoted by $\vdash_L \Phi$.

Theorem 7.5.8 (Deduction Theorem for L^0). Let Γ be a set of external L-formulas and Φ and Ψ be external L-formulas. Then $\Gamma \cup \{\Phi\} \vdash_L \Psi$ implies $\Gamma \vdash_L \Phi \to \Psi$.

Proof. Proof is similar to the proof of the Deduction Theorem for classical propositional calculus [Mendelson, 1997]. □

Lemma 7.5.9. Any axiom of group (A) of L^0 is an L-tautology.

Proof. The truth tables for $\neg$ and $\to$ can be used to show that axioms of the group (A) (i.e. axioms obtained from schemata A1–A3) are L-tautologies similarly to classical logic. □

Lemma 7.5.10. Let Γ be a set of external L-formulas and Φ be an external L-formula. Suppose that each axiom of group (B) is an L-tautology. Then $\Gamma \vdash_L \Phi$ implies $\Gamma \models_L \Phi$.

Proof. By induction over the length of an L^0-inference of Φ (from Γ), using Lemma 7.5.9. This proof is similar to that of the Soundness Theorem for classical propositional calculus (see, e.g. [Mendelson, 1997]). □

Lemma 7.5.11. Let L be an FPJ logic. Then every axiom of group (B) is an L-tautology.

Proof. Consider axioms obtained from (B1) and (B2). Let $p \in \mathscr{P}(L, \mathsf{I})$. Take an arbitrary L-valuation v. Then $v(p)$ can take exactly one value from $\mathscr{V}(L, \mathsf{I})$. Therefore, $v(\mathrm{J}_\varepsilon\, p)$ takes value $\mathbf{t}$ for one and only one $\varepsilon \in \mathscr{V}(L)$. Clearly, in this case

$$v\left(\bigvee_{\varepsilon \in \mathscr{V}(L,\mathsf{I})} \mathrm{J}_\varepsilon(p)\right) = \mathbf{t}$$

and

$$v\,(\mathrm{J}_\varepsilon(p) \to \neg \mathrm{J}_\delta(p)) = \mathbf{t},$$

where $\varepsilon, \delta \in \mathscr{V}(L, \mathsf{I})$, $\varepsilon \neq \delta$. □

Lemma 7.5.12. Let L be an FPJ logic, $\tau \in \mathscr{V}(L, \mathsf{I})$, $I = \mathrm{I}^\tau(L)$. Every axiom of group (B) is an I-tautology.

Proof. By direct checking. Take an arbitrary I-valuation v. Then for each $n \in \omega$, $v(p)$ may belong only to exactly one of the sets $\mathrm{Ch}(\varepsilon, n)$, where $\varepsilon \in \mathscr{V}(L, \mathsf{I})$. Hence,

$$v\left(\bigvee_{\varepsilon \in \mathscr{V}(L,\mathsf{I})} \mathrm{J}_{(\varepsilon,n)}(p)\right) = \mathbf{t}$$

and

$$v\left(\mathrm{J}_{(\varepsilon,n)}(p) \to \neg \mathrm{J}_{(\delta,n)}(p)\right) = \mathbf{t},$$

where $\varepsilon, \delta \in \mathscr{V}(L, \mathsf{I})$, $\varepsilon \neq \delta$.

Therefore axioms obtained from (B1) and (B2) are I-tautologies.

Let us suppose now that $k > n$. Let

$$v\,(p) \in \mathrm{Ch}\,(\varepsilon, n) \subseteq \mathrm{Ch}\,(\varepsilon, k)$$

(see 7.4.12). Consequently $v(\mathrm{J}_{(\varepsilon,k)}(p)) = \mathbf{t}$. Thus the axiom schema ($\mathrm{B}^{\mathbf{It}}3$) has been checked. Analogously by the use of 7.4.12 it can be shown that all axioms obtained from the schema (B4) are **It**-tautologies. □

Theorem 7.5.13 (Soundness Theorem for FPJ logics). Let L be an FPJ logic, Γ be a set of external L-formulas and Φ be an external L-formula. Then

$$\Gamma \vdash_L \Phi \quad \text{implies} \quad \Gamma \models_L \Phi.$$

Proof. This statement directly follows from Lemmas 7.5.9, 7.5.10 and 7.5.11. □

Theorem 7.5.14 (Soundness Theorem for iterative versions of FPJ logics). Let L be an FPJ logic, $\tau \in \mathscr{V}(L, \mathsf{I})$, $I = \mathrm{I}^\tau(L)$, Γ be a set of external I-formulas and Φ be an external I-formula. Then

$$\Gamma \vdash_I \Phi \quad \text{implies} \quad \Gamma \models_I \Phi.$$

Proof. This statement directly follows from Lemmas 7.5.9, 7.5.10 and 7.5.12. □

Below we will define some notions that are necessary for further work and will prove a series of auxiliary statements necessary to prove the Completeness Theorem. First we will define a one-to-one correspondence between the set of external L-formulas and the set of **C**-formulas. We will prove the statement that links derivability of external L-formulas and derivability of their images (**C**-formulas). Then we will define a correspondence between L-valuations and **C**-valuations and the important notion of **C**-*expressible PJ logic*. In the Main Lemma we prove that for a PJ logic to be complete it is sufficient to be **C**-expressible. Then we prove that any FPJ logic L and its iterative version I are **C**-expressible. As corollaries we have the completeness theorems for these logics.

Definition 7.5.15. We say that an external formula of logic L is *basic* if it is either an external propositional variable or a J-atomic formula.

The notion of a *basic external L-formula* is important. Namely, basic external formulas are considered in the basis of induction in various inductive definitions and inductive proofs.

Definition 7.5.16. Note that both the set of basic external L-formulas and the set of propositional variables of classical logic are countable. Hence, there is a bijection between these two sets. Choose one such bijection and denote it by $\hat{}$. For brevity we write $\hat{p}_\alpha$ instead of $\widehat{\mathrm{J}_\alpha p}$, where $p \in \mathscr{P}(L,\mathsf{I})$. Then obviously,

$$\mathscr{P}(\mathbf{C}) = \{\hat{p}_\alpha \mid \mathrm{J}_\alpha \in \mathscr{J}(L), p \in \mathscr{P}(L,\mathsf{I})\} \cup \{\hat{q} \mid q \in \mathscr{P}(L,\mathsf{E})\}.$$

We recursively define the mapping C of the set of external formulas of propositional logic L into the set of formulas of classical propositional logic as follows:

(i) $\mathrm{C}(q) = \hat{q} \quad (q \in \mathscr{P}(L,\mathsf{E}))$;
(ii) $\mathrm{C}(\mathrm{J}_\alpha p) = \hat{p}_\alpha \quad (p \in \mathscr{P}(L,\mathsf{I}), \mathrm{J}_\alpha \in \mathscr{J}(L))$;
(iii) $\mathrm{C}(\neg\Phi) = (\neg\mathrm{C}(\Phi))$,
$\mathrm{C}(\Phi \to \Psi) = (\mathrm{C}(\Phi) \to \mathrm{C}(\Psi))$, where Φ and Ψ are external formulas of L.

Let Φ be an external L-formula. Then the **C**-formula $\mathrm{C}(\Phi)$ is called the C-*image* of Φ.

Lemma 7.5.17. Mapping C defined above is a *bijection* of the set of external L-formulas ($\mathscr{F}(L,\mathsf{E})$) onto the set of **C**-formulas ($\mathscr{F}(\mathbf{C})$).

Proof. First we prove that C is a *surjection*. To do this, we show that, for any **C**-formula φ, there is an external L-formula Φ such that $\mathrm{C}(\Phi) = \varphi$. The proof is trivial by induction over formulas of classical logic.

BASIS. For $\hat{q}$, where $q \in \mathscr{P}(L,\mathsf{E})$, an inverse image of C is p. For $\hat{p}_\alpha$, where $p \in \mathscr{P}(L,\mathsf{I})$, such a required external L-formula is $(\mathrm{J}_\alpha p)$.

INDUCTION STEP. If external L-formulas Φ and Ψ are inverse images of $\mathbf{C}$-formulas φ and ψ, respectively, then it is obvious that $(\neg\Phi)$ and $(\Phi \to \Psi)$ are inverse images of formulas $(\neg\varphi)$ and $(\varphi \to \psi)$, respectively.

We define recursively the mapping

$$\mathrm{P}\colon \mathscr{F}(\mathbf{C}) \to \mathscr{F}(L,\mathsf{E})$$

as follows:

(i) $\mathrm{P}(\hat{q}) = q \quad (q \in \mathscr{P}(L,\mathsf{E}))$;
(ii) $\mathrm{P}(\hat{p}_\alpha) = (\mathrm{J}_\alpha\, p) \quad (p \in \mathscr{P}(L,\mathsf{I}))$;
(iii) $\mathrm{P}(\neg\varphi) = (\neg\mathrm{P}(\varphi))$,
$\mathrm{P}(\varphi \to \psi) = (\mathrm{P}(\varphi) \to \mathrm{P}(\psi)) \quad (\varphi, \psi \in \mathscr{F}(\mathbf{C}))$.

It can be proved by induction over external L-formulas that $\mathrm{P}(\mathrm{C}(\Phi)) = \Phi$ for any external L-formula Φ. Hence, C is an *injection*.

Thus, C is a bijection and P is its inverse mapping. So, P is also a bijection. □

Definition 7.5.18. Let Γ be a set of external L-formulas. By definition let

$$\mathrm{C}(\Gamma) = \{\mathrm{C}(\Phi) \mid \Phi \in \Gamma\}.$$

$\mathrm{C}(\Gamma)$ is called the C-*image* of Γ. By $\mathrm{C}(\mathrm{B}^L)$ we denote the set of C-images of axioms of the group (B^L).

Remark 7.5.19. Let L be an FPJ logic. Then the set $\mathrm{C}(\mathrm{B}^L)$ contains formulas of the following form:

(i) $\bigvee\limits_{\varepsilon\in\mathscr{V}(L,\mathsf{I})} \hat{p}_\varepsilon$,
(ii) $\hat{p}_\varepsilon \to \neg\hat{p}_\delta \quad (\varepsilon, \delta \in \mathscr{V}(L,\mathsf{I}), \varepsilon \neq \delta)$,

where $p \in \mathscr{P}(L,\mathsf{I})$.

Remark 7.5.20. Let $I = \mathrm{I}^\tau(L)$, where L is an FPJ logic, $\tau \in \mathscr{V}(L,\mathsf{I})$. Then the set $\mathrm{C}(\mathrm{B}^I)$ contains formulas of the following form:

(i) $\bigvee\limits_{\varepsilon\in\mathscr{V}(L,\mathsf{I})} \hat{p}_{(\varepsilon,n)} \quad (n \in \omega)$,
(ii) $\hat{p}_{(\varepsilon,n)} \to \neg\hat{p}_{(\delta,n)} \quad (\varepsilon, \delta \in \mathscr{V}(L,\mathsf{I}), \varepsilon \neq \delta, n \in \omega)$,
(iii) $\hat{p}_{(\varepsilon,n)} \to \hat{p}_{(\varepsilon,k)} \quad (\varepsilon \in \mathscr{D}^\tau(L), n, k \in \omega, k > n)$,
(iv) $\hat{p}_{(\tau,k)} \to \hat{p}_{(\tau,n)} \quad (n, k \in \omega, k > n)$,

where $p \in \mathscr{P}(I)$.

Lemma 7.5.21. Let Γ be a set of external L-formulas and Φ be an external L-formula. Then

$$\mathrm{C}(\Gamma) \cup \mathrm{C}(\mathrm{B}^L) \vdash_{\mathbf{C}} \mathrm{C}(\Phi) \quad \text{implies} \quad \Gamma \vdash_L \Phi.$$

Proof. Trivial induction over the length of inference in classical propositional calculus.

If $\mathrm{C}(\Phi) \in \mathrm{C}(\Gamma)$, then $\Phi \in \Gamma$. If $\mathrm{C}(\Phi) \in \mathrm{C}(\mathrm{B}^L)$, then Φ is an axiom of group (B). If $\mathrm{C}(\Phi)$ follows from $\mathrm{C}(\Psi)$ and $\mathrm{C}(\Psi) \to \mathrm{C}(\Phi)$ by *modus ponens* for classical propositional calculus, then Φ follows from Ψ and $\Psi \to \Phi$ by *modus ponens* for L^0.

The inverse assertion, i.e. the fact that

$$\Gamma \vdash_L \Phi \quad \text{implies} \quad \mathrm{C}(\Gamma) \cup \mathrm{C}(\mathrm{B}^L) \vdash_{\mathrm{C}} \mathrm{C}(\Phi),$$

also holds. This can be easily proved by induction over the length of inference in L^0. □

Definition 7.5.22. Let v be an L-valuation. The C-*image of the L-valuation v* is a mapping

$$\hat{v}\colon \mathscr{F}(\mathbf{C}) \to \{\mathbf{t}, \mathbf{f}\}$$

recursively defined as follows:

(i) $\hat{v}(\hat{q}) = v(q) \quad (q \in \mathscr{P}(L, \mathsf{E}));$
(ii) $\hat{v}(\hat{p}_\alpha) = v(\mathrm{J}_\alpha p) \quad (p \in \mathscr{P}(L, \mathsf{I}));$
(iii) $\hat{v}(\neg\varphi) = (\neg\hat{v}(\varphi),$
$\hat{v}(\varphi \to \psi) = (\hat{v}(\varphi) \to \hat{v}(\psi)) \quad (\varphi, \psi \in \mathscr{F}(\mathbf{C})).$

Clearly, the C-image of the L-valuation v is a **C**-valuation.

Definition 7.5.23. Let u be a **C**-valuation (see Definition 7.2.8). We call u *L-coherent* if it is a **C**-model of the set $\mathrm{C}(\mathrm{B}^L)$.

Lemma 7.5.24. Suppose that v is an L-valuation. Then, for every external formula Φ of propositional logic L:

$$v(\Phi) = \hat{v}(\mathrm{C}(\Phi)).$$

Proof. Proof is trivial by induction over external L-formulas. □

Corollary 7.5.25. For any L-valuation v, the map $\hat{v}$ is an L-coherent **C**-valuation.

Definition 7.5.26. Let L be a PJ logic. We say that L is **C**-*expressible* if for each L-coherent **C**-valuation u there is an L-valuation v such that $u = \hat{v}$.

Lemma 7.5.27 (Main Lemma). Suppose PJ logic L is **C**-expressible. Let Γ be a set of external L-formulas and Φ be an external L-formula. Then

$$\Gamma \models_L \Phi \quad \text{implies} \quad \Gamma \vdash_L \Phi.$$

Proof. Suppose this is not true. Let $\Gamma \models_L \Phi$, but it is not true that $\Gamma \vdash_L \Phi$. By Lemma 7.5.21, $\mathrm{C}(\Gamma) \cup \mathrm{C}(\mathrm{B}^L) \vdash_{\mathrm{C}} \mathrm{C}(\Phi)$ does not hold either. By the Completeness Theorem for classical propositional logic (see 7.3.7), $\mathrm{C}(\Gamma) \cup \mathrm{C}(\mathrm{B}^L) \not\models_{\mathrm{C}} \mathrm{C}(\Phi)$. Then, there is a **C**-valuation u such that u is a **C**-model of the set $\mathrm{C}(\Gamma) \cup \mathrm{C}(\mathrm{B}^L)$, but u is not a **C**-model of the formula $\mathrm{C}(\Phi)$. The **C**-valuation u is L-coherent (u is a **C**-model of $\mathrm{C}(\mathrm{B}^L)$). In this case, there is an L-valuation v such that $u = \hat{v}$ by the assumption (and Definition 7.5.26). By Lemma 7.5.24, for any external L-formula Ψ it holds that $v(\Psi) = \hat{v}\mathrm{C}(\Psi))$. Then L-valuation v is a model Γ, but is not a model of Φ. Therefore, $\Gamma \not\models_L \Phi$, which results in contradiction. □

Lemma 7.5.28. Every FPJ logic L is **C**-expressible, i.e. for any L-coherent **C**-valuation u there is an L-valuation v such that $u = \hat{v}$.

Proof. Let $p \in \mathscr{P}(L,\mathsf{I})$. Since u is an L-coherent **C**-valuation, u is a model of C-images of axioms obtained by (B1) and (B2) (see (i) and (ii) in Definition 7.5.19).

Then, there is a unique $\varepsilon \in \mathscr{V}(L,\mathsf{I})$ such that $u(\hat{p}_\varepsilon) = \mathbf{t}$. For any $\hat{p} \in \mathscr{P}(L,\mathsf{I})$ let $v(p) = \varepsilon$ if $u(\hat{p}_\varepsilon) = \mathbf{t}$.

If $q \in \mathscr{P}(L,\mathsf{E})$ then let $v(q) = u(\hat{q})$.

Obviously, $v\colon \mathscr{P}(L) \to \mathscr{V}(L)$ is well defined. We recursively extend the mapping v to the entire set of L-formulas in a standard way (see Definition 7.4.19).

Now we show that $u = \hat{v}$, i.e. for any formula $\varphi \in \mathscr{F}(\mathbf{C})$, it holds that $u(\varphi) = \hat{v}(\varphi)$. The proof is by induction over the formulas of classical propositional logic.

Each propositional variable of **C** can be represented in one and only one of the following forms: it is $\hat{q}$, wehere $q \in \mathscr{V}(L,\mathsf{E})$, or it is $\hat{p}_\alpha$, where $p \in \mathscr{V}(L,\mathsf{I})$.

It is evident that

$$u(\hat{q}) = \hat{v}(\hat{q})$$

for any $q \in \mathscr{V}(L,\mathsf{E})$.

Let us prove that

$$u(\hat{p}_\alpha) = \hat{v}(\hat{p}_\alpha)$$

for any $p \in \mathscr{V}(L,\mathsf{I})$. For this purpose, it suffices to show that $u(\hat{p}_\alpha) = v(\mathrm{J}_\alpha\, p)$ for any $p \in \mathscr{P}(L)$ and $\mathrm{J}_\alpha \in \mathscr{J}(L)$ (see Definition 7.5.22) .

Let $v(p) = \varepsilon \in \mathscr{V}(L,\mathsf{I})$. Then $u(\hat{p}\varepsilon) = \mathbf{t}$. If $\alpha = \varepsilon$ then $u(\hat{p}_\alpha) = v(\mathrm{J}_\alpha\, p) = \mathbf{t}$. If $\alpha \neq \varepsilon$ then $u(\hat{p}_\alpha) = v(\mathrm{J}_\alpha\, p) = \mathbf{f}$.

Thus, we have proved the statement for propositional variables of classical logic. The induction step is trivial. □

Lemma 7.5.29. Let $I = \mathrm{I}^\tau(L)$ for an FPJ logic L, where $\tau \in \mathscr{V}(L,\mathsf{I})$. Then I is **C**-expressible, i.e. for any I-coherent **C**-valuation u, there is an I-valuation v such that $u = \hat{v}$.

Proof. Let $q \in \mathscr{P}(I,\mathsf{E})$. Then let $v(q) = u(\hat{q})$. It is evident that in this case

$$u(\hat{q}) = \hat{v}(\hat{q})\,.$$

Let $p \in \mathscr{P}(I,\mathsf{I})$. Since u is an I-coherent **C**-valuation, u is a model of **C**-images of axioms obtained from (B1–B4) (see Remark 7.5.20). Then:

(i) For each $n \in \omega$ there exists a unique $\varepsilon \in \mathscr{V}(L,\mathsf{I})$ such that $u(\hat{p}_{(\varepsilon,n)}) = \mathbf{t}$,
(ii) If $u(\hat{p}_{(\varepsilon,n)}) = \mathbf{t}$, where $\varepsilon \in \mathscr{D}^\tau(L)$ then $u(\hat{p}_{(\varepsilon,m)}) = \mathbf{t}$ for any $m > n$,
(iii) If $u(\hat{p}_{(\tau,n)}) = \mathbf{t}$ then $u(\hat{p}_{(\tau,k)}) = \mathbf{t}$ for any $k < n$.

For any $p \in \mathscr{P}(I,|)$ let by definition

$$v(p) = \begin{cases} \langle \varepsilon, n \rangle & \text{if } n = \min\{m \in \omega \mid u(\hat{p}_{(\varepsilon,m)}) = \mathbf{t}\}, \\ \tau & \text{if } u(\hat{p}_{(\tau,m)}) = \mathbf{t} \text{ for any } m \in \omega, \end{cases}$$

where $\varepsilon \in \mathscr{D}^{\tau}(L)$, $n \in \omega$.

It follows from (i) and (ii) that there exists no more than one $\varepsilon \in \mathscr{D}^{\tau}(L)$ such that $u(\hat{p}_{(\varepsilon,m)}) = \mathbf{t}$ for some $m \in \omega$. Therefore the definition of the mapping

$$v\colon \mathscr{P}(I) \to \mathscr{V}(I)$$

is correct. Let us continue the mapping v onto the entire set of the formulas of the iterative logic by induction in the usual way (see definition of the I-valuation that is analogous to the Definition 7.4.19).

Now we must prove that $u = \hat{v}$, i.e. $u(\varphi) = \hat{v}(\varphi)$ holds for any formula $\varphi \in \mathscr{F}(\mathbf{C})$. Proof is by induction over the formulas of the classical propositional logic. First we must show that $u(\hat{p}_{\alpha}) = \hat{v}(\hat{p}_{\alpha})$ for any $\hat{p}_{\alpha} \in \mathscr{P}(\mathbf{C})$. For this it is enough to show that

$$u(\hat{p}_{(\delta,m)}) = \mathrm{J}_{(\delta,m)}(v(p)) \tag{7.1}$$

for any $p \in \mathscr{P}(I,|)$, $\delta \in \mathbf{St}$, $m \in \omega$.

Above it has been shown that by (i) and (ii) there exists not more than one $\varepsilon \in \mathscr{D}^{\tau}(L)$ such that $u(\hat{p}_{(\varepsilon,m)}) = \mathbf{t}$ for some $m \in \omega$. Let us consider the possible cases.

CASE 1. There exists $m \in \omega$ such that $u(\hat{p}_{(\varepsilon,m)}) = \mathbf{t}$, where $\varepsilon \in \mathscr{D}^{\tau}(L)$. Then $v(p) = \langle \varepsilon, n \rangle$, where $n = \min\{m \in \omega \mid u(\hat{p}_{(\varepsilon,m)}) = \mathbf{t}\}$. In this case the following statements hold:

(a) For arbitrary $k < n$,

$$u(\hat{p}_{(\delta,k)}) = \begin{cases} \mathbf{t} \text{ if } \delta = \tau, \\ \mathbf{f} \text{ if } \delta \neq \tau; \end{cases}$$

(b) For any $m \geq n$,

$$u(\hat{p}_{(\delta,m)}) = \begin{cases} \mathbf{t} \text{ if } \delta = \varepsilon, \\ \mathbf{f} \text{ if } \delta \neq \varepsilon. \end{cases}$$

On the other hand, because of $v(p) = \langle \varepsilon, n \rangle$, the following statements hold:

(a′) For any $k < n$, $v(p) \in \mathrm{Ch}(\tau,k)$ and, consequently, $v(p) \notin \mathrm{Ch}(\delta,k)$, if $\delta \neq \tau$;
(b′) For any $m \geq n$, $v(p) \in \mathrm{Ch}(\varepsilon,m)$ and, consequently, $v(p) \notin \mathrm{Ch}(\delta,m)$, if $\delta \neq \varepsilon$.

It follows from (a), (b), (a′), and (b′) that (7.1) holds for arbitrary $p \in \mathscr{P}(I,|)$, $\delta \in \mathbf{St}$, $m \in \omega$.

CASE 2. For arbitrary $m \in \omega$ and $\varepsilon \in \mathscr{V}(L,|)$, $u(\hat{p}_{(\varepsilon,m)}) = \mathbf{f}$. Then due to (i), $u(\hat{p}_{(\tau,m)}) = \mathbf{f}$ for any $m \in \omega$. In this case $v(p) = \tau$ and the following statements hold:

(c) For any $m \in \omega$

$$u\left(\hat{p}_{(\delta,m)}\right) = \begin{cases} \mathbf{t} \text{ if } \delta = \tau, \\ \mathbf{f} \text{ if } \delta \neq \tau; \end{cases}$$

(c′) $v(p) \in \mathrm{Ch}(\tau,k)$ for any $m \in \omega$.

It follows from (c) and (c′) that (7.1) holds for any $p \in \mathscr{P}(I,|)$, $\delta \in \mathbf{St}$, $m \in \omega$.

Thus (7.1) holds in each possible case for any $p \in \mathscr{P}(I,|)$, $\delta \in \mathbf{St}$, $m \in \omega$. Therefore, $u = \hat{v}$, as was to be proved. □

Theorem 7.5.30 (Completeness Theorem for FPJ logics). Let L be an FPJ logic, Γ be a set of external L-formulas and Φ be an external L-formula. Then

$$\Gamma \models_L \Phi \quad \text{implies} \quad \Gamma \vdash_L \Phi.$$

Proof. This statement directly follows from the Main Lemma (7.5.27) and Lemma 7.5.28. □

Theorem 7.5.31 (Completeness Theorem for iterative versions of FPJ logics). Let $I = \mathrm{I}^{\tau}(L)$, where L is an FPJ logics, $\tau \in \mathscr{V}(L,|)$. Let Γ be a set of external I-formulas and Φ be an external I-formula. Then

$$\Gamma \models_I \Phi \quad \text{implies} \quad \Gamma \models_I \Phi.$$

Proof. This statement directly follows from the Main Lemma (7.5.27) and Lemma 7.5.29. □

Definition 7.5.32. Set Γ of external L-formulas is called *inconsistent* if there is an external L-formula Φ such that $\Gamma \vdash_L \Phi \wedge \neg\Phi$. Otherwise, set Γ is called *consistent.*

Proposition 7.5.33. If a set of external L-formulas is inconsistent, then it contains a finite inconsistent subset.

Proof. Let Γ be an inconsistent set of external L-formulas. Then, the set of all elements of Γ contained in some inference of the formula $\Phi \wedge \neg\Phi$ from Γ in L^0 is a finite inconsistent subset of Γ. □

Proposition 7.5.34. Let Γ be a set of external L-formulas. Then, Γ is consistent if and only if it has a model.

Theorem 7.5.35 (Compactness Theorem). Let Γ be a set of external formulas of propositional logic L. Suppose that every finite subset of Γ has a model. Then, the entire set Γ has a model.

Proof. Proof is by application of Propositions 7.5.33 and 7.5.34. □

Chapter 8
First-Order Logics

In this chapter we introduce the first-order logics that correspond to the propositional ones discussed in the previous section. We will consider many-sorted first-order logics where many-sortedness is significant for the applications further to be discussed.

8.1 Terms and Notation

First the term "string", the finite sequence of elements of some set, will be introduced, then the basic properties of strings will be analysed and operations over strings will be defined and special notations will be introduced. These notations will allow a much shorter description of the definitions and statements, e.g. we will write $\vec{x}$ instead of $x_1, \ldots, x_n$ and thus we can avoid the long expressions with dots. Obviously, a part of the present section that deals with the strings is auxiliary but is very important since it provides technical tools that allow a more transparent discussion.

Definition 8.1.1. Let U be a non-empty set. Let U^* denote the *set of all strings* of elements of U. To generate all the strings it is enough to have the *empty string* (denoted by $\mathbf{e}$) and the *right-concatenation operation* (denoted by $\bullet$). Then a *string over* U can be inductively defined in the following way:

(i) The empty string is a string over U;
(ii) If ξ is a string over U, $x \in U$, then $\xi \bullet x$ is also a string over U.

The operation $\bullet$ adds a member of U to the right side of a string earlier constructed. This operation can be defined in various ways. E.g. if we put

$$\xi \bullet x = \langle \xi, x \rangle ,$$

then the strings over U are exactly the ordered n-tuples (as they are defined in abstract set theory).

O. Anshakov, T. Gergely, *Cognitive Reasoning*, Cognitive Technologies,
DOI 10.1007/978-3-540-68875-4_8,

Definition 8.1.2. By the use of operation $\bullet$ the *concatenation* operation on the set of strings can be defined. This operation is denoted by $+$ (similarly to the character $+$ used in programming languages). This definition is by induction over the strings. Let η be a string. Then:

(i) $\eta + \mathbf{e} = \eta$;
(ii) Let the value of $\eta + \xi$ be known for $\xi \in U^*$ and let $x \in U$, then $\eta + (\xi \bullet x) = (\eta + \xi) \bullet x$.

Proposition 8.1.3. It can be proved that for any $\xi, \eta, \zeta \in U$ the following holds:

(a) $(\xi + \eta) + \zeta = \xi + (\eta + \zeta)$,
(b) $\xi + \mathbf{e} = \mathbf{e} + \xi = \xi$,
(c) $\xi + \eta = \xi + \zeta$ implies $\eta = \xi$,
(d) $\eta + \xi = \zeta + \xi$ implies $\eta = \xi$,

i.e. the algebra $\langle U^*; +, \mathbf{e} \rangle$ is a (non-commutative) monoid with right and left reductions. Further on we will sometimes use these properties of the strings implicitly, since they are obvious.

Agreement 8.1.4. Let x, y, z, ... be variables for elements of U. Then we use $\vec{x}$, $\vec{y}$, $\vec{z}$, ... as *variables for elements of* U^*; i.e. $\vec{x}$, $\vec{y}$, $\vec{z}$, ... will denote strings over U. We assume that x, y, z, ... and $\vec{x}$, $\vec{y}$, $\vec{z}$, ... form different name spaces. This allows us to consider expressions such as $\vec{x} \bullet x$ (where $x \in U$ and $\vec{x} \in U^*$).

Notation 8.1.5. Let $f \colon U \to V$. Then f^* denotes a mapping from U^* to V^*, defined recursively in the following way:

(i) $f^*(\mathbf{e}) = \mathbf{e}$,
(ii) $f^*(\xi \bullet x) = f^*(\xi) \bullet f(x)$.

In this case, if $\vec{x}$ is the sequence x_1, ..., x_n, then $f^*(\vec{x})$ is the sequence $f(x_1)$, ..., $f(x_n)$. By the use of angle brackets we have

$$f^*(\langle x_1, \ldots, x_n \rangle) = \langle f(x_1), \ldots, f(x_n) \rangle .$$

Let the sequence x_1, ..., x_n be denoted by $\vec{x}$. Then in the case of the postfix notation for the function symbol f (i.e. if one uses x^f instead of $f(x)$), we denote the sequence $x_1^f, \ldots, x_n^f$ by $\vec{x}^{f^*}$.

Definition 8.1.6. We denote the *set of all items occurring in* $\vec{x} \in U^*$ by $\mathrm{Set}(\vec{x})$. The set $\mathrm{Set}(\vec{x})$ can be defined by induction in the following way:

(i) $\mathrm{Set}(\mathbf{e}) = \emptyset$;
(ii) Let the value of $\mathrm{Set}(\xi)$ be already known, $\xi \in U^*$, $x \in U$; then

$$\mathrm{Set}(\xi \bullet x) = \mathrm{Set}(\xi) \cup \{x\} .$$

Definition 8.1.7. Let $k,n \in \omega$. Then the set $\{i \in \omega \mid k \leq i \leq n\}$ is said to be *an interval of the natural sequence from k to n* and is denoted by $[k\,..\,n]$. It is obvious that $[k\,..\,n] = \emptyset$ if $k > n$.

Notation 8.1.8. Note that we can mean any string over U as a function from a segment of natural sequence to U. In this context, we say that elements of this segment are *indices* of elements of the string. Unless otherwise indicated, the first element of the segment is equal to 1. $\xi\,[i]$ denotes the member of ξ indexed by i.

Len (ξ) denotes the *length* of ξ.

IS (ξ) denotes the *indexing set* of ξ. Unless otherwise indicated

$$\mathrm{IS}\,(\xi) = [1\,..\,\mathrm{Len}\,(\xi)]\,.$$

Last (ξ) denotes the *last element* of ξ, while First (ξ) denotes the first element of ξ. Values Last (ξ) and First (ξ) are undefined if $\xi = \mathbf{e}$.

Definition 8.1.9. A $\xi \in U^*$ is called *regular*, if it does not contain repeating components, i.e. if

$$\mathrm{Len}\,(\xi) = \mathrm{Card}\,(\mathrm{Set}\,(\xi))\,.$$

Here and hereafter Card (M) denotes the cardinality of the set M.

The set of all regular strings over U is denoted by $U^\diamond$.

Often it is important that strings do not contain identical elements. E.g. when a substitution is defined it is required that the string of arguments of the substitution will not contain repeating elements, i.e. it would be regular in the sense of Definition 8.1.9. Below we will define a substitution by using the regular strings explicitly. Note that regular strings of variables will be used in the definition of *pure atomic formulas*. By the use of these formulas the important term *locality* will be defined for the plausible inference rules (modification rules); this term being one of the main notions for the theory of cognitive reasoning (CR theories).

Definition 8.1.10. Let $\xi \in U^\diamond$, $\eta \in V^*$. Let a mapping of the set Set (ξ) onto the set Set (η), denoted by $\binom{\xi}{\eta}$ or simply by $\frac{\xi}{\eta}$, be such that, for any $i \in \mathrm{IS}\,(\xi)$,

$$\tfrac{\xi}{\eta}\,(\xi\,[i]) = \eta\,[i]\,.$$

Then this mapping is called a *substitution* of Set (ξ) to Set (η).

Evidently $\frac{\xi}{\eta}$ is a standard substitution, which is usually denoted by

$$\begin{pmatrix} \xi_1 & \dots & \xi_n \\ \eta_1 & \dots & \eta_n \end{pmatrix} \quad \text{or by} \quad \begin{matrix} \xi_1 & \dots & \xi_n \\ \eta_1 & \dots & \eta_n \end{matrix}\,.$$

Thus the part dealing with strings is completed. The terms and notations necessary for further use have been introduced. Now we turn to introduce the language of the *many-sorted* predicate logic. First the terms and notations specific for the many-sorted case will be introduced. Further on we will consider first-order many-sorted logics.

Notation 8.1.11. The notion of sort is similar to the notion of data type in programming languages. Variables and constants for the sorts are denoted by characters of sans serif font. The names of the sets of sorts are written by bold sans serif fonts.

(i) Denote by IS the set of *sorts of individuals* (intividual sorts). This set is not empty, at most countable and fixed furtheron.
(ii) Denote by TS the set of *sorts of truth values* (truth sorts). $\mathsf{TS} = \{\mathsf{I}, \mathsf{E}\}$, where E is the sort of external truth values and I is the sort of internal truth values. The notations E and I have already been used in this work.

Definition 8.1.12. *Signature* is a quadruple

$$\sigma = \langle \mathsf{Sorts}_\sigma, \mathrm{Fu}_\sigma, \mathrm{Pr}_\sigma, \mathrm{Ar}_\sigma, \mathrm{Sort}_\sigma \rangle,$$

where $\mathsf{Sorts}_\sigma \subseteq \mathsf{IS}$ is the set *of sorts of signature* σ, Fu_σ is the set of *function symbols of signature* σ, Pr_σ is the set of *predicate symbols of signature* σ, $\mathrm{Fu}_\sigma \cap \mathrm{Pr}_\sigma = \emptyset$, mappings

$$\mathrm{Ar}_\sigma : \mathrm{Fu}_\sigma \cup \mathrm{Pr}_\sigma \to \mathsf{Sorts}_\sigma{}^*,$$

$$\mathrm{Sort}_\sigma : \mathrm{Fu}_\sigma \cup \mathrm{Pr}_\sigma \to \mathsf{Sorts}_\sigma \cup \mathsf{TS},$$

satisfy $\mathrm{Sort}_\sigma(\mathrm{Fu}_\sigma) \subseteq \mathsf{Sorts}_\sigma$, $\mathrm{Sort}_\sigma(\mathrm{Pr}_\sigma) \subseteq \mathsf{TS}$.

The value $\mathrm{Ar}_\sigma(\varphi)$ will be called the *arity of* φ and the value $\mathrm{Sort}_\sigma(\varphi)$ is called the *sort of* φ ($\varphi \in \mathrm{Fu}_\sigma \cup \mathrm{Pr}_\sigma$).

If the arity of a function symbol is an empty string of sorts, then this symbol is *nullary* and is called an *individual constant*. Predicate symbols whose arity is an empty sequence of sorts are called *logical constants*.

Denote by Const_σ the *set of individual constants of signature* σ.

If the sort of a predicate symbol p is E, then p is called *external*; if the sort of this p is I, then p is called *internal*.

The *cardinality of signature* σ is the cardinality of $\mathrm{Fu}_\sigma \cup \mathrm{Pr}_\sigma$. We assume that the cardinality of this set is at most countable. Fu_σ, and Pr_σ, both separately and together, may be empty.

Signature σ is called *finite* if the cardinality of σ is finite.

From here on we will often use the same symbol as notation for different entities in different contexts. Recall that in the previous section we proposed to denote the corresponding operations of different algebras by the same symbols. The functional metasymbols can denote different entities when the sets of the argument types differ. For example, above we have defined Fu_σ and Pr_σ as sets of functional and predicate symbols of the signature σ, respectively. In this case each of the symbols from Fu_σ and Pr_σ denotes a concrete permanent object through which they can be considered as metaconstants, i.e. *zeroary* functional metasymbols. We agree to allow *overloading for similar metasymbols* as is common in some widespread programming languages (e.g. C++, C#). Thus for another collection of argument types Fu_σ

and Pr_σ will have another interpretation, e.g. they will be considered as binary or unary metasymbols.

Now we continue to introduce definitions and give an example for overloading of functional metasymbols.

Let:

$$\mathrm{Fu}_\sigma(\vec{\mathsf{a}},\mathsf{s}) = \{f \in \mathrm{Fu}_\sigma \mid \mathrm{Ar}_\sigma(f) = \vec{\mathsf{a}}, \mathrm{Sort}_\sigma(f) = \mathsf{s}\},$$

$$\mathrm{Pr}_\sigma(\vec{\mathsf{a}},\mathsf{s}) = \{p \in \mathrm{Pr}_\sigma \mid \mathrm{Ar}_\sigma(p) = \vec{\mathsf{a}}, \mathrm{Sort}_\sigma(p) = \mathsf{s}\}.$$

There are situations where only the sorts of functional and predicate symbols will interest us. In these cases we put

$$\mathrm{Fu}_\sigma(\mathsf{s}) = \{f \in \mathrm{Fu}_\sigma \mid \mathrm{Sort}_\sigma(f) = \mathsf{s}\},$$

$$\mathrm{Pr}_\sigma(\mathsf{s}) = \{p \in \mathrm{Pr}_\sigma \mid \mathrm{Sort}_\sigma(p) = \mathsf{s}\}.$$

Moreover, we introduce a notation for the set of individual constants of sort s:

$$\mathrm{Const}_\sigma(\mathsf{s}) = \{c \in \mathrm{Const}_\sigma \mid \mathrm{Sort}_\sigma(c) = \mathsf{s}\}.$$

Definition 8.1.13. A signature σ is called *classical* if all predicate symbols of this signature are external, i.e. if $\mathrm{Pr}_\sigma = \mathrm{Pr}_\sigma(\mathsf{E})$.

In this section we study only those notions that are *common* for classical logic **C**, and any PJ logics. The notion of signature as defined above is an example of such notions.

We proceed with the consideration of the first-order logics. Logics **C** and PJ logics have different sets of propositional connectives. Therefore, the first-order languages for these logics are also different. However, these languages can have the same set of symbols. For example, for a fixed signature the first-order languages for logics **C** and a PJ logic can have common sets of predicate and function symbols.

The following definitions will introduce the syntactical notions: variable, term and atomic formula. At present the definition of formulas will not yet be introduced since it will not be general for the hereby discussed logics.

We assume that the first-order languages for the logics **C** and for any PJ logic have the same set of individual variables.

Definition 8.1.14. *The set of individual variables* is denoted by Var. We assume that the following mapping assigns a sort to each variable:

$$\mathrm{Sort}\colon \mathrm{Var} \to \mathsf{IS}.$$

By definition let

$$\mathrm{Var}(\mathsf{s}) = \{x \in \mathrm{Var} \mid \mathrm{Sort}(x) = \mathsf{s}\}.$$

For each signature σ,

$$\mathrm{Var}_\sigma = \{x \in \mathrm{Var} \mid \mathrm{Sort}(x) \in \mathsf{Sorts}_\sigma\}.$$

The set Var_σ is called *the set of individual variables of signature* σ.

Definition 8.1.15. We simultaneously recursively define the *terms of signature* σ and the mapping Sort_σ from the set of this terms to Sorts_σ:

(i) If $x \in \mathrm{Var}_\sigma$ then x is a term of signature σ and $\mathrm{Sort}_\sigma(x) = \mathrm{Sort}(x)$.
(ii) If $c \in \mathrm{Const}_\sigma(\mathsf{s})$ then c is a term of signature σ and $\mathrm{Sort}_\sigma(c) = \mathsf{s}$.
(iii) Let f be a functional symbol of signature σ such that $\mathrm{Ar}_\sigma(f) \neq \mathbf{e}$ and $\vec{t}$ be a term string of signature σ such that $\mathrm{Sort}_\sigma{}^*(\vec{t}) = \mathrm{Ar}_\sigma(f)$. Then $f(\vec{t})$ is a term of signature σ and $\mathrm{Sort}_\sigma(f(\vec{t})) = \mathrm{Sort}_\sigma(f)$.

Let Term_σ denote the set of all terms of signature σ. By definition let

$$\mathrm{Term}_\sigma(\mathsf{s}) = \{t \in \mathrm{Term}_\sigma \mid \mathrm{Sort}_\sigma(t) = \mathsf{s}\}.$$

Let CTerm_σ denote the set of *closed (constant) terms*, i.e. of those terms that do not contain any variables. Obtain the recursive definition of these terms if we omit (i). By definition let

$$\mathrm{CTerm}_\sigma(\mathsf{s}) = \{t \in \mathrm{CTerm}_\sigma \mid \mathrm{Sort}_\sigma(t) = \mathsf{s}\}.$$

Definition 8.1.16. Now we simultaneously define the *atomic formulas of signature* σ and extend the function Sort_σ to the set of atomic formulas: Let t_1 and t_2 be terms of signature σ such that $\mathrm{Sort}_\sigma(t_1) = \mathrm{Sort}_\sigma(t_2)$. Then expression $t_1 = t_2$ is an *atomic formula* of signature σ and $\mathrm{Sort}_\sigma(t_1 = t_2) = \mathsf{E}$.

Let p be a predicate symbol of signature σ and $\vec{t}$ be a finite sequence of terms of signature σ such that $\mathrm{Sort}_\sigma{}^*(\vec{t}) = \mathrm{Ar}_\sigma(p)$. Then $p(\vec{t})$ is an atomic formula of signature σ and $\mathrm{Sort}_\sigma(p(\vec{t})) = \mathrm{Sort}_\sigma(p)$.

An atomic formula of sort E is called an *external* atomic formula; an *atomic* formula of sort I is called an *internal* atomic formula.

Let AF_σ denote the *set of all atomic formulas of signature* σ. By definition let

$$\mathrm{AF}_\sigma(\mathsf{s}) = \{\varphi \in \mathrm{AF}_\sigma \mid \mathrm{Sort}_\sigma(\varphi) = \mathsf{s}\}.$$

Now we define the basic semantic terms considered general for all logics in this book. One of them is the term *universe frame*. A universe frame determines the collection of universes of different sorts. On the basis of a universe frame we can define *structures* which will play the role of models for the collections of formulas (theories) in all logics discussed here.

Definition 8.1.17. An ordered pair $U = \langle \mathrm{Dom}_U, \mathrm{Sort}_U \rangle$, where Dom_U is a non-empty set,

$$\mathrm{Sort}_U : \mathrm{Dom}_U \to \mathsf{IS},$$

is called a *universe frame (U-frame).*

Suppose $\mathsf{s} \in \mathsf{IS}$. By definition, let

$$\mathrm{Dom}_U(\mathsf{s}) = \{x \in \mathrm{Dom}_U \mid \mathrm{Sort}_U(x) = \mathsf{s}\}.$$

The set $\mathrm{Dom}_U(\mathsf{s})$ is called *the domain of sort* s *in* U.

By definition let

$$\begin{aligned}\mathsf{Sorts}_U &= \mathrm{Range}(\mathrm{Sort}_U) \\ &= \{\mathsf{s} \in \mathsf{IS} \mid \mathrm{Sort}_U(x) = \mathsf{s} \text{ for some } x \in \mathrm{Dom}_U\}.\end{aligned}$$

Then the following statements hold:

(i) $\mathrm{Dom}_U(\mathsf{s}) \neq \emptyset$ for any $\mathsf{s} \in \mathsf{Sorts}_U$,
(ii) $\mathrm{Dom}_U = \bigcup_{\mathsf{s} \in \mathsf{Sorts}_U} \mathrm{Dom}_U(\mathsf{s})$,
(iii) $\mathrm{Dom}_U(\mathsf{s}) \cap \mathrm{Dom}_U(\mathsf{t}) = \emptyset$ for any $\mathsf{s}, \mathsf{t} \in \mathsf{Sorts}_U$ such that $\mathsf{s} \neq \mathsf{t}$.

Definition 8.1.18. A U-frame $U = \langle \mathrm{Dom}_U, \mathrm{Sort}_U \rangle$ is called *finite* if Dom_U is a finite set.

Definition 8.1.19. Let $U = \langle \mathrm{Dom}_U, \mathrm{Sort}_U \rangle$ be a U-frame and $\vec{\mathsf{s}} \in \mathsf{Sorts}_U{}^*$. By definition let

$$\mathrm{Dom}_U(\vec{\mathsf{s}}) = \{\vec{x} \in \mathrm{Dom}_U{}^* \mid \mathrm{Sort}_U{}^*(\vec{x}) = \vec{\mathsf{s}}\}.$$

It is not difficult to show that

(i) $\mathrm{Dom}_U(\vec{\mathsf{s}}) \neq \emptyset$ for any $\vec{\mathsf{s}} \in \mathsf{Sorts}_U{}^*$,
(ii) $\mathrm{Dom}_U{}^* = \bigcup_{\vec{\mathsf{s}} \in \mathsf{Sorts}_U{}^*} \mathrm{Dom}_U{}^*(\vec{\mathsf{s}})$,
(iii) $\mathrm{Dom}_U(\vec{\mathsf{s}}) \cap \mathrm{Dom}_U(\vec{\mathsf{t}}) = \emptyset$, if $\vec{\mathsf{s}} \neq \vec{\mathsf{t}}$.

Definition 8.1.20. Let $U = \langle \mathrm{Dom}_U, \mathrm{Sort}_U \rangle$ be a universe frame, $\vec{\mathsf{a}} \in \mathsf{Sorts}_U{}^*$ and $\mathsf{s} \in \mathsf{Sorts}_U$. The mapping

$$f\colon \mathrm{Dom}_U(\vec{\mathsf{a}}) \to \mathrm{Dom}_U(\mathsf{s})$$

will be called an *operation* over the universe frame U.

The sequence $\vec{\mathsf{a}}$ of the sorts will be called the *arity of the operation* f in the U-frame U and will be denoted as $\mathrm{Ar}_U(f)$.

The sort s will *be called the sort of the operation* f in the U-frame U and it will be denoted as $\mathrm{Sort}_U(f)$.

Denote by Fu_U the set of all operations (functions) in U-frame U.

By definition let

$$\begin{aligned}\mathrm{Fu}_U(\vec{\mathsf{a}}, \mathsf{s}) &= \{f \in \mathrm{Fu}_U \mid \mathrm{Ar}_U(f) = \vec{\mathsf{a}}, \mathrm{Sort}_U(f) = \mathsf{s}\}, \\ \mathrm{Fu}_U(\mathsf{s}) &= \{f \in \mathrm{Fu}_U \mid \mathrm{Sort}_U(f) = \mathsf{s}\}.\end{aligned}$$

The terms to be introduced below will, to some extent, depend on the logic. They will also be general for the logics discussed in the book but they will contain a reference as a parameter to the discussed logic. One of the main terms to be introduced is the term L-structure, where L is one of the logics considered here. Namely, L-structures will play the role of models of theories constructed over the logic L.

Definition 8.1.21. Let L be a one- or two-sorted logic, e.g. **C**, or a PJ logic. Let $\mathsf{s} \in \mathsf{TS}$. By definition let:

$$\mathrm{Dom}_L(\mathsf{s}) = \begin{cases} \text{undefined, if } L \text{ is a one-sorted logic and } \mathsf{s} = \mathsf{I}, \\ \mathscr{V}(L,\mathsf{s}), \quad \text{otherwise.} \end{cases}$$

Definition 8.1.22. Let $U = \langle \mathrm{Dom}_U, \mathrm{Sort}_U \rangle$ be a universe frame. L is either a one-sorted or two-sorted logic. Let $\vec{\mathsf{a}} \in \mathsf{Sorts}_U{}^*$, $\mathsf{s} \in \mathsf{TS}$. Moreover let us suppose that the conditions "L is one-sorted" and "$\mathsf{s} = \mathsf{I}$" do not hold simultaneously. Then the mapping

$$p\colon \mathrm{Dom}_U(\vec{\mathsf{a}}) \to \mathscr{V}(L,\mathsf{s})$$

is called an *L-predicate in U*, where

(i) $\vec{\mathsf{a}}$ is called the *arity of p in U* and denoted by $\mathrm{Ar}_U(p)$;
(ii) s is called the *sort of p in U* and denoted by $\mathrm{Sort}_U(p)$.

An L-predicate p in U is called *internal* if $\mathrm{Sort}_U(p) = \mathsf{I}$.
An L-predicate p in U is called *external* if $\mathrm{Sort}_U(p) = \mathsf{E}$.
$\mathrm{Pr}_U(L)$ denotes the *set of all L-predicates in U*. By definition let

$$\mathrm{Pr}_U(L)(\vec{\mathsf{a}},\mathsf{s}) = \{p \in \mathrm{Pr}_U(L) \mid \mathrm{Ar}_U(p) = \vec{\mathsf{a}}, \mathrm{Sort}_U(p) = \mathsf{s}\},$$
$$\mathrm{Pr}_U(L)(\mathsf{s}) = \{p \in \mathrm{Pr}_U(L) \mid \mathrm{Sort}_U(p) = \mathsf{s}\}.$$

Definition 8.1.23. Let σ be an arbitrary signature, U be a universe frame and L be a one- or two-sorted logic. An ordered pair $\mathfrak{L} = \langle U, I \rangle$ is called the *L-structure of signature σ over U*, if the following conditions hold:

(a) The mapping
$$I\colon \mathrm{Fu}_\sigma \cup \mathrm{Pr}_\sigma \to \mathrm{Fu}_U \cup \mathrm{Pr}_U(L)$$
is such that $I(\mathrm{Fu}_\sigma) \subseteq \mathrm{Fu}_U$, $I(\mathrm{Pr}_\sigma) \subseteq \mathrm{Pr}_U(L)$,
(b) $\mathrm{Ar}_U(I(f)) = \mathrm{Ar}_\sigma(f)$ and $\mathrm{Sort}_U(I(f)) = \mathrm{Sort}_\sigma(f)$ for any $f \in \mathrm{Fu}_\sigma$,
(c) $\mathrm{Ar}_U(I(p)) = \mathrm{Ar}_\sigma(p)$ and $\mathrm{Sort}_U(I(p)) = \mathrm{Sort}_\sigma(p)$ for any $p \in \mathrm{Pr}_\sigma$.

For the L-structure defined above:

(i) The U-frame $U = \langle \mathrm{Dom}_U, \mathrm{Sort}_U \rangle$ is called the *frame of* $\mathfrak{L}$ and is denoted by $\mathrm{Frame}_{\mathfrak{L}}$;
(ii) The set Dom_U is called the *domain of* $\mathfrak{L}$ and is denoted by $\mathrm{Dom}_{\mathfrak{L}}$;
(iii) The mapping Sort_U is denoted by $\mathrm{Sort}_{\mathfrak{L}}$;
(iv) For any $\mathsf{s} \in \mathsf{Sorts}_U$ the set $\mathrm{Dom}_U(\mathsf{s})$ is called the *domain of sort* s *in* $\mathfrak{L}$ and is denoted by $\mathrm{Dom}_{\mathfrak{L}}(\mathsf{s})$ (see Definition 8.1.17);
(v) The mapping I is called the *interpretation function of* $\mathfrak{L}$ and is denoted by $\mathrm{I}_{\mathfrak{L}}$;

(vi) For the sake of conciseness we write $\varphi^{\mathfrak{L}}$ instead of $\mathrm{I}_{\mathfrak{L}}(\varphi)$ for any $\varphi \in \mathrm{Fu}_\sigma \cup \mathrm{Pr}_\sigma$.

By definition let:

(vii) $\mathrm{Fu}_{\mathfrak{L}} = \{f^{\mathfrak{L}} \mid f \in \mathrm{Fu}_\sigma\}$,
(viii) $\mathrm{Pr}_{\mathfrak{L}}(L) = \{p^{\mathfrak{L}} \mid p \in \mathrm{Pr}_\sigma\}$,
(ix) $\mathrm{Fu}_{\mathfrak{L}}(\vec{\mathsf{a}}, \mathsf{s}) = \{f^{\mathfrak{L}} \mid f \in \mathrm{Fu}_\sigma(\vec{\mathsf{a}}, \mathsf{s})\}$,
(x) $\mathrm{Fu}_{\mathfrak{L}}(\mathsf{s}) = \{f^{\mathfrak{L}} \mid f \in \mathrm{Fu}_\sigma(\mathsf{s})\}$,
(xi) $\mathrm{Pr}_{\mathfrak{L}}(L)(\vec{\mathsf{a}}, \mathsf{s}) = \{p^{\mathfrak{L}} \mid p \in \mathrm{Pr}_\sigma(\vec{\mathsf{a}}, \mathsf{s})\}$,
(xii) $\mathrm{Pr}_{\mathfrak{L}}(L)(\mathsf{s}) = \{p^{\mathfrak{L}} \mid p \in \mathrm{Pr}_\sigma(\mathsf{s})\}$.

Remark 8.1.24. Let σ be a classical signature (see 8.1.13). Then, obviously, any L-structure of signature σ, where L is a PJ logic, over an arbitrary U-frame will actually be **C**-structure.

Notation 8.1.25. Let σ be a signature and U be a U-frame such that $\mathsf{Sorts}_U = \mathsf{Sorts}_\sigma$. Denote by

$$\mathbf{Struct}_\sigma(L, U)$$

the set of all L-structures of signature σ over U. By definition let

$$\mathrm{Var}_U = \{x \in \mathrm{Var} \mid \mathrm{Sort}(x) \in \mathsf{Sorts}_U\}.$$

Now we introduce a few definitions related to valuations in L-structures.

Definition 8.1.26. Let U be a U-frame and $\mathfrak{L}$ be an L-structure.

(i) A *valuation of individual variables in U* is a mapping

$$v\colon \mathrm{Var}_U \to \mathrm{Dom}_U$$

such that $v(\mathrm{Var}_U(\mathsf{s})) \subseteq \mathrm{Dom}_U(\mathsf{s})$ for any $\mathsf{s} \in \mathsf{Sorts}_U$.

(ii) *A valuation of individual variables in* $\mathfrak{L}$ is a valuation of individual variables in the U-frame $\mathrm{Frame}_{\mathfrak{L}}$.

(iii) Let v be a valuation of individual variables in the U-frame U, $x \in \mathrm{Var}_U$, $a \in \mathrm{Dom}_U$, $\mathrm{Sort}(x) = \mathrm{Sort}_U(a)$. Let $v[x \leftarrow a]$ denote the valuation of individual variables in the U-frame U defined in the following way:

$$v[x \leftarrow a](y) = \begin{cases} a, & \text{if } y = x, \\ v(y), & \text{if } y \neq x, \end{cases}$$

where $y \in \mathrm{Var}_U$.

Definition 8.1.27. Let φ be a functional or predicate symbol of signature σ. By definition let

$$\mathrm{Dom}_{\sigma}(\varphi) = \left\{\vec{t} \in \mathrm{Term}_{\sigma}{}^{*} \mid \mathrm{Sort}_{\sigma}{}^{*}(\vec{t}) = \mathrm{Ar}_{\sigma}(\varphi)\right\}.$$

$\mathrm{Dom}_{\sigma}(\varphi)$ is called the *domain of signature σ for the symbol φ.*

Definition 8.1.28. Let $\mathfrak{L}$ be an L-structure of signature σ and v be a valuation in $\mathfrak{L}$. Suppose t is a term of signature σ. The *value of the term t under the valuation v in $\mathfrak{L}$* (written: $t^{\mathfrak{L},v}$) is recursively defined as follows:

(i) Let x be an individual variable. Then $x^{\mathfrak{L},v} = v(x)$.
(ii) Let $f \in \mathrm{Fu}_{\sigma}$, $\vec{t} \in \mathrm{Dom}_{\sigma}(f)$. Then $\left(f(\vec{t})\right)^{\mathfrak{L},v} = f^{\mathfrak{L}}\left(\vec{t}^{\mathfrak{L},v^{*}}\right)$.

Definition 8.1.29. Let $\mathfrak{L}$ be an L-structure of signature σ and v be a valuation in $\mathfrak{L}$. Suppose φ is an atomic formula of signature σ. The *value of φ under the valuation v in $\mathfrak{L}$* (written: $\varphi^{\mathfrak{L},v}$) is recursively defined as follows:

(i) Let t and s be terms of signature σ such that $\mathrm{Sort}_{\sigma}(t) = \mathrm{Sort}_{\sigma}(s)$. Then

$$(t = s)^{\mathfrak{L},v} = \begin{cases} \mathbf{t} \text{ if } t^{\mathfrak{L},v} = s^{\mathfrak{L},v}, \\ \mathbf{f} \text{ otherwise.} \end{cases}$$

(ii) Let $p \in \mathrm{Pr}_{\sigma}$, $\vec{t} \in \mathrm{Dom}_{\sigma}(p)$. Then $\left(p(\vec{t})\right)^{\mathfrak{L},v} = p^{\mathfrak{L}}\left(\vec{t}^{\mathfrak{L},v^{*}}\right)$.

Obviously, the value of each term (atomic formula) under a valuation in an L-structure of the corresponding signature has the same sort as that of the corresponding term (atomic formula).

This section is completed with a number of definitions and statements connected with the term *isomorphism.* We consider isomorphism of L-structures, but it can be seen that this term coincides with the usual term of isomorphism of many-sorted structures.

Definition 8.1.30. Let $\mathfrak{J}$ and $\mathfrak{K}$ be L-structures of signature σ. An *isomorphism* of *L-structure $\mathfrak{J}$ onto the L-structure* $\mathfrak{K}$ is a mapping

$$\varphi\colon \mathrm{Dom}_{\mathfrak{J}} \to \mathrm{Dom}_{\mathfrak{K}}$$

such that:

(i) φ is a bijection;
(ii) $\varphi(\mathrm{Dom}_{\mathfrak{J}}(\mathsf{s})) = \mathrm{Dom}_{\mathfrak{K}}(\mathsf{s})$ for any $\mathsf{s} \in \mathsf{Sorts}_{\sigma}$ (φ preserves the sorts of the objects);
(iii) If $f \in \mathrm{Fu}_{\mathfrak{J}}$, $\vec{a} \in \mathrm{Dom}\left(f^{\mathfrak{J}}\right)$ then $\varphi\left(f^{\mathfrak{J}}(\vec{a})\right) = f^{\mathfrak{K}}(\varphi^{*}(\vec{a}))$;
(iv) If $q \in \mathrm{Pr}_{\mathfrak{J}}(L)$, $\vec{a} \in \mathrm{Dom}\left(q^{\mathfrak{J}}\right)$ then $q^{\mathfrak{J}}(\vec{a}) = q^{\mathfrak{K}}(\varphi^{*}(\vec{a}))$.

Proposition 8.1.31. Let $\mathfrak{J}$ be an L-structure of the signature σ. Let ι be an identity mapping of the set $\mathrm{Dom}_{\mathfrak{J}}$ onto itself, i.e. $\iota(a) = a$ for any $a \in \mathrm{Dom}_{\mathfrak{J}}$. Then ι is an *isomorphism* of $\mathfrak{J}$ onto $\mathfrak{J}$.

Proof. It is obvious. □

Proposition 8.1.32. Let $\mathfrak{J}$ and $\mathfrak{K}$ be L-structures of signature σ. Let φ be an isomorphism $\mathfrak{J}$ onto $\mathfrak{K}$. Let $\psi = \varphi^{-1}$ (the mapping ψ is inverse to φ, i.e.

$$\psi\colon \mathrm{Dom}_{\mathfrak{K}} \to \mathrm{Dom}_{\mathfrak{J}},$$

where $\psi(\varphi(a)) = a$ for any $a \in \mathrm{Dom}_{\mathfrak{J}}$).

Then ψ is an isomorphism $\mathfrak{K}$ onto $\mathfrak{J}$.

Proof. Similarly to the proof of the corresponding statement for the many-sorted classical structures. □

Proposition 8.1.33. Let $\mathfrak{J}$, $\mathfrak{K}$ and $\mathfrak{L}$ be L-structures of the signature σ. Let φ and ψ be isomorphisms of $\mathfrak{J}$ onto $\mathfrak{K}$ and of $\mathfrak{K}$ onto $\mathfrak{L}$, respectively. Let $\theta = \varphi \circ \psi$ (θ is the composition of the mappings φ and ψ, i.e.

$$\theta\colon \mathrm{Dom}_{\mathfrak{J}} \to \mathrm{Dom}_{\mathfrak{L}}$$

is such that $\theta(a) = \psi(\varphi(a))$ for any $a \in \mathrm{Dom}_{\mathfrak{J}}$).

Then θ is an isomorphism of $\mathfrak{J}$ onto $\mathfrak{L}$.

Proof. Similarly to the proof of the corresponding assertions for the classical many-sorted structures. □

Definition 8.1.34. Let $\mathfrak{J}$ and $\mathfrak{K}$ be L-structures of the signature σ. L-structures $\mathfrak{J}$ and $\mathfrak{K}$ are called *isomorphic* if there exists an isomorphism of $\mathfrak{J}$ onto $\mathfrak{K}$. Let $\mathfrak{J} \cong \mathfrak{K}$ denote the fact that the L-structures $\mathfrak{J}$ and $\mathfrak{K}$ are isomorphic.

Proposition 8.1.35. Let $\mathfrak{J}$, $\mathfrak{K}$ and $\mathfrak{L}$ be L-structures of the signature σ. Then:

(i) $\mathfrak{J} \cong \mathfrak{J}$,
(ii) $\mathfrak{J} \cong \mathfrak{K}$ implies $\mathfrak{K} \cong \mathfrak{J}$,
(iii) $\mathfrak{J} \cong \mathfrak{K}$ and $\mathfrak{K} \cong \mathfrak{L}$ implies $\mathfrak{J} \cong \mathfrak{L}$.

Proof. Similarly to the proof of the corresponding propositions for the many-sorted structures. Use Propositions 8.1.31, 8.1.32 and 8.1.33. □

Lemma 8.1.36. Let $\mathfrak{J}$ and $\mathfrak{K}$ be two L-structures of signature σ and φ be an isomorphism of $\mathfrak{J}$ onto $\mathfrak{K}$. Let v be a valuation in $\mathfrak{J}$. Let us define a valuation u of the individual variables in the L-structure $\mathfrak{K}$ as follows: for any $x \in \mathrm{Var}_{\mathfrak{K}}$, let

$$u(x) = \varphi(v(x)).$$

The well-definedness of the valuation u follows from the fact that an isomorphism, as well as a valuation, preserves the sort of the entities to be mapped.

Then for arbitrary term t of the signature σ,

$$\varphi\left(t^{\mathfrak{J},v}\right) = t^{\mathfrak{K},u}.$$

Proof. By a standard induction over the terms.

BASIS. For any individual variable x,

$$\begin{aligned}\varphi\left(x^{\mathfrak{J},v}\right) &= \varphi\left(v\left(x\right)\right) && \text{(by Definition 8.1.28)}\\ &= u\left(x\right) && \text{(by the condition)}\\ &= x^{\mathfrak{K},u} && \text{(by Definition 8.1.28).}\end{aligned}$$

INDUCTION STEP. Let us suppose that the statement holds for any term of $\vec{t} \in \mathrm{Dom}_{\sigma}(f)$. Then

$$\varphi^{*}\left(\vec{t}^{\mathfrak{J},v^{*}}\right) = \vec{t}^{\mathfrak{K},u^{*}},$$

consequently,

$$\begin{aligned}\varphi\left(\left(f\left(\vec{t}\right)\right)^{\mathfrak{J},v}\right) &= \varphi\left(f^{\mathfrak{J}}\left(\vec{t}^{\mathfrak{J},v^{*}}\right)\right) && \text{(by Definition 8.1.28)}\\ &= f^{\mathfrak{K}}\left(\varphi^{*}\left(\vec{t}^{\mathfrak{J},v^{*}}\right)\right) && \text{(by Definition 8.1.30)}\\ &= f^{\mathfrak{K}}\left(\vec{t}^{\mathfrak{K},u^{*}}\right). && \text{(by the induction hypothesis)}\end{aligned}$$

□

Lemma 8.1.37. Let $\mathfrak{J}$ and $\mathfrak{K}$ be two L-structures of signature σ and φ be an isomorphism of $\mathfrak{J}$ onto $\mathfrak{K}$. Let v be a valuation in $\mathfrak{J}$. Let us define a valuation u in $\mathfrak{K}$ similarly to the previous lemma; assume

$$u\left(x\right) = \varphi\left(v\left(x\right)\right)$$

for any $x \in \mathrm{Var}_{\mathfrak{K}}$.

Then for any atomic formula θ of signature σ it holds that

$$\theta^{\mathfrak{J},v} = \theta^{\mathfrak{K},u}.$$

Proof. Consider the possible cases.

CASE 1. Let θ be the formula $t_1 = t_2$, where t_1 and t_2 are terms of signature σ that have the same sort.

$$\begin{aligned}(t_1 = t_2)^{\mathfrak{J},v} = \mathbf{t} \quad &\text{implies} \quad t_1^{\mathfrak{J},v} = t_2^{\mathfrak{J},v} && \text{(by 8.1.29 (i))}\\ &\text{implies} \quad \varphi\left(t_1^{\mathfrak{J},v}\right) = \varphi\left(t_2^{\mathfrak{J},v}\right) &&\\ &\text{implies} \quad t_1^{\mathfrak{K},u} = t_2^{\mathfrak{K},u} && \text{(by 8.1.36)}\\ &\text{implies} \quad (t_1 = t_2)^{\mathfrak{K},u} = \mathbf{t}. && \text{(by 8.1.29 (i))}\end{aligned}$$

On the other hand,

$$\begin{aligned}
(t_1 = t_2)^{\mathfrak{J},v} = \mathbf{f} \quad &\text{implies} \quad t_1^{\mathfrak{J},v} \neq t_2^{\mathfrak{J},v} && \text{(by 8.1.29 (i))}\\
&\text{implies} \quad \varphi\left(t_1^{\mathfrak{J},v}\right) \neq \varphi\left(t_2^{\mathfrak{J},v}\right) && \text{(by 8.1.30 (i))}\\
&\text{implies} \quad t_1^{\mathfrak{K},u} \neq t_2^{\mathfrak{K},u} && \text{(by 8.1.36)}\\
&\text{implies} \quad (t_1 = t_2)^{\mathfrak{K},u} = \mathbf{f}. && \text{(by 8.1.29 (i))}
\end{aligned}$$

CASE 2. Let θ be the formula $p(\vec{t})$, where $p \in \mathrm{Pr}_\sigma(L)$, $\vec{t} \in \mathrm{Dom}_\sigma(p)$. Then by Definitions 8.1.29 and 8.1.30 (iv) and Lemma 8.1.36, we obtain the consequence of this lemma. □

8.2 Classical First-Order Logic (Syntax and Semantics)

In this section we consider syntax and semantics of classical first-order many-sorted logics, which we denote as we did in the propositional case by **C**. Similarly we will deal with other logics also considered in the book. This will not result in any confusion since in the forthcoming sections where the formalism related to cognitive reasoning is introduced we will refer only to the results related to first-order logics.

Below we define the formulas of the **C** logics and the values of the formulas by valuation in the **C**-structures. We also define the truth, general validity and the term of (semantic) *consequence* in the **C**-structures. We conclude the section with some well-known statements on the meaning of terms and formulas and truth in the isomorphic **C**-structures.

Note that analogues of the statements of this section will hold also for PJ logics with similar proof.

Definition 8.2.1. Let σ be a classical signature (see Definition 8.1.13). Then the *alphabet of signature σ for the first-order many-sorted classical logic* contains the following groups of symbols:

(i) The set $\mathrm{Fu}_\sigma \cup \mathrm{Pr}_\sigma$ (see Definition 8.1.12)
(ii) The equality character $=$
(iii) The set of individual variables Var (see Definition 8.1.14)
(iv) Classical propositional connectives $\neg$ and $\rightarrow$
(v) Classical universal quantifier $\forall$
(vi) Comma ',' and parentheses '(' and ')'

Definition 8.2.2. Let σ be a classical signature (see Definition 8.1.13). Then (*first-order*) **C**-*formulas of signature* σ are recursively defined in the following way:

(i) Each atomic formula of signature σ is a **C**-formula of signature σ;
(ii) If φ and ψ are **C**-formulas of signature σ and x is an individual variable then the expressions $(\varphi \rightarrow \psi)$, $(\neg\varphi)$, and $(\forall x\, \varphi)$ are also **C**-formulas of signature σ.

The set of all **C**-*formulas of signature* σ is denoted by $\mathscr{F}_\sigma(\mathbf{C})$.

Agreement 8.2.3. Propositional connectives $\wedge$, $\vee$, $\leftrightarrow$, and the quantifier $\exists$ are expressed in the usual way through $\neg$, $\rightarrow$ and $\forall$; for propositional connectives see Definitions 7.2.6 and 7.2.2. As for the existential quantifier $\exists$, we agree that the expression $(\exists x\ \varphi)$ is an abbreviation for $(\neg(\forall x\ (\neg\varphi)))$. By analogy with 7.2.3 we accept the following *agreement on the priority of propositional connectives and quantifiers*:

- The unary connective $\neg$ and expressions $\forall x$ and $\exists x$, where $x \in \mathrm{Var}$, have the highest priority (equal for all of them);
- The operations $\wedge$ and $\vee$ have (equal) lower priority;
- The operations $\rightarrow$ and $\leftrightarrow$ both have the lowest priority.

By analogy with 7.2.7 we make the following *agreement on omission of parentheses*:

- The outer parentheses of a formula are omitted,
- Parentheses are omitted if this is allowed by the agreement on the precedence of propositional connectives and quantifiers stated above.

Definition 8.2.4. Let θ be a **C**-formula of signature σ and $\mathfrak{C}$ be a **C**-structure of the same signature. Denote by $\theta^{\mathfrak{C},v}$ the value of θ under v in $\mathfrak{C}$. *The values of* **C**-*formulas under the valuation v in* $\mathfrak{C}$ are recursively defined in the following way:

(i) The values of atomic formulas are defined above;

(ii) $(\neg\varphi)^{\mathfrak{C},v} = \neg\left(\varphi^{\mathfrak{C},v}\right)$,
$(\varphi \rightarrow \psi)^{\mathfrak{C},v} = \left(\varphi^{\mathfrak{C},v}\right) \rightarrow \left(\psi^{\mathfrak{C},v}\right)$;

(iii) $$(\forall x\ \varphi)^{\mathfrak{C},v} = \begin{cases} \mathbf{t}, \text{ if } \varphi^{\mathfrak{C},v[x\leftarrow c]} = \mathbf{t} \text{ for all } c \in \mathrm{Dom}_{\mathfrak{C}}(\mathsf{s}), \\ \mathbf{f}, \text{ otherwise,} \end{cases}$$

where φ and ψ are **C**-formulas of signature σ, $x \in \mathrm{Var}(\mathsf{s})$.

Definition 8.2.5. Let φ be a **C**-formula, Γ be a set of **C**-formulas of signature σ and $\mathfrak{C}$ be a **C**-structure of the same signature.

(i) We say that a **C**-formula φ is *valid in* $\mathfrak{C}$ (written: $\mathfrak{C} \models_{\mathbf{C}} \varphi$) if $\varphi^{\mathfrak{C},v} = \mathbf{t}$ for any valuation v in this **C**-structure. In this case, $\mathfrak{C}$ is called a **C**-*model of* φ.

(ii) We say that $\mathfrak{C}$ is a **C**-*model of* Γ (written: $\mathfrak{C} \models_{\mathbf{C}} \Gamma$) if it is a **C**-model of each **C**-formula of Γ.

(iii) We say that φ *is a* **C**-*consequence of* Γ (written: $\Gamma \models_{\mathbf{C}} \varphi$) if each **C**-model of Γ is also a **C**-model of φ.

(iv) We say that φ *is* **C**-*valid* (written: $\models_{\mathbf{C}} \varphi$) if it is a **C**-consequence of the empty set of **C**-formulas, i.e. if it is valid in any **C**-structure of the corresponding signature.

The forthcoming two statements are well-known facts connecting values under valuation and truth in the isomorphic **C**-structures. These results can be easily transferred to the case of PJ logics, which will be done below.

Lemma 8.2.6. Let $\mathfrak{J}$ and $\mathfrak{K}$ be two **C**-structures of signature σ, φ be a isomorphism of $\mathfrak{J}$ onto $\mathfrak{K}$ and v be a valuation in $\mathfrak{J}$. Let u be a valuation in $\mathfrak{K}$ defined in the following way:

$$u(x) = \varphi(v(x)).$$

Note that the valuation u is well defined, because of the fact that isomorphism, as well as valuation, preserves the sorts.

Then for any term t of signature σ and for any **C**-formula Φ of the same signature the following holds:

(i) $\varphi\left(t^{\mathfrak{J},v}\right) = t^{\mathfrak{K},u}$,
(ii) $\Phi^{\mathfrak{J},v} = \Phi^{\mathfrak{K},u}$.

Proof. (i) follows from Lemma 8.1.36. (ii) can be proved by induction over the **C**-formulas.

BASIS. By Lemma 8.1.36, (ii) holds for the atomic formulas.

INDUCTION STEP. Let us suppose that (ii) holds for the formulas Φ and Ψ. Then

$$\begin{aligned}
(\Phi \to \Psi)^{\mathfrak{J},v} &= \Phi^{\mathfrak{J},v} \to \Psi^{\mathfrak{J},v} && \text{(by Definition 8.2.4 (ii))}\\
&= \Phi^{\mathfrak{K},u} \to \Psi^{\mathfrak{K},u} && \text{(by the induction hypothesis)}\\
&= (\Phi \to \Psi)^{\mathfrak{J},u} && \text{(by Definition 8.2.4 (ii))}
\end{aligned}$$

The fact that

$$(\neg\Phi)^{\mathfrak{J},v} = (\neg\Phi)^{\mathfrak{K},u}$$

can be established similarly.

Let us prove that

$$(\forall x\, \Phi)^{\mathfrak{J},v} = (\forall x \Phi)^{\mathfrak{K},u}.$$

Let $\mathrm{Sort}(x) = \mathsf{s}$. Suppose that $(\forall x\, \Phi)^{\mathfrak{J},v} = \mathbf{t}$. Then $\Phi^{\mathfrak{J},v[x\leftarrow a]} = \mathbf{t}$ for any $a \in \mathrm{Dom}_{\mathfrak{J}}(\mathsf{s})$. By the inductive assumption,

$$\Phi^{\mathfrak{J},v[x\leftarrow a]} = \Phi^{\mathfrak{K},u[x\leftarrow\varphi(a)]},$$

since

$$u[x \leftarrow \varphi(a)](y) = \varphi(v[x \leftarrow a](y)).$$

But φ is a bijection that preserves the sorts. Therefore for any $b \in \mathrm{Dom}_{\mathfrak{K}}(\mathsf{s})$ there exits an $a \in \mathrm{Dom}_{\mathfrak{J}}(\mathsf{s})$ such that $b = \varphi(a)$. Then for any b,

$$\Phi^{\mathfrak{K},u[x\leftarrow b]} = \mathbf{t}.$$

Consequently, $(\forall x\Phi)^{\mathfrak{K},u} = \mathbf{t}$.

Similarly, it can be shown that

$$(\forall x\, \Phi)^{\mathfrak{K},u} = \mathbf{t} \quad \text{implies} \quad (\forall x\, \Phi)^{\mathfrak{J},v} = \mathbf{t}.$$

□

Corollary 8.2.7. Let $\mathfrak{J}$ and $\mathfrak{K}$ be two **C**-structures of signature σ and Φ be a **C**-formula of the same signature. Let $\mathfrak{J} \cong \mathfrak{K}$. Then

$$\mathfrak{J} \models_{\mathbf{C}} \Phi \quad \text{iff} \quad \mathfrak{K} \models_{\mathbf{C}} \Phi.$$

Proof. ($\Leftarrow$) Let $\mathfrak{K} \models_{\mathbf{C}} \Phi$. Then $\Phi^{\mathfrak{K},u} = \mathbf{t}$ holds for any valuation u in $\mathfrak{K}$. Let v be an arbitrary valuation in the **C**-structure $\mathfrak{J}$. Let φ be an isomorphism of $\mathfrak{J}$ onto $\mathfrak{K}$. Then by Lemma 8.2.6

$$\Phi^{\mathfrak{J},v} = \Phi^{\mathfrak{K},u} = \mathbf{t}.$$

($\Rightarrow$) Similarly. □

8.3 Classical First-Order Logic (Calculus)

In this section we formulate axiom schemata and inference rules for the classical many-sorted first-order calculi, we define the terms of inference, proof, deducibility from hypothesis and provability in a calculus. We conclude with the well-known theorems on the classical first-order calculus: the deduction theorem, correctness theorem and completeness theorem.

We permit ourselves not to define some syntactic constructions. For example, we will not give any strict definition of the notions of *subterm* and *subformula*. We will be satisfied with explanations such as "a part of a term that is itself a term" and "a part of a formula that is itself a formula". We also will not give a strict definition of *an occurrence of a variable* (or a term) in a term or in a formula. (For formulas, it also makes sense to speak of occurrences of subformulas.)

In this section we consider formulas of classical first-order logic (**C**-formulas). However, everything said above holds also for formulas of any logic discussed in this work. Therefore, we used above the general term "formula", but did not mention particular kinds of formulas, e.g. **C**-formulas, **St**-formulas or **It**-formulas.

Note that a variable can occur several times in a term or in a formula. Therefore, we can refer to *different* occurrences (of the same) variable in a term or in a formula, to *all* or *some* occurrences of a variable in a term or in a formula, etc. Similarly, one can speak of occurrences of terms and formulas.

We assume that the notion of occurrence is fairly intuitive. We can, of course, give a strict definition of this notion in terms of string algebra with concatenation operation, but we think that in this case the exposition would have been too formal and cumbersome.

We will also informally consider the notion of *scope of quantifier*. The formula $(\forall x\ \varphi)$ is *the scope of the universal quantifier over the variable x*. The formula $(\forall x\ \varphi)$ can be considered as the scope of the quantifier even if it is a subformula of any other formula. The scope of the existential quantifier is defined similarly.

We underline that in this definition of the scope of a quantifier we do not take into account the agreement of omission of parentheses; so, we assume that a formula contains all parentheses that it should have by definition. Therefore, when parentheses are omitted in a formula, they are also omitted for the scope of a quantifier. For example, upon omitting parentheses, the scope of a quantifier will look like $\forall x\, \varphi$.

Now, using the informal notions of occurrence and scope of quantifier, we give a definition of free and bound variable occurrences.

Definition 8.3.1. (i) An occurrence of a variable x in a **C**-formula φ is called *bound* if it is within the scope of a quantifier over x. Otherwise, the occurrence of the variable is called *free*.

(ii) The set of individual variables occurring in the term t is denoted by $\mathrm{Var}(t)$.

(iii) The set of individual variables freely occurring in the formula φ is denoted by $\mathrm{FVar}(\varphi)$.

(iv) A first-order **C**-formula φ is called *closed* (*or a* **C**-*sentence*) if it does not contain free variable occurrences.

(v) We say that a term t is *admissible for substituting it for* x *in* φ if $\mathrm{Sort}_\sigma(t) = \mathrm{Sort}(x)$ and no occurrence of x in φ lies within the scope of quantifiers over variables occurring in the term t.

(vi) Let x be an individual variable, t be a term and φ be a first-order **C**-formula. By φ_t^x we denote the result of substituting t for x in φ in all free occurrences of x.

If we do not mention a signature when referring to terms and/or first-order **C**-formulas in a definition or in a statement, it means that the notion is defined (or statement holds) for terms and/or formulas *of any signature*.

Now we introduce *the axiom schemata of the first-order many-sorted classical calculus*. Here and below this calculus is denoted by $\mathbf{C}^1$.

Axiom Schemata 8.3.2 (Axiom schemata of $\mathbf{C}^1$). Let φ, ψ and θ be formulas of classical many-sorted first-order logic; x, y and z be individual variables; and t be a term. Then the following formulas are *axioms of* $\mathbf{C}^1$:

(A1) $\varphi \to (\psi \to \varphi)$,
(A2) $(\varphi \to (\psi \to \theta)) \to ((\varphi \to \psi) \to (\varphi \to \theta))$,
(A3) $(\neg\varphi \to \psi) \to ((\neg\varphi \to \neg\psi) \to \varphi)$.

(Q1) $\forall x\, \varphi \to \varphi_t^x$ (t is admissible for substitution for x in φ),
(Q2) $\forall x\, (\varphi \to \psi) \to (\varphi \to \forall x\, \psi)$ (φ does not contain free occurrences of x);

(E1) $x = x$,
(E2) $y = z \to (\varphi_y^x \to \varphi_z^x)$ (y and z are admissible for substitution for x in φ and $\mathrm{Sort}x = \mathrm{Sort}y = \mathrm{Sort}z$).

Inference Rules 8.3.3 (Inference rules of $\mathbf{C}^1$). The following two inference rules are used in classical many-sorted calculus:

$$\frac{\varphi, \quad \varphi \to \psi}{\psi} \qquad (\textit{modus ponens})$$

$$\frac{\varphi}{\forall x\, \varphi} \qquad (\textit{generalization rule for variable } x)$$

where φ and ψ are formulas of classical many-sorted first-order logic and x is an individual variable.

Definition 8.3.4. Let Γ be a set of **C**-formulas. An *inference in* $\mathbf{C}^1$ ($\mathbf{C}^1$*-inference*) *from* Γ is a finite sequence of **C**-formulas such that every its element:

- Either is an axiom of $\mathbf{C}^1$,
- Or belongs to Γ,
- Or follows from preceding formulas by one of the inference rules of $\mathbf{C}^1$.

An inference from the empty set of hypotheses in the $\mathbf{C}^1$ is called *a proof in* $\mathbf{C}^1$ ($\mathbf{C}^1$*-proof*).

Definition 8.3.5. Let Γ be a set of **C**-formulas and φ be a **C**-formula. We say that φ is *derivable in* $\mathbf{C}^1$ *from* Γ (written: $\Gamma \vdash_{\mathbf{C}} \varphi$) if there exists an inference in $\mathbf{C}^1$ from Γ containing φ. An external **C**-formula φ is called *provable in* $\mathbf{C}^1$ or a *theorem of* $\mathbf{C}^1$ (written: $\vdash_{\mathbf{C}} \Phi$) if φ is derivable in $\mathbf{C}^1$ from the empty set of **C**-formulas.

Now we formulate three well-known theorems on the classical first-order calculi.

Theorem 8.3.6 (Deduction Theorem for $\mathbf{C}^1$). Let Γ be a set of **C**-formulas and φ and ψ be **C**-formulas. Suppose that $\Gamma \cup \{\varphi\} \vdash_{\mathbf{C}} \psi$ and there exists a $\mathbf{C}^1$-inference of ψ from $\Gamma \cup \{\varphi\}$ that contains no application of the generalisation rule for variables freely occurring in φ. Then $\Gamma \vdash_{\mathbf{C}} \varphi \to \psi$.

Proof. The proof is similar to that in [Mendelson, 1997]. □

Theorem 8.3.7 (Soundness Theorem for $\mathbf{C}^1$). Let Γ be a set of **C**-formulas and φ be a **C**-formula. Then $\Gamma \vdash_{\mathbf{C}} \varphi$ implies $\Gamma \models_{\mathbf{C}} \varphi$.

Proof. The proof is similar to that in [Mendelson, 1997]. □

Theorem 8.3.8 (Completeness Theorem for $\mathbf{C}^1$). Let Γ be a set of **C**-formulas and φ be a **C**-formula. Then $\Gamma \models_{\mathbf{C}} \varphi$ implies $\Gamma \vdash_{\mathbf{C}} \varphi$.

Proof. The proof is similar to that in [Mendelson, 1997]. □

8.4 First-Order PJ Logics (Syntax and Semantics)

In this section we consider the syntax and semantics of the first-order many-sorted FPJ logics and their iterative versions. The discussion will follow the structure of the presentation of the syntax and semantics of the classical first-order logic (see Sect. 8.2). The proofs of the statements below will be similar to those of the corresponding statements of the classical first-order logic.

Definition 8.4.1. Let σ be an arbitrary signature (see Definition 8.1.12). Then, the *alphabet of signature σ for the first-order many-sorted PJ logic* contains the same groups of symbols as the alphabet of classical first-order many-sorted logic, i.e. groups of symbols (i)–(vi) from Definition 8.2.1, as well as the set of symbols of J-operators of PJ logic L that form the group:

(vii^L) $\{\mathrm{J}_\alpha \mid \alpha \in \mathrm{JI}(L)\}$,

e.g. for any FPJ logic L we have

(vii^L) $\{\mathrm{J}_\varepsilon \mid \varepsilon \in \mathscr{V}(L, \mathsf{I})\}$

and for the logic $I = \mathrm{I}^\tau(L)$, where L is an FPJ logic, $\tau \in \mathscr{D}^\tau(L)$, we have

(vii^I) $\left\{\mathrm{J}_{(\varepsilon,n)} \mid \varepsilon \in \mathscr{V}(L, \mathsf{I}), n \in \omega\right\}$.

In 8.1.16, we gave the definition of an atomic formula of signature σ and the extension of the function Sort_σ on the set of atomic formulas of signature σ. This definition was not related to a particular logic, but was suited for classical first-order logic. Obviously, all atomic formulas of the first-order classical logic have sort E, i.e. they are external atomic formulas. This follows from the fact that atomic formulas are formulas of the classical signature all predicate symbols of which have sort E, i.e. these predicate symbols should be external.

If the signature σ is not classical, then atomic formulas may be both external and internal. In the PJ logic L there is no constraint on the signature, therefore, *atomic formulas of signature σ of logic L* are exactly atomic formulas of signature σ in the sense of Definition 8.1.16. These formulas may be either external or internal.

Definition 8.4.2. Let L be a first-order many-sorted PJ logic. Expressions of the form $(\mathrm{J}_\alpha\, \varphi)$, where φ is an internal atomic L-formula of signature σ, $\mathrm{J}_\alpha \in \mathscr{J}(L)$, are called *J-atomic L-formulas of signature σ*. We state by definition that, for any J-atomic L-formula Φ of signature σ, $\mathrm{Sort}_\sigma(\Phi) = \mathsf{E}$.

Definition 8.4.3. Let L be a first-order many-sorted PJ logic. We recursively define *external L-formulas of signature σ* in the following way:

(i) Every external atomic formula is an external L-formula;
(ii) Every J-atomic formula is an external L-formula;
(iii) If Φ and Ψ are external L-formulas then expressions $(\Phi \to \Psi)$ and $(\neg\Phi)$ are the also external L-formulas;
(iv) If Φ is an external L-formula and x is an individual variable then the expression $(\forall x\, \Phi)$ is an external L-formula.

Note that in this definition we omitted the term "signature σ", since it is clear from the context.

We state by definition that

$$\mathrm{Sort}_\sigma(\Phi) = \mathsf{E}$$

for any external L-formula Φ.

L-formulas of signature σ are either internal atomic L-formulas or external L-formulas of this signature.

Denote by $\mathscr{F}_\sigma(L)$ the set of all L-formulas of signature σ. By definition let

$$\mathscr{F}_\sigma(L, \mathsf{s}) = \{\Theta \in \mathscr{F}_\sigma(L) \mid \mathrm{Sort}_\sigma(\Theta) = \mathsf{s}\}.$$

Agreement 8.4.4. As in classical logic, we assume that propositional connectives $\wedge$, $\vee$, $\leftrightarrow$, and quantifier $\exists$ are expressed in the usual way by $\neg$, $\rightarrow$ and $\forall$ (see Agreement 8.2.3) . We add the agreement on the priority of J-operators to Agreement 8.2.3. We consider that J-operators have the same priority as the connective $\neg$ (negation).

Further we will assume that by L some PJ logic is denoted.

Definition 8.4.5. Let Ξ be an external L-formula of signature σ, $\mathfrak{S}$ be an L-structure of the same signature and v be an valuation in $\mathfrak{S}$. Denote by $\Xi^{\mathfrak{S},v}$ the value of Ξ under v in $\mathfrak{S}$. *The value of external formulas of signature σ under v in $\mathfrak{S}$* is recursively defined as follows:

(i) The values of external atomic formulas are defined above (see 8.1.26);
(ii) $(\mathrm{J}_\alpha\,\varphi)^{\mathfrak{S},v} = \mathrm{J}_\alpha\left(\varphi^{\mathfrak{S},v}\right)$ (φ is an internal atomic L-formula, $\mathrm{J}_\alpha \in \mathscr{J}(L)$);
(iii) $(\neg\Phi)^{\mathfrak{S},v} = \neg\left(\Phi^{\mathfrak{S},v}\right)$,
$(\Phi \rightarrow \Psi)^{\mathfrak{S},v} = \left(\Phi^{\mathfrak{S},v}\right) \rightarrow \left(\Psi^{\mathfrak{S},v}\right)$ (Φ and Ψ are external L-formulas);
(iv) $(\forall x\,\Phi)^{\mathfrak{S},v} = \begin{cases} \mathbf{t}, \text{ if } \Phi^{\mathfrak{S},v[x\leftarrow a]} = \mathbf{t} \text{ for any } a \in \mathrm{Dom}_{\mathfrak{S}}(\mathsf{s}), \\ \mathbf{f}, \text{ otherwise}, \end{cases}$

where Φ is an external L-formula and $x \in \mathrm{Var}(\mathsf{s})$.

Definition 8.4.6. Let Φ be an external L-formula, Γ be a set of external L-formulas and $\mathfrak{S}$ be an L-structure of the same signature σ.

(i) An external L-formula Φ is called *valid in an L-structure* $\mathfrak{S}$ if $\Phi^{\mathfrak{S},v} = \mathbf{t}$ for every valuation in this L-structure. In this case, we say that $\mathfrak{S}$ *is an L-model of* Φ (written: $\mathfrak{S} \models_L \Phi$).
(ii) We say that $\mathfrak{S}$ *is an L-model of Γ* (written: $\mathfrak{S} \models_L \Gamma$) if it is an model of each formula of Γ.
(iii) We say that Φ is an *L-consequence* of Γ (written: $\Gamma \models_L \Phi$) if every model of Γ is also a model of Φ.
(iv) An external L-formula Φ is called *L-valid* (written: $\models_L \Phi$) if it is an L-consequence of the empty set of L-formulas, i.e. if it is valid in any L-structure of the corresponding signature.

We end this section by describing a particular three-sorted **St**-structure, which will be used later to illustrate our deductive technique.

Example 8.4.7. Now we consider a structure that formalises the semantic aspects of the example from Section 5.2, where we spoke about Mick the robot, which is on an alien planet and tries to accommodate itself to the local dynamic environment.

Let the sorts of individuals be as follows: P is the sort of phenomena and D is the sort of dates.

First let us define a U-frame W.

For the elements of $\mathrm{Dom}_W(\mathsf{P})$ we introduce in Table 8.1 the abbreviations, which are the same as were introduced in Section 5.2. In this table we also introduce the abbreviations for the elements of $\mathrm{Dom}_W(\mathsf{D})$.

We consider this example for the logic **Si** (see Definition 7.4.7). Now we define a signature for our **Si**-structure. Let signature σ be such that:

Table 8.1 Abbreviations

P	
Phenomenon	**Abbreviation**
Purple clouds	PC
Storm	S
Large moon	LM
Rising tide	RT

D	
Date	**Abbreviation**
08.01	D1
08.02	D2
08.04	D4
08.05	D5
08.07	D7

Table 8.2 Predicates $C^{\mathfrak{I}}$, $F^{\mathfrak{I}}$ and $A^{\mathfrak{I}}$

$C^{\mathfrak{I}}$		
P	P	I
PC	S	?
LM	RT	?

$F^{\mathfrak{I}}$			
D	P	P	I
D1	PC	S	!
D2	LM	RT	!
D4	PC	S	!
D5	LM	RT	!
D7	PC	S	?

$A^{\mathfrak{I}}$		
D	P	I
D1	PC	!
D1	S	!
D2	LM	!
D2	RT	!
D4	PC	!
D4	S	!
D5	LM	!
D5	RT	!
D7	PC	!
D7	S	?

(i) $\mathrm{Fu}_\sigma = \emptyset$, $\mathrm{Pr}_\sigma = \{\mathrm{C}, \mathrm{F}, \mathrm{A}\}$,
(ii) $\mathrm{Ar}_\sigma(\mathrm{C}) = \langle \mathsf{P}, \mathsf{P} \rangle$, $\mathrm{Ar}_\sigma(\mathrm{F}) = \langle \mathsf{D}, \mathsf{P}, \mathsf{P} \rangle$, $\mathrm{Ar}_\sigma(\mathrm{A}) = \langle \mathsf{P}, \mathsf{P} \rangle$,
(iii) $\mathrm{Sort}_\sigma(\mathrm{C}) = \mathsf{I}$, $\mathrm{Sort}_\sigma(\mathrm{F}) = \mathsf{I}$, $\mathrm{Sort}_\sigma(\mathrm{A}) = \mathsf{I}$.

The informal interpretation of the predicate symbols of signature σ is as follows:

- $\mathrm{C}(e_1, e_2)$ means that "e_1 is a phenomenological cause of e_2", where

$$e_1, e_2 \in \mathrm{Dom}_W(\mathsf{P}),$$

  i.e. e_1 and e_2 are phenomena (events);
- $\mathrm{F}(d, e_1, e_2)$ means that "On d within a short time interval after e_1 there occurred e_2", where

$$d \in \mathrm{Dom}_W(\mathsf{D}) \quad \text{and} \quad e_1, e_2 \in \mathrm{Dom}_W(\mathsf{P}),$$

  i.e. d is a day, and e_1 and e_2 are phenomena;

- $\mathrm{A}(d,e)$ means that "On d there occurred e", where

$$d \in \mathrm{Dom}_W(\mathsf{D}) \quad \text{and} \quad e \in \mathrm{Dom}_W(\mathsf{P}),$$

  i.e. d is a day and e is a phenomenon.

To complete the definition of **Si**-structure (which is denoted by $\mathfrak{J}$), we define the **Si**-predicates $\mathrm{C}^{\mathfrak{J}}$, $\mathrm{F}^{\mathfrak{J}}$ and $\mathrm{A}^{\mathfrak{J}}$. The tables of these **Si**-predicates are informally represented in Sect. 5.2 (Tables 5.1 and 5.2). Using the abbreviation of this example, we reperesent the discussed predicates in Table 8.2. Note that for short we include in these table the only rows that correspond the rows of tables from Sect. 5.2. For other combinations of arguments, these predicates take the value "?".

Lemma 8.4.8. Let $\mathfrak{J}$ and $\mathfrak{K}$ be two L-structures of the signature σ and φ be an isomorphism of $\mathfrak{J}$ onto $\mathfrak{K}$. Let v be a valuation in $\mathfrak{J}$. Let u be a valuation in $\mathfrak{K}$ defined in the following way:

$$u(x) = \varphi(v(x)).$$

Note that the valuation u is well defined, because of the fact that isomorphism, similarly to valuation, preserves the sort of the mapped entities.

Then for any term t of the signature σ and for any L-formula Φ of the same signature the following hold:

(i) $\varphi\left(t^{\mathfrak{J},v}\right) = t^{\mathfrak{K},u}$,
(ii) $\Phi^{\mathfrak{J},v} = \Phi^{\mathfrak{K},u}$.

Proof. (i) follows from Lemma 8.1.36. (ii) can be proved by induction over the L-formulas.

BASIS. By Lemma 8.1.36, (ii) holds for the atomic formulas. Let us check the validity of (ii) for the J-atomic formulas. Obviously

$$(\mathrm{J}_\alpha\,\theta)^{\mathfrak{J},v} = \mathrm{J}_\alpha\left(\theta^{\mathfrak{J},v}\right) = \mathrm{J}_\alpha\left(\theta^{\mathfrak{K},u}\right) = (\mathrm{J}_\alpha\,\theta)^{\mathfrak{K},u}.$$

INDUCTION STEP. It can be proved similarly to 8.2.6. □

Corollary 8.4.9. Let $\mathfrak{J}$ and $\mathfrak{K}$ be two L-structures of the signature σ and Φ be an L-formula of the same signature. Let $\mathfrak{J} \cong \mathfrak{K}$. Then

$$\mathfrak{J} \models_L \Phi \quad \text{iff} \quad \mathfrak{K} \models_L \Phi.$$

Proof. It is similar to 8.2.7. □

8.5 First-Order PJ Logic (Calculus)

We extend some notions and agreements formulated above for **C** to any PJ logic L. First, we note that the remark at the beginning of Sect. 8.3 has universal meaning.

Therefore, we can freely use the notions of *subterm, subformula, occurrence of a variable or subterm in a term or formula, occurrence of a subformula in a formula* and *scope of quantifier* in the context of any PJ logic L.

Replacing the term "first-order **C**-formula φ" by "external first-order L-formula φ" in Definition 8.3.1, we obtain definitions of the following notions:

- Free and bound variable occurrences in external L-formulas
- Closed external L-formulas (L-sentences)
- Terms admissible for substitution for an individual variable in an external L-formula

We also introduce the notation Φ^x_t for the result of substituting term t for variable x in an external L-formula Φ in all free occurrences of this variable. Moreover, we agree that in a certain context we do not mention the signature of a term or L-formula, since we assume that the corresponding statement (context) holds for any signature.

Let L be a PJ logic. The first-order many-sorted calculus for L is denoted by L^1.

Axiom Schemata 8.5.1 (Axiom schemata of L^1). Let Φ, Ψ and Θ be external L-formulas; x,y, z be individual variables; and t be a term. Then, the following formulas are *axioms of* L^1:

(A1) $\Phi \to (\Psi \to \Phi)$,
(A2) $(\Phi \to (\Psi \to \Theta)) \to ((\Phi \to \Psi) \to (\Phi \to \Theta))$,
(A3) $(\neg\Phi \to \Psi) \to ((\neg\Phi \to \neg\Psi) \to \Phi)$.

The calculus L^1 should include axiom schemata of the groups (B), which describe the relationship of J-defined subsets of the set $\mathscr{V}(L, \mathsf{I})$).

(Q1) $\forall x\, \Phi \to \Phi^x_t$ (t is admissible for substitution for x in Φ),
(Q2) $\forall x\, (\Phi \to \Psi) \to (\Phi \to \forall x\, \Psi)$ (Φ does not contain free occurrences of x);

(E1) $x = x$,
(E2) $y = z \to (\Phi^x_y \to \Phi^x_z)$ (y and z are admissible for substitution for x in Φ and $\mathrm{Sort}x = \mathrm{Sort}y = \mathrm{Sort}z$).

If an axiom is a formula of signature σ, we will refer to it as an *axiom of signature* σ. By $\mathrm{B}^{L,\sigma}$ we denote the *set of all axioms of group* (B^L) of signature σ, i.e. the set of axioms of signature σ obtained by schemata of group (B^L). Similar notation may also be introduced for other axiom groups (A^σ, Q^σ, E^σ), but we do not need them.

Axiom Schemata 8.5.2 (Axiom schemata of group (B) for an FPJ logic). Let L be an FPJ logic and φ be an internal atomic L-formula. Then the following formulas are axioms of L^1:

(B1) $\bigvee\limits_{\varepsilon \in \mathscr{V}(L,\mathsf{I})} \mathrm{J}_\varepsilon\, \varphi$,
(B2) $\mathrm{J}_\varepsilon\, \varphi \to \neg \mathrm{J}_\delta\, \varphi$ $(\varepsilon, \delta \in \mathscr{V}(L, \mathsf{I}),\ \varepsilon \neq \delta)$.

Axiom Schemata 8.5.3 (Axiom schemata of group (B) for an iterative version of an FPJ logic). Let $I = \mathrm{I}^{\tau}(L)$, where L is an FPJ logic, $\tau \in \mathscr{D}^{\tau}(L)$ and φ be an internal atomic I-formula. Then the following formulas are axioms of I^1:

(B1) $\bigvee_{\varepsilon \in \mathscr{V}(L,\mathsf{I})} \mathrm{J}_{(\varepsilon,n)}\,\varphi \quad (n \in \omega)$,
(B2) $\mathrm{J}_{(\varepsilon,n)}\,\varphi \to \neg \mathrm{J}_{(\delta,n)}\,\varphi \quad (\varepsilon, \delta \in \mathscr{V}(L,\mathsf{I}),\ \varepsilon \neq \delta,\ n \in \omega)$,
(B3) $\mathrm{J}_{(\varepsilon,n)}\,\varphi \to \mathrm{J}_{(\varepsilon,k)}\,\varphi \quad (\varepsilon \in \mathscr{D}^{\tau}(L),\ n,k \in \omega,\ k > n)$,
(B4) $\mathrm{J}_{(\tau,k)}\,\varphi \to \mathrm{J}_{(\tau,n)}\,\varphi \quad (n,k \in \omega,\ k > n)$.

Inference Rules 8.5.4 (Inference rules of L^1). L^1 has two rules:

$$\frac{\Phi, \quad \Phi \to \Psi}{\Psi} \qquad (modus\ ponens)$$

$$\frac{\Phi}{\forall x\, \Phi} \qquad (generalisation\ rule\ for\ variable\ x)$$

where Φ and Ψ are external first-order formulas and x is an individual variable.

Now we define derivation from a collection of hypothesis in the calculus L^1, proofs in L^1, deducibility and provability in L^1. These definitions will be similar to the corresponding definitions for the logic $\mathbf{C}^1$. The theorem on deduction for L^1 will also be formulated and its proof will be similar to the proof of the corresponding theorem for $\mathbf{C}^1$.

Definition 8.5.5. Let Γ be a set of external first-order L-formulas of signature σ. An *inference from Γ in L^1* is a finite sequence of *external* first-order L-formulas such that every one of its element:

- Either is an axiom of L^1,
- Or belongs to Γ,
- Or follows from preceding formulas by one of the inference rules of L^1.

An inference from the empty set of hypotheses in L^1 is called a *proof in L^1* (*L^1-proof*).

Definition 8.5.6. Let Γ be a set of external L-formulas and Φ be an external L-formula. We say that Φ *is derivable in L^1 from Γ* (written: $\Gamma \vdash_L \Phi$) if there exists an inference from Γ in L^1 that contains Φ. Φ is called *provable in L^1* or *a theorem of L^1* (written: $\vdash_L \Phi$) if Φ is derivable in L^1 from the empty set of hypotheses.

Theorem 8.5.7 (Deduction Theorem for L^1). Let Γ be a set of external first-order L-formulas and let Φ and Ψ be external first-order L-formulas. Suppose that $\Gamma \cup \{\Phi\} \vdash_L \Psi$ and there exists an L^1-inference of Ψ from $\Gamma \cup \{\Phi\}$ that contains no application of the generalisation rule for variables that have free occurrences in Ψ. Then, $\Gamma \vdash_L \Phi \to \Psi$.

Proof. Proof is similar to the proof of Deduction Theorem for classical $\mathbf{C}^1$ [Mendelson, 1997]. □

Now we formulate a number of statements that will end with theorems on correctness for the calculi L^1 and I^1, where L is an FPJ logic and I is an iterative version of an FPJ logic. The proof of these statements will be analogous to the proof of the corresponding statements for the logic $\mathbf{C}^1$.

Lemma 8.5.8. Let L be a first-order PJ logic. Then its all axioms of the groups (A), (Q) and (E) are L-valid.

Proof. It is similar to the proof of $\mathbf{C}$-validity of the corresponding axioms of $\mathbf{C}^1$. □

Lemma 8.5.9. Let L be a first-order PJ logic. Assume that its all axioms of the group (B) are L-valid. Then $\Gamma \vdash_L \Phi$ implies $\Gamma \models_L \Phi$.

Proof. Due to Lemma 8.5.8, it is enough to show that the rules *modus ponens* and *generalisation* preserve L-validity. This can be done similarly to the case of $\mathbf{C}^1$. □

Lemma 8.5.10. The following statements hold:

(i) Let L be a first-order FPJ logic. Then axioms obtained by the schemata of the group (B) are **St**-valid.
(ii) Let $I = \mathrm{I}^\tau(L)$, where L is an FPJ logic, $\tau \in \mathscr{D}^\tau(L)$. Then axioms obtained by the schemas of the group (B) are I-valid.

Proof. (i) can be proved similarly to Lemma 7.5.11. (ii) can be proved similarly to Lemma 7.5.12. □

Theorem 8.5.11 (Soundness Theorem for FPJ logics). Let L be a first-order FPJ logic, Γ be a set of first-order external L-formulas and Φ be an external first-order L-formula. Then $\Gamma \vdash_L \Phi$ implies $\Gamma \models_L \Phi$.

Proof. This statement directly follows from 8.5.8–8.5.10. □

Theorem 8.5.12 (Soundness Theorem for iterative versions of FPJ logics). Let $I = \mathrm{I}^\tau(L)$, where L is an FPJ logic, $\tau \in \mathscr{D}^\tau(L)$, Γ be a set of first-order external I-formulas and Φ be an external first-order I-formula. Then $\Gamma \vdash_I \Phi$ implies $\Gamma \models_I \Phi$.

Proof. This statement follows directly from 8.5.8–8.5.10. □

We start with a number of definitions and statements which prepare the proof of the theorem on completeness for the calculi L^1 and I^1, where L is an FPJ logic and I is an iterative version of some FPJ logic. These statements will hold for an arbitrary PJ logic. First we establish a one-to-one correspondence between the sets of signatures for the classical logic for an arbitrary PJ logic L. Then we define a bijection of external L-formulas into the set of $\mathbf{C}$-formulas. Moreover, we define a one-to-one correspondence between the classes of L-structures and $\mathbf{C}$-structures. We also establish a connection between the truths in the corresponding L- and $\mathbf{C}$-structures.

Moreover, we define the very important term **C**-*expressible PJ logic* and prove the Main Lemma. The latter claims that if a PJ logic is **C**-expressible then the calculus L^1 will be complete w.r.t. general L-validity. I.e. **C**-expressibility is a sufficient condition for completeness. Then we will prove that any FPJ logic L and its iterative version I are **C**-expressible. In this case the completeness theorems for the calculi L^1 and I^1 are consequences of the Main Lemma.

Definition 8.5.13. Let σ be an arbitrary signature. The C-*image of signature* σ is the classical signature σ' defined in the following way:

(i) $\mathrm{Fu}_{\sigma'} = \mathrm{Fu}_{\sigma}$;
(ii) $\mathrm{Pr}_{\sigma'} = \mathrm{Pr}_{\sigma}(\mathsf{E}) \cup (\mathrm{Pr}_{\sigma}(\mathsf{I}) \times \mathscr{J}(L))$
(in particular, since symbols may be of any nature, as symbols we will use ordered pairs of the form $\langle p, \mathrm{J}_\alpha \rangle$, where p is an internal predicate symbol of signature σ and J_α is a J-operator from the set $\mathscr{J}(L)$);
(iii) If $\theta \in \mathrm{Fu}_{\sigma} \cup \mathrm{Pr}_{\sigma}$ then

$$\mathrm{Ar}_{\sigma'}(\theta) = \mathrm{Ar}_{\sigma}(\theta) \quad \text{and} \quad \mathrm{Sort}_{\sigma'}(\theta) = \mathrm{Sort}_{\sigma}(\theta);$$

(iv) If $\theta \in \mathrm{Pr}_{\sigma}(\mathsf{I})$ then

$$\mathrm{Ar}_{\sigma'}(\langle \theta, \mathrm{J}_\alpha \rangle) = \mathrm{Ar}_{\sigma}(\theta), \quad \mathrm{Sort}_{\sigma'}(\langle \theta, \mathrm{J}_\alpha \rangle) = \mathsf{E},$$

for any $\mathrm{J}_\alpha \in \mathscr{J}(L)$.

Obviously, every signature has a unique C-image. The C-image of a signature σ will be denoted by $\mathrm{C}(\sigma)$. We agree to use the *abridged notation* p_α for the ordered pair $\langle p, \mathrm{J}_\alpha \rangle$ ($p \in \mathrm{Pr}_{\sigma}(\mathsf{I})$, $\mathrm{J}_\alpha \in \mathscr{J}(L)$) in the context of the C-image of signature σ.

Remark 8.5.14. Obviously, if σ is a classical signature, then $\mathrm{C}(\sigma) = \sigma$.

Agreement 8.5.15. Without loss of generality we assume that we have two sets of signatures: $\Sigma_{\mathbf{C}}$ and Σ_L. We take signatures from $\Sigma_{\mathbf{C}}$ for **C**-formulas and **C**-structures. From Σ_L we take signatures for L-formulas and L-structures. We assume that the sets of signatures $\Sigma_{\mathbf{C}}$ and Σ_L satisfy the following conditions:

(i) No signature of Σ_L contains symbols of the form $\langle p, \mathrm{J}_\alpha \rangle$, where p is an internal predicate symbol of a signature from Σ_L, $\mathrm{J}_\alpha \in \mathscr{J}(L)$;
(ii) Any signature of $\Sigma_{\mathbf{C}}$ is a C-image of a signature from Σ_L.

Now we explain the relation between $\Sigma_{\mathbf{C}}$ and Σ_L in more detail. Let us take Σ_L. Then $\Sigma_{\mathbf{C}}$ is obtained as follows:

- We put into $\Sigma_{\mathbf{C}}$ all classical signatures of Σ_L (see Remark 8.5.14);
- For every nonclassical signature σ of Σ_L (i.e. signatures that contains at least one internal predicate symbol), we add $\mathrm{C}(\sigma)$ constructed by Definition 8.5.13 to the set $\Sigma_{\mathbf{C}}$.

The fact that our assumptions do not violate generality is explained as follows:

- In this section we use a small collection of well-known statements about classical many-sorted first-order logic (Deduction Theorem, Soundness Theorem, and Completeness Theorem are particular cases of the corresponding statements about classical many-sorted first-order logic);
- In the proof of statements listed in the above section, signature is considered fixed.

Lemma 8.5.16. The mapping $\mathrm{C}\colon \Sigma_L \to \Sigma_{\mathrm{C}}$ that takes each signature σ from Σ_L to its C-image (see Definition 8.5.13) is a bijection.

Proof. The mapping C is surjective due to Agreement 8.5.15. It is obvious that C is injective. □

Definition 8.5.17. Let σ be a signature and $\mathrm{C}(\sigma)$ be a C-image of signature σ. We define the C-image of terms and external L-formulas of signature σ. If χ is a term or an external formula of signature σ, then C-image of χ is denoted by $\mathrm{C}(\chi)$.

(i) For each term t of signature σ, we put $\mathrm{C}(t) = t$.
(ii) For each external atomic L-formula φ of signature σ, we put $\mathrm{C}(\varphi) = \varphi$.
(iii) For each internal predicate symbol p of signature σ and each J-operator $\mathrm{J}_\alpha \in \mathscr{J}(L)$, we put $\mathrm{C}\left(\mathrm{J}_\alpha\left(p\left(\vec{t}\right)\right)\right) = p_\alpha\left(\vec{t}\right)$, if $\vec{t} \in \mathrm{Dom}_\sigma(p)$ (see Definition 8.5.13).
(iv) For any external L-formulas Φ and Ψ and any individual variable x we put

$$\mathrm{C}(\Phi \to \Psi) = (\mathrm{C}(\Phi) \to \mathrm{C}(\Psi)),$$
$$\mathrm{C}(\neg\Phi) = (\neg\mathrm{C}(\Phi)),$$
$$\mathrm{C}(\forall x\, \Phi) = (\forall x\, \mathrm{C}(\Phi)).$$

Lemma 8.5.18. The mapping C defined above is a bijection of the set of external L-formulas of signature σ onto the set of **C**-formulas of signature $\mathrm{C}(\sigma)$ (C is the identity mapping for terms).

Proof. Proof is similar to the proof of Lemma 7.5.17. □

Definition 8.5.19. Let Γ be a set of external first-order L-formulas of signature σ. The set $\{\mathrm{C}(\Phi) \mid \Phi \in \Gamma\}$ is called the C-*image of the set* Γ and is denoted by $\mathrm{C}(\Gamma)$.

Definition 8.5.20. Let $\mathfrak{S}$ be an L-structure of signature σ. The C-*image* of the L-structure $\mathfrak{S}$ is a **C**-structure $\mathfrak{C}$ of signature $\mathrm{C}(\sigma)$ defined in the following way:

(i) $\mathrm{Frame}_{\mathfrak{C}} = \mathrm{Frame}_{\mathfrak{S}}$;
(ii) $f^{\mathfrak{C}} = f^{\mathfrak{S}}$ for each function symbol f of signature σ;
(iii) $q^{\mathfrak{C}} = q^{\mathfrak{S}}$ for each external predicate symbol q of the signature σ;
(iv) $p_\alpha^{\mathfrak{C}}(\vec{x}) = \mathrm{J}_\alpha\left(p^{\mathfrak{S}}(\vec{x})\right)$ for each internal predicate symbol p, J-operator $\mathrm{J}_\alpha \in \mathscr{J}(L)$ and sequence $\vec{x} \in {\mathrm{Dom}_{\mathfrak{C}}}^* = {\mathrm{Dom}_{\mathfrak{S}}}^*$ such that

$$\mathrm{Ar}_\sigma(p) = {\mathrm{Sort}_\sigma}^*(\vec{x}).$$

The C-image of the L-structure $\mathfrak{S}$ is denoted by $\mathrm{C}(\mathfrak{S})$.

Remark 8.5.21. Obviously, if $\mathfrak{S}$ is a classical L-structure then

$$\mathrm{C}(\mathfrak{S}) = \mathfrak{S}.$$

Lemma 8.5.22. Let $\mathfrak{S}$ be an L-structure of signature σ and v be a valuation in $\mathfrak{S}$. Then

(i) $t^{\mathfrak{S},v} = (\mathrm{C}(t))^{\mathrm{C}(\mathfrak{S}),v}$ for any term t of signature σ;
(ii) $\left(q\left(\vec{t}\right)\right)^{\mathfrak{S},v} = \left(\mathrm{C}\left(q\left(\vec{t}\right)\right)\right)^{\mathrm{C}(\mathfrak{S}),v}$ for any external predicate symbol q of signature σ and any sequence of terms $\vec{t} \in \mathrm{Dom}_{\sigma}(q)$ of signature σ;
(iii) $\left(\mathrm{J}_{\alpha}\left(p\left(\vec{t}\right)\right)\right)^{\mathfrak{S},v} = \left(\mathrm{C}\left(\mathrm{J}_{\alpha}\left(p\left(\vec{t}\right)\right)\right)\right)^{\mathrm{C}(\mathfrak{S}),v}$ for any $p \in \mathrm{Pr}_{\sigma}(\mathsf{I})$, $\mathrm{J}_{\alpha} \in \mathscr{J}(L)$ and any sequence of terms $\vec{t} \in \mathrm{Dom}_{\sigma}(p)$;
(iv) $\Phi^{\mathfrak{S},v} = (\mathrm{C}(\Phi))^{\mathrm{C}(\mathfrak{S}),v}$ for any external L-formula Φ of signature σ.

Proof. (i) By trivial induction over the terms. The fact that $t = \mathrm{C}(t)$ for any term t is used.
(ii) (i) and Definition 8.5.20(iii) are used.
(iii) (i) and Definition 8.5.20(iv) are used.
(iv) By obvious induction over internal formulas of signature σ; (ii) and (iii) are used.

□

Corollary 8.5.23. Let Γ be a set of external L-formulas of signature σ, Φ be an external L-formula of the same signature and $\mathfrak{S}$ be an L-structure of signature σ. Then:

(i) $\mathfrak{S} \vDash_L \Phi$ iff $\mathrm{C}(\mathfrak{S}) \vDash_{\mathrm{C}} \mathrm{C}(\Phi)$;
(ii) $\mathfrak{S} \vDash_L \Gamma$ iff $\mathrm{C}(\mathfrak{S}) \vDash_{\mathrm{C}} \mathrm{C}(\Gamma)$.

Definition 8.5.24. Let σ be a signature of Σ_L. A **C**-structure $\mathfrak{C}$ of signature $\mathrm{C}(\sigma)$ is called *L-coherent* if C-images of all axioms of group (B) of signature σ are valid in it, i.e. if $\mathfrak{C} \vDash_{\mathrm{C}} \mathrm{C}\left(\mathrm{B}^{L,\sigma}\right)$.

Lemma 8.5.25. Let Γ be a set of external L-formulas and Φ be an external L-formula. Then

$$\mathrm{C}(\Gamma) \cup \mathrm{C}\left(\mathrm{B}^{L}\right) \vdash_{\mathrm{C}} \mathrm{C}(\Phi) \quad \text{implies} \quad \Gamma \vdash_L \Phi.$$

Proof. By an obvious induction over the length of inferences in $\mathbf{C}^1$. □

The inverse of Lemma 8.5.25, i.e. the statement

$$\Gamma \vdash_L \Phi \quad \text{implies} \quad \mathrm{C}(\Gamma) \cup \mathrm{C}\left(\mathrm{B}^{L}\right) \vdash_{\mathrm{C}} \mathrm{C}(\Phi),$$

also holds. This can be easily proved by standard induction over the length of inferences in the first-order calculus L.

Below an important property, the **C**-expressiveness of PJ logics, will be established. Then the Main Lemma will be proved, which states that **C**-expressiveness of a PJ logic L is a sufficient condition for the completeness of the calculus L^1 w.r.t. general L-validity.

Definition 8.5.26. Let L be a PJ logic. We say that L is **C**-expressible if the following statements hold:

(i) Let $\mathfrak{S}$ be an L-structure of signature $\sigma \in \Sigma_L$. Then $\mathrm{C}(\mathfrak{S})$ is an L-coherent classical structure.
(ii) Let $\mathfrak{C}$ be an L-coherent **C**-structure of signature $\mathrm{C}(\sigma)$, where σ is a signature from Σ_L. Then there is an L-structure $\mathfrak{S}$ of signature σ such that $\mathfrak{C} = \mathrm{C}(\mathfrak{S})$.

Lemma 8.5.27 (Main Lemma). Let L be a **C**-expressible PJ logic. Then

$$\Gamma \vDash_L \Phi \quad \text{implies} \quad \Gamma \vdash_L \Phi.$$

Proof. It is similar to the proof of the Main Lemma for L^0 (7.5.9). By the use of the Soundness Theorem and Completeness Theorem for $\mathbf{C}^1$ (8.3.7, 8.3.8) and Lemma 8.5.25. □

Lemma 8.5.28. Any FPJ logic L is **C**-expressible.

Proof. We must prove that for for any FPJ logic L, the statements (i) and (ii) of Definition 8.5.26 hold.

(i) Axioms of group (B) are L-valid (see 8.5.10). Then for each L-structure $\mathfrak{S}$ of signature σ, $\mathfrak{S} \vDash_L \mathrm{B}^{L,\sigma}$ holds. Therefore

$$\mathrm{C}(\mathfrak{S}) \vDash_{\mathbf{C}} \mathrm{C}\left(\mathrm{B}^{L,\sigma}\right)$$

also holds (see Corollary 8.5.23) .

(ii) Let us define an L-structure $\mathfrak{S}$ of signature σ as follows:

(a) $\mathrm{Frame}_{\mathfrak{S}} = \mathrm{Frame}_{\mathfrak{C}}$;
(b) $f^{\mathfrak{S}} = f^{\mathfrak{C}}$ for each function symbol f of signature σ;
(c) $q^{\mathfrak{S}} = q^{\mathfrak{C}}$ for each external predicate symbol q of signature σ;
(d) Define an internal predicate $p^{\mathfrak{S}}$ for each internal predicate symbol p of signature σ similarly to how **St**-valuation is defined in the proof of Lemma 7.5.28; namely, let:

$$p^{\mathfrak{S}}(\vec{x}) = \varepsilon, \quad \text{if} \quad p_{\varepsilon}^{\mathfrak{C}}(\vec{x}) = \mathbf{t},$$

where $\vec{x} \in \mathrm{Dom}_{\mathfrak{S}}{}^* = \mathrm{Dom}_{\mathfrak{C}}{}^*$, $\mathrm{Ar}_\sigma(p) = \mathrm{Sort}_\sigma{}^*(\vec{x})$.

The accuracy of (a)–(c) is obvious. Correctness of (d) can be proved similarly to the corresponding statement from the proof of Lemma 7.5.28. (We use the fact that $\mathfrak{C}$ is an L-coherent **C**-structure.)

Then, by analogy with the proof of Lemma 7.5.28, we state that

$$\mathrm{J}_\alpha\left(p^{\mathfrak{S}}(\vec{x})\right) = p_\alpha^{\mathfrak{C}}(\vec{x})$$

for every internal predicate symbol p, J-operator $\mathrm{J}_\alpha \in \mathscr{J}(L)$ and sequence $\vec{x} \in \mathrm{Dom}_{\mathfrak{S}}{}^* = \mathrm{Dom}_{\mathfrak{C}}{}^*$ such that $\mathrm{Ar}_\sigma(p) = \mathrm{Sort}_\sigma{}^*(\vec{x})$. Obviously, this implies the equality $\mathfrak{C} = \mathrm{C}(\mathfrak{S})$. □

Lemma 8.5.29. Let $I = \mathrm{I}^\tau(L)$, where L is an FPJ logic, $\tau \in \mathscr{D}^\tau(L)$. Then I is **C**-expressible.

Proof. We must prove that, for the logic I, the statements (i) and (ii) of Definition 8.5.26 hold.

(i) Axioms of group (B) are I-valid (see 8.5.10). Then for each I-structure $\mathfrak{S}$ of signature σ, $\mathfrak{S} \vDash_I \mathrm{B}^{I,\sigma}$ holds. Therefore,

$$\mathrm{C}(\mathfrak{S}) \vDash_{\mathbf{C}} \mathrm{C}\left(\mathrm{B}^{I,\sigma}\right)$$

also holds (see Corollary 8.5.23).

(ii) Let us define an I-structure $\mathfrak{I}$ of signature σ as follows:

(a) $\mathrm{Frame}_{\mathfrak{I}} = \mathrm{Frame}_{\mathfrak{C}}$;
(b) $f^{\mathfrak{I}} = f^{\mathfrak{C}}$ for each function symbol f of signature σ;
(c) $q^{\mathfrak{I}} = q^{\mathfrak{C}}$ for each external predicate symbol q of signature σ;
(d) Define an internal predicate $p^{\mathfrak{I}}$ for each internal predicate symbol p of signature σ similarly to the definition of I-valuation in the proof of Lemma 7.5.29; namely, let:

$$p^{\mathfrak{I}}(\vec{x}) = \begin{cases} \langle \varepsilon, n\rangle & \text{if } n = \min\left\{m \in \omega \mid p^{\mathfrak{C}}_{(\varepsilon,m)}(\vec{x}) = \mathbf{t}\right\}, \\ \tau & \text{if } p^{\mathfrak{C}}_{(\tau,m)}(\vec{x}) = \mathbf{t} \text{ for any } m \in \omega, \end{cases}$$

where $\vec{x} \in \mathrm{Dom}_{\mathfrak{I}}{}^* = \mathrm{Dom}_{\mathfrak{C}}{}^*$, $\mathrm{Ar}_\sigma(p) = \mathrm{Sort}_\sigma{}^*(\vec{x})$.

Correctness of (a)–(c) of the definition of the I-structure $\mathfrak{I}$ is obvious. Correctness of (d) will be proved similarly to the corresponding statement from the proof of Lemma 7.5.29. (We use the fact that $\mathfrak{I}$ is an I-coherent **C**-structure.)

Then, by analogy with the proof of Lemma 7.5.29, we state that

$$\mathrm{J}_\alpha\left(p^{\mathfrak{I}}(\vec{x})\right) = p^{\mathfrak{C}}_\alpha(\vec{x})$$

for every internal predicate symbol p, J-operator $\mathrm{J}_\alpha \in \mathscr{J}(I)$ and sequence $\vec{x} \in \mathrm{Dom}_{\mathfrak{I}}{}^* = \mathrm{Dom}_{\mathfrak{C}}{}^*$ such that $\mathrm{Ar}_\sigma(p) = \mathrm{Sort}_\sigma{}^*(\vec{x})$. Obviously, this implies the equality $\mathfrak{C} = \mathrm{C}(\mathfrak{I})$. □

Theorem 8.5.30 (Completeness Theorem for FPJ logics). Let L be an FPJ logic, Γ be a set of external L-formulas and Φ be an L-formula. Then

$$\Gamma \vDash_L \Phi \quad \text{implies} \quad \Gamma \vdash_L \Phi.$$

Proof. The theorem follows directly from Lemmas 8.5.27 (Main Lemma) and 8.5.28. □

Theorem 8.5.31 (Completeness Theorem for iterative version of FPJ logics). Let $I = \mathrm{I}^{\tau}(L)$, where L is an FPJ logic, $\tau \in \mathscr{D}^{\tau}(L)$, Γ be a set of external I-formulas and Φ be an external I-formula. Then

$$\Gamma \models_I \Phi \quad \text{implies} \quad \Gamma \vdash_I \Phi.$$

Proof. The theorem follows directly from Lemmas 8.5.27 (Main Lemma) and 8.5.29. □

Definition 8.5.32. A set of external L-formulas Γ is called *L-inconsistent* if there exists an external L-formula Φ such that $\Gamma \vdash_L \Phi \wedge \neg \Phi$. Otherwise the set Γ is called *L-consistent*.

Proposition 8.5.33. If the set of external L-formulas is L-inconsistent, then it contains a finite inconsistent L-subset.

Proposition 8.5.34. Let Γ be a set of external first-order L-formulas. Then Γ is L-consistent if and only if it has an model.

Theorem 8.5.35 (Compactness Theorem for first-order logic L). Let Γ be a set of external first-order L-formulas. We assume that any finite subset of the set Γ has model. Then the whole set Γ has an L-model.

Propositions 8.5.33 and 8.5.34, and Theorem 8.5.35 can be proved similarly to the corresponding propositions for propositional logic L (see 7.5.33–7.5.35).

Part III
Formal CR Framework

Chapter 9
Introductory Explanation

Our objective in the present part is the development of the main instrument of our formal approach, which will provide appropriate formalism for the representation, simulation and generation of the cognitive reasoning processes of cognizing agents, i.e. the processes that allow cognizing subjects to gain new knowledge. The proposed instrument, the formal CR framework, will provide appropriate tools for

- The description of the observed events and objects
- The description and representation of knowledge
- The cognitive reasoning processes from observation to extracting information and knowledge
- The representation and construction of the cognitive reasoning processes permitting reasoning operators of various types to be combined into one process

Moreover, the formal CR framework will be able to adequately represent and handle the cognitive reasoning processes. For this we have to take into account that the reasoning processes fundamentally differ from the logical proofs at least in the following (cf. [Joinet, 2001]):

(i) Reasoning processes are of a dynamic nature in two regards: informational and representational. Reasoning, from the informational point of view, can produce substantively new information, whereas the representational aspect emphasises the possibility to modify the history of a reasoning process. In contrast, proofs, represented as discourses and especially as texts, are static, i.e. they do not modify the informational content significantly from the point of view of the informational aspect.

(ii) Reasoning processes are indeterminate in the sense that they remain reasoning processes even if they seem incorrect (whereas a wrong or incorrect proof is not a proof).

(iii) Reasoning processes are referential, as they presuppose semantic character connected with meaning, i.e. with the internal representation of objects and their structures. (As for proofs, they are of inferential nature, articulating statements inferentially, according only to their shape and regardless of the reference.)

O. Anshakov, T. Gergely, *Cognitive Reasoning*, Cognitive Technologies,
DOI 10.1007/978-3-540-68875-4_9, © Springer-Verlag Berlin Heidelberg 2010

The formal CR framework to be proposed will provide certain solutions to handle these features. Below these three features will be explained from the point of view of the proposed approach.

(i) The informationally static character of logical proofs means that we cannot obtain any unexpected new information by simply applying deduction only. For example, in mathematics, deduction is a tool for verification and representation of the results of thinking and not a tool for searching for new mathematical *entities* and *statements*. Especially scientific researchers' way of thinking cannot be reduced to deduction only.
On the contrary, we can discover principially new relationships by the use of nondeductive techniques of thinking. However, this discovery has its price, that is, the non-correctness of non-deductive reasoning, the possibility of obtaining non-correct conclusions even from valid premises and the possibility of obtaining contradictions.
In the present part we will develop a model of *dynamics of cognitive reasoning processes*. This model is based on the representation of a cognitive reasoning process as a *motion from ignorance to knowledge*. In the course of this motion the uncertainty domain is permanently narrowing. The proposed formalism deals with a specific syntactic structure, the so-called *modification inference*. Applying traditional inference rules we can only add formulas to an inference but we cannot do anything with formulas which are already in the inference. In the proposed formalism, adding formulas to an inference by non-deductive rules we may cause modifications in those constituents which are already in the inference. Therefore, non-deductive rules will be called modification rules in our formalism.
Of course, any formalism imposes certain restrictions and puts the cognitive processes into a certain frame. For example, our view of dynamics of cognitive processes is *discrete* and is rather rigidly regulated. This cognitive process is split into an unlimited number of stages. Each stage is further split into modules (the number of which is finite and is the same for all stages). Each module is divided into two phases: the *rapid one*, where plausible reasoning appears, and the *torpid one*, where only deductive reasoning appears. Thus the structure representing dynamics has three levels of embedding: stage, module and phase.

(ii) In spite of the rigid structure of the representation of reasoning our approach permits the applification of plausible rules and even formal logical contradictions to be obtained. An interesting effect of the proposed formalism is the temporal character of the obtained contradictions. A cognitive process may be free of contradiction at the initial stage. Contradiction may appear at one of the internal stages and then it may disappear. At the subsequent stages of this cognitive process a contradiction may also appear but for a different reason.

(iii) The proposed formalism has a dual (semantic–syntactic) nature. Formally it can be considered as a method of constructing calculus of a special form, the so-called *modification calculus*. By the use of such calculi, syntactic objects, i.e. inferences, may be generated. The state description is an important constituent of each calculus. It is a collection of axioms which *entirely and uniquely describes the semantic structure*, the latter of which is considered a model of our initial knowledge about the subject domain.

Note that the completion of the rapid phase of each module of a cognitive reasoning process results in a new collection of axioms. These axioms describe a new semantic structure as a model of the knowledge state about the subject domain. The sequence of such structures can be considered as a model of the modification calculus itself.

The modification calculi are used to analyse facts obtained as a result of interiorisation of experiments and observations. The aim of the analysis is the discovery of regularities. This analysis of facts is carried out by the use of formal rules.

The set of facts, the axioms of initial knowledge and the set of formal (deductive and plausible) rules form the so-called *modification theory*. This theory in general, is temporarily open, which means that a cognizing subject as the carrier of the considered cognitive processes is open to its environment, from where facts may be received at any time. However, we suppose that new facts from the environment arrive through a special sluice, which accumulates the new facts and advances them in the form of batches.

Chapter 10
Modification Calculi

We introduce a formalism that has a dual semantical–syntactical character. Modification calculi contain a collection of statements, practically playing the role of axioms, which identically define a certain canonical model. The representation of this model implicitly affects certain inference rules. Note that, besides some usual inference rules of first-order calculus (*modus ponens* and generalisation), we will also use the plausible inference rules, i.e. the modification rules. Application of these rules is interpreted as a move from uncertainty to definiteness, i.e. from ignorance to knowing. Application of these rules creates a new knowledge state.

Let us consider the application of modification rules. From the semantic point of view their application modifies the canonical model, while from the syntactic point of view the inference itself is modified. Note that the notion of inference, as usually regarded in formal logic, is traditionally considered as a written text and, at first glance, it lacks true dynamic dimension. The inference rules allow us to add new elements (formulas, labelled formulas, sequences etc.) to the text, but they do not allow modification of anything in the already written parts of the inference. Thus, indeed, one occasionally finds the absense of dynamics used against the formalist perspective in logic, which is then typically blamed for being only occupied with static things and thus incomparable to human reasoning.

On the contrary, an inference in the modification calculi is a dynamic, changing object. The inference rules in the modification calculi allow not only addition of new elements to the inference but they also allow the modification of the earlier added elements. I.e. if a traditional proof is a written text that is allowed to be extended, then an inference in the modification calculi can be represented as a collection of records in the cells of an electronic table (or records in a relational database). Here one can both add new records and modify some of the already existing ones.

Note that the mentioned modifications are far from being arbitrary and they are subject to strict rules. We can only block some records of the inference, making them inactive and invisible for the rules. However, the deactivation of statements earlier obtained already brings dynamics into the inference. Addition of new elements has always been possible, but now we are able to hide old elements too. Thus a com-

O. Anshakov, T. Gergely, *Cognitive Reasoning*, Cognitive Technologies,
DOI 10.1007/978-3-540-68875-4_10,

bination of addition and deactivation of inference elements allows us to modify the state of knowledge more comparably to in human reasoning.

Note that the cognitive process simulated in the modification calculi by inference is rather rigidly regulated. This becomes necessary so as to realise such an inference on a computer, and to minimise the risk of the appearance of contradictions even though we assume the application of plausible inference rules that do not guarantee reliable consequences. However, the regulation may not save us from the appearance of such contradictions. An interesting property of an inference in the modification calculi is that it can lead to *temporal* contradictions. After several inference steps there may appear a situation where the inference will contain a contradictory set of active (visible) formulas. Then by applying modification rules the contradiction may be eliminated.

Let us return to the semantic aspect. As we have already mentioned, inference allows the modification of the canonical model. Of course, this modification is not produced by deduction which, in our formalism, is represented by *modus ponens* and generalisation, but by plausible inference rules which change the knowledge state. In general, our cognitive strategy is rather rigidly given (otherwise computer realisation would be out of the question). In the framework of this strategy there are certain cycles in which the combination of deductive and plausible reasoning occurs. The completion of each of these cycles is accompanied by a new knowledge state. The cognitive procedure generates a sequence of knowledge states. Each consistent state has its own corresponding canonical model.

The next section of this chapter will be devoted to the investigation of the semantic properties of our formalism.

10.1 State Descriptions over Sets of Constants

In this section we introduce the notion of *state description* with the meaning of a knowledge state as discussed above. Since our formalism supposes the possibility of computer realisation we represent knowledge in a certain finite universe. The initial state can be considered as a system of observational data about a finite fragment of a subject domain. State description uniquely defines a canonical model that will be called a *syntactic* model.

Note that we will use the apparatus of models constructed from constants proposed by Henkin [1949].

We recall that the individual constants have been defined in 8.1.12 as functional symbols of empty arity. The notations Const_σ and $\mathrm{Const}_\sigma(\mathsf{s})$ have also been defined there for the set of all individual constants of signature σ and for the set of all individual constants of sort s, respectively.

Definition 10.1.1. Let σ be a signature, $D \subseteq \mathrm{Const}_\sigma$. The set D is called a σ-*vocabulary* if the following conditions hold:

(i) D is finite,

(ii) For any $\mathsf{s} \in \mathsf{Sorts}_\sigma$,
$$D \cap \mathrm{Const}_\sigma(\mathsf{s}) \neq \emptyset,$$
i.e. D contains constants of all sorts.

Definition 10.1.2. Let σ be a signature and D be a σ-vocabulary.

(i) By definition let
$$\mathrm{Frame}_\sigma(D) = \langle D, \mathrm{Sort}_\sigma|_D \rangle .$$
We recall that $\mathrm{Sort}_\sigma|_D$ denotes the restriction of the function Sort_σ to D. $\mathrm{Frame}_\sigma(D)$ is called the *D-frame of signature σ*.

(ii) By definition let
$$\mathrm{Dom}_\sigma(D, \mathsf{s}) = \mathrm{Dom}_\sigma(\mathsf{s}) \cap D = \{c \in D \mid \mathrm{Sort}_\sigma(c) = \mathsf{s}\} .$$
$\mathrm{Dom}_\sigma(D, \mathsf{s})$ is called the *D-s-domain of signature σ*.

Proposition 10.1.3. Let σ be a signature and D be a σ-vocabulary. Then $\mathrm{Frame}_\sigma(D)$ is a U-frame such that $\mathsf{Sorts}_U = \mathsf{Sorts}_\sigma$.

Proof. Since D is a σ-vocabulary, the map $\mathrm{Sort}_\sigma|_D$ is surjective. Therefore, $\mathsf{Sorts}_U = \mathsf{Sorts}_\sigma$ (see Definition 8.1.17). □

Corollary 10.1.4. Let σ be a signature and D be a σ-vocabulary. Then

(i) $\mathrm{Dom}_\sigma(D, \mathsf{s}) \neq \emptyset$ for any $\mathsf{s} \in \mathsf{Sorts}_\sigma$,
(ii) $\mathrm{Dom}_\sigma(D) = \bigcup_{\mathsf{s} \in \mathsf{Sorts}_\sigma} \mathrm{Dom}_\sigma(D, \mathsf{s})$,
(iii) $\mathrm{Dom}_\sigma(D, \mathsf{s}) \cap \mathrm{Dom}_\sigma(D, \mathsf{t}) = \emptyset$ for any $\mathsf{s}, \mathsf{t} \in \mathsf{Sorts}_\sigma$ such that $\mathsf{s} \neq \mathsf{t}$.

Definition 10.1.5. Let L be an FPJ logic, σ be a signature and D be a σ-vocabulary.

(i) Suppose $\varphi \in \mathrm{Fu}_\sigma \cup \mathrm{Pr}_\sigma$. By definition let:
$$\mathrm{Dom}_\sigma(D, \varphi) = \mathrm{Dom}_\sigma(\varphi) \cap D^* = \{\vec{c} \in D^* \mid \mathrm{Ar}_\sigma(\varphi) = \mathrm{Sort}_\sigma{}^*(\vec{c})\} .$$
$\mathrm{Dom}_\sigma(D, \varphi)$ is called *the D-domain of signature σ for φ*.

(ii) Suppose $f \in \mathrm{Fu}_\sigma$. An arbitrary map
$$F\colon \mathrm{Dom}_\sigma(D, f) \to \mathrm{Dom}_\sigma(D, \mathsf{s}) ,$$
where $\mathsf{s} = \mathrm{Sort}_\sigma(f)$, is said to be a *syntactic f-operation over D*.

(iii) Suppose $p \in \mathrm{Pr}_\sigma$. An arbitrary map
$$P\colon \mathrm{Dom}_\sigma(D, p) \to \mathscr{V}(L, \mathsf{s}) ,$$
where $\mathsf{s} = \mathrm{Sort}_\sigma(p)$, is said to be a *syntactic p-predicate over D in L*.

(iv) We say that F is a *syntactic operation of signature σ over D*, if F is a *syntactic f-operation over D* for some $f \in \mathrm{Fu}_\sigma$.

(v) We say that P is a *syntactic predicate of signature* σ *over* D *in* L, if P is a *syntactic p-predicate over* D *in* L for some $p \in \mathrm{Pr}_\sigma$.

In this case:

- If $\mathrm{Sort}_\sigma(p) = \mathsf{E}$, then P is called an *external syntactic predicate of signature* σ *over* D *in* L (ES_σ-predicate over D in L);
- If $\mathrm{Sort}_\sigma(p) = \mathsf{I}$, then P is called an *internal syntactic predicate of signature* σ *over* D *in* L (IS_σ-predicate over D in L).

We will consider the above-defined σ-vocabulary D as a finite universum of some L-structure. Below we will call such L-structures syntactic L-structures over D.

Let Γ be a set of external L-formulas of the signature σ. If a syntactic L-structure over a σ-vocabulary D is an L-model for Γ, then such a Γ will be of particular interest. Our prime interest here is with those sets of L-formulas which uniquely define the corresponding syntactic L-structure. We also would like that these sets of L-formulas could be built up by fairly simple rules.

For an FPJ-logic L, such simply constructed sets of L-formulas (which uniquely define their own syntactical L-model) will be further called *state descriptions*. This denomination has the following intuitive motivation: each syntactical L-structure, where L is an FPJ logic, is considered as a model of a *knowledge state*. Sequences of such L-structures represent the history of the cognitive process.

Besides the state descriptions we are interested in the sets of external L-formulas designated to solve more particular tasks. Namely, we are interested in the sets that would describe internal syntactic predicates of the signature σ over the σ-vocabulary D. Below they will be called *IS_σ-predicate descriptions*.

In order to define an L-structure we must define internal L-predicates. Moreover, internal syntactic L-predicates will play a special role in our formalism. They will be modified as a result of the cognitive process, and their *uncertaity domains* (i.e. the subsets of their domains, for the elements of which predicates have the undefined value τ) will drop. The cognitive process will modify them and decrease their uncertaity domains.

Thus we formalise the situation of moving from ignorance to knowledge in the cognitive process.

Definition 10.1.6. In this definition we will introduce a notation for various kinds of sets of J-atomic formulas.

Let L be a PJ logic, σ be a signature and D be a σ-vocabulary. By definition let:

(i) $\mathrm{JA}_\sigma(L,p,D) = \{\mathrm{J}_\alpha\, p(\vec{c}) \mid \vec{c} \in \mathrm{Dom}_\sigma(D,p), \alpha \in \mathrm{JI}(L)\}$,
where $p \in \mathrm{Pr}_\sigma(\mathsf{I})$;

(ii) $\mathrm{JA}_\sigma(L,D) = \{\mathrm{J}_\alpha\, p(\vec{c}) \mid p \in \mathrm{Pr}_\sigma(\mathsf{I}), \vec{c} \in \mathrm{Dom}_\sigma(D,p), \alpha \in \mathrm{JI}(L)\}$.

Suppose Γ to be a set of external L-formulas of signature σ. By definition let:

(iii) $\mathrm{JA}_\sigma(\Gamma,p,D) = \mathrm{JA}_\sigma(L,p,D) \cap \Gamma$,
where $p \in \mathrm{Pr}_\sigma(\mathsf{I})$;

(iv) $\mathrm{JA}_\sigma(\Gamma,D) = \mathrm{JA}_\sigma(L,D) \cap \Gamma$.

Agreement 10.1.7. Further, for the sake of brevity in the case when L is known from context, we will write $\mathrm{JA}_\sigma(p,D)$ and $\mathrm{JA}_\sigma(D)$ instead of $\mathrm{JA}_\sigma(L,p,D)$ and $\mathrm{JA}_\sigma(L,p,D)$, respectively. We agree also to omit L in other similar cases.

Definition 10.1.8. Let L be an FPJ logic, σ be a signature, D be a σ-vocabulary, and

$$\Gamma \subseteq \mathrm{JA}_\sigma(L,D).$$

Then Γ is called an *internal state predicate description of signature σ over D in L* if, for any $p \in \mathrm{Pr}_\sigma(\mathsf{I})$, $\vec{a} \in \mathrm{Dom}_\sigma(D,p)$, there exists a unique $\varepsilon \in \mathscr{V}(L,\mathsf{I})$ such that $\mathrm{J}_\varepsilon\, p(\vec{a}) \in \Gamma$.

The set of all internal state predicate descriptions of signature σ over D in L is denoted by $\mathrm{ISPD}_\sigma(D,L)$.

Notation 10.1.9. Let L be a PJ logic and Γ be a finite set of external L-formulas. For the conjuctions and disjunctions of all L-formulas of Γ the following notations are used

$$\bigwedge_{\Phi\in\Gamma} \Phi \quad \text{and} \quad \bigvee_{\Phi\in\Gamma} \Phi.$$

Definition 10.1.10. Let L be an FPJ logic and D be a σ-vocabulary. A set Δ of external L-formulas of signature σ is called a *state description of signature σ over D in L* if it meets the following conditions:

(i) Δ contains all the formulas of the form $\neg(a=b)$, where a and b are two distinct constants from D of the same sort,

(ii) For each $s \in \mathsf{Sorts}_\sigma$, Δ contains the formula

$$\forall x \left(\bigvee_{a\in \mathrm{Dom}_\sigma(\mathsf{s})} x=a \right),$$

where $\mathrm{Sort}(x) = \mathsf{s}$;

(iii) For any $f \in \mathrm{Fu}_\sigma$, $\vec{a} \in \mathrm{Dom}_\sigma(D,f)$ there exists a unique constant

$$b \in \mathrm{Dom}_\sigma(D, \mathrm{Sort}_\sigma(f))$$

such that the formula $f(\vec{a}) = b$ belongs to Δ;

(iv) For any $p \in \mathrm{Pr}_\sigma(\mathsf{E})$, $\vec{a} \in \mathrm{Dom}_\sigma(D,p)$ either $p(\vec{a}) \in \Delta$ or $\neg p(\vec{a}) \in \Delta$;

(v) For any $p \in \mathrm{Pr}_\sigma(\mathsf{I})$, $\vec{a} \in \mathrm{Dom}_\sigma(D,p)$ there exists a unique $\varepsilon \in \mathscr{V}(L,\mathsf{I})$ such that $\mathrm{J}_\varepsilon\, p(\vec{a}) \in \Delta$;

(vi) The set Δ does not contain any other formulas different from the ones given in (i) – (v),

The *set of all state descriptions of signature σ over D* in L is denoted by $\mathbf{SD}_\sigma(L,D)$.

Example 10.1.11. Take **Si** as an FPJ logic (see Definition 7.4.7). Consider Example 8.4.7 as an information source for constructing an example of a state description in **Si**.

Let σ be the signature and D be the set of abreviations from Example 8.4.7, i.e.

$$D = \{\mathrm{PC}, \mathrm{S}, \mathrm{LM}, \mathrm{RT}, \mathrm{D1}, \mathrm{D2}, \mathrm{D4}, \mathrm{D5}, \mathrm{D7}\}.$$

Enrich σ by adding D as a set of individual constants. Denote the obtained signature by σ'. Let

$$\mathrm{Sort}_{\sigma'}(\mathrm{PC}) = \mathrm{Sort}_{\sigma'}(\mathrm{S}) = \mathrm{Sort}_{\sigma'}(\mathrm{LM}) = \mathrm{Sort}_{\sigma'}(\mathrm{RT}) = \mathsf{P},$$

(i.e. the sort of these constants is the sort of phenomena) and

$$\mathrm{Sort}_{\sigma'}(\mathrm{D1}) = \mathrm{Sort}_{\sigma'}(\mathrm{D2}) = \mathrm{Sort}_{\sigma'}(\mathrm{D4}) = \mathrm{Sort}_{\sigma'}(\mathrm{D5}) = \mathrm{Sort}_{\sigma'}(\mathrm{D7}) = \mathsf{D},$$

(i.e. the sort of the above constants is the sort of days).

Then D is finite and for any $\mathsf{s} \in \mathsf{Sorts}_{\sigma'}$,

$$D \cap \mathrm{Const}_{\sigma'}(\mathsf{s}) \neq \emptyset,$$

i.e. D contains constants of all sorts. Therefore, D is a σ'-vocabulary in the sense of Definition 10.1.1.

Let Δ be a set of **Si**-formulas that contains:

(i) All the formulas of the form $\neg(e_1 = e_2)$, where $e_1, e_2 \in \mathrm{Dom}_{\sigma'}(D, \mathsf{P})$, and all the formulas of the form $\neg(d_1 = d_2)$, where $d_1, d_2 \in \mathrm{Dom}_{\sigma'}(D, \mathsf{P})$;

(ii) The formulas

$$\forall e\,(e = \mathrm{PC} \vee e = \mathrm{S} \vee e = \mathrm{LM} \vee e = \mathrm{RT})$$

and

$$\forall d\,(d = \mathrm{D1} \vee d = \mathrm{D2} \vee d = \mathrm{D4} \vee d = \mathrm{D5} \vee d = \mathrm{D5} \vee d = \mathrm{D7}),$$

where $\mathrm{Sort}(e) = \mathsf{P}$, $\mathrm{Sort}(d) = \mathsf{D}$;

(iii) The formulas $!\mathrm{F}(\mathrm{D1}, \mathrm{PC}, \mathrm{S})$, $!\mathrm{F}(\mathrm{D2}, \mathrm{LM}, \mathrm{RT})$, $!\mathrm{F}(\mathrm{D4}, \mathrm{PC}, \mathrm{S})$, $!\mathrm{F}(\mathrm{D5}, \mathrm{LM}, \mathrm{RT})$, (see Table 8.2 from Example 8.4.7)

(iv) All the formulas of the form $?\mathrm{F}(d, e_1, e_2)$, where

$$d \in \mathrm{Dom}_{\sigma'}(D, \mathsf{D}) \quad \text{and} \quad e_1, e_2 \in \mathrm{Dom}_{\sigma'}(D, \mathsf{P}),$$

and the argument tuple $\langle d, e_1, e_2 \rangle$ differs from any such tuple of (iii);

(v) The formulas $!\mathrm{A}(\mathrm{D1}, \mathrm{PC})$, $!\mathrm{A}(\mathrm{D1}, \mathrm{S})$, $!\mathrm{A}(\mathrm{D2}, \mathrm{LM})$, $!\mathrm{A}(\mathrm{D2}, \mathrm{RT})$, $!\mathrm{A}(\mathrm{D4}, \mathrm{PC})$, $!\mathrm{A}(\mathrm{D4}, \mathrm{S})$, $!\mathrm{A}(\mathrm{D5}, \mathrm{LM})$, $!\mathrm{A}(\mathrm{D5}, \mathrm{RT})$, $!\mathrm{A}(\mathrm{D7}, \mathrm{PC})$, (see Table 8.2 from Example 8.4.7);

(vi) All the formulas of the form $?\mathrm{A}(d, e)$, where

$$d \in \mathrm{Dom}_{\sigma'}(D, \mathsf{D}) \quad \text{and} \quad e \in \mathrm{Dom}_{\sigma'}(D, \mathsf{P}),$$

and the argument pair $\langle d, e \rangle$ differs from any such pair of (v);

(vii) All the formulas of the form $?\mathrm{C}(e_1, e_2)$, where $e_1, e_2 \in \mathrm{Dom}_{\sigma'}(D, \mathsf{P})$;

(viii) The set $\varDelta$ does not contain any other formulas different from the ones given in (i)–(vii).

Recall that the signature σ' from Example 8.4.7 contains no functional symbols and no external predicate symbols. Then, it is easy to prove that $\varDelta$ is a state description over D in **Si** (see Definition 10.1.10).

Remark 10.1.12. Let L be an FPJ logic, σ be a signature, D be a σ-vocabulary and $\varDelta$ be a state description of signature σ over D in L. Then

(i) $\mathrm{JA}_\sigma(\varDelta, D)$ is the set of formulas whose elements are defined in Definition 10.1.10 (v),
(ii) $\mathrm{JA}_\sigma(\varDelta, D) \in \mathrm{ISPD}_\sigma(D, L)$ (see Definition 10.1.8).

Notation 10.1.13. Let us introduce the notation for the result of replacing one element with another in a set. I.e. by definition let:

$$M[a \leftarrow b] = (M \setminus \{a\}) \cup \{b\}.$$

Similarly, let us define the replacement of a subset in a set:

$$M[K \leftarrow L] = (M \setminus K) \cup L.$$

Proposition 10.1.14. Let L be an FPJ-logic, D be σ-vocabulary, $\varDelta \in \mathbf{SD}_\sigma(L, D)$, i.e. $\varDelta$ is a state description of signature σ over D in L. Let $\mathrm{J}_\varepsilon\, p\left(\vec{b}\right) \in \varDelta$. Then for any $\delta \in \mathscr{V}(L, \mathsf{I})$,

$$\varDelta\left[\mathrm{J}_\varepsilon\, p\left(\vec{b}\right) \leftarrow \mathrm{J}_\delta\, p\left(\vec{b}\right)\right] \in \mathbf{SD}_\sigma(L, D).$$

Proof. By direct checking of the conditions (i)–(vi) of Definition 10.1.10. □

Proposition 10.1.15. Let L be an FPJ logic, $\varDelta \in \mathbf{SD}_\sigma(L, D)$, $\Xi \subseteq \mathrm{JA}_\sigma(D)$. Then

$$\varDelta[\mathrm{JA}_\sigma(\varDelta, D) \leftarrow \Xi] \in \mathbf{SD}_\sigma(L, D)$$

if and only if $\Xi \in \mathrm{ISPD}_\sigma(D, L)$.

Proof. The statement directly follows from Definitions 10.1.6, 10.1.8 and 10.1.10. □

Definition 10.1.16. Let L be an FPJ logic, D be a σ-vocabulary and $\varDelta \in \mathbf{SD}_\sigma(L, D)$. An L-structure $\mathfrak{J}$ over the U-frame $\mathrm{Frame}_\sigma(D)$ is called the *syntactic L-structure of signature σ for $\varDelta$ over D* if it meets the following conditions:

(i) $a^{\mathfrak{J}} = a$ for each $a \in D$;
(ii) For any $f \in \mathrm{Fu}_\sigma$, $\vec{a} \in \mathrm{Dom}_\sigma(D, f)$ and $b \in \mathrm{Dom}_\sigma(D, \mathrm{Sort}_\sigma(f))$,

$$f^{\mathfrak{J}}(\vec{a}) = b,$$

if and only if the formula $f(\vec{a}) = b$ belongs to $\varDelta$;

(iii) For any $p \in \mathrm{Pr}_\sigma(Ext)$, $\vec{a} \in \mathrm{Dom}_\sigma(D,p)$,

$$p^{\mathfrak{J}}(\vec{a}) = \begin{cases} \mathbf{t} \text{ if } p(\vec{a}) \in \Delta, \\ \mathbf{f} \text{ otherwise;} \end{cases}$$

(iv) For any $p \in \mathrm{Pr}_\sigma(\mathsf{I})$, $\vec{a} \in \mathrm{Dom}_\sigma(D,p)$, $\varepsilon \in \mathscr{V}(L,\mathsf{I})$,

$$p^{\mathfrak{J}}(\vec{a}) = \varepsilon \quad \text{iff} \quad \mathrm{J}_\varepsilon\, p(\vec{a}) \in \Delta.$$

Example 10.1.17. Consider the **Si**-structure $\mathfrak{J}$ from Example 8.4.7. Enrich the signature σ from this example to the signature σ' from Example 10.1.11. Take D from Example 10.1.11. Define the **Si**-structure $\mathfrak{K}$ of signatire σ' as follows:

- Let $\mathrm{Frame}_{\mathfrak{K}} = \mathrm{Frame}_{\mathfrak{J}}$;
- Let $p^{\mathfrak{K}} = p^{\mathfrak{J}}$ for any $p \in \mathrm{Pr}_\sigma(\mathsf{I})$;
- Let $c^{\mathfrak{K}} = c$ for any $c \in D$.

It is easy to prove that the above **Si**-structure $\mathfrak{K}$ is a syntactic **Si**-structure of signature σ' over D in **Si**.

Proposition 10.1.18. Let L be an FPJ logic, D be a σ-vocabulary and

$$\Delta \in \mathbf{SD}_\sigma(L,D).$$

Then there exists a unique syntactic L-structure of signature σ for Δ over D.

Proof. Existence follows from Definition 10.1.10. Uniqueness can be easily proved by induction also with the help of Definition 10.1.10. □

Notation 10.1.19. Let L be an FPJ logic, D be a σ-vocabulary and $\Delta \in \mathbf{SD}_\sigma(L,D)$. Let $\mathrm{Sint}_L(\Delta,D)$ denote the syntactic L-structure of signature σ for Δ over D.

Proposition 10.1.20. Let L be an FPJ logic, D be a σ-vocabulary and

$$\Delta \in \mathbf{SD}_\sigma(L,D).$$

Then the L-structure

$$\mathfrak{J} = \mathrm{Sint}_L(\Delta,D)$$

is an L-model of Δ.

Proof. By Definition 10.1.10, the set Δ contains five groups of formulas. Let us consider each of these groups separately.

(i) Δ contains all formulas of the form $\neg(a = b)$, where a and b are two distinct constants from D of the same sort.
By definion 10.1.16 we have $a^{\mathfrak{J}} = a$ for each constant $a \in D$. Hence all formulas of the given group are valid in $\mathfrak{J}$.

(ii) For each $s \in \mathsf{Sorts}_\sigma$ the set Δ contains the formula

$$\forall x \left(\bigvee_{a \in \mathrm{Dom}_\sigma(s)} x = a \right),$$

where $\mathrm{Sort}(x) = \mathsf{s}$.
Formulas of group (ii) are valid in $\mathfrak{J}$ on the same basis as for the formulas of the previous group.

(iii) For any $f \in \mathrm{Fu}_\sigma$, $\vec{a} \in \mathrm{Dom}_\sigma(D, f)$ there exists a unique constant

$$b \in \mathrm{Dom}_\sigma(D, \mathrm{Sort}_\sigma(f))$$

such that the formula $f(\vec{a}) = b$ belongs to Δ.
By Definition 10.1.16, $f^{\mathfrak{J}}(\vec{a}) = b$ if the formula $f(\vec{a}) = b$ belongs to Δ. Consequently all formulas from this group are valid in $\mathfrak{J}$.

(iv) For any $p \in \mathrm{Pr}_\sigma(\mathsf{E})$, $\vec{a} \in \mathrm{Dom}_\sigma(D, p)$ either $p(\vec{a}) \in \Delta$, or $\neg p(\vec{a}) \in \Delta$.
By Definition 10.1.16,

$$p^{\mathfrak{J}}(\vec{a}) = \begin{cases} \mathbf{t} \text{ if } p(\vec{a}) \in \Delta, \\ \mathbf{f} \text{ otherwise.} \end{cases}$$

If $p(\vec{a}) \in \Delta$, then $p^{\mathfrak{J}}(a) = \mathbf{t}$. In the opposite case $p^{\mathfrak{J}}(\vec{a}) = \mathbf{f}$. Here $\neg p(\vec{a})$ is valid in $\mathfrak{J}$. In this very case $\neg p(\vec{a}) \in \Delta$. Therefore any formula of the given group is valid in $\mathfrak{J}$.

(v) For any $p \in \mathrm{Pr}_\sigma(\mathsf{I})$, $\vec{a} \in \mathrm{Dom}_\sigma(D, p)$ there exists a unique $\varepsilon \in \mathscr{V}(L, \mathsf{I})$ such that $\mathrm{J}_\varepsilon\, p(\vec{a}) \in \Delta$.
By Definition 10.1.16, for any $\varepsilon \in \mathscr{V}(L, \mathsf{I})$, $p^{\mathfrak{J}}(\vec{a}) = \varepsilon$, if $\mathrm{J}_\varepsilon\, p(\vec{a}) \in \Delta$. Then obviously, $\mathrm{J}_\varepsilon\, p(\vec{a})$ is valid in $\mathfrak{J}$.

By Definition 10.1.10 (vi), the set Δ does not contain formulas others than the ones enumerated in (i)–(v). Therefore all the formulas from Δ are valid in $\mathfrak{J}$. □

Corollary 10.1.21. Let L be an FPJ logic, D be a σ-vocabulary and $\Delta \in \mathbf{SD}_\sigma(L, D)$. Then the set Δ is L-consistent.

Proof. It follows from the Correctness Theorem (8.5.11) that an L-inconsistent set cannot have any model. Therefore, the set Δ is L-consistent. □

Proposition 10.1.22. Let L be an FPJ logic, D be a σ-vocabulary, $\Delta \in \mathbf{SD}_\sigma(L, D)$. Then any two L-models of the set Δ are isomorphic.

Proof. By Proposition 8.1.35 it is enough to show that any model of Δ is isomorphic to $\mathrm{Sint}_L(\Delta, D)$.

Let $\mathfrak{K}$ be an arbitrary model of Δ, $\mathfrak{J} = \mathrm{Sint}_L(\Delta, D)$. Let us define the map

$$\varphi\colon D \to \mathrm{Dom}_{\mathfrak{K}}$$

in the following way: Suppose

$$\varphi(a) = a^{\mathfrak{K}} \tag{10.1}$$

for any constant $a \in D$.

Now we show that φ is an isomorphism of $\mathfrak{J}$ onto $\mathfrak{K}$.

The map φ is a bijection because the formulas of Δ given in (i) and (ii) of Definition 10.1.10 are valid in $\mathfrak{K}$. Clearly, the map φ preserves the sorts.

Let $f \in \mathrm{Fu}_\sigma$, $\vec{a} \in \mathrm{Dom}_\sigma(D, f)$. By (10.1) it is enough to show that

$$\varphi\left(f^{\mathfrak{J}}(\vec{a})\right) = f^{\mathfrak{K}}\left(\vec{a}^{\mathfrak{K}^*}\right).$$

By Definition 10.1.10 (iii), there exists a unique constant $b \in \mathrm{Dom}_\sigma(D, \mathrm{Sort}_\sigma(f))$ such that the formula $f(\vec{a}) = b$ belongs to Δ. By Definition 10.1.16 (ii), if the formula $f(\vec{a}) = b$ belongs to Δ, then $f^{\mathfrak{J}}(\vec{a}) = b$. Since all the formulas from Δ are valid in $\mathfrak{K}$ we obtain

$$f^{\mathfrak{K}}\left(\vec{a}^{\mathfrak{K}^*}\right) = b^{\mathfrak{K}} = \varphi(b) = \varphi\left(f^{\mathfrak{J}}(a)\right).$$

Let $p \in \mathrm{Pr}_\sigma(\mathsf{E})$, $\vec{a} \in \mathrm{Dom}_\sigma(D, p)$. By (10.1), it is enough to show that

$$p^{\mathfrak{J}}(\vec{a}) = p^{\mathfrak{K}}\left(\vec{a}^{\mathfrak{K}^*}\right).$$

It is not difficult to prove the correctness of the latter using the following:

- Δ satisfies condition (iv) of Definition 10.1.10,
- $\mathfrak{J}$ satisfies condition (iii) of Definition 10.1.16,
- All formulas from Δ are valid in $\mathfrak{K}$.

The case when $p \in \mathrm{Pr}_\sigma(\mathsf{I})$ may be considered in a similar way by using Definition 10.1.10 (v) and Definition 10.1.16 (iv). □

Proposition 10.1.23. Let L be an FPJ logic, D be a σ-vocabulary, $\Delta \in \mathbf{SD}_\sigma(L, D)$, $\mathfrak{J} = \mathrm{Sint}_L(\Delta, D)$ and Φ be an external L-formula of signature σ. Then

$$\Delta \vdash_L \Phi \quad \text{iff} \quad \mathfrak{J} \models_L \Phi.$$

Proof. ($\Rightarrow$). According to the Soundness Theorem for L^1 (Theorem 8.5.11), $\Delta \vdash_L \Phi$ implies $\mathfrak{J} \models_L \Phi$.

($\Leftarrow$) Let $\mathfrak{J} \models_L \Phi$. By Proposition 10.1.22, all models of the set Δ are isomorphic to $\mathfrak{J}$. Then a formula that is valid in $\mathfrak{J}$ will be also valid in any model of the set Δ (see 8.4.9) . Therefore $\Delta \models_L \Phi$ and by Completeness Theorem for L^1 (Theorem 8.5.30) $\Delta \vdash_L \Phi$. □

Corollary 10.1.24. Let L be an FPJ logic, D be a σ-vocabulary, $\Delta \in \mathbf{SD}_\sigma(L, D)$ and Γ be an L^1-consistent set of external L-formulas, $\Delta \subseteq \Gamma$. Then

$$\Gamma \vdash_L \Phi \quad \text{iff} \quad \Delta \vdash_{\mathbf{St}} \Phi.$$

Proof. ($\Rightarrow$) Let $\Gamma \vdash_L \Phi$. By the assumption, Γ is L^1-consistent. Therefore it has an L^1-model. Since $\Delta \subseteq \Gamma$, any L^1-model of Γ is at the same time an L^1-model of Δ. By Proposition 10.1.22, all models of Δ are isomorphic to $\mathfrak{J} = \mathrm{Sint}_L(\Delta, D)$. Hence $\mathfrak{J} \models_L \Phi$, whence by Proposition 10.1.23 we obtain $\Delta \vdash_L \Phi$.

($\Leftarrow$) It is evident. □

10.2 Inference

The most important syntactic construction in our formalism is the type of inferences that will be appropriate to imitate the cognitive process. In this section we introduce all the syntactic notions necessary for defining the required notion of inference. The first series of the necessary notions to be introduced is as follows: *modification rule, modification rule system, modification calculus*. The latter is the most complex of these. In our formalism it is an *approximate* analogue of calculus or theory. However, as has already been said, although a modification calculus is a pure syntactic object, it still implies a reference to the canonical model of knowledge state.

Definition 10.2.1. Let L be an FPJ-logic, σ be a signature, $p \in \mathrm{Pr}_\sigma(\mathsf{I})$,

$$\vec{x} \in \mathrm{Var}_\sigma{}^\diamond \cap \mathrm{Dom}_\sigma(p),$$

and $\varepsilon \in \mathscr{V}(L, \mathsf{I})$. Then the ordered triple

$$M = \langle \mathrm{J}_\tau\, p(\vec{x}), \Phi; \mathrm{J}_\varepsilon\, p(\vec{x})\rangle,$$

where Φ is an external L-formula such that $\mathrm{FVar}(\Phi) \subseteq \mathrm{Set}(\vec{x})$, is called a *modification ε-rule for the predicate symbol p.*

The formula $\mathrm{J}_\tau\, p(\vec{x})$ is called the *minor premise of the rule M*, Φ the *major premise* of this rule and $\mathrm{J}_\varepsilon\, p(\vec{x})$ its *conclusion.*

The rule M is written as either

$$\mathrm{J}_\tau\, p(\vec{x}), \Phi \Vdash \mathrm{J}_\varepsilon\, p(\vec{x})$$

or

$$\frac{\mathrm{J}_\tau\, p(\vec{x}), \quad \Phi}{\mathrm{J}_\varepsilon\, p(\vec{x})}.$$

The intuitive meaning of the modification rule is as follows: if in a record indexed by the tuple $\vec{x}$ there occurs the undefined value τ and the condition Φ holds, then the value ε can be written into this record.

Definition 10.2.2. Let L be an FPJ-logic, σ be a signature, $p \in \mathrm{Pr}_\sigma(\mathsf{I})$,

$$\vec{x} \in \mathrm{Var}_\sigma{}^\diamond \cap \mathrm{Dom}_\sigma(p),$$

for any $\varepsilon \in \mathscr{V}(L,\mathsf{I})$ and Φ_ε be an external L-formula such that $\mathrm{FVar}(\Phi_\varepsilon) \subseteq \mathrm{Set}(\vec{x})$. Then any finite set

$$S = \{\mathrm{J}_\tau\, p(\vec{x}), \Phi_\varepsilon \Vdash \mathrm{J}_\varepsilon\, p(\vec{x}) \mid \varepsilon \in \mathscr{V}(L,\mathsf{I})\}$$

is called a *complete modification set* (CMC) for p in L. If

$$\vdash_L \Phi_\varepsilon \to \neg \Phi_\delta$$

for any $\varepsilon, \delta \in \mathscr{V}(L,\mathsf{I})$ such that $\varepsilon \neq \delta$ then the CMC S is called *disjoint*.

Below is the definition of the modification rule system, which is an important constituent of a modification calculi. The modification rule system can be considered as an analogue of a set of own (non-logical) axioms for traditional first-order theories.

Definition 10.2.3. Let L be an FPJ-logic. The collection of sets

$$R = \{S_p \mid p \in \mathrm{Pr}_\sigma(\mathsf{I})\}$$

is called a *modification rule system of signature* σ in L if, for each $p \in \mathrm{Pr}_\sigma(\mathsf{I})$, S_p is a disjoint CMC for p in L.

The set of all modification rule systems of signature σ in L is denoted by $\mathrm{MRS}_\sigma(L)$

Definition 10.2.4. Let L be an FPJ-logic and $R \in \mathrm{MRS}_\sigma(L)$. Clearly for any $p \in \mathrm{Pr}_\sigma(\mathsf{I})$ and $\varepsilon \in \mathscr{V}(L,\mathsf{I})$ there exists a unique modification ε-rule for the predicate symbol p, which belongs to an element of this collection. The rule, its major premise and its conclusion are denoted by $R[p,\varepsilon]$, $\mathrm{MPr}[R,p,\varepsilon]$ and $\mathrm{Con}[R,p,\varepsilon]$, respectively.

The set $\{R[p,\varepsilon] \mid \varepsilon \in \mathbf{ITS}\}$ is called the *p-projection of* R and is denoted by $R[p]$.

Definition 10.2.5. Let L be an FPJ logic, $R \in \mathrm{MRS}_\sigma(L)$, $p \in \mathrm{Pr}_\sigma(\mathsf{I})$, $\varepsilon \in \mathscr{V}(L,\mathsf{I})$. Suppose that the rule $R[p,\varepsilon]$ is of the form

$$\mathrm{J}_\tau\, p(\vec{x}), \Phi(\vec{x}) \Vdash \mathrm{J}_\varepsilon\, p(\vec{x}),$$

where $\vec{x} \in \mathrm{Var}_\sigma{}^\diamond \cap \mathrm{Dom}_\sigma(p)$. Let $\vec{t} \in \mathrm{Dom}_\sigma(p)$. Then we say that the formula $\mathrm{J}_\varepsilon\, p(\vec{t})$ *is obtained from* $\mathrm{J}_\tau\, p(\vec{t})$ *and* $\Phi(\vec{t})$ *by the rule* $R[p,\varepsilon]$.

Now we turn to the definition of the notion of modification calculus. As was mentioned above, a modification calculus can be considered as an *approximate* analogue of a calculus or a theory where we emphasise the approximateness.

A modification calculus consists of four components. The first one is a modification rule system, which is an analogue of a set of non-logical axioms. The next two components are the universe and the state description over this universe; these uniquely define the canonical model of the initial state of knowledge. We can also say that these components implicitly define the standard semantics for the modification calculus. The state description can also be considered as a collection of additional non-logical axioms. Note that a state description includes the *diagram* (see e.g. [Nourani, 1991]) of the canonical model. The last component of modification calculi defines the ordering on the finite set of internal predicate symbols. This component defines the strategy of the cognitive process.

Definition 10.2.6. Let L be an FPJ logic and σ be a signature. Suppose that the ordered quadruple

$$\mathscr{M} = \langle R, D, \Delta, O \rangle$$

satisfies the following conditions:

(i) $R \in \mathrm{MRS}_\sigma(L)$,
(ii) D is a σ-vocabulary,
(iii) $\Delta \in \mathbf{SD}_\sigma(L, D)$,
(iv) O is a map of the segment $[1 \mathinner{..} Q]$ to $\mathrm{Pr}_\sigma(\mathsf{I})$, where $Q \geq 1$; i.e. the map O forms a finite sequence (string) of internal predicate symbols of signature σ.

Then the quadruple $\mathscr{M} = \langle R, D, \Delta, O \rangle$ is called a *modification calculus of signature* σ in L.

The set of all modification calculi of signature σ in L is denoted by $\mathrm{MC}_\sigma(L)$.

Let us introduce notation for the components of the modification calculus $\mathscr{M}$. The modification rule system R is denoted by $\mathrm{MR}_{\mathscr{M}}$. The set of constants D is denoted by $\mathrm{Univ}_{\mathscr{M}}$, the state description Δ by $\mathrm{Descr}_{\mathscr{M}}$ and the map O by $\mathrm{Ord}_{\mathscr{M}}$. This map establishes an order of traversal for some string of internal predicate symbols.

For brevity let us write $\mathrm{P}_i^{\mathscr{M}}$ instead of $\mathrm{Ord}_{\mathscr{M}}(i)$, i.e. by $\mathrm{P}_i^{\mathscr{M}}$ we denote the predicate symbol $p \in \mathrm{Pr}_\sigma(\mathsf{I})$ that is the i-th member of the sequence formed by the map $\mathrm{Ord}_{\mathscr{M}}$. The length of this sequence (which is equal to Q from the segment $[1 \mathinner{..} Q]$) is denoted by $\mathrm{Q}_{\mathscr{M}}$.

Let us denote the major premise of the rule $\mathrm{MR}_{\mathscr{M}}[p, \varepsilon]$, where $p \in \mathrm{Pr}_\sigma(\mathsf{I})$, $\varepsilon \in \mathscr{V}(L, \mathsf{I})$, by $\mathrm{MPr}_{\mathscr{M}}[p, \varepsilon]$ (or by $\mathrm{MPr}_{\mathscr{M}}[p, \varepsilon](\vec{x})$ if we need to show explicitly what variables the formula $\mathrm{MPr}_{\mathscr{M}}[p, \varepsilon]$ depends on).

Let us denote by $\mathrm{JA}_\sigma(\mathscr{M})$ the set $\mathrm{JA}_\sigma(\mathrm{Descr}_{\mathscr{M}}, \mathrm{Univ}_{\mathscr{M}})$ and by $\mathrm{JA}_\sigma(p, \mathscr{M})$ the set $\mathrm{JA}_\sigma(\mathrm{Descr}_{\mathscr{M}}, p, \mathrm{Univ}_{\mathscr{M}})$, where $p \in \mathrm{Pr}_\sigma(\mathsf{I})$.

Let us also denote by $\mathrm{Dom}_{\mathscr{M}}(p)$ the set $\mathrm{Dom}_\sigma(\mathrm{Univ}_{\mathscr{M}}, p)$ and by $\mathrm{Dom}_{\mathscr{M}}(s)$ the set $\mathrm{Dom}_\sigma(\mathrm{Univ}_{\mathscr{M}}, s)$, where $p \in \mathrm{Pr}_\sigma(\mathsf{I})$, $s \in \mathsf{Sorts}_\sigma$.

Denote by $\mathrm{Sint}_L(\mathscr{M})$ the L-structure $\mathrm{Sint}_L(\mathrm{Descr}_{\mathscr{M}}, \mathrm{Univ}_{\mathscr{M}})$.

Example 10.2.7. Take **Si** as an FPJ logic and σ' from Example 10.1.11 as a signature. Take D and Δ from Example 10.1.11 as a σ'-vocabulary and a state description of signature σ' over D in **Si**, respectively.

Let the map O take each $p \in \mathrm{Pr}_{\sigma'}(\mathsf{I})$ to p's ordinal number in the tuple $\langle \mathrm{C}, \mathrm{F}, \mathrm{A} \rangle$.

Recall that in **Si** the value "uncertain" is denoted by ?, i.e. we use '?' instead of 'τ' in this example. Recall that in **Si** we denote $\mathrm{J}_?$ and $\mathrm{J}_!$ by '?' and '!', respectively.

Let S_C be the set of the following modification rules:

$$\frac{?\mathrm{C}(e_1,e_2),\ \exists d_1 \exists d_2 (d_1 \neq d_2 \wedge !\mathrm{F}(d_1,e_1,e_2) \wedge !\mathrm{F}(d_2,e_1,e_2))}{!\mathrm{C}(e_1,e_2)}, \tag{10.2}$$

$$\frac{?\mathrm{C}(e_1,e_2),\ \neg(\exists d_1 \exists d_2 (d_1 \neq d_2 \wedge !\mathrm{F}(d_1,e_1,e_2) \wedge !\mathrm{F}(d_2,e_1,e_2)))}{?\mathrm{C}(e_1,e_2)}. \tag{10.3}$$

Let S_{F} be the set of the following modification rules:

$$\frac{?\mathrm{F}(d,e_1,e_2),\ !\mathrm{A}(d,e_1) \wedge !\mathrm{C}(e_1,e_2)}{!\mathrm{F}(d,e_1,e_2)}, \tag{10.4}$$

$$\frac{?\mathrm{F}(d,e_1,e_2),\ \neg(!\mathrm{A}(d,e_1) \wedge !\mathrm{C}(e_1,e_2))}{?\mathrm{F}(d,e_1,e_2)}. \tag{10.5}$$

Let S_{A} be the set of the following modification rules:

$$\frac{?\mathrm{A}(d,e_2),\ !\mathrm{A}(d,e_1) \wedge !\mathrm{C}(e_1,e_2)}{!\mathrm{A}(d,e_2)}, \tag{10.6}$$

$$\frac{?\mathrm{A}(d,e_2),\ \neg(!\mathrm{A}(d,e_1) \wedge !\mathrm{C}(e_1,e_2))}{?\mathrm{A}(d,e_2)}. \tag{10.7}$$

Let

$$R = \{S_{\mathrm{C}}, S_{\mathrm{F}}, S_{\mathrm{A}}\} \quad \text{and} \quad \mathscr{M} = \langle R, D, \Delta, O \rangle .$$

Then it is easy to prove that $R \in \mathrm{MRS}_{\sigma'}(\mathbf{Si})$ and $\mathscr{M}$ is a modification calculus of signature σ' in **Si**.

Below we consider such formal constructions, which we will call inference in the modification calculus. These constructions will be defined similarly to inferences in the usual Hilbert-type calculi. A peculiarity of the introduced formalism is that it allows reconsideration of the already proved statements.

An inference will be defined as a string of *records*. Each record of an inference is a tuple that includes the formula, the reason for the inclusion of this record in the inference and some additional information.

For an inference, it is important to know the reason why its records are included. Now we introduce a formal notation that indicates the reason for such inclusion.

Definition 10.2.8. Let L be an FPJ logic, σ be a signature, i, j, k be any positive integers, $p \in \mathrm{Pr}_\sigma(\mathsf{I})$, $\varepsilon \in \mathscr{V}(L,\mathsf{I})$, and Hyp, LAx, DAx, OAx, $\mathrm{MP}(i,j)$, $\mathrm{Gen}(i)$, $\mathrm{MR}[p,\varepsilon](j,k)$ and $\mathrm{Conf}(i)$ be formal expressions any, of which is called a *reason*. Each reason indicates why a record is included into an inference. Namely

- Hyp denotes that a formula from a record is a hypothesis,

- LAx denotes that a formula from a record is a logical axiom,
- DAx means that a formula from a record is an axiom that belongs to the set $\mathrm{JA}_\sigma(\mathscr{M})$ for some $\mathscr{M} \in \mathrm{MC}_\sigma(L)$,
- OAx means that a formula from a record is an axiom that belongs to the set $\mathrm{Descr}_{\mathscr{M}} \setminus \mathrm{JA}_\sigma(\mathscr{M})$ for some $\mathscr{M} \in \mathrm{MC}_\sigma(L)$ (any other axiom).
- $\mathrm{MP}(i,j)$ means that a formula from a record of an inference is obtained by the *modus ponens* rule (MP) from the formulas of the records with indices i and j in this inference.
- $\mathrm{Gen}(i)$ means that a formula from a record of an inference is obtained by the generalisation rule (Gen) from the formula of the record with index i in this inference.
- $\mathrm{MR}[p,\varepsilon](j,k)$ means that a formula from a record of an inference is obtained from the formulas of the records with indices i and j by the rule, where $\mathscr{M} \in \mathrm{MC}_\sigma(L)$, with respect to which the inference is considered.
- $\mathrm{Conf}(i)$ means that a formula from a record of an inference is obtained from a formula contained in the record with index i by the confirmation rule. This rule simply transfers a J-atomic formula without any changes onto the next stage of the inference.

Definition 10.2.9. The expressions Hyp, LAx, DAx, OAx, MP, Gen, MR and Conf are called *reason types*. The *reason type of r* is denoted by $\mathrm{ToR}(r)$. By definition let

$$\mathrm{ToR}(r) = \begin{cases} \mathrm{MP} & \text{if } r = \mathrm{MP}(i,j) \text{ for some } i,j, \\ \mathrm{Gen} & \text{if } r = \mathrm{Gen}(i) \text{ for some } i, \\ \mathrm{Conf} & \text{if } r = \mathrm{Conf}(i) \text{ for some } i, \\ \mathrm{MR} & \text{if } r = \mathrm{MR}[p,\varepsilon](j,k) \text{ for some } p,\varepsilon,j,k, \\ r & \text{otherwise.} \end{cases}$$

The *extended reason type of r* is denoted by $\mathrm{ERT}(r)$. By definition put

$$\mathrm{ERT}(r) = \begin{cases} \mathrm{MR}[p,\varepsilon] & \text{if } r = \mathrm{MR}[p,\varepsilon](j,k) \text{ for some } p,\varepsilon,j,k, \\ \mathrm{ToR}(r) & \text{otherwise.} \end{cases}$$

Below is the definition of the main structural component of an inference. From now on we will stick to the term *record.* Note that we do not regard a record as indivisible; it has its own structure that contains five components as defined below.

Definition 10.2.10. A *record* of signature σ in L, where L is an FPJ logic, is an ordered quintuple

$$\mathfrak{a} = \langle \Phi, r, s, m, h \rangle,$$

where:

- Φ is an external L-formula of signature σ, which we call *the formula of the record* $\mathfrak{a}$ and denote by $\mathscr{F}(\mathfrak{a})$,

- r is a reason, which we call *the reason of the record* $\mathfrak{a}$ and denote by Reason$(\mathfrak{a})$,
- s, m, h are natural numbers interpreted as follows:
  - s is the index of the stage at which the record is added to the inference. This stage is called the *stage of the record* $\mathfrak{a}$ and is denoted by Stage$(\mathfrak{a})$.
  - m is the index of the module (substage) at which the given record is added to the inference. This index will be equal to the index of the predicate symbol processed while the given record is being added. We call m the *module of the record* $\mathfrak{a}$ and denote it by Module$(\mathfrak{a})$.
  - h is the index of the record that deactivates (hides) the given record $\mathfrak{a}$; h is denoted by DNum$(\mathfrak{a})$.

The ordered pair $\langle s, m\rangle$ is called *rank* of the record $\mathfrak{a}$ and is denoted by Rank$(\mathfrak{a})$.

Definition 10.2.11. A record $\mathfrak{a}$ is called *active* if DNum$(\mathfrak{a}) = 0$. Otherwise $\mathfrak{a}$ is said to be *inactive*. For a record string $\vec{\mathfrak{a}}$ let by definition

$$\mathrm{ASet}(\vec{\mathfrak{a}}) = \{\mathfrak{a} \in \mathrm{Set}(\vec{\mathfrak{a}}) \mid \mathrm{DNum}(\mathfrak{a}) = 0\}.$$

ASet$(\vec{\mathfrak{a}})$ is the set of active records of the string $\vec{\mathfrak{a}}$. Often it will be important to know the stage and module of a record. Therefore let:

$$\mathrm{ASet}(\vec{\mathfrak{a}}, s) = \{\mathfrak{a} \in \mathrm{Set}(\vec{\mathfrak{a}}) \mid \mathrm{DNum}(\mathfrak{a}) = 0, \mathrm{Stage}(\mathfrak{a}) = s\}.$$

Similarly we can define ASet$(\vec{\mathfrak{a}}, s, m)$, where s is a stage index and m is a module index.

Sometimes it is more convenient to use the index set instead of the set of records of a string. Let us introduce the following notation:

$$\mathrm{AIS}(\vec{\mathfrak{a}}) = \{i \in \mathrm{IS}(\vec{\mathfrak{a}}) \mid \mathrm{DNum}(\mathfrak{a}[i]) = 0\}.$$

AIS$(\vec{\mathfrak{a}})$ is the index set of active elements of the string $\vec{\mathfrak{a}}$. The set AIS$(\vec{\mathfrak{a}})$ will be called the *active index set of the string* $\vec{\mathfrak{a}}$. Similarly we can define the sets AIS$(\vec{\mathfrak{a}}, s)$ and AIS$(\vec{\mathfrak{a}}, s, m)$ for the stage s and module m, respectively.

Below two imporant types of records, the deductively and the non-deductively grounded ones, will be defined. In the following we often need the record to be grounded non-deductively. Formulas of non-deductively grounded records describe the current state of the fact base, also including those facts that can be interpreted as knowledge. The transfer to the next stage of an inference is possible only after having checked the entire fact base.

As for the deductively grounded records, they are necessary for the definition of the correspondence between the inferences in modification calculi and proofs in the traditional logical calculi. This correspondence will be studied in Sect. 11.4.

Definition 10.2.12. A record $\mathfrak{a}$ of signature σ is called *deductively grounded*, if ToR(Reason$(\mathfrak{a})$) takes one of the following values: Hyp, OAx, LAx, MP or Gen. If otherwise, i.e., ToR(Reason$(\mathfrak{a})$)takes the value DAx, MR or Conf, the record is called *non-deductively grounded*.

The set of deductively grounded active records of the string $\vec{\mathfrak{a}}$ will be denoted by DASet$(\vec{\mathfrak{a}})$, and the set of non-deductively grounded active records by NASet$(\vec{\mathfrak{a}})$. Let, by definition,

$$\mathrm{DASet}(\vec{\mathfrak{a}}, s) = \{\mathfrak{c} \in \mathrm{DASet}(\vec{\mathfrak{a}}) \mid \mathrm{Stage}(\mathfrak{c}) = s\},$$

$$\mathrm{NASet}(\vec{\mathfrak{a}}, s) = \{\mathfrak{c} \in \mathrm{NASet}(\vec{\mathfrak{a}}) \mid \mathrm{Stage}(\mathfrak{c}) = s\}.$$

The sets DASet$(\vec{\mathfrak{a}}, s)$ and NASet$(\vec{\mathfrak{a}}, s)$ can be defined similarly for the stage s and the module m.

Let us introduce notations for the index set of active deductively (non-deductively) grounded elements:

$$\mathrm{DAIS}(\vec{\mathfrak{a}}) = \{i \in \mathrm{AIS}(\vec{\mathfrak{a}}) \mid \mathfrak{a}[i] \text{ is deductively grounded}\},$$

$$\mathrm{NAIS}(\vec{\mathfrak{a}}) = \{i \in \mathrm{AIS}(\vec{\mathfrak{a}}) \mid \mathfrak{a}[i] \text{ is non-deductively grounded}\}.$$

Similarly we can define the sets

$$\mathrm{DAIS}(\vec{\mathfrak{a}}, s),\ \mathrm{NAIS}(\vec{\mathfrak{a}}, s),\ \mathrm{DAIS}(\vec{\mathfrak{a}}, s, m) \text{ and } \mathrm{NAIS}(\vec{\mathfrak{a}}, s, m)$$

for the stage s and the module m.

Definition 10.2.13. Often it is more convenient to consider a set of formulas occurring in a record string instead of sets of records or indices. Moreover, we will only be interested in the active records of the string. Let, by definition,

$$\mathrm{FSet}(\vec{\mathfrak{a}}) = \{\mathscr{F}(\mathfrak{a}) \mid \mathfrak{a} \in \mathrm{ASet}(\vec{\mathfrak{a}})\},$$

$$\mathrm{DFSet}(\vec{\mathfrak{a}}) = \{\mathscr{F}(\mathfrak{a}) \mid \mathfrak{a} \in \mathrm{DASet}(\vec{\mathfrak{a}})\},$$

$$\mathrm{NFSet}(\vec{\mathfrak{a}}) = \{\mathscr{F}(\mathfrak{a}) \mid \mathfrak{a} \in \mathrm{NASet}(\vec{\mathfrak{a}})\}.$$

FSet$(\vec{\mathfrak{a}})$ is called the *formula set* of the string $\vec{\mathfrak{a}}$. DFSet$(\vec{\mathfrak{a}})$ is called the *deductive*, and NFSet$(\vec{\mathfrak{a}})$ the *non-deductive* formula set of the string $\vec{\mathfrak{a}}$. Similarly to the two previous definitions we can define FSet$(\vec{\mathfrak{a}}, s)$, DFSet$(\vec{\mathfrak{a}}, s)$, NFSet$(\vec{\mathfrak{a}}, s)$, FSet$(\vec{\mathfrak{a}}, s, m)$, DFSet$(\vec{\mathfrak{a}}, s, m)$ and NFSet$(\vec{\mathfrak{a}}, s, m)$. Let us consider an example. Let by definition

$$\mathrm{NFSet}(\vec{\mathfrak{a}}, s) = \{\mathscr{F}(\mathfrak{a}) \mid \mathfrak{a} \in \mathrm{NASet}(\vec{\mathfrak{a}}, s)\}.$$

Definition 10.2.14. Let L be an FPJ logic, σ be a signature, D be a σ-vocabulary and $\vec{\mathfrak{a}}$ be a record string of signature σ. Let by definition:

$$\mathrm{JA}_\sigma(\vec{\mathfrak{a}}, D) = \mathrm{JA}_\sigma(\mathrm{NFSet}(\vec{\mathfrak{a}}), D)$$

(see 10.1.6, 10.2.13).

Clearly, $\mathrm{JA}_\sigma(\vec{\mathfrak{a}}, D)$ can be defined by induction as follows

(i) $\mathrm{JA}_\sigma(\mathbf{e}, D) = \emptyset$,

$$\text{(ii) } \mathrm{JA}_{\sigma}(\vec{\mathfrak{a}}+\mathfrak{b},D)=\begin{cases}\mathrm{JA}_{\sigma}(\vec{\mathfrak{a}},D)+\mathscr{F}(\mathfrak{b}) & \text{if } \mathscr{F}(\mathfrak{b})\in\mathrm{JA}_{\sigma}(D) \text{ and}\\ & \mathfrak{b} \text{ is non-deductively grounded,}\\ \mathrm{JA}_{\sigma}(\vec{\mathfrak{a}},D) & \text{otherwise.}\end{cases}$$

This inductive definition is convenient to use in proofs by induction on the length of record strings.

Let $\mathscr{M}\in\mathrm{MC}_{\sigma}(L)$. By definition put

$$\mathrm{JA}_{\sigma}(\vec{\mathfrak{a}},\mathscr{M})=\mathrm{JA}_{\sigma}(\vec{\mathfrak{a}},\mathrm{Univ}_{\mathscr{M}}).$$

Definition 10.2.15. Let L be an FPJ logic, σ be a signature, D be a σ-vocabulary and $\vec{\mathfrak{a}}$ be a record string of signature σ, $p\in\mathrm{Pr}_{\sigma}(\mathsf{I})$. Let by definition

$$\mathrm{JA}_{\sigma}(\vec{\mathfrak{a}},p,D)=\mathrm{JA}_{\sigma}(\mathrm{NFSet}(\vec{\mathfrak{a}}),p,D)$$

(see 10.1.6 and 10.2.13).

Clearly, $\mathrm{JA}_{\sigma}(\vec{\mathfrak{a}},p,D)$ can be inductively defined similarly to the inductive version of Definition 10.2.14.

It is also clear that

$$\mathrm{JA}_{\sigma}(\vec{\mathfrak{a}},D)=\bigcup_{p\in\mathrm{Pr}_{\sigma}(\mathsf{I})}\mathrm{JA}_{\sigma}(\vec{\mathfrak{a}},p,D),$$

and $\mathrm{J}_{\varepsilon}\,p(\vec{\mathfrak{a}})\in\mathrm{JA}_{\sigma}(\vec{\mathfrak{a}},D)$ if and only if $\mathrm{J}_{\varepsilon}\,p(\vec{\mathfrak{a}})\in\mathrm{JA}_{\sigma}(\vec{\mathfrak{a}},p,D)$.

Let $\mathscr{M}\in\mathrm{MC}_{\sigma}(L)$. By definition put

$$\mathrm{JA}_{\sigma}(\vec{\mathfrak{a}},p,\mathscr{M})=\mathrm{JA}_{\sigma}(\vec{\mathfrak{a}},p,\mathrm{Univ}_{\mathscr{M}}).$$

Definition 10.2.16. Let L be an FPJ logic, σ be a signature, D be a σ-vocabulary, $\vec{\mathfrak{a}}$ be a record string of signature σ and s be a natural number. Let by definition:

$$\mathrm{JA}_{\sigma}(\vec{\mathfrak{a}},D,s)=\mathrm{JA}_{\sigma}(\mathrm{NFSet}(\vec{\mathfrak{a}},s),D)$$

(see 10.1.6 and 10.2.13).

The set $\mathrm{JA}_{\sigma}(\vec{\mathfrak{a}},p,D,s)$ can be defined similarly.

It is clear that inductive definitions of $\mathrm{JA}_{\sigma}(\vec{\mathfrak{a}},D,s)$ and $\mathrm{JA}_{\sigma}(\vec{\mathfrak{a}},p,D,s)$ can be given similarly to the definition of $\mathrm{JA}_{\sigma}(\vec{\mathfrak{a}},D)$.

Let $\mathscr{M}\in\mathrm{MC}_{\sigma}(L)$. By definition put

$$\mathrm{JA}_{\sigma}(\vec{\mathfrak{a}},\mathscr{M},s)=\mathrm{JA}_{\sigma}(\vec{\mathfrak{a}},\mathrm{Univ}_{\mathscr{M}},s).$$

The set $\mathrm{JA}_{\sigma}(\vec{\mathfrak{a}},p,\mathscr{M},s)$ can be defined similarly.

Definition 10.2.17. Let L be an FPJ logic, σ be a signature, D be a σ-vocabulary and $\vec{\mathfrak{a}}$ be a record string of signature σ, $p \in \mathrm{Pr}_\sigma(\mathsf{I})$. Let by definition

$$\mathrm{Dom}_\sigma(D,p,\vec{\mathfrak{a}}) = \{\vec{\mathfrak{a}} \in \mathrm{Dom}_\sigma(D,p) \mid \mathrm{J}_\varepsilon\, p(\vec{\mathfrak{a}}) \in \mathrm{JA}_\sigma(\vec{\mathfrak{a}},D) \text{ for some } \varepsilon \in \mathrm{ITS}\}.$$

$\mathrm{Dom}_\sigma(D,p,\vec{\mathfrak{a}})$ is called the *D-domain of p in* $\vec{\mathfrak{a}}$.

Let $\mathscr{M} \in \mathrm{MC}_\sigma(L)$. By definition put

$$\mathrm{Dom}_{\mathscr{M}}(p,\vec{\mathfrak{a}}) = \mathrm{Dom}_\sigma(\mathrm{Univ}_{\mathscr{M}},p,\vec{\mathfrak{a}}).$$

Definition 10.2.18. Let L be an FPJ logic, σ be a signature, D be a σ-vocabulary, $\vec{\mathfrak{a}}$ be a record string of signature σ and s be a natural number, $p \in \mathrm{Pr}_\sigma(\mathsf{I})$. Let by definition

$$\mathrm{Dom}_\sigma(D,p,\vec{\mathfrak{a}},s) = \{\vec{\mathfrak{a}} \in \mathrm{Dom}_\sigma(D,p) \mid \mathrm{J}_\varepsilon\, p(\vec{\mathfrak{a}}) \in \mathrm{JA}_\sigma(\vec{\mathfrak{a}},D,s) \text{ for some } \varepsilon \in \mathscr{V}(L,\mathsf{I})\}. \quad (10.8)$$

$\mathrm{Dom}_\sigma(D,p,\vec{\mathfrak{a}},s)$ is called the *D-domain of p in* $\vec{\mathfrak{a}}$ *at the stage s*.

Let $\mathscr{M} \in \mathrm{MC}_\sigma(L)$. By definition put

$$\mathrm{Dom}_{\mathscr{M}}(p,\vec{\mathfrak{a}},s) = \mathrm{Dom}_\sigma(\mathrm{Univ}_{\mathscr{M}},p,\vec{\mathfrak{a}},s).$$

Lemma 10.2.19. Let D be a σ-vocabulary, $\vec{\mathfrak{a}}$, $\vec{\mathfrak{b}}$ be record strings of signature σ and s be a natural number, $p \in \mathrm{Pr}_\sigma(\mathsf{I})$. Then:

(i) $\mathrm{JA}_\sigma(\vec{\mathfrak{a}},D,s) \subseteq \mathrm{JA}_\sigma(\vec{\mathfrak{a}},D)$,
(ii) $\mathrm{Dom}_\sigma(D,p,\vec{\mathfrak{a}},s) \subseteq \mathrm{Dom}_\sigma(D,p,\vec{\mathfrak{a}})$,
(iii) $\mathrm{JA}_\sigma(\vec{\mathfrak{a}},D) \subseteq \mathrm{JA}_\sigma\left(\vec{\mathfrak{a}}+\vec{\mathfrak{b}},D\right)$,
(iv) $\mathrm{JA}_\sigma(\vec{\mathfrak{a}},p,D) \subseteq \mathrm{JA}_\sigma\left(\vec{\mathfrak{a}}+\vec{\mathfrak{b}},p,D\right)$
(v) $\mathrm{Dom}_\sigma(D,p,\vec{\mathfrak{a}}) \subseteq \mathrm{Dom}_\sigma\left(D,p,\vec{\mathfrak{a}}+\vec{\mathfrak{b}}\right)$,
(vi) $\mathrm{JA}_\sigma(\vec{\mathfrak{a}},D,s) \subseteq \mathrm{JA}_\sigma\left(\vec{\mathfrak{a}}+\vec{\mathfrak{b}},D,s\right)$,
(vii) $\mathrm{Dom}_\sigma(D,p,\vec{\mathfrak{a}},s) \subseteq \mathrm{Dom}_\sigma\left(D,p,\vec{\mathfrak{a}}+\vec{\mathfrak{b}},s\right)$.

Proof. Clearly, $\mathrm{ASet}(\vec{\mathfrak{a}},s) \subseteq \mathrm{ASet}(\vec{\mathfrak{a}})$. From this by Definitions 10.2.14 and 10.2.16 we obtain (ii).

It is also clear that $\mathrm{ASet}(\vec{\mathfrak{a}}) \subseteq \mathrm{ASet}\left(\vec{\mathfrak{a}}+\vec{\mathfrak{b}}\right)$. Then by 10.2.14 and 10.2.16 we obtain (iii) and (iv), but from (iii) by Definitions 10.2.17 and 10.2.18 we obtain (v).

Statements (vi) and (vii) can be considered similarly □

Corollary 10.2.20. Let D be a σ-vocabulary, $\vec{\mathfrak{a}}$, $\vec{\mathfrak{b}}$ be record strings of signature σ and s be a natural number, $p \in \mathrm{Pr}_\sigma(\mathsf{I})$. Suppose that

$$\mathrm{Dom}_\sigma(D,p,\vec{\mathfrak{a}}) = \mathrm{Dom}_\sigma(D,p).$$

Then

$$\mathrm{Dom}_\sigma\left(D,p,\vec{\mathfrak{a}}+\vec{\mathfrak{b}}\right) = \mathrm{Dom}_\sigma(D,p).$$

Proof. Clearly,

$$\begin{aligned}
\mathrm{Dom}_\sigma(D,p) &= \mathrm{Dom}_\sigma(D,p,\vec{\mathfrak{a}}) && \text{(by hypothesis)}\\
&\subseteq \mathrm{Dom}_\sigma\left(D,p,\vec{\mathfrak{a}}+\vec{\mathfrak{b}}\right) && \text{(by 10.2.19(v))}\\
&\subseteq \mathrm{Dom}_\sigma(D,p) && \text{(by 10.2.17)}
\end{aligned}$$

□

Definition 10.2.21. Let σ be a signature, $\vec{\mathfrak{a}}$ be a record atring of signature σ and Φ be an external L-formula of the same signature. Let us inductively define the records of $\vec{\mathfrak{a}}$ that *deductively depend on* Φ *in* $\vec{\mathfrak{a}}$.

BASIS. If $a \in \mathrm{ASet}(\vec{\mathfrak{a}})$ and $\mathscr{F}(a) = \Phi$, then $\mathfrak{a}$ *deductively depends on* Φ *in* $\vec{\mathfrak{a}}$.

INDUCTION STEP.

(i) Let:

- $i, j, k \in \mathrm{AIS}(\vec{\mathfrak{a}})$,
- $i < k, \quad j < k$,
- $\mathscr{F}(\vec{\mathfrak{a}}[j]) = (\mathscr{F}(\vec{\mathfrak{a}}[i]) \to \mathscr{F}(\vec{\mathfrak{a}}[k]))$,
- $\mathrm{Reason}(\vec{\mathfrak{a}}[k]) = \mathrm{MP}(i,j)$,
- At least one of the records $\vec{\mathfrak{a}}[i]$, $\vec{\mathfrak{a}}[j]$ *deductively depends on* Φ *in* $\vec{\mathfrak{a}}$.

Then $\vec{\mathfrak{a}}[k]$ *is deductively depends on* Φ *in* $\vec{\mathfrak{a}}$.

(ii) Let:

- $i, k \in \mathrm{AIS}(\vec{\mathfrak{a}})$,
- $i < k$,
- $\mathscr{F}(\vec{\mathfrak{a}}[k]) = \forall x \mathscr{F}(\vec{\mathfrak{a}}[i])$,
- $\mathrm{Reason}(\vec{\mathfrak{a}}[k]) = \mathrm{Gen}(i)$,
- The record $\vec{\mathfrak{a}}[i]$ *deductively depends on* Φ *in* $\vec{\mathfrak{a}}$.

Then $\vec{\mathfrak{a}}[k]$ *deductively depends on* Φ *in* $\vec{\mathfrak{a}}$.

If a record does not depend deductively on the formula Φ in the string $\vec{\mathfrak{a}}$, *then it is called deductively* Φ*-free in* $\vec{\mathfrak{a}}$.

Proposition 10.2.22. Let $\mathfrak{c}$ be a record, $\vec{\mathfrak{a}}$ be a record string and Φ be an external L-formula. It is clear that $\mathfrak{c}$ deductively depends on Φ in $\vec{\mathfrak{a}}$ if and only if $\mathfrak{c} = \vec{\mathfrak{a}}[k]$ for some $k \in \mathrm{AIS}(\vec{\mathfrak{a}})$ and at least one of the following conditions hold:

(i) $\mathscr{F}(\mathfrak{c}) = \Phi$;

(ii) Reason $(\mathfrak{c}) = \mathrm{MP}(i, j)$ and there exist $i, j \in \mathrm{AIS}(\vec{\mathfrak{a}})$ such that $i < k$, $j < k$; $\mathscr{F}(\mathfrak{c})$ follows from $\mathscr{F}(\vec{\mathfrak{a}}[i])$ and $\mathscr{F}(\vec{\mathfrak{a}}[j])$ by the *modus ponens* rule and at least one of the records $\vec{\mathfrak{a}}[i]$, $\vec{\mathfrak{a}}[j]$ deductively depends on Φ in $\vec{\mathfrak{a}}$;

(iii) Reason $(\mathfrak{c}) = \mathrm{Gen}(i)$ and there exists an $i \in \mathrm{AIS}(\vec{\mathfrak{a}})$ such that $i < k$; $\mathscr{F}(c)$ follows from $\mathscr{F}(\vec{\mathfrak{a}}[i])$ by the generalisation rule and $\vec{\mathfrak{a}}[i]$ deductively depends on Φ in $\vec{\mathfrak{a}}$.

Now two important operations named Deactivate and Activate will be defined on the records. The application of the operation Deactivate on a record makes it invisible for the inference rules. The operation Activate makes the record visible again.

Definition 10.2.23. Let

$$\mathfrak{a} = \langle \Phi, r, s, m, h \rangle$$

be a record. Then Deactivate $(\mathfrak{a}, k)$, where k is a natural number, denotes the record $\langle \Phi, r, s, m, k \rangle$, and Activate $(\mathfrak{a})$ denotes the record $\langle \Phi, r, s, m, 0 \rangle$. That is, the function Deactivate establishes the deactivation index (and thus deactivates the record), and the function Activate makes the record active.

We will now define a very important operation named Suspend on the record strings. This operation allows the deactivation of all the records that deductively depend on a certain formula. This is the operation that makes our inference dynamic and that permits the minimisation of the risk of appearance of contradictions.

Definition 10.2.24. Let σ be a signature, $\vec{\mathfrak{a}}$ be a finite record string of signature σ and Φ be an external L-formula of the same signature. Suspend $(\vec{\mathfrak{a}}, \Phi, k)$ denotes a string $\vec{\mathfrak{b}}$ such that $\mathrm{Len}\left(\vec{\mathfrak{b}}\right) = \mathrm{Len}(\vec{\mathfrak{a}})$ and for any $i\,(1 \le i \le \mathrm{Len}(\vec{\mathfrak{a}}))$,

$$\vec{\mathfrak{b}}[i] = \begin{cases} \vec{\mathfrak{a}}[i] & \vec{\mathfrak{a}}[i] \text{ is deductively } \Phi\text{-free in } \vec{\mathfrak{a}}, \\ \text{Deactivate}\left(\vec{\mathfrak{a}[i]}, k\right) & \vec{\mathfrak{a}}[i] \text{ deductively depends on } \Phi \text{ in } \vec{\mathfrak{a}}. \end{cases}$$

The operation Suspend makes hidden all those and only those records of $\vec{\mathfrak{a}}$ which deductively depend on Φ and sets the deactivation indices of these records.

We can also give an inductive definition of the function Suspend, which will be more convenient for the proofs.

BASIS. Suspend $(\mathbf{e}, \Phi, k) = \mathbf{e}$.

INDUCTION STEP. Suspend $(\vec{\mathfrak{a}} \bullet \mathfrak{a}, \Phi, k) = $ Suspend $(\vec{\mathfrak{a}}, \Phi, k) \bullet \mathfrak{b}$,
where

$$\mathfrak{b} = \begin{cases} \mathfrak{a} & \mathfrak{a} \text{ is deductively } \Phi\text{-free in } \vec{\mathfrak{a}} \bullet \mathfrak{a}, \\ \text{Deactivate}(\mathfrak{a}, k) & \mathfrak{a} \text{ deductively depends on } \Phi \text{ in } \vec{\mathfrak{a}} \bullet \mathfrak{a}. \end{cases}$$

Lemma 10.2.25. Let σ be a signature, $\vec{\mathfrak{a}}$ and $\vec{\mathfrak{b}}$ be record strings of signature σ and Φ be an external L-formula of the same signature, s, m be natural numbers. Then there exists a record string $\vec{\mathfrak{c}}$ such that

(i) $\mathrm{Len}(\vec{\mathfrak{c}}) = \mathrm{Len}\left(\vec{\mathfrak{b}}\right)$,

(ii) $\mathrm{Suspend}\left(\vec{\mathfrak{a}}+\vec{\mathfrak{b}}, \Phi, k\right) = \mathrm{Suspend}(\vec{\mathfrak{a}}, \Phi, k) + \vec{\mathfrak{c}}$.

Proof. By induction on the length of the string $\vec{\mathfrak{b}}$.

BASIS. Let $\mathrm{Len}\left(\vec{\mathfrak{b}}\right) = 0$. Then for $\vec{\mathfrak{c}} = \mathbf{e}$ the conditions (i) and (ii) hold.

INDUCTION STEP. Let the statement be true if the length of $\vec{\mathfrak{b}}$ is n. Suppose that the length of $\vec{\mathfrak{b}}$ is $n+1$. Then for some record $\mathfrak{d}$, $\vec{\mathfrak{b}} = \vec{\mathfrak{d}} \bullet \mathfrak{d}$, where $\mathrm{Len}\left(\vec{\mathfrak{d}}\right) = n$. By the inductive assumption there exists $\vec{\mathfrak{k}}$ such that

(i) $\mathrm{Len}\left(\vec{\mathfrak{k}}\right) = \mathrm{Len}\left(\vec{\mathfrak{d}}\right) = n$;

(ii) $\mathrm{Suspend}\left(\vec{\mathfrak{a}}+\vec{\mathfrak{d}}, \Phi, k\right) = \mathrm{Suspend}(\vec{\mathfrak{a}}, \Phi, k) + \vec{\mathfrak{k}}$.

Then

$$\begin{aligned}\mathrm{Suspend}\left(\vec{\mathfrak{a}}+\vec{\mathfrak{b}}, \Phi, k\right) &= \mathrm{Suspend}\left(\vec{\mathfrak{a}}+\vec{\mathfrak{d}}\bullet\mathfrak{d}, \Phi, k\right)\\ &= \mathrm{Suspend}\left(\vec{\mathfrak{a}}+\vec{\mathfrak{d}}, \Phi, k\right)\bullet\mathfrak{c}\\ &= \mathrm{Suspend}(\vec{\mathfrak{a}}, \Phi, k) + \vec{\mathfrak{k}}\bullet\mathfrak{c},\end{aligned}$$

where

$$\mathfrak{c} = \begin{cases}\mathfrak{d}, & \mathfrak{d} \text{ is deductively } \Phi\text{-free in } \vec{\mathfrak{a}}+\vec{\mathfrak{b}},\\ \mathrm{Deactivate}(\mathfrak{d}, k), & \mathfrak{d} \text{ deductively depends on } \Phi \text{ in } \vec{\mathfrak{a}}+\vec{\mathfrak{b}}.\end{cases}$$

Let $\vec{\mathfrak{c}} = \vec{\mathfrak{k}}\bullet\mathfrak{c}$. Then, clearly (i) and (ii) hold. □

Lemma 10.2.26. Let σ be a signature, $\vec{\mathfrak{a}}$ be a record string of signature σ, k,s be natural numbers and Φ be an external L-formula of the same signature. Suppose that the following conditions hold:

(i) $\vec{\mathfrak{a}}$ does not contain a record $\mathfrak{c}$, for which simultaneously

$$\mathscr{F}(\mathfrak{c}) = \Phi \quad \text{and} \quad \mathrm{Stage}(\mathfrak{c}) = s,$$

(ii) $\vec{\mathfrak{b}} = \mathrm{Suspend}(\vec{\mathfrak{a}}, \Phi, k)$.

Then for any $p \in \mathrm{Pr}_{\sigma}(\mathsf{I})$

$$\mathrm{Dom}_{\sigma}\left(D, p, \vec{\mathfrak{b}}, s\right) = \mathrm{Dom}_{\sigma}(D, p, \vec{\mathfrak{a}}, s).$$

Proof. The operation Suspend can make inaccessible only those records that are deductively grounded or contain formula Φ. Taking into account the restrictions on the formula Φ given by the conditions we obtain

$$\mathrm{NFSet}\left(\vec{\mathfrak{b}}, s\right) = \mathrm{NFSet}(\vec{\mathfrak{a}}, s).$$

Therefore for any $p \in \mathrm{Pr}_\sigma(\mathsf{I})$

$$\mathrm{Dom}_\sigma\left(D, p, \vec{\mathfrak{b}}, s\right) = \mathrm{Dom}_\sigma(D, p, \vec{a}, s)$$

(see 10.2.18, 10.2.16 and 10.2.13). □

Definition 10.2.27. Let $\mathrm{Len}(\vec{\mathfrak{a}}) + 1$ be the default value for the parameter k. By definition, put

$$\mathrm{Suspend}(\vec{\mathfrak{a}}, \Phi) = \mathrm{Suspend}(\vec{\mathfrak{a}}, \Phi, \mathrm{Len}(\vec{\mathfrak{a}}) + 1).$$

How is an inference built up and how can an element be added to it? To answer this question requires the introduction of a few notions with reasons, on the basis of which new records can be added to an inference.

Definition 10.2.28. Let $\vec{\mathfrak{a}}$ is a record string of signature σ and Φ be an external L-formula of the same signature.

(i) We say that Φ may be added to $\vec{\mathfrak{a}}$ *with the reason* Ax if Φ is an axiom of L^1.
(ii) Let $i, j \in \mathrm{AIS}(\vec{\mathfrak{a}})$. We say that Φ may be added to $\vec{\mathfrak{a}}$ *with the reason* $\mathrm{MP}(i, j)$ if

$$\mathscr{F}(\vec{\mathfrak{a}}[j]) = (\mathscr{F}(\vec{a}[i]) \to \Phi).$$

(iii) Let $i \in \mathrm{AIS}(\vec{\mathfrak{a}})$. We say that Φ may be added to $\vec{\mathfrak{a}}$ *with the reason* $\mathrm{Gen}(i)$ if

$$\Phi = \forall x \mathscr{F}(\vec{\mathfrak{a}}[i])$$

for some $x \in \mathrm{Var}_\sigma$.

Definition 10.2.29. Let L be an FPJ logic, σ be a signature, $\mathscr{M} \in \mathrm{MC}_\sigma(L)$ and Γ be a set of external L-formulas of signature σ. We say that Γ *is approprate to* $\mathscr{M}$ if $\Gamma \subseteq \mathscr{F}_\sigma(L, \mathsf{E})$. Denote by $\mathrm{AFS}(\mathscr{M})$ the collection of all sets of external L-formulas that are appropriate to $\mathscr{M}$.

Definition 10.2.30. Let $\mathscr{M} \in \mathrm{MC}_\sigma(L)$, $\Gamma \in \mathrm{AFS}(\mathscr{M})$, $\vec{\mathfrak{a}}$ be a record string of signature σ and Φ is an external L-formula of the same signature, $\vec{a} \in (\mathrm{Univ}_{\mathscr{M}})^*$, $i \in \mathrm{NAIS}(\vec{\mathfrak{a}})$, $j \in \mathrm{AIS}(\vec{\mathfrak{a}})$.

(i) We say that Φ may be added to $\vec{\mathfrak{a}}$ *with the reason* Hyp w.r.t. Γ, if $\Phi \in \Gamma$.
(ii) We say that Φ may be added to $\vec{\mathfrak{a}}$ *with the reason* OAx w.r.t. $\mathscr{M}$, if

$$\Phi \in \mathrm{Descr}_{\mathscr{M}} \backslash \mathrm{JA}_\sigma(\mathscr{M}).$$

(iii) We say that Φ may be added to $\vec{\mathfrak{a}}$ *with the reason* DAx w.r.t. $\mathscr{M}$ and $\vec{a}$, if there exist $p \in \mathrm{Pr}_\sigma(\mathsf{I})$ and $\varepsilon \in \mathscr{V}(L, \mathsf{I})$ such that

$$\vec{\mathfrak{a}} \in \mathrm{Dom}_\sigma(D, p) \quad \text{and} \quad \Phi = \mathrm{J}_\varepsilon\, p(\vec{a}) \in \mathrm{JA}_\sigma(\mathscr{M}).$$

(iv) We say that Φ may be added to $\vec{\mathfrak{a}}$ *with the reason* $\mathrm{MR}[p, \varepsilon](i, j)$ w.r.t. $\mathscr{M}$ and $\vec{a}$, if there exist $p \in \mathrm{Pr}_\sigma(\mathsf{I})$ and $\varepsilon \in \mathscr{V}(L, \mathsf{I})$ such that:

- $\Phi = \mathrm{J}_{\varepsilon}\, p(\vec{a})$,
- $\mathscr{F}(\vec{\mathfrak{a}}[i]) = \mathrm{J}_{\tau}\, p(\vec{a})$,
- $\mathscr{F}(\vec{\mathfrak{a}}[j]) = \Theta(\vec{a})$, where $\Theta(\vec{x})$ is a major premise of the rule $\mathrm{MR}_{\mathscr{M}}[p,\varepsilon]$.

(v) We say that Φ may be added to $\vec{\mathfrak{a}}$ *with the reason* $\mathrm{Conf}(i)$ w.r.t. $\mathscr{M}$ and $\vec{\mathfrak{a}}$ if there exist $p \in \mathrm{Pr}_{\sigma}(\mathsf{I})$ and $\varepsilon \in \mathscr{D}^{\tau}(L)$ such that

$$\Phi = \mathrm{J}_{\varepsilon}\, p(\vec{\mathfrak{a}}) = \mathscr{F}(\vec{\mathfrak{a}}[i]).$$

Now we are ready to introduce the main notion of the present section, i.e. the notion of inference in modification calculi. Not only can new elements (records) be added to this inference but also the records added to it earlier can be modified.

Definition 10.2.31. Let L be an FPJ logic, σ be a signature, $\mathscr{M} \in \mathrm{MC}_{\sigma}(L)$, $\Gamma \in \mathrm{AFS}(\mathscr{M})$. Below we inductively define the *$\mathscr{M}$-inference from Γ*.

To each $\mathscr{M}$-inference corresponds an ordered triple $\langle s,m,d\rangle$, which is said to be its *rank* (s, m and d are called the *stage*, the *module* and the *regime* of the inference, respectively). For short an $\mathscr{M}$-inference of rank $\langle s,m,d\rangle$ is called an $\mathscr{M}(s,m,d)$-inference. As usual, s and m are natural numbers (the stage index and the module index). There are two regimes of an inference: **D** the deduction regime and **M** the modification regime.

BASIS.

The empty records string is an $\mathscr{M}(s,m,\mathbf{D})$-inference from Γ.

INDUCTION STEP.

For the deductive ground (Ded)

Let $\vec{\mathfrak{a}}$ be an $\mathscr{M}(s,m,\mathbf{D})$-*inference* from Γ and Φ be an external L-formula. Suppose that Φ can be added to $\vec{a}$ with the reason r, where r is one of the following reasons:

- LAx
- Hyp w.r.t. Γ
- OAx w.r.t. $\mathscr{M}$
- $\mathrm{MP}(i,j)$ for some $i,j \in \mathrm{AIS}(\vec{\mathfrak{a}})$
- $\mathrm{Gen}(i)$ for some $i \in \mathrm{AIS}(\vec{\mathfrak{a}})$

Then the string

$$\vec{\mathfrak{a}} \bullet \langle \Phi, r, s, m, 0\rangle$$

is an $\mathscr{M}(s,m,\mathbf{D})$-*inference* from Γ.

For the transfer onto a new stage (NS)

Let $\vec{\mathfrak{a}}$ be an $\mathscr{M}(s,\mathrm{Q}_{\mathscr{M}},d)$-*inference* from Γ. Then $\vec{\mathfrak{a}}$ is an $\mathscr{M}(s+1,m,\mathbf{D})$-*inference* from Γ.

For the non-deductive ground (NDed)

Let $\vec{\mathfrak{a}}$ be an $\mathscr{M}(s,m,d)$-*inference* from Γ. Let

$$\Phi = \mathrm{J}_{\varepsilon}\, p(\vec{a}),$$

where $p \in \mathrm{Pr}_\sigma(\mathsf{I})$, $\varepsilon \in \mathcal{V}(L,\mathsf{I})$, $\vec{a} \in \mathrm{Dom}_\sigma(D,p)$. Suppose that for m and d one and only one of the following conditions holds:

(a) $d = \mathbf{M}$ and $p = \mathrm{P}_m^{\mathcal{M}}$;
(b) $d = \mathbf{D}$, $m < \mathrm{Q}_{\mathcal{M}}$ and $p = \mathrm{P}_{m+1}^{\mathcal{M}}$.

Also suppose that for the stage s and reason r one and only one of the following conditions holds:

(c) $s = 0, r = \mathrm{DAx}$
(d) $s > 0, r = \mathrm{MR}[p,\varepsilon](i,j)$ or $r = \mathrm{Conf}(i)$ for some $i \in \mathrm{NAIS}(\vec{\mathfrak{a}})$, $j \in \mathrm{AIS}(\vec{\mathfrak{a}})$

Let it be that Φ can be added to $\vec{\mathfrak{a}}$ with the reason r w.r.t. $\mathcal{M}$ and $\vec{a}$. Then the string

$$\vec{\mathfrak{b}} \bullet \langle \Phi, r, s, m', 0, 0 \rangle,$$

where

$$\vec{\mathfrak{b}} = \begin{cases} \mathrm{Suspend}(\vec{\mathfrak{a}}, \mathrm{J}_\tau\, p(\vec{a})) & \text{if } r = \mathrm{MR}[p,\varepsilon](i,j) \text{ and } \varepsilon \neq \tau, \\ \vec{\mathfrak{a}} & \text{otherwise,} \end{cases}$$

is an $\mathcal{M}(s,m',d')$-*inference* from Γ, where

$$m' = \begin{cases} m & \text{if (a) is true,} \\ m+1 & \text{if (b) is true,} \end{cases} \qquad d' = \begin{cases} \mathbf{D} & \text{if } \mathrm{Dom}_\sigma(D,p,\vec{\mathfrak{a}},s) \cup \{\vec{a}\} = \mathrm{Dom}_\sigma(D,p), \\ \mathbf{M} & \text{otherwise.} \end{cases}$$

Definition 10.2.32. Let L be an FPJ logic, σ be a signature, $\mathcal{M} \in \mathrm{MC}_\sigma(L)$, $\Gamma \in \mathrm{AFS}(\mathcal{M})$, $s \in \omega$, $m \in [0..\mathrm{Q}_{\mathcal{M}}]$. We say that a record string $\vec{\mathfrak{a}}$ is an $\mathcal{M}(s,m)$-*inference* from Γ if there exists a $d \in \{\mathbf{D},\mathbf{M}\}$ such that $\vec{\mathfrak{a}}$ is an $\mathcal{M}(s,m,d)$-inference from Γ. Similarly, we say that a record string $\vec{\mathfrak{a}}$ is an $\mathcal{M}(s)$-*inference* from Γ if $\vec{\mathfrak{a}}$ is an $\mathcal{M}(s,m)$-inference from Γ for some $m \in [0..\mathrm{Q}_{\mathcal{M}}]$.

Similarly, we can define the notions: $\mathcal{M}(s,,d)$-*inference*, $\mathcal{M}(,m,d)$-*inference*, $\mathcal{M}(,m)$-*inference*, $\mathcal{M}(,d)$-*inference* and $\mathcal{M}$-*inference* from Γ. For example, we say that a record string $\vec{\mathfrak{a}}$ is an $\mathcal{M}(s,,d)$-*inference* from Γ if $\vec{\mathfrak{a}}$ is an $\mathcal{M}(s,m,d)$-inference from Γ for some $m \in [0..\mathrm{Q}_{\mathcal{M}}]$; $\vec{\mathfrak{a}}$ is an $\mathcal{M}$-*inference* from Γ if there exist s, m and d such that $\vec{\mathfrak{a}}$ is an $\mathcal{M}(s,m,d)$-inference.

We say that a record string $\vec{\mathfrak{a}}$ is an $\mathcal{M}(s,m,d)$-*inference* if there exists a $\Gamma \in \mathrm{AFS}(\mathcal{M})$ such that $\vec{\mathfrak{a}}$ is an $\mathcal{M}(s,m,d)$-inference from Γ. Similarly, we can define the notions: $\mathcal{M}(s,m)$-*inference*, $\mathcal{M}(s,,d)$-*inference*, $\mathcal{M}(,m,d)$-*inference*, $\mathcal{M}(s)$-*inference*, $\mathcal{M}(,m)$-*inference*, $\mathcal{M}(,d)$-*inference* and $\mathcal{M}$-*inference*.

Definition 10.2.33. Let L be an FPJ logic, σ be a signature, $\mathcal{M} \in \mathrm{MC}_\sigma(L)$, $s \in \omega$, $m \in [0..\mathrm{Q}_{\mathcal{M}}]$, $d \in \{\mathbf{D},\mathbf{M}\}$. We say that a record string $\vec{\mathfrak{a}}$ is a *pure* $\mathcal{M}(s,m,d)$-*inference* if $\vec{\mathfrak{a}}$ is an $\mathcal{M}(s,m,d)$-inference from $\emptyset$.

Similarly, we can define the notions: *pure* $\mathcal{M}(s,m)$-*inference*, *pure* $\mathcal{M}(s,,d)$-*inference*, *pure* $\mathcal{M}(,m,d)$-*inference*, *pure* $\mathcal{M}(s)$-*inference*, *pure* $\mathcal{M}(,m)$-*inference*, *pure* $\mathcal{M}(,d)$-*inference* and *pure* $\mathcal{M}$-*inference*.

We use the terms $\mathcal{M}(s,m,d)$-*quasiproof*, $\mathcal{M}(s,m)$-*quasiproof* etc. as synonyms for the terms *pure* $\mathcal{M}(s,m,d)$-*inference*, *pure* $\mathcal{M}(s,m)$-*inference* etc.

The notion of pure $\mathscr{M}$-inference (quasiproof) is also important. Primarily we will define the corresponding semantics for the pure $\mathscr{M}$-inference. Many results in the forthcoming investigation will be obtained only for this type of inference. Generally speaking the set $\mathrm{Descr}_{\mathscr{M}}$ contains all the necessary information about the model of the subject domain. Formulas from $\Gamma \setminus \mathrm{Descr}_{\mathscr{M}}$, where $\Gamma \in \mathrm{AFS}(\mathscr{M})$, do not allow one to obtain any new types of consequences that could not be obtained otherwise by using only $\mathrm{Descr}_{\mathscr{M}}$ (see Corollary 10.1.24). They only help to obtain these consequences somewhat faster than in the case of using only $\mathrm{Descr}_{\mathscr{M}}$. We can get something new from $\Gamma \in \mathrm{AFS}(M)$ only if Γ is inconsistent. Moreover, Corollary 10.1.24 shows the specific status of pure $\mathscr{M}$-inferences in our syntactic and semantic constructions. This specific status will be comfirmed below in a long series of results.

Agreement 10.2.34. An $\mathscr{M}$-inference, where $\mathscr{M}$ is a modification calculus, is a rather lengthy construction. To shorten the formulation of an $\mathscr{M}$-inference, we agree to replace a series of aplicatioins of logical axioms and deduction rules (MP and Gen) with one application of generalised deduction rule (written: Ded).

We illustrate the efficiency of this agreement in Example 10.2.35.

Example 10.2.35. Consider Example 10.2.7, which formalises the example from Sect. 5.2. Denote the rules from Example 10.2.7 in accordance with Definition 10.2.6. Then:

- (10.2) is denoted by $\mathrm{MR}_{\mathscr{M}}[\mathrm{C}, !]$,
- (10.3) is denoted by $\mathrm{MR}_{\mathscr{M}}[\mathrm{C}, ?]$,
- (10.4) is denoted by $\mathrm{MR}_{\mathscr{M}}[\mathrm{F}, !]$,
- (10.5) is denoted by $\mathrm{MR}_{\mathscr{M}}[\mathrm{F}, ?]$,
- (10.6) is denoted by $\mathrm{MR}_{\mathscr{M}}[\mathrm{A}, !]$,
- (10.7) is denoted by $\mathrm{MR}_{\mathscr{M}}[\mathrm{A}, ?]$.

Using Agreement 10.2.34 and the above notation we can formulate the inference from Sect. 5.2 in the form represented in Table 10.1.

Table 10.1 $\mathscr{M}$-inference in shortened notation

#	Formula	Reason	Stage	Module	DN
1	$D1 \neq D4$	OAx	0	0	
2	$D2 \neq D5$	OAx	0	0	
3	$?C(PC,S)$	DAx	0	1	12
4	$?C(LM,RT)$	DAx	0	1	13
5	$!F(D1,PC,S)\,,\,!F(D4,PC,S)$	DAx	0	2	
6	$!F(D2,LM,RT)\,,\,!F(D5,LM,RT)$	DAx	0	2	
7	$?F(D7,PC,S)$	DAx	0	2	15
8	$!A(D7,PC)$	DAx	0	3	
9	$?A(D7,S)$	DAx	0	3	16
10	$\exists d_1 \exists d_2\,(d_1 \neq d_2 \wedge !F(d_1,PC,S) \wedge !F(d_2,PC,S))$	Ded (1, 5)	1	0	
11	$\exists d_1 \exists d_2\,(d_1 \neq d_2 \wedge !F(d_1,LM,RT) \wedge !F(d_2,LM,RT))$	Ded (2, 6)	1	0	
12	$!C(PC,S)$	MR [C, !] (3, 10)	1	1	
13	$!C(LM,RT)$	MR [C, !] (4, 11)	1	1	
14	$!A(D7,PC) \wedge !C(PC,S)$	Ded (8, 12)	1	1	
15	$!F(D7,PC,S)$	MR [F, !] (7, 14)	1	2	
16	$!A(D7,S)$	MR [A, !] (9, 14,)	1	3	

Chapter 11
Derivability in Modification Calculi and L^1

11.1 Cuts of Record Strings: the General Case

Here we introduce the cut of arbitrary strings and we discuss a few basic properties of the cuts. In this case we are not interested in the nature of the elements of a string. Obviously we will use the tools introduced above for working with the record strings.

Definition 11.1.1. Let $U \neq \emptyset$, $\xi \in U^*$ and $M \subseteq \mathrm{IS}(\xi)$. Let us define a string generated from ξ by the use of a subset of indices of M by induction over the cardinality of M. This string is denoted by $\xi(\mathscr{M})$.

(i) If $M = \emptyset$, then $\xi(M) = e$,
(ii) Assume that we can find $\xi(S)$ for any $S \subseteq \mathrm{IS}(\xi)$, the cardinality of which equals n. Assume that the cardinality of M is $n+1$. Let m be the maximal index of M. Then $M = S \cup \{m\}$, where $\mathrm{Card}(S) = n$. In this case, by definition, put

$$\xi(M) = \xi(S) \bullet \xi[m].$$

Example 11.1.2. Let $\xi =$ "qwertyuiop" be a string of letters of the Roman alphabet. Then $\xi(\{2,4,8,9\}) =$ "wrio".

Definition 11.1.3. Let $U \neq \emptyset$, $\xi, \eta \in U^*$. We say that ξ is *an initial segment of* η if there exist a $\zeta \in U^*$ such that $\eta = \xi + \zeta$.

Lemma 11.1.4. Let $U \neq \emptyset$, $\xi \in U^*$, $K = \mathrm{IS}(\xi)$, $M \subseteq K$. Assume that for any $i, j \in K$ such that $i \leq j$, $j \in M$ implies $i \in M$.

Then the set $\xi(M)$ is an initial segment of ξ, i.e. there exists $\eta \in U^*$ such that $\xi = \xi(M) + \eta$.

Proof. If $K \backslash M = \emptyset$, then $\xi(M) = \xi$ and consequently is an initial segment of ξ.

If $K \backslash M \neq \emptyset$, then it contains the least number m. Let us show that in this case $M = \{k \in K \mid k < m\}$. Indeed, let $k < m$. Then $k \notin K \backslash M$ (otherwise k would be less

O. Anshakov, T. Gergely, *Cognitive Reasoning*, Cognitive Technologies,
DOI 10.1007/978-3-540-68875-4_11, © Springer-Verlag Berlin Heidelberg 2010

than the least number from $K \backslash M$. Therefore $k \in M$. Now let $k \in M$. Suppose that $m \leq k$. Then by the assumption of the lemma: $m \in M$). This is a contradiction.

Thus, $M = \{k \in K \mid k < m\}$, $K \backslash M = \{k \in K \mid k \geq m\}$. Obviously in this case: $\xi = \xi(M) + \xi(K \backslash M)$. □

Lemma 11.1.5. Let $U \neq \emptyset$, $\xi \in U^*$, $K = \mathrm{IS}(\xi)$, $M \subseteq K$. Suppose that for any $i, j \in K$, such that $i \leq j$,

$$i \in \mathscr{M} \quad \text{implies} \quad j \in M.$$

Then the set $\xi(M)$ is a final segment of ξ, i.e., there exists $\eta \in U^*$ such that $\xi = \eta + \xi(M)$.

Proof. The proof is similar to the proof of the previous lemma. □

Definition 11.1.6. Let $\xi, \eta, \zeta \in U^*$ for a non-empty set U. The ordered pair

$$S = \langle \eta, \zeta \rangle$$

is called a *cut* of the string ξ if $\xi = \eta + \zeta$. The string η is called the *upper segment* of the cut S and denoted by $\mathrm{Lo}(S)$, while the sting ζ is called the *lower segment* of the cut S and is denoted by $\mathrm{Hi}(S)$.

Proposition 11.1.7. Let $\xi \in U^*$, $U \neq \emptyset$ and S be a cut of the string ξ. Then:

(i) $\mathrm{Lo}(S) = e$ iff $\mathrm{Hi}(S) = \xi$,
(ii) $\mathrm{Hi}(S) = e$ iff $\mathrm{Lo}(S) = \xi$,
(iii) For any $\kappa \in U^*$, $\langle \mathrm{Lo}(S), \mathrm{Hi}(S) + \kappa \rangle$ is a cut of the string $\xi + \kappa$.

Remark 11.1.8. Let $\xi, \eta, \zeta, \kappa \in U^*$, $U \neq \emptyset$. Suppose that

(i) $\mathrm{Len}(\xi) = \mathrm{Len}(\zeta)$,
(ii) $\mathrm{Len}(\eta) = \mathrm{Len}(\kappa)$,
(iii) $\xi + \eta = \zeta + \kappa$.

Then $\xi = \zeta$, $\eta = \kappa$.

Although this obvious statement does not use the notion of cut in an explicit way it will be useful while dealing with cuts.

Remark 11.1.9. Let $\xi, \eta, \zeta \in U^*$, $U \neq \emptyset$. Suppose that

$$\mathrm{Len}(\xi + \eta) = \mathrm{Len}(\zeta).$$

Then there exist $\kappa, \nu \in U^*$ such that

(i) $\zeta = \kappa + \nu$,
(ii) $\mathrm{Len}(\xi) = \mathrm{Len}(\kappa)$,
(iii) $\mathrm{Len}(\eta) = \mathrm{Len}(\nu)$.

This trivial statement will also be very useful for working with cuts.

Lemma 11.1.10. Let $U \neq \emptyset$, $\xi \in U^*$, $K = \mathrm{IS}(\xi)$, $M \subseteq K$. Suppose that for any $i, j \in K$ such that $i \leq j$,

$$j \in \mathscr{M} \quad \text{implies} \quad i \in M.$$

Then the ordered pair

$$\langle \xi(M), \xi(K \backslash M) \rangle$$

is a cut of the string ξ.

Proof. This lemma follows directly from Lemma 11.1.4. □

Lemma 11.1.11. Let $U \neq \emptyset$, $\xi \in U^*$, $K = \mathrm{IS}(\xi)$, $M \subseteq K$. Let us suppose that, for any $i, j \in K$ such that $i \leq j$,

$$i \in \mathscr{M} \quad \text{implies} \quad j \in M.$$

Then the ordered pair

$$\langle \xi(K \backslash M), \xi(M) \rangle$$

is a cut of the string ξ.

Proof. This lemma follows directly from Lemma 11.1.5. □

Before considering the cut of record strings we will define an operation over record strings which is an inverse one to the operation Suspend and we will discuss some basic properties of this operation.

Definition 11.1.12. Let $\vec{\mathfrak{a}}$ be a record string of signature σ and k be a natural number. Let $\mathrm{Resume}(\vec{\mathfrak{a}}, k)$ denote a string $\vec{\mathfrak{b}}$ such that $\mathrm{Len}\left(\vec{\mathfrak{b}}\right) = \mathrm{Len}(\vec{\mathfrak{a}})$ and for any $i \in [1..\mathrm{Len}(\vec{\mathfrak{a}})]$

$$\vec{\mathfrak{b}}[i] = \begin{cases} \mathrm{Activate}(\vec{\mathfrak{a}}[i]) \text{ if } \mathrm{DNum}(\vec{\mathfrak{a}}[i]) > k, \\ \vec{\mathfrak{a}}[i] \qquad\qquad\quad \text{otherwise.} \end{cases}$$

The operation Resume makes visible those and only those records of $\vec{a}$ which were hidden by records whose indicies were greater than k.

We define the "default" value of k and its short form similarly to the definition of the dual operation Suspend. The "default" value of k is $\mathrm{Len}(\vec{\mathfrak{a}})$. By definition let

$$\mathrm{Resume}(\vec{\mathfrak{a}}) = \mathrm{Resume}(\vec{\mathfrak{a}}, \mathrm{Len}(\vec{\mathfrak{a}})).$$

Lemma 11.1.13. Let L be an FPJ logic, σ be a signature, $\mathscr{M} \in \mathrm{MC}_\sigma(L)$, $\Gamma \in \mathrm{AFS}(\mathscr{M})$, $\vec{\mathfrak{a}}$ be an $\mathscr{M}(s,m,d)$-inference from a family of hypotheses Γ and Φ be an external L-formula. Then

$$\mathrm{Resume}(\mathrm{Suspend}(\vec{\mathfrak{a}}, \Phi)) = \vec{\mathfrak{a}}.$$

Proof. The records that deductively depend on Φ are deactivated in the string $\mathrm{Suspend}(\vec{\mathfrak{a}},\Phi)$. These deactivated records have the deactivation index $\mathrm{Len}(\vec{\mathfrak{a}})+1$. But the operation Resume will reactivate the records whose deactivation index is more than $\mathrm{Len}(\vec{\mathfrak{a}})$. Among them are those records whose deactivation index is equal to $\mathrm{Len}(\vec{\mathfrak{a}})+1$. It is easy to show (by a trivial induction) that deactivation indices of the records in the inference do not exceed the length of the inference. Therefore $\mathrm{Len}(\vec{\mathfrak{a}})+1$ is the only deactivation index greater than $\mathrm{Len}(\vec{\mathfrak{a}})$ in the string $\mathrm{Suspend}(\vec{\mathfrak{a}},\Phi)$. It was set in some records by Suspend. In fact Resume cancels the results of Suspend. □

Lemma 11.1.14. Let L be an FPJ logic, σ be a signature, $\mathscr{M}\in\mathrm{MC}_\sigma(L)$, $\Gamma\in\mathrm{AFS}(\mathscr{M})$ and $\vec{\mathfrak{a}}$ be an $\mathscr{M}(s,m,d)$-inference from Γ, $k,l\in\omega$, $k\le l$. Then

$$\mathrm{Resume}(\mathrm{Resume}(\vec{\mathfrak{a}},l),k)=\mathrm{Resume}(\vec{\mathfrak{a}},k).$$

Proof. The operation Resume in the expession $\mathrm{Resume}(\vec{\mathfrak{a}},l)$ reactivates the records in the string $\vec{\mathfrak{a}}$ whose deactivation index is greater than l. This operation in the expression $\mathrm{Resume}(\mathrm{Resume}(\vec{\mathfrak{a}},l),k)$ activates the records in the string $\mathrm{Resume}(\vec{\mathfrak{a}},l)$, whose deactivation index is greater than $k\le l$. Obviously, we obtain the same result that we would obtain if the records with deactivation index greater than k were deactivated initially.

More formally: let $\mathrm{ND}(\vec{\mathfrak{r}})$ denote the set of deactivation indices in the string $\vec{\mathfrak{r}}$. Evidently

(i) $\mathrm{ND}(\mathrm{Resume}(\vec{\mathfrak{a}},l))=\mathrm{ND}(\vec{\mathfrak{a}})\setminus[l+1..]$,
(ii) $\mathrm{ND}(\mathrm{Resume}(\mathrm{Resume}(\vec{\mathfrak{a}},l),k))=(\mathrm{ND}(\vec{\mathfrak{a}})\setminus[l+1..])\setminus[k+1..]$,
(iii) $\mathrm{ND}(\mathrm{Resume}(\vec{\mathfrak{a}},k))=\mathrm{ND}(\vec{\mathfrak{a}})\setminus[k+1..]$.

It is easy to show that

$$(\mathrm{ND}(\vec{\mathfrak{a}})\setminus[l+1..])\setminus[k+1..]=\mathrm{ND}(\vec{\mathfrak{a}})\setminus[k+1..],$$

since $[l+1..]\subseteq[k+1..]$. □

Remark 11.1.15. It is easy to show that, for any record strings $\vec{\mathfrak{a}}$ and $\vec{\mathfrak{b}}$,

$$\mathrm{Resume}\left(\vec{\mathfrak{a}}+\vec{\mathfrak{b}},k\right)=\mathrm{Resume}(\vec{\mathfrak{a}},k)+\mathrm{Resume}\left(\vec{\mathfrak{b}},k\right).$$

Therefore the operation Resume could be inductively defined similarly to Definition 10.2.24.

BASIS. $\mathrm{Resume}(\mathbf{e},k)=\mathbf{e}$.

INDUCTION STEP. $\mathrm{Resume}(\vec{\mathfrak{a}}\bullet\mathfrak{a},k)=\mathrm{Resume}(\vec{\mathfrak{a}},k)\bullet\mathfrak{b}$, where

$$\mathfrak{b}=\begin{cases}\mathrm{Activate}(\mathfrak{a}) & \text{if } \mathrm{DNum}(\mathfrak{a})>k,\\ \mathfrak{a} & \text{otherwise.}\end{cases}$$

Lemma 11.1.16. Let L be an FPJ logic, σ be a signature, $\mathscr{M} \in \mathrm{MC}_\sigma(L)$, $\Gamma \in \mathrm{AFS}(\mathscr{M})$ and $\vec{\mathfrak{a}}$ be an $\mathscr{M}(s,m,d)$-inference from Γ. Then $\vec{\mathfrak{a}}$ is an $\mathscr{M}(s',m',d)$-inference from Γ, where

$$\langle s',m'\rangle = \mathrm{Rank}(\mathrm{Last}(\vec{\mathfrak{a}})).$$

Proof. It is obvious (see 10.2.31). Each inference of rank $\langle s, \mathrm{Q}_{\mathscr{M}}\rangle$ is at the same time an inference of rank $\langle s+1,0\rangle$ (see 10.2.31), which can be further considered as inference of rank $\langle s, \mathrm{Q}_{\mathscr{M}}\rangle$ until a record of rank $\langle s+1,0\rangle$ would be added to it. If such a record was added then we obtain an inference of rank $\langle s+1,0\rangle$ and not of $\langle s, \mathrm{Q}_{\mathscr{M}}\rangle$. □

Lemma 11.1.17. Let L be an FPJ logic, σ be a signature, $\mathscr{M} \in \mathrm{MC}_\sigma(L)$, $\Gamma \in \mathrm{AFS}(\mathscr{M})$ and $\vec{\mathfrak{a}}$ be an $\mathscr{M}(s,m,d)$-inference from Γ. Then, for any $k \leq l$,

$$\mathrm{Dom}_\sigma\left(D, \mathrm{P}_k^{\mathscr{M}}, \vec{\mathfrak{a}}, s\right) = \mathrm{Dom}_\sigma\left(D, \mathrm{P}_k^{\mathscr{M}}\right),$$

where

$$l = \begin{cases} m & \text{if } d = \mathbf{D}, \\ m-1 & \text{if } d = \mathbf{M}. \end{cases}$$

Proof. By induction on the length of the inference $\vec{\mathfrak{a}}$.

BASIS. If $\vec{\mathfrak{a}} = \mathbf{e}$, then $m = 0$ and $d = \mathbf{D}$. The assumption $k \leq 0$ always holds, because the ordering of predicate symbols starts with 1. Then the lemma trivially holds.

INDUCTION STEP. Suppose the lemma holds for the inference of length n. Let $n+1$ be the length of $\vec{\mathfrak{a}}$. Then $\vec{\mathfrak{a}} = \vec{\mathfrak{b}} \bullet \mathfrak{c}$. Consider the possible cases.

CASE 1. $\mathfrak{c}$ is added to the inference by a rule of Definition 10.2.31 (Ded). Then $\vec{\mathfrak{b}}$ is an $\mathscr{M}(s,m,d)$-inference. By the inductive assumption

$$\mathrm{Dom}_\sigma\left(D, \mathrm{P}_k^{\mathscr{M}}, \vec{\mathfrak{b}}, s\right) = \mathrm{Dom}_\sigma\left(D, \mathrm{P}_k^{\mathscr{M}}\right)$$

for any $k \leq m$. But in the considered case

$$\mathrm{Dom}_\sigma\left(D, \mathrm{P}_k^{\mathscr{M}}, \vec{\mathfrak{a}}, s\right) = \mathrm{Dom}_\sigma\left(D, \mathrm{P}_k^{\mathscr{M}}, \vec{\mathfrak{b}}, s\right)$$

for any $k \in [1..\mathrm{Q}_{\mathscr{M}}]$. Then for any $k \leq m$,

$$\mathrm{Dom}_\sigma\left(D, \mathrm{P}_k^{\mathscr{M}}, \vec{\mathfrak{a}}, s\right) = \mathrm{Dom}_\sigma\left(D, \mathrm{P}_k^{\mathscr{M}}\right), \tag{11.1}$$

holds as was to be proved.

Case 2. $\mathfrak{c}$ is added to the inference by a rule of Definition 10.2.31 (NDed). Then $m > 0$, $\mathscr{F}(c) = \mathrm{J}_\varepsilon \mathrm{P}_m^{\mathscr{M}}(\vec{a})$ for some $\varepsilon \in \mathscr{V}(L, \mathsf{I})$ and there exists an n-long $\mathscr{M}(s,m',d')$-inference $\vec{\mathfrak{d}}$ such that

$$\vec{\mathfrak{b}} = \vec{\mathfrak{d}} \quad \text{or} \quad \vec{\mathfrak{b}} = \mathrm{Suspend}\left(\vec{\mathfrak{d}}, \mathrm{J}_\tau \mathrm{P}_m^{\mathscr{M}}(\vec{a})\right).$$

As for m' and d', they, according to 10.2.31, should satisfy one and only one of the following conditions:

(a) $d' = \mathbf{M}$ and $m' = m$
(b) $d' = \mathbf{D}$ and $m' = m - 1$

Note that, in $\vec{\mathfrak{b}}$, $J_\tau P^{\mathcal{M}}_k(\vec{a})$ does not occur in the records with stage s. (A formula with the subformula $P^{\mathcal{M}}_k(\vec{a})$ was only added to $\vec{\mathfrak{a}}$ at the stage s, and earlier similar formulas did not occur in the inference.) Then by Lemma 10.2.26 for any $p \in \mathrm{Pr}_\sigma(\mathsf{l})$

$$\mathrm{Dom}_\sigma\left(D, p, \vec{\mathfrak{b}}, s\right) = \mathrm{Dom}_\sigma\left(D, p, \vec{\mathfrak{d}}, s\right).$$

Hence by the inductive assumption we obtain

$$\mathrm{Dom}_\sigma\left(D, P^{\mathcal{M}}_k, \vec{\mathfrak{b}}, s\right) = \mathrm{Dom}_\sigma\left(D, P^{\mathcal{M}}_k\right) \tag{11.2}$$

for any $k \leq l$, where

$$l = \begin{cases} m' & \text{if } d' = \mathbf{D}, \\ m' - 1 & \text{if } d' = \mathbf{M}. \end{cases} \tag{11.3}$$

Note also that, for any $k \leq m - 1$,

$$\mathrm{Dom}_\sigma\left(D, P^{\mathcal{M}}_k, \vec{\mathfrak{a}}, s\right) = \mathrm{Dom}_\sigma\left(D, P^{\mathcal{M}}_k, \vec{\mathfrak{b}}, s\right). \tag{11.4}$$

It is also true that

$$\mathrm{Dom}_\sigma\left(D, P^{\mathcal{M}}_m, \vec{\mathfrak{a}}, s\right) = \mathrm{Dom}_\sigma\left(D, P^{\mathcal{M}}_m, \vec{\mathfrak{b}}, s\right) \cup \{\vec{a}\}. \tag{11.5}$$

Consider the possible subcases:

SUBCASE 2.1. $d' = \mathbf{M}$ and $d = \mathbf{M}$. By (a) these equalities are valid only if $m' = m$. Then by the induction hypothesis we obtain that (11.1) holds for all $k \leq m - 1$ (see (11.2), (11.3) and (11.4)).

SUBCASE 2.2. $d' = \mathbf{D}$ and $d = \mathbf{M}$. By (b) these equalities are valid only if $m' = m - 1$. Then also by the induction hypothesis we obtain that (11.1) holds for all $k \leq m - 1$ (see (11.2), (11.3) and (11.4)).

SUBCASE 2.3. $d' = \mathbf{M}$ and $d = \mathbf{D}$. By (a) we obtain $m' = m$. By Definition 10.2.31 these equalities are valid only if

$$\mathrm{Dom}_\sigma\left(D, P^{\mathcal{M}}_m, \vec{\mathfrak{b}}, s\right) \cup \{\vec{a}\} = \mathrm{Dom}_\sigma\left(D, P^{\mathcal{M}}_m\right). \tag{11.6}$$

From (11.5) and (11.6) we obtain

$$\mathrm{Dom}_\sigma\left(D, P^{\mathcal{M}}_m, \vec{\mathfrak{a}}, s\right) = \mathrm{Dom}_\sigma\left(D, P^{\mathcal{M}}_m\right). \tag{11.7}$$

Besides, by using the induction hypothesis it can be shown, like in the two previous cases, that (11.1) holds for any $k \leq m-1$ (see (11.2), (11.3) and (11.4)). Hence from (11.7) we obtain that (7.1) holds for any $k \leq m$, as was to be proved.

SUBCASE 2.4. $d' = \mathbf{D}$ and $d = \mathbf{D}$. By (b) we obtain $m' = m - 1$. These equalities are valid only if

$$\mathrm{Dom}_\sigma\left(D, \mathrm{P}_m^{\mathscr{M}}\right) = \{\vec{a}\}.$$

It can be considered similarly to the previous subcase. □

Lemma 11.1.18. Let L be an FPJ logic, σ be a signature, $\mathscr{M} \in \mathrm{MC}_\sigma(L)$, $\Gamma \in \mathrm{AFS}(\mathscr{M})$ and $\vec{\mathfrak{a}}$ be an $\mathscr{M}(s,m,d)$-inference from Γ such that

$$\mathrm{Dom}_\sigma\left(D, \mathrm{P}_m^{\mathscr{M}}, \vec{\mathfrak{a}}, s\right) = \mathrm{Dom}_\sigma\left(D, \mathrm{P}_m^{\mathscr{M}}\right).$$

Then $d = \mathbf{D}$.

Proof. By induction on the length of the inference $\vec{\mathfrak{a}}$.

BASIS. If $\vec{\mathfrak{a}} = \mathbf{e}$, then $\vec{\mathfrak{a}}$ is an $\mathscr{M}(0,0,\mathbf{D})$-inference from Γ.

INDUCTION STEP. Suppose the lemma holds for the n-long inferences. Let the length of $\vec{\mathfrak{a}}$ be $n+1$. Then $\vec{\mathfrak{a}} = \vec{\mathfrak{b}} \bullet \mathfrak{c}$. Consider the possible cases:

CASE 1. $\mathfrak{c}$ is added to the inference by a rule of Definition 10.2.31 (Ded). Then $\vec{\mathfrak{b}}$ is also an $\mathscr{M}(s,m,d)$-inference. By the induction hypothesis the validity of

$$\mathrm{Dom}_\sigma\left(D, \mathrm{P}_m^{\mathscr{M}}, \vec{\mathfrak{b}}, s\right) = \mathrm{Dom}_\sigma\left(D, \mathrm{P}_m^{\mathscr{M}}\right) \tag{11.8}$$

implies $d = \mathbf{D}$. Note that

$$\mathrm{Dom}_\sigma\left(D, \mathrm{P}_m^{\mathscr{M}}, \vec{\mathfrak{a}}, s\right) = \mathrm{Dom}_\sigma\left(D, \mathrm{P}_m^{\mathscr{M}}, \vec{\mathfrak{b}}, s\right). \tag{11.9}$$

Now suppose that

$$\mathrm{Dom}_\sigma\left(D, \mathrm{P}_m^{\mathscr{M}}, \vec{\mathfrak{a}}, s\right) = \mathrm{Dom}_\sigma\left(D, \mathrm{P}_m^{\mathscr{M}}\right). \tag{11.10}$$

From (11.10) and (11.9) it follows (11.8); (11.8) by the inductive assumption implies $d = \mathbf{D}$, as was to be proved.

CASE 2. $\mathfrak{c}$ is added to the inference by a rule of Definition 10.2.31 (NDed). Then $m > 0$ and $\mathscr{F}(\mathfrak{c}) = \mathrm{J}_\varepsilon \mathrm{P}_m^{\mathscr{M}}(\vec{a})$ for some $\varepsilon \in \mathscr{V}(L, \mathsf{I})$. Suppose (11.10) holds. Then

$$\mathrm{Dom}_\sigma\left(D, \mathrm{P}_m^{\mathscr{M}}, \vec{\mathfrak{b}}, s\right) \cup \{\vec{a}\} = \mathrm{Dom}_\sigma\left(D, \mathrm{P}_m^{\mathscr{M}}, \vec{\mathfrak{a}}, s\right) = \mathrm{Dom}_\sigma\left(D, \mathrm{P}_m^{\mathscr{M}}\right).$$

By Definition 10.2.31, $d = \mathbf{D}$, as was to be proved. □

Corollary 11.1.19. Let L be an FPJ logic, σ be a signature, $\mathscr{M} \in \mathrm{MC}_\sigma(L)$, $\Gamma \in \mathrm{AFS}(\mathscr{M})$ and $\vec{\mathfrak{a}}$ be an $\mathscr{M}(s,m,d)$-inference from Γ. Then

$$d = \begin{cases} \mathbf{D} \text{ if } \mathrm{Dom}_\sigma\left(D, \mathrm{P}_k^{\mathcal{M}}, \vec{\mathfrak{a}}, s\right) = \mathrm{Dom}_\sigma\left(D, \mathrm{P}_k^{\mathcal{M}}\right) \text{ for any } k \leq m, \\ \mathbf{M} \text{ otherwise.} \end{cases}$$

Proof. Let $\mathrm{Dom}_\sigma\left(D, \mathrm{P}_k^{\mathcal{M}}, \vec{\mathfrak{a}}, s\right) = \mathrm{Dom}_\sigma\left(D, \mathrm{P}_k^{\mathcal{M}}\right)$ for all $k \leq m$. Then

$$\mathrm{Dom}_\sigma\left(D, \mathrm{P}_m^{\mathcal{M}}, \vec{\mathfrak{a}}, s\right) = \mathrm{Dom}_\sigma\left(D, \mathrm{P}_m^{\mathcal{M}}\right).$$

Hence by Lemma 11.1.18, $d = \mathbf{D}$.

Let $d = \mathbf{D}$. Then, by Lemma 11.1.17, $\mathrm{Dom}_\sigma\left(D, \mathrm{P}_k^{\mathcal{M}}, \vec{\mathfrak{a}}, s\right) = \mathrm{Dom}_\sigma\left(D, \mathrm{P}_k^{\mathcal{M}}\right)$ for any $k \leq m$. □

Corollary 11.1.20. Let L be an FPJ logic, σ be a signature, $\mathcal{M} \in \mathrm{MC}_\sigma(L)$, $\Gamma \in \mathrm{AFS}(\mathcal{M})$ and $\vec{\mathfrak{a}}$ be an $\mathcal{M}(s,m,d)$-inference from Γ. Then

$$d = \begin{cases} \mathbf{D} \text{ if } \mathrm{Dom}_\sigma\left(D, \mathrm{P}_m^{\mathcal{M}}, \vec{\mathfrak{a}}, s\right) = \mathrm{Dom}_\sigma\left(D, \mathrm{P}_m^{\mathcal{M}}\right), \\ \mathbf{M} \text{ otherwise.} \end{cases}$$

Proof. Similarly to the proof of the previous corollary. □

Lemma 11.1.21. Let L be an FPJ logic, σ be a signature, $\mathcal{M} \in \mathrm{MC}_\sigma(L)$, $\Gamma \in \mathrm{AFS}(\mathcal{M})$ and $\vec{\mathfrak{a}} = \vec{\mathfrak{b}} \bullet c$ be an $\mathcal{M}(s,m,d)$-inference from Γ. Then $\mathrm{Resume}\left(\vec{\mathfrak{b}}\right)$ is an $\mathcal{M}(s',m',d')$- inference from Γ, where

(i) $\langle s', m' \rangle = \mathrm{Rank}\left(\mathrm{Last}\left(\vec{\mathfrak{b}}\right)\right)$,

(ii) $d' = \begin{cases} \mathbf{D} \text{ if } \mathrm{Dom}_\sigma\left(D, \mathrm{P}_m^{\mathcal{M}}, \mathrm{Resume}\left(\vec{\mathfrak{b}}\right), s\right) = \mathrm{Dom}_\sigma\left(D, \mathrm{P}_m^{\mathcal{M}}\right), \\ \mathbf{M} \text{ otherwise.} \end{cases}$

Proof. By considering the possible cases.

CASE 1. Let record $\mathfrak{c}$ be added to the inference by a rule of Definition 10.2.31 (Ded). Then $\mathrm{Resume}\left(\vec{\mathfrak{b}}\right) = \vec{\mathfrak{b}}$, and $\vec{\mathfrak{b}}$ is an $\mathcal{M}(s',m',d')$-inference. The conditions (i) and (ii) hold for s', m' and d' by Lemma 11.1.16 and Corollary 11.1.20, respectively.

CASE 2. Let record $\mathfrak{c}$ be added to the inference by a rule of Definition 10.2.31 (NDed). Then by Definition 10.2.9, $\mathcal{F}(\mathfrak{c}) = \mathrm{J}_\varepsilon \mathrm{P}_m^{\mathcal{M}}(\vec{a})$ for some $\varepsilon \in \mathcal{V}(L, \mathsf{I})$ and there exists an $\mathcal{M}(s,m',d')$-inference $\vec{\mathfrak{d}}$ such that

$$\vec{\mathfrak{b}} = \vec{\mathfrak{d}} \quad \text{or} \quad \vec{\mathfrak{b}} = \mathrm{Suspend}\left(\vec{\mathfrak{d}}, \mathrm{J}_\tau \mathrm{P}_m^{\mathcal{M}}(\vec{a})\right).$$

Then in any case $\mathrm{Resume}\left(\vec{\mathfrak{b}}\right) = \vec{\mathfrak{d}}$ (see 11.1.13), while $\vec{\mathfrak{d}}$ is an $\mathcal{M}(s',m',d')$-inference. The fact that the conditions (i) and (ii) hold for s', m' and d' can be proved similarly to CASE 1. □

For the traditional notion of proof in a Hilbert-type calculus the following statement holds: any initial segment of a proof is a proof. This fact permits us to develop proofs by induction on the length of proofs in the Hilbert-type calculi. Such proofs are widely known. With the help of induction on the length of proofs many important theorems will be obtained, such as the Deduction Theorem, Correctness Theorem etc.

Unfortunately, the initial segment of an inference in a modification calculus is not necessarily an inference. The reason is that modification rules require the application of the operation Suspend to the inferences in modification calculi and this operation does not preserve the property of record strings "to be an inference".

Now we establish an important property of cuts. We will see that if the operation Resume is applied to an initial segment of an inference in a modification calculus then its result will be an inference. This result will allow us to use induction on the length of inferences in modification calculi almost as freely as in the case of traditional proofs in the Hilbert-type calculi.

Theorem 11.1.22. Let L be an FPJ logic, σ be a signature, $\mathscr{M} \in \mathrm{MC}_\sigma(L)$, $\Gamma \in \mathrm{AFS}(\mathscr{M})$ and $S = \left\langle \vec{\mathfrak{b}}, \vec{\mathfrak{c}} \right\rangle$ be a cut of $\vec{\mathfrak{a}}$, where $\vec{\mathfrak{a}}$ is an $\mathscr{M}(s,m,d)$-inference from Γ. Then $\mathrm{Resume}\left(\vec{\mathfrak{b}}\right)$ is an $\mathscr{M}(s',m',d')$-inference from Γ, where:

(i) $\langle s',m' \rangle = \mathrm{Rank}\left(\mathrm{Last}\left(\vec{\mathfrak{b}}\right)\right)$,

(ii) $d' = \begin{cases} \mathbf{D} \text{ if } \mathrm{Dom}_\sigma\left(D, \mathrm{P}_m^{\mathscr{M}}, \mathrm{Resume}\left(\vec{\mathfrak{b}}\right), s\right) = \mathrm{Dom}_\sigma\left(D, \mathrm{P}_m^{\mathscr{M}}\right), \\ \mathbf{M} \text{ otherwise.} \end{cases}$

Proof. By induction on the length of $\vec{\mathfrak{c}}$ (on the length of the upper segment of the cut S).

BASIS. If $\vec{\mathfrak{c}} = \mathbf{e}$ then $\vec{\mathfrak{b}} = \vec{\mathfrak{a}}$. Therefore $\vec{\mathfrak{b}}$ is an $\mathscr{M}(s',m',d')$-inference from Γ. By 11.1.16 and 11.1.20, we obtain (i) and (ii), respectively.

INDUCTION STEP. Suppose the statement holds for the length of $\vec{\mathfrak{c}}$ (the length of the upper segment of the corresponding cut) equal to n. Let the length of $\vec{\mathfrak{c}}$ be $n+1$. Then $\vec{\mathfrak{c}} = \vec{\mathfrak{d}} \bullet \mathfrak{e}$ for some record string $\vec{\mathfrak{d}}$ and record $\mathfrak{e}$. In this case

$$\vec{\mathfrak{a}} = \vec{\mathfrak{b}} + \vec{\mathfrak{c}} = \left(\vec{\mathfrak{b}} + \vec{\mathfrak{d}}\right) \bullet \mathfrak{e}.$$

Recall that $\vec{\mathfrak{a}}$ is an $\mathscr{M}$-inference. Then by Lemma 11.1.21, $\mathrm{Resume}\left(\vec{\mathfrak{b}} + \vec{\mathfrak{d}}\right)$ is also an $\mathscr{M}$-inference. Let $l = \mathrm{Len}\left(\vec{\mathfrak{b}} + \vec{\mathfrak{d}}\right)$. Then

$$\begin{aligned} \mathrm{Resume}\left(\vec{\mathfrak{b}} + \vec{\mathfrak{d}}\right) &= \mathrm{Resume}\left(\vec{\mathfrak{b}} + \vec{\mathfrak{d}}, l\right) && \text{(by 11.1.12)} \\ &= \mathrm{Resume}\left(\vec{\mathfrak{b}}, l\right) + \mathrm{Resume}\left(\vec{\mathfrak{d}}, l\right) && \text{(by 11.1.15)} \end{aligned}$$

Let

$$\vec{\mathfrak{f}} = \text{Resume}\left(\vec{\mathfrak{b}}, l\right), \tag{11.11}$$

$$\vec{\mathfrak{g}} = \text{Resume}\left(\vec{\mathfrak{d}}, l\right). \tag{11.12}$$

Then

$$\vec{\mathfrak{f}} + \vec{\mathfrak{g}} = \text{Resume}\left(\vec{\mathfrak{b}} + \vec{\mathfrak{d}}\right)$$

is an $\mathscr{M}$-inference. By (11.12), $\text{Len}\left(\vec{\mathfrak{g}}\right) = \text{Len}\left(\vec{\mathfrak{d}}\right) = n$. Then by the inductive assumption, $\text{Resume}\left(\vec{\mathfrak{f}}\right)$ is an $\mathscr{M}$-inference. Let $k = \text{Len}\left(\vec{\mathfrak{f}}\right) = \text{Len}\left(\vec{\mathfrak{b}}\right)$. Then

$$\begin{aligned}
\text{Resume}\left(\vec{\mathfrak{f}}\right) &= \text{Resume}\left(\vec{\mathfrak{f}}, k\right) && \text{(by 11.1.12)} \\
&= \text{Resume}\left(\text{Resume}\left(\vec{\mathfrak{b}}, l\right), k\right) && \text{(by (11.11))} \\
&= \text{Resume}\left(\vec{\mathfrak{b}}, k\right) && \text{(by 11.1.14)} \\
&= \text{Resume}\left(\vec{\mathfrak{b}}\right) && \text{(by 11.1.12)}
\end{aligned}$$

Thus, $\text{Resume}\left(\vec{\mathfrak{f}}\right) = \text{Resume}\left(\vec{\mathfrak{b}}\right)$. Then $\text{Resume}\left(\vec{\mathfrak{b}}\right)$ is an $\mathscr{M}(s', m', d')$-inference from Γ, for some s', and d', as was to be proved. By 11.1.16 and 11.1.20, we obtain (i) and (ii), respectively.

The above statement completes this section, which is devoted to arbitrary cuts of any strings and any record strings. Below we will investigate cuts of some special types. □

11.2 (m, s)-Cuts

The section will be started with the definition of two operations on the set $\omega \times M$, where M is an initial segment of ω. The first argument of these operations is regarded as the stage number, and the second one as the module number of the inference in a modification calculus. The operations are called *levelling-up* and *levelling-down*. These operations are required due to the *ambiguity of numeral parameters* of inference in our formalism. Thus by Definition 10.2.31, each $\mathscr{M}(s, \mathrm{Q}_{\mathscr{M}}, \mathbf{D})$-inference is at the same time an $\mathscr{M}(s+1, 0, \mathbf{D})$-inference. Resolution of this ambiguity is an *important technique* which further on will be widely used in inductive proofs.

The levelling operations eliminate this ambiguity. They give one and the same result for the numerical characteristics of the same inference regardless of whether we mean an $\mathscr{M}(s, \mathrm{Q}_{\mathscr{M}}, d)$-inference or an $\mathscr{M}(s+1, m, \mathbf{D})$-inference. This elimination of ambiguity is very important for some notions considered below.

These operations will be investigated below. They will prove to have very interesting properties. The operations will be defined in a simple way and so the proof of their properties will also be simple. School knowledge of mathematics will be sufficient to carry out the proofs.

Agreement 11.2.1. Suppose that a lexicographic order relation is given on the set $\omega \times \omega$, i.e. $\langle k,l\rangle < \langle m,n\rangle$ if $k<m$ or $k=m$ and $l<n$. The non-strict order is defined as usual: $\langle k,l\rangle \leq \langle m,n\rangle$ if $\langle k,l\rangle < \langle m,n\rangle$ or $\langle k,l\rangle = \langle m,n\rangle$.

Definition 11.2.2. Let $\vec{\mathfrak{a}}$ be a record string. The string $\vec{\mathfrak{a}}$ will be called *regular w.r.t. rank* (or simply *ranked*), if for any $i,j \in \mathrm{IS}(\vec{\mathfrak{a}})$,

$$i \leq j \quad \text{implies} \quad \mathrm{Rank}(\vec{\mathfrak{a}}[i]) \leq \mathrm{Rank}(\vec{\mathfrak{a}}[j]).$$

Remark 11.2.3. Let $\vec{\mathfrak{a}}$ and $\vec{\mathfrak{b}}$ be record strings. Then if $\vec{\mathfrak{a}} + \vec{\mathfrak{b}}$ is a ranked string then $\vec{\mathfrak{a}}$ and $\vec{\mathfrak{b}}$ are also ranked strings.

Lemma 11.2.4. Let $\vec{\mathfrak{a}}$ be a ranked record string and k, l be natural numbers. Let

$$M = \{i \in \mathrm{IS}(\vec{\mathfrak{a}}) \mid \mathrm{Rank}(\vec{\mathfrak{a}}[i]) < \langle k,l\rangle\}.$$

Then, for any $i,j \in \mathrm{IS}(\vec{\mathfrak{a}})$ such that $i \leq j$,

$$j \in \mathcal{M} \quad \text{implies} \quad i \in M.$$

Proof. It is simple by the use of Definition 11.2.2. □

Definition 11.2.5. Let σ be a signature. Define the map

$$\overline{\Delta}_\sigma \colon \omega \times [0..\mathrm{Q}_{\mathcal{M}}] \to \omega \times [0..\mathrm{Q}_{\mathcal{M}}]$$

as follows: let

$$\overline{\Delta}_\sigma(s,m) = \begin{cases} \langle s+1,0\rangle & \text{if } m = \mathrm{Q}_{\mathcal{M}}, \\ \langle s,m\rangle & \text{otherwise.} \end{cases}$$

The map $\overline{\Delta}_\sigma$ is called *levelling-up* (of signature σ).

$$\overline{\Delta}_\sigma(s,m) = \begin{cases} \langle s+1,0\rangle & \text{if } m = \mathrm{Q}_{\mathcal{M}}, \\ \langle s,m\rangle & \text{otherwise.} \end{cases}$$

Definition 11.2.6. Let σ be a signature. Define the map

$$\underline{\nabla}_\sigma \colon \omega \times [0..\mathrm{Q}_{\mathcal{M}}] \to \omega \times [0..\mathrm{Q}_{\mathcal{M}}]$$

as follows: let

$$\underline{\nabla}_\sigma(s,m) = \begin{cases} \langle s-1,\mathrm{Q}_{\mathcal{M}}\rangle & \text{if } m = 0 \text{ and } s > 0, \\ \langle s,m\rangle & \text{otherwise.} \end{cases}$$

The map $\underline{\nabla}_\sigma$ is called *levelling-down* (of signature σ).

Remark 11.2.7. Let σ be a signature. Then for any $s \in \omega$:

(i) $\overline{\Delta}_\sigma(s+1,0) = \overline{\Delta}_\sigma(s,Q_{\mathcal{M}}) = \langle s+1,0\rangle$,
(ii) $\underline{\nabla}_\sigma(s+1,0) = \underline{\nabla}_\sigma(s,Q_{\mathcal{M}}) = \langle s,Q_{\mathcal{M}}\rangle$.

Proof. It is straightforward. □

Lemma 11.2.8. Let σ be a signature. Then for any $s \in \omega$, $m \in [0..Q_{\mathcal{M}}]$,

(i) $\underline{\nabla}_\sigma(s,m) \leq \langle s,m\rangle \leq \overline{\Delta}_\sigma(s,m)$,
(ii) $\underline{\nabla}_\sigma(s,m) < \langle s,m\rangle$ iff $m = 0$ and $s > 0$,
(iii) $\langle s,m\rangle < \overline{\Delta}_\sigma(s,m)$ iff $m = Q_{\mathcal{M}}$,
(iv) $\overline{\Delta}_\sigma(\overline{\Delta}_\sigma(s,m)) = \overline{\Delta}_\sigma(s,m)$,
(v) $\underline{\nabla}_\sigma(\underline{\nabla}_\sigma(s,m)) = \underline{\nabla}_\sigma(s,m)$.

Proof. It is straightforward. □

Lemma 11.2.9. Let σ be a signature. Then for any $s_1,s_2 \in \omega$, $m_1,m_2 \in [0..Q_{\mathcal{M}}]$:

(i) $\langle s_1,Q_{\mathcal{M}}\rangle < \langle s_2,m_2\rangle$ implies $\langle s_1+1,0\rangle \leq \langle s_2,m_2\rangle$,
(ii) $\langle s_1,m_1\rangle < \langle s_2,0\rangle$ implies $s_2 > 0$ and $\langle s_1,m_1\rangle \leq \langle s_2-1,Q_{\mathcal{M}}\rangle$.

Proof. (i) Obviously, $\langle s_1,Q_{\mathcal{M}}\rangle < \langle s_2,m_2\rangle$ implies $s_2 \geq s_1+1$. Furthermore, $m_2 \geq 0$ is always true.
(ii) Obviously, $\langle s_1,m_1\rangle < \langle s_2,0\rangle$ implies $s_2 > 0$ and $s_1 \leq s_2-1$. Besides, $m_1 \leq 0$ always hold. □

Corollary 11.2.10. Let σ be a signature. Then for any $s_1,s_2 \in \omega$, $m_1,m_2 \in [0..Q_{\mathcal{M}}]$:

(i) $\langle s_1,m_1\rangle < \langle s_2,m_2\rangle$ implies $\overline{\Delta}_\sigma(s_1,m_1) \leq \langle s_2,m_2\rangle$,
(ii) $\langle s_1,m_1\rangle < \langle s_2,m_2\rangle$ implies $\langle s_1,m_1\rangle \leq \underline{\nabla}_\sigma(s_2,m_2)$.

Proof. It directly follows from 11.2.5, 11.2.6 and 11.2.9. □

Lemma 11.2.11. Let σ be a signature. Then for any $s_1,s_2 \in \omega$, $m_1,m_2 \in [0..Q_{\mathcal{M}}]$:

(i) $\langle s_1,m_1\rangle \leq \langle s_2,m_2\rangle$ implies $\overline{\Delta}_\sigma(s_1,m_1) \leq \overline{\Delta}_\sigma(s_2,m_2)$,
(ii) $\langle s_1,m_1\rangle \leq \langle s_2,m_2\rangle$ implies $\underline{\nabla}_\sigma(s_1,m_1) \leq \underline{\nabla}_\sigma(s_2,m_2)$.

Proof. It is clear that $\langle s_1,m_1\rangle = \langle s_2,m_2\rangle$ implies $\overline{\Delta}_\sigma(s_1,m_1) \leq \overline{\Delta}_\sigma(s_2,m_2)$ and $\underline{\nabla}_\sigma(s_1,m_1) \leq \underline{\nabla}_\sigma(s_2,m_2)$. Therefore we consider the case

$$\langle s_1,m_1\rangle < \langle s_2,m_2\rangle\,. \tag{11.13}$$

Then, obviously,

$$\begin{aligned} \overline{\Delta}_\sigma(s_1,m_1) &\leq \langle s_2,m_2\rangle && \text{(by (11.13) and 11.2.10(i))} \\ &\leq \overline{\Delta}_\sigma(s_2,m_2) && \text{(by 11.2.8(i))} \end{aligned}$$

Also,

$$\begin{aligned}\underline{\nabla}_\sigma(s_1,m_1) &\leq \langle s_2,m_2\rangle && \text{(by 11.2.8(i))}\\ &\leq \underline{\nabla}_\sigma(s_2,m_2) && \text{(by (11.13) and 11.2.10(ii))}\end{aligned}$$

□

Corollary 11.2.12. Let σ be a signature. Then for any $s_1,s_2 \in \omega$, $m_1,m_2 \in [0..\mathrm{Q}_{\mathcal{M}}]$

(i) $\overline{\Delta}_\sigma(s_1,m_1) > \overline{\Delta}_\sigma(s_2,m_2)$ implies $\langle s_1,m_1\rangle > \langle s_2,m_2\rangle$,
(ii) $\underline{\nabla}_\sigma(s_1,m_1) > \underline{\nabla}_\sigma(s_2,m_2)$ implies $\langle s_1,m_1\rangle > \langle s_2,m_2\rangle$.

Lemma 11.2.13. Let σ be a signature. Then for any $s \in \omega$, $m \in [0..\mathrm{Q}_{\mathcal{M}}]$

$$\underline{\nabla}_\sigma\left(\overline{\Delta}_\sigma(s,m)\right) \leq \langle s,m\rangle .$$

Proof. Consider the possible cases:

CASE 1. Let $m \neq 0$, $m \neq \mathrm{Q}_{\mathcal{M}}$. Then by 11.2.5 and 11.2.6,

$$\underline{\nabla}_\sigma\left(\overline{\Delta}_\sigma(s,m)\right) = \underline{\nabla}_\sigma(s,m) = \langle s,m\rangle .$$

CASE 2. Let $m = \mathrm{Q}_{\mathcal{M}}$. Then by 11.2.5 and 11.2.6,

$$\underline{\nabla}_\sigma\left(\overline{\Delta}_\sigma(s,\mathrm{Q}_{\mathcal{M}})\right) = \underline{\nabla}_\sigma(s+1,0) = \langle s,\mathrm{Q}_{\mathcal{M}}\rangle .$$

CASE 3. Let $m = 0$. Then by 11.2.5 and 11.2.6,

$$\underline{\nabla}_\sigma\left(\overline{\Delta}_\sigma(s,0)\right) = \underline{\nabla}_\sigma(s,0) = \langle s-1,\mathrm{Q}_{\mathcal{M}}\rangle < \langle s,0\rangle .$$

□

Lemma 11.2.14. Let σ be a signature. Then for any $s \in \omega$, $m \in [0..\mathrm{Q}_{\mathcal{M}}]$

$$\overline{\Delta}_\sigma(\underline{\nabla}_\sigma(s,m)) \geq \langle s,m\rangle .$$

Proof. Consider the possible cases:

CASE 1. Let $m \neq 0$, $m \neq \mathrm{Q}_{\mathcal{M}}$. Then by 11.2.5 and 11.2.6,

$$\overline{\Delta}_\sigma(\underline{\nabla}_\sigma(s,m)) = \overline{\Delta}_\sigma(s,m) = \langle s,m\rangle .$$

CASE 2. Let $m = 0$, $s > 0$. Then by 11.2.5 and 11.2.6,

$$\overline{\Delta}_\sigma(\underline{\nabla}_\sigma(s,0)) = \overline{\Delta}_\sigma(s-1,\mathrm{Q}_{\mathcal{M}}) = \langle s,0\rangle .$$

Subcase $s = 0$ is similar to CASE 1.

CASE 3. Let $m = Q_{\mathcal{M}}$. Then by 11.2.5 and 11.2.6,

$$\overline{\Delta}_\sigma(\underline{\nabla}_\sigma(s, Q_{\mathcal{M}})) = \overline{\Delta}_\sigma(s, Q_{\mathcal{M}}) = \langle s+1, 0\rangle > \langle s, Q_{\mathcal{M}}\rangle .$$

□

Lemma 11.2.15. Let σ be a signature. Then for any $s_1, s_2 \in \omega$, $m_1, m_2 \in [0..Q_{\mathcal{M}}]$

$$\langle s_1, m_1\rangle \leq \overline{\Delta}_\sigma(s_2, m_2) \quad \text{iff} \quad \underline{\nabla}_\sigma(s_1, m_1) \leq \langle s_2, m_2\rangle .$$

Proof. ($\Rightarrow$) Let

$$\langle s_1, m_1\rangle \leq \overline{\Delta}_\sigma(s_2, m_2) . \tag{11.14}$$

Then

$$\begin{aligned} \underline{\nabla}_\sigma(s_1, m_1) &\leq \underline{\nabla}_\sigma\left(\overline{\Delta}_\sigma(s_2, m_2)\right) && \text{(by (11.14) and 11.2.11(ii))} \\ &\leq \langle s_2, m_2\rangle && \text{(by 11.2.13)} \end{aligned}$$

($\Leftarrow$) Let

$$\underline{\nabla}_\sigma(s_1, m_1) \leq \langle s_2, m_2\rangle . \tag{11.15}$$

Then

$$\begin{aligned} \langle s_1, m_1\rangle &\leq \overline{\Delta}_\sigma(\underline{\nabla}_\sigma(s_1, m_1)) && \text{(by 11.2.14)} \\ &\leq \overline{\Delta}_\sigma(s_2, m_2) && \text{(by (11.15) and 11.2.11(i))} \end{aligned}$$

□

Corollary 11.2.16. Let σ be a signature. Then for any $s_1, s_2 \in \omega$, $m_1, m_2 \in [0..Q_{\mathcal{M}}]$

$$\langle s_1, m_1\rangle > \overline{\Delta}_\sigma(s_2, m_2) \quad \text{iff} \quad \underline{\nabla}_\sigma(s_1, m_1) > \langle s_2, m_2\rangle .$$

Lemma 11.2.17. Let σ be a signature. Then for any $s_1, s_2 \in \omega$, $m_1, m_2 \in [0..Q_{\mathcal{M}}]$:

(i) $\overline{\Delta}_\sigma(s_1, m_1) \leq \overline{\Delta}_\sigma(s_2, m_2)$ implies $\langle s_1, m_1\rangle \leq \overline{\Delta}_\sigma(s_2, m_2)$,
(ii) $\underline{\nabla}_\sigma(s_1, m_1) \leq \underline{\nabla}_\sigma(s_2, m_2)$ implies $\underline{\nabla}_\sigma(s_1, m_1) \leq \langle s_2, m_2\rangle$.

Proof. (i) Assume the converse. Let

$$\overline{\Delta}_\sigma(s_2, m_2) < \langle s_1, m_1\rangle . \tag{11.16}$$

Then,

$$\begin{aligned} \overline{\Delta}_\sigma(s_2, m_2) &< \langle s_1, m_1\rangle && \text{(by (11.16))} \\ &\leq \overline{\Delta}_\sigma(s_1, m_1) && \text{(by 11.2.8(i))} \end{aligned}$$

which contradicts the condition.

(ii) Assume the converse. Let

$$\langle s_2, m_2\rangle < \underline{\nabla}_\sigma (s_1, m_1)\,. \tag{11.17}$$

Then,

$$\begin{aligned} \underline{\nabla}_\sigma (s_2, m_2) &\le \langle s_2, m_2\rangle && \text{(by 11.2.8(i))}\\ &< \underline{\nabla}_\sigma (s_1, m_1) && \text{(by (11.17))} \end{aligned}$$

which contradicts the condition.

□

Lemma 11.2.18. Let σ be a signature. Then for any $s_1, s_2 \in \omega$, $m_1, m_2 \in [0..Q_{\mathscr{M}}]$

$$\overline{\Delta}_\sigma (s_1, m_1) \le \overline{\Delta}_\sigma (s_2, m_2) \quad \text{iff} \quad \underline{\nabla}_\sigma (s_1, m_1) \le \underline{\nabla}_\sigma (s_2, m_2)\,.$$

Proof. $(\Rightarrow)$

$$\begin{aligned} &\overline{\Delta}_\sigma (s_1, m_1) \le \overline{\Delta}_\sigma (s_2, m_2) && \\ &\qquad \text{implies} \quad \langle s_1, m_1\rangle \le \overline{\Delta}_\sigma (s_2, m_2) && \text{(by 11.2.17(i))}\\ &\qquad \text{implies} \quad \underline{\nabla}_\sigma (s_1, m_1) \le \langle s_2, m_2\rangle && \text{(by 11.2.15)}\\ &\qquad \text{implies} \quad \underline{\nabla}_\sigma (\underline{\nabla}_\sigma (s_1, m_1)) \le \underline{\nabla}_\sigma (s_2, m_2) && \text{(by 11.2.11(ii))}\\ &\qquad \text{implies} \quad \underline{\nabla}_\sigma (s_1, m_1) \le \underline{\nabla}_\sigma (s_2, m_2), && \text{(by 11.2.8(v))} \end{aligned}$$

as was to be proved.

$(\Leftarrow)$ Suppose $\underline{\nabla}_\sigma (s_1, m_1) \le \underline{\nabla}_\sigma (s_2, m_2)$. Arguing backwards we can get

$$\overline{\Delta}_\sigma (s_1, m_1) \le \overline{\Delta}_\sigma (s_2, m_2)$$

(see Lemmas 11.2.17(ii), 11.2.15, 11.2.11(i) and 11.2.8(iv)). □

Corollary 11.2.19. Let σ be a signature. Then for any $s_1, s_2 \in \omega$, $m_1, m_2 \in [0..Q_{\mathscr{M}}]$

$$\overline{\Delta}_\sigma (s_1, m_1) > \overline{\Delta}_\sigma (s_2, m_2) \quad \text{iff} \quad \underline{\nabla}_\sigma (s_1, m_1) > \underline{\nabla}_\sigma (s_2, m_2)\,.$$

The important notion we introduce now is the (s,m)-cut of ranked record strings, where s is the stage index and m is the module index. Naturally, we are interested mainly in (s,m)-cuts of inferences in modification calculi. The lower segment of an (s,m)-cut of such an inference includes all the records added to it until the moment when a non-deductive rule of the next module was used for the first time.

The lower segments of an (s,m)-cut are the basic tools in proving the fact that the sequences of L-structures to be defined in the next section could be considered as adequate semantics for our formalism. Further on, induction on the length of upper segment of (s,m)-cuts will be a frequently used technique.

Definition 11.2.20. Let $\vec{\mathfrak{a}}$ be a ranked record string and s and m be arbitrary natural numbers. The ordered pair $\left\langle \vec{\mathfrak{b}}, \vec{\mathfrak{c}} \right\rangle$ is called the $(s,\, m)$*-cut of* $\vec{\mathfrak{a}}$, if

$$\vec{\mathfrak{b}} = \vec{\mathfrak{a}}\,(\mathscr{M}), \quad \vec{\mathfrak{c}} = \vec{\mathfrak{a}}\,(\mathrm{IS}\,(\vec{\mathfrak{a}})\setminus M),$$

where

$$M = \left\{ i \in \mathrm{IS}\,(\vec{\mathfrak{a}}) \mid \mathrm{Rank}\,(\vec{\mathfrak{a}}\,[i]) \leq \overline{\Delta}_{\sigma}\,(s,m) \right\}.$$

By Lemmas 11.2.4 and 11.1.11, the (s,m)-cut of the string $\vec{\mathfrak{a}}$ is really its cut in the sense of Definition 11.1.6. Moreover, it is clear that, for any ranked record string, and any natural numbers s and m there exists a unique (s,m)-cut.

$\mathrm{Sec}\,(\vec{\mathfrak{a}},s,m)$ denotes the (s,m)-cut of $\vec{\mathfrak{a}}$. $\mathrm{Lo}\,(\vec{\mathfrak{a}},s,m)$ denotes the lower segment and $\mathrm{Hi}\,(\vec{\mathfrak{a}},s,m)$ the upper segment of this cut.

Lemma 11.2.21. Let $\vec{\mathfrak{a}}$ be a ranked record string and s be a natural number. Then $\mathrm{Sec}\,(\vec{\mathfrak{a}},s,\mathrm{Q}_{\mathscr{M}}) = \mathrm{Sec}\,(\vec{\mathfrak{a}},s+1,0)$.

Proof. By 11.2.7 (i), $\overline{\Delta}_{\sigma}\,(s,\mathrm{Q}_{\mathscr{M}}) = \overline{\Delta}_{\sigma}\,(s+1,0)$. Then by Definition 11.2.20,

$$\mathrm{Sec}\,(\vec{\mathfrak{a}},s,\mathrm{Q}_{\mathscr{M}}) = \mathrm{Sec}\,(\vec{\mathfrak{a}},s+1,0)\,.$$

□

Lemma 11.2.22. Let $\vec{\mathfrak{a}}$ be a ranked record string and s and m be any natural numbers. Then $\mathrm{Lo}\,(\vec{\mathfrak{a}},s,m) = \vec{\mathfrak{a}}$ if and only if for any $i \in \mathrm{IS}\,(\vec{\mathfrak{a}})$, $\mathrm{Rank}\,(\vec{\mathfrak{a}}\,[i]) \leq \overline{\Delta}_{\sigma}\,(s,m)$.

Proof. By direct use of Definition 11.2.20. □

Lemma 11.2.23. Let $\vec{\mathfrak{a}} = \vec{\mathfrak{b}} \bullet \mathfrak{c}$ be a ranked record string and s and m be any natural numbers. Let

$$\left\langle \vec{\mathfrak{k}}, \vec{\mathfrak{l}} \bullet \mathfrak{c} \right\rangle = \mathrm{Sec}\,(\vec{\mathfrak{a}},s,m)\,.$$

Then $\left\langle \vec{\mathfrak{k}}, \vec{\mathfrak{l}} \right\rangle = \mathrm{Sec}\left(\vec{\mathfrak{b}},s,m\right)$.

Proof. Let $M_{\vec{\mathfrak{a}}}$ denote the index set for the lower segment of the (s,m)-cut of $\vec{\mathfrak{a}}$, and $M_{\vec{\mathfrak{b}}}$ denote the index set for the lower segment of the (s,m)-cut of $\vec{\mathfrak{b}}$. By the condition, the record $\mathfrak{c}$ belongs to the upper segment of the (s,m)-cut of the string $\vec{\mathfrak{a}}$. Then it is clear that $M_{\vec{\mathfrak{a}}} = M_{\vec{\mathfrak{b}}}$, and consequently $\mathrm{Lo}\,(\vec{\mathfrak{a}},s,m) = \mathrm{Lo}\left(\vec{\mathfrak{b}},s,m\right) = \vec{\mathfrak{k}}$. From $\vec{\mathfrak{k}} + \vec{\mathfrak{l}} \bullet \mathfrak{c} = \vec{\mathfrak{b}} \bullet \mathfrak{c}$ it follows that $\vec{\mathfrak{k}} + \vec{\mathfrak{l}} = \vec{\mathfrak{b}}$. Then $\mathrm{Hi}\left(\vec{\mathfrak{b}},s,m\right) = \vec{\mathfrak{l}}$. □

Lemma 11.2.24. Let L be an FPJ logic, σ be a signature, $\mathscr{M} \in \mathrm{MC}_{\sigma}\,(L)$, $\vec{\mathfrak{a}}$ be an $\mathscr{M}\,(s,m,d)$-inference from a set of external L-formulas. Then one and only one of the following holds:

(i) $\vec{\mathfrak{a}} = \mathbf{e}$,
(ii) $\underline{\nabla}_\sigma(s,m) \leq \mathrm{Rank}(\mathrm{Last}(\vec{\mathfrak{a}})) \leq \langle s,m \rangle$,

Proof. By straightforward induction on $\mathcal{M}$-inferences. □

Lemma 11.2.25. Let L be an FPJ logic, σ be a signature, $\mathcal{M} \in \mathrm{MC}_\sigma(L)$, $\vec{\mathfrak{a}}$ be an $\mathcal{M}$-inference from some set of external L-formulas. Then $\vec{\mathfrak{a}}$ is a ranked record string (in the sense of Definition 11.2.2), i.e. for any $i, j \in \mathrm{IS}(\vec{\mathfrak{a}})$,

$$i \leq j \quad \text{implies} \quad \mathrm{Rank}(\vec{\mathfrak{a}}[i]) \leq \mathrm{Rank}(\vec{\mathfrak{a}}[j]).$$

Proof. By induction on $\mathcal{M}$-inferences. The statement is evident for the empty $\mathcal{M}$-inference. By Lemma 11.2.24 and Definition 10.2.31 we can conclude that the rank of a record added to the inference is not lower than the rank of the previous record. Hence by the inductive assumption we obtain what was desired. □

Lemma 11.2.26. Let L be an FPJ logic, σ be a signature, $\mathcal{M} \in \mathrm{MC}_\sigma(L)$, $\vec{\mathfrak{a}}$ be an $\mathcal{M}(s,m,d)$-inference from some set of external L-formulas. Suppose

$$\langle s',m' \rangle \in \omega \times [1..\mathrm{Q}_{\mathcal{M}}], \quad \langle s',m' \rangle \leq \langle s,m \rangle.$$

Then there exists a $\mathfrak{b} \in \mathrm{Set}(\vec{\mathfrak{a}})$ such that $\mathrm{Rank}(\mathfrak{b}) = \langle s',m' \rangle$.

Proof. By induction on the length of the inference $\vec{\mathfrak{a}}$.

BASIS. Let $\vec{\mathfrak{a}} = \mathbf{e}$. Then $\vec{\mathfrak{a}}$ is an $\mathcal{M}(0,0,d)$-inference. In this case the premise of the statement to be proved is false.

INDUCTION STEP. Let the statement hold for all $\mathcal{M}$-inferences of length n. Suppose that the length of the $\mathcal{M}(s,m,d)$-inference $\vec{\mathfrak{a}}$ is $n+1$. Then $\vec{\mathfrak{a}} = \vec{\mathfrak{c}} \bullet \mathfrak{d}$, where $\vec{\mathfrak{e}} = \mathrm{Resume}(\vec{\mathfrak{c}})$ is an $\mathcal{M}$-inference of length n. Consider the possible cases.

CASE 1. Let $\mathfrak{d}$ be added by a rule from Definition 10.2.31 (Ded). Then $\vec{\mathfrak{c}}$ is an $\mathcal{M}(s,m,d)$-inference of length n, and by the induction hypothesis for any $\langle s',m' \rangle \in \omega \times [1..\mathrm{Q}_{\mathcal{M}}]$ ($\langle s',m' \rangle \leq \langle s,m \rangle$), there exists a record $\mathfrak{b} \in \mathrm{Set}(\vec{\mathfrak{c}}) \subseteq \mathrm{Set}(\vec{\mathfrak{a}})$ such that $\mathrm{Rank}(\mathfrak{b}) = \langle s',m' \rangle$.

CASE 2. Let $\mathfrak{d}$ be added by a rule from Definition 10.2.31 (NDed). Then $m > 0$. Consider the subcases.

SUBCASE 2.1. Let $\vec{\mathfrak{e}} = \mathrm{Resume}(\vec{\mathfrak{c}})$ be an $\mathcal{M}(s,m,\mathbf{M})$-inference. Let

$$\langle s',m' \rangle \in \omega \times [1..\mathrm{Q}_{\mathcal{M}}], \quad \langle s',m' \rangle \leq \langle s,m \rangle.$$

By the induction hypothesis there exists a record $\mathfrak{f} \in \mathrm{Set}(\vec{\mathfrak{e}})$ such that $\mathrm{Rank}(\mathfrak{f}) = \langle s',m' \rangle$. Let $\mathfrak{f} = \vec{\mathfrak{e}}[i]$. Take $\mathfrak{b} = \vec{\mathfrak{c}}[i]$. Then $\mathfrak{b} \in \mathrm{Set}(\vec{\mathfrak{c}}) \subseteq \mathrm{Set}(\vec{\mathfrak{a}})$ and $\mathrm{Rank}(\mathfrak{b}) = \mathrm{Rank}(\mathfrak{f}) = \langle s',m' \rangle$ (see Definition 11.1.12).

SUBCASE 2.2. Let $\vec{\mathfrak{e}} = \mathrm{Resume}(\vec{\mathfrak{c}})$ be an $\mathcal{M}(s,m-1,\mathbf{D})$-inference. Then $\vec{\mathfrak{a}} = \vec{\mathfrak{c}} \bullet \mathfrak{d}$ is an $\mathcal{M}(s,m,\mathbf{M})$-inference and $\mathrm{Rank}(\mathfrak{d}) = \langle s,m \rangle$. By the induction hypothesis, like in the previous subcase it can be shown that for any $\langle s',m' \rangle \in \omega \times [1..\mathrm{Q}_{\mathcal{M}}]$

$(\langle s',m'\rangle \le \langle s,m-1\rangle)$ there exists a record $\mathfrak{b} \in \mathrm{Set}(\vec{\mathfrak{c}})$ such that $\mathrm{Rank}(\mathfrak{b}) = \langle s',m'\rangle$. Since $\mathrm{Rank}(\mathfrak{d}) = \langle s,m\rangle$, we can conclude that for an arbitrary $\langle s',m'\rangle \in \omega \times [1..\mathrm{Q}_{\mathscr{M}}]$ $(\langle s',m'\rangle \le \langle s,m\rangle)$ there exists a record of rank $\langle s',m'\rangle$ that belong to $\vec{\mathfrak{a}}$ of rank $\langle s',m'\rangle$.

SUBCASE 2.3. Let $\vec{\mathfrak{e}} = \mathrm{Resume}(\vec{\mathfrak{c}})$ be an $\mathscr{M}(s-1,0,\mathbf{D})$-inference. Then $m=1$, and $\vec{\mathfrak{a}} = \vec{\mathfrak{c}} \bullet \mathfrak{d}$ is an $\mathscr{M}(s,1,\mathbf{M})$-inference and $\mathrm{Rank}(\mathfrak{d}) = \langle s,1\rangle$. This subcase can be considered analogously to the previous subcase. □

Lemma 11.2.27. Let L be an FPJ logic, σ be a signature, $\mathscr{M} \in \mathrm{MC}_{\sigma}(L)$, $\vec{\mathfrak{a}}$ be an $\mathscr{M}(s,m,d)$-inference from a set of external L-formulas. Let $k \in \omega$, $l \in [0..\mathrm{Q}_{\mathscr{M}}]$. Then

$$\mathrm{Lo}(\vec{\mathfrak{a}},k,l) = \vec{\mathfrak{a}} \quad \text{iff} \quad \overline{\Delta}_{\sigma}(s,m) \le \overline{\Delta}_{\sigma}(k,l).$$

Proof. By the hypothesis for any $i \in \mathrm{IS}(\vec{\mathfrak{a}})$, $\mathrm{Rank}(\vec{\mathfrak{a}}[i]) \le \langle s,m\rangle$.

($\Leftarrow$) Suppose $\overline{\Delta}_{\sigma}(s,m) \le \overline{\Delta}_{\sigma}(k,l)$. Then by 11.2.17(i), $\langle s,m\rangle \le \overline{\Delta}_{\sigma}(k,l)$. Therefore $\mathrm{Rank}(\vec{\mathfrak{a}}[i]) \le \overline{\Delta}_{\sigma}(k,l)$ holds for any $i \in \mathrm{IS}(\vec{\mathfrak{a}})$. In this case by Lemma 11.2.22 we get $\mathrm{Lo}(\vec{\mathfrak{a}},k,l) = \vec{\mathfrak{a}}$.

($\Rightarrow$)Suppose $\mathrm{Lo}(\vec{\mathfrak{a}},k,l) = \vec{\mathfrak{a}}$. If $\vec{\mathfrak{a}} = \mathbf{e}$, then obviously $\overline{\Delta}_{\sigma}(s,m) \le \overline{\Delta}_{\sigma}(k,l)$. Let $\vec{\mathfrak{a}} \ne \mathbf{e}$. Then by Lemma 11.2.24, there exists $i \in \mathrm{IS}(\vec{\mathfrak{a}})$ such that

$$\underline{\nabla}_{\sigma}(s,m) \le \mathrm{Rank}(\vec{a}[i]). \tag{11.18}$$

Le tus prove that $\overline{\Delta}_{\sigma}(s,m) \le \overline{\Delta}_{\sigma}(k,l)$. Suppose the converse. Let

$$\overline{\Delta}_{\sigma}(s,m) > \overline{\Delta}_{\sigma}(k,l). \tag{11.19}$$

From (11.19) by Corollary 11.2.19,

$$\underline{\nabla}_{\sigma}(s,m) > \underline{\nabla}_{\sigma}(k,l). \tag{11.20}$$

Then (see (11.18) and (11.20)) there exists $i \in \mathrm{IS}(\vec{\mathfrak{a}})$ such that

$$\underline{\nabla}_{\sigma}(k,l) < \underline{\nabla}_{\sigma}(s,m) \le \mathrm{Rank}(\vec{\mathfrak{a}}[i]). \tag{11.21}$$

Hence by Lemma 11.2.22 we obtain $\mathrm{Lo}(\vec{\mathfrak{a}},k,l) \ne \vec{\mathfrak{a}}$, which is a contradiction. □

Lemma 11.2.28. Let L be an FPJ logic, σ be a signature, $\mathscr{M} \in \mathrm{MC}_{\sigma}(L)$, $\vec{\mathfrak{a}}$ be an $\mathscr{M}(s,m,\mathbf{D})$-inference from a set of external L-formulas Γ, $\langle s',m'\rangle = \overline{\Delta}_{\sigma}(s,m)$. Then $\vec{\mathfrak{a}}$ is an $\mathscr{M}(s',m',\mathbf{D})$-inference from Γ.

Proof. By simple case analysis. □

Theorem 11.1.22 has been proved above. It states that the result of application of the operation Resume to a lower segment of any cut of an inference in a modification calculus will also be an inference. This theorem allows us to prove statements about inferences in modification calculi by induction *over the ranks of inference*, i.e. by embedded induction on the stages and modules of the inference. In this case it is important that the inference obtained as a result of applying the operation Resume has a definite rank that may differ from the rank of the last record of the inference. The following theorem exactly justifies such a proof technique.

Combination of the induction techniques on the length of inference, on the ranks of inference and on the length of upper segments of cuts of various types allows us to obtain most of the results concerning inferences in modification calculi and possessing syntactic character.

Theorem 11.2.29 (Theorem on (k,l)-cut). Let L be an FPJ logic, σ be a signature, $\mathscr{M} \in \mathrm{MC}_\sigma(L)$, $\vec{\mathfrak{a}}$ be an $\mathscr{M}(s,m,d)$-inference from a set of external L-formulas Γ. Then for any $k \in \omega$, $l \in [0..\mathrm{Q}_{\mathscr{M}}]$ such that

$$\overline{\Delta}_\sigma(k,l) < \overline{\Delta}_\sigma(s,m),$$

the string $\mathrm{Resume}(\mathrm{Lo}(\vec{\mathfrak{a}},k,l))$ is an $\mathscr{M}(s',m',\mathbf{D})$-inference from Γ, where

$$\langle s',m'\rangle = \overline{\Delta}_\sigma(k,l).$$

Proof. Let $\mathrm{Sec}(\vec{\mathfrak{a}},s,m) = \left\langle \vec{\mathfrak{b}},\vec{\mathfrak{c}} \right\rangle$, where $\vec{\mathfrak{b}} = \mathrm{Lo}(\vec{\mathfrak{a}},k,l)$, $\vec{\mathfrak{c}} = \mathrm{Hi}(\vec{\mathfrak{a}},k,l)$. It is easy to show that $\vec{\mathfrak{c}} \neq \mathbf{e}$. In the converse case, $\mathrm{Lo}(\vec{\mathfrak{a}},k,l) = \vec{\mathfrak{a}}$. This implies $\overline{\Delta}_\sigma(s,m) \leq \overline{\Delta}_\sigma(k,l)$ by Lemma 11.2.27, which contradicts the conditions.

Let $n = \mathrm{Len}\left(\vec{\mathfrak{b}}\right) + 1$. Then $\vec{\mathfrak{a}}[n]$ is the first element of the string $\vec{\mathfrak{c}}$. Let us consider the possible cases.

CASE 1. $l < \mathrm{Q}_{\mathscr{M}}$. Then

$$\overline{\Delta}_\sigma(k,l) = \langle k,l\rangle. \tag{11.22}$$

By Definition 11.2.20, $\mathrm{Rank}\left(\mathrm{Last}\left(\vec{\mathfrak{b}}\right)\right) \leq \langle k,l\rangle$. On the other hand, by Lemma 11.2.26, there exists a $\mathfrak{d} \in \mathrm{Set}(\vec{\mathfrak{a}})$ such that $\mathrm{Rank}(\mathfrak{d}) = \langle k,l\rangle$. By Definition 11.2.20 $\mathfrak{d} \in \mathrm{Set}\left(\vec{\mathfrak{b}}\right)$. Since $\vec{\mathfrak{a}}$ is ranked,

$$\langle k,l\rangle = \mathrm{Rank}(d) \leq \mathrm{Rank}\left(\mathrm{Last}\left(\vec{\mathfrak{b}}\right)\right) \leq \langle k,l\rangle.$$

Thus,

$$\mathrm{Rank}\left(\mathrm{Last}\left(\vec{\mathfrak{b}}\right)\right) = \langle k,l\rangle. \tag{11.23}$$

Let us determine the rank of $\vec{\mathfrak{a}}[n]$. Obviously, $\mathrm{Rank}(\vec{\mathfrak{a}}[n]) \geq \langle k,l+1\rangle$. Furthermore, by Lemma 11.2.26, there exists $\mathfrak{e} \in \mathrm{Set}(\vec{\mathfrak{a}})$ such that $\mathrm{Rank}(\mathfrak{e}) = \langle k,l+1\rangle \leq \langle s,m\rangle$. Then

$$\langle k,l+1\rangle \leq \mathrm{Rank}(\vec{\mathfrak{a}}[n]) \leq \mathrm{Rank}(\mathfrak{e}) = \langle k,l+1\rangle.$$

Thus,

$$\mathrm{Rank}(\vec{\mathfrak{a}}[n]) = \langle k,l+1\rangle. \tag{11.24}$$

By Theorem 11.1.22, $\mathrm{Resume}\left(\vec{\mathfrak{b}}\right)$ is an $\mathscr{M}(s',m',d')$-inference from the set of external L-formulas Γ, where

$$\begin{aligned}\langle s',m'\rangle &= \text{Rank}\left(\text{Last}\left(\vec{\mathfrak{b}}\right)\right) && \text{(by 11.1.22)}\\ &= \langle k,l\rangle && \text{(by (11.23))}\\ &= \overline{\Delta}_\sigma(k,l). && \text{(by (11.22))}\end{aligned}$$

Thus, it was proved that Resume$\left(\vec{\mathfrak{b}}\right)$ is an $\mathscr{M}(s',m',d')$-inference from Γ, where $\langle s',m'\rangle = \overline{\Delta}_\sigma(k,l)$. It remains to prove that $d' = \mathbf{D}$.

By Theorem 11.1.22 and (11.24), Resume$\left(\vec{\mathfrak{b}} \bullet \vec{\mathfrak{a}}[n]\right)$ is an $\mathscr{M}(k,l+1,d'')$-inference. According to Definition 10.2.31 (NDed, b) changing of a module index is possible only if the inference to which an element is added is in the regime **D**. In the given case $\vec{\mathfrak{a}}[n]$ is added to the $\mathscr{M}(s',m',d')$-inference Resume$\left(\vec{\mathfrak{b}}\right)$.

Therefore $d' = \mathbf{D}$, which was to be proved.

CASE 2. $l = \mathrm{Q}_{\mathscr{M}}$. Then
$$\overline{\Delta}_\sigma(k,l) = \langle k+1,0\rangle.$$

Then by the assumption,
$$\langle k+1,0\rangle < \overline{\Delta}_\sigma(s,m). \tag{11.25}$$

Here two subcases are possible.

SUBCASE 2.1. $\overline{\Delta}_\sigma(s,m) = \langle s,m\rangle$. Then due to (11.25): $\langle k+1,0\rangle < \langle s,m\rangle$. In this case it is obvious:
$$\langle k+1,1\rangle \le \langle s,m\rangle. \tag{11.26}$$

SUBCASE 2.2. $\overline{\Delta}_\sigma(s,m) = \langle s+1,0\rangle$. Then $m = \mathrm{Q}_{\mathscr{M}}$ and due to (11.25), $k < s$. It is easy to see that in this case (11.26) also holds.

By Lemma 11.2.26 and (11.26) (similarly to CASE 1) we obtain that, in CASE 2,
$$\text{Rank}(\vec{\mathfrak{a}}[n]) = \langle k+1,1\rangle. \tag{11.27}$$

It is easy to show that in the considered case Rank$\left(\text{Last}\left(\vec{\mathfrak{b}}\right)\right)$ is equal either to $\langle k,\mathrm{Q}_{\mathscr{M}}\rangle$ or to $\langle k+1,0\rangle$. In any case by 11.2.28, Resume$\left(\vec{\mathfrak{b}}\right)$ can be considered as an $\mathscr{M}(k+1,0)$-inference. (Resume$\left(\vec{\mathfrak{b}}\right)$ is an $\mathscr{M}$-inference by 11.1.22).

Analoguously to CASE 1 the fact that the regime of Resume$\left(\vec{\mathfrak{b}}\right)$ is equal to **D** can be justified by the use of (11.27). □

Corollary 11.2.30. Let L be an FPJ logic, σ be a signature, $\mathscr{M} \in \mathrm{MC}_\sigma(L)$, $\vec{\mathfrak{a}}$ be an $\mathscr{M}(s,m,\mathbf{D})$-inference from a set of external L-formulas Γ. Then for any $k \in \omega$, $l \in [0..\mathrm{Q}_{\mathscr{M}}]$
$$\text{Resume}(\text{Lo}(\vec{\mathfrak{a}},k,l))$$
is an $\mathscr{M}(s',m',\mathbf{D})$-inference from Γ, where

$$\langle s', m' \rangle = \min\left(\overline{\Delta}_\sigma(k,l), \overline{\Delta}_\sigma(s,m)\right).$$

Proof. Consider the possible cases.

CASE 1. $\overline{\Delta}_\sigma(k,l) < \overline{\Delta}_\sigma(s,m)$. Then $\min\left(\overline{\Delta}_\sigma(k,l), \overline{\Delta}_\sigma(s,m)\right) = \overline{\Delta}_\sigma(k,l)$. The fact that $\text{Resume}(\text{Lo}(\vec{\mathfrak{a}}, k, l))$ is an $\mathscr{M}(s', m', \mathbf{D})$-inference from Γ, where $\langle s', m' \rangle = \overline{\Delta}_\sigma(k,l)$, is obtained by Theorem 11.2.29.

CASE 2. $\overline{\Delta}_\sigma(s,m) \leq \overline{\Delta}_\sigma(k,l)$. Then $\min\left(\overline{\Delta}_\sigma(k,l), \overline{\Delta}_\sigma(s,m)\right) = \overline{\Delta}_\sigma(s,m)$. The fact that $\text{Resume}(\text{Lo}(\vec{\mathfrak{a}}, k, l))$ is an $\mathscr{M}(s', m', \mathbf{D})$-inference from Γ, where $\langle s', m' \rangle = \overline{\Delta}_\sigma(s,m)$, is obtained by Lemmas 11.2.27 and 11.2.28. □

Corollary 11.2.31. Let L be an FPJ logic, σ be a signature, $\mathscr{M} \in \text{MC}_\sigma(L)$ and $\vec{\mathfrak{a}}$ be an $\mathscr{M}(0, m, d)$-inference from a set of external L-formulas Γ. Suppose l is a natural number such that

$$\overline{\Delta}_\sigma(0,l) < \overline{\Delta}_\sigma(0,m).$$

Let $\left\langle \vec{\mathfrak{b}}, \vec{\mathfrak{c}} \right\rangle = \text{Sec}(\vec{\mathfrak{a}}, 0, l)$. Then $\vec{\mathfrak{b}}$ is an $\mathscr{M}(0, l, \mathbf{D})$-inference from Γ.

Proof. By 11.2.29, $\text{Resume}\left(\vec{\mathfrak{b}}, 0, l\right)$ is an $\mathscr{M}(s', m', \mathbf{D})$-inference from Γ, where $\langle s', m' \rangle = \overline{\Delta}_\sigma(0,l)$. But $\overline{\Delta}_\sigma(0,l) < \overline{\Delta}_\sigma(0,m)$ implies $\langle 0, l \rangle < \langle 0, m \rangle$. Then $l < m \leq \text{Q}_{\mathscr{M}}$. In this case $\overline{\Delta}_\sigma(0,l) = \langle 0, l \rangle$. By Definition 11.1.12: $\text{Resume}\left(\vec{\mathfrak{b}}, 0, l\right) = \vec{\mathfrak{b}}$, since the records cannot be deactivated at stage 0. □

11.3 Deductive Cuts and Their Applications

In reasoning processes, it is important to be able to deal with deductive and non-deductive parts separately. Deductive cuts that we intend to introduce in the present section will allow us the required separate handling. Namely, deductive cuts will have the following important property. The upper segment of these cuts contains only deductively grounded records, while the last record of the lower segment will be non-deductively grounded. (If the chain does not contain non-deductively grounded records then the lower segment is empty.) Thus by the use of deductive cuts we can extract from the chain a fragment which does not contain the application of modification rules and which is situated at the end of the string

Deductive cuts of inferences in the modification calculi help to determine the moment of stabilisation of knowledge. No changes in the upper segment of the deductive cuts generate new knowledge, since we assume that new knowledge can be generated in a mechanical reasoning process only with the help of plausible inference. More precisely, new knowledge may be extracted from data with the help of such rules.

Deductive cuts allow the extension of our derivation technique through induction on the length of upper segments of deductive cuts, i.e. on the length of final fragment of the chain that does not contain non-deductively grounded records. Below we will find a number of examples using this technique.

Lemma 11.3.1. Let $\vec{\mathfrak{a}}$ be a record string, $K = \mathrm{IS}(\vec{\mathfrak{a}})$. Let $M \subseteq K$ be defined as follows: let $i \in M$ if and only if for any $k \geq i$ the record $\vec{\mathfrak{a}}[k]$ is deductively grounded. Then the ordered pair $\langle \vec{\mathfrak{a}}(K \backslash M), \vec{\mathfrak{a}}(M) \rangle$ is a cut of the string $\vec{\mathfrak{a}}$.

Proof. By direct checking we can see that, for any $i, j \in K$ such that $i \leq j$,

$$i \in M \quad \text{implies} \quad j \in M.$$

Then by Lemma 11.1.9 we obtain what was desired. □

Definition 11.3.2. Let $\vec{\mathfrak{a}}$ be a record string. An ordered pair $\left\langle \vec{\mathfrak{b}}, \vec{\mathfrak{c}} \right\rangle$ is called a *deductive cut* of $\vec{\mathfrak{a}}$, if $\vec{\mathfrak{b}} = \vec{\mathfrak{a}}(\mathrm{IS}(\vec{\mathfrak{a}}) \backslash M)$, $\vec{\mathfrak{c}} = \vec{\mathfrak{a}}(M)$, where

$$M = \{ i \in \mathrm{IS}(\vec{\mathfrak{a}}) \mid \text{ for any } k \geq i,\ \vec{\mathfrak{a}}[k] \text{ is deductively grounded} \}.$$

By Lemma 11.3.1, a deductive cut of $\vec{\mathfrak{a}}$ is really its cut in the sense of Definition 11.1.6. Moreover it is clear that for any record string there exists a unique deductive cut.

$\mathrm{Ded}(\vec{\mathfrak{a}})$ denotes a deductive cut of $\vec{\mathfrak{a}}$. $\mathrm{Lo}^{\mathrm{Ded}}(\vec{\mathfrak{a}})$ and $\mathrm{Hi}^{\mathrm{Ded}}(\vec{\mathfrak{a}})$ denote the lower and upper segments of this cut, respectively.

It is interesting to analyse inferences in modification calculi that are *completed at the zeroth stage*. At this stage only hypotheses obtained from the initial knowledge state will become non-deductively grounded records. It is induction on the length of the upper segment of deductive cuts with the help of which some of the statements below will be proved

Note that the statements below will concern the sets of J-atomic formulas contained in the inferences and domains of internal predicate symbols. Despite the intuitive clarity of the statements below we provide their strict proof or, at least, the idea of the proof.

Lemma 11.3.3. Let L be an FPJ logic, σ be a signature, $\mathscr{M} \in \mathrm{MC}_\sigma(L)$, $p \in \mathrm{Pr}_\sigma(\mathrm{I})$ and $\vec{\mathfrak{a}}$ be an $\mathscr{M}(0,m,d)$-inference. Then

$$\mathrm{JA}_\sigma(\vec{\mathfrak{a}}, p, M) \subseteq \mathrm{JA}_\sigma(p, M) \quad \text{and} \quad \mathrm{JA}_\sigma(\vec{\mathfrak{a}}, M) \subseteq \mathrm{JA}_\sigma(\mathscr{M}).$$

Proof. By straightforward induction on the length of the $\mathscr{M}(0,m,d)$-inference. By Definitions 10.2.14 and 10.2.15 (inductive version). □

Lemma 11.3.4. Let L be an FPJ logic, σ be a signature, $\mathscr{M} \in \mathrm{MC}_\sigma(L)$, $m \in [0..\mathrm{Q}_{\mathscr{M}}]$ and $\vec{\mathfrak{a}}$ be an $\mathscr{M}(0,m,\mathbf{D})$-inference. Then for any $i \leq m$,

$$\mathrm{Dom}_{\mathscr{M}}\left(\mathrm{P}_i^{\mathscr{M}}, \vec{\mathfrak{a}}\right) = \mathrm{Dom}_{\mathscr{M}}\left(\mathrm{P}_i^{\mathscr{M}}\right).$$

Proof. By induction on m.

BASIS. For $m = 0$ the statement is obvious, since ordering of the internal predicate symbols starts with 1.

INDUCTION STEP. Suppose that the statement is true for $m = n$. Let us prove it for $m = n+1$. The proof will be conducted by induction on the length of the upper segment of the deductive cut of the string $\vec{\mathfrak{a}}$ (i.e. on the length of $\mathrm{Hi}^{\mathrm{Ded}}(\vec{\mathfrak{a}})$).

Basis. Let $\mathrm{Hi}^{\mathrm{Ded}}(\vec{\mathfrak{a}}) = \mathbf{e}$. Since $n+1 > 1$, the last record of the string $\vec{\mathfrak{a}}$ should be non-deductively grounded. Therefore this record was added to the inference $\vec{\mathfrak{a}}$ by a rule of 10.2.31 (NDed). Since $\vec{\mathfrak{a}}$ has the regime $\mathbf{D}$ (although the last record of $\vec{\mathfrak{a}}$ is non-deductively grounded) only one case is possible:

$$\mathrm{Dom}_{\mathscr{M}}\left(\mathrm{P}_{n+1}^{\mathscr{M}}, \vec{\mathfrak{a}}, 0\right) = \mathrm{Dom}_{\mathscr{M}}\left(\mathrm{P}_{n+1}^{\mathscr{M}}\right). \tag{11.28}$$

By 11.2.31, $\vec{\mathfrak{b}} = \mathrm{Lo}(\vec{\mathfrak{a}}, 0, n)$ is an $\mathscr{M}(0, n, \mathbf{D})$-inference. Then by the inductive assumption and Corollary 10.2.20 for any $i \leq n$ the following holds:

$$\mathrm{Dom}_{\mathscr{M}}\left(\mathrm{P}_{i}^{\mathscr{M}}, \vec{\mathfrak{a}}\right) = \mathrm{Dom}_{\mathscr{M}}\left(\mathrm{P}_{i}^{\mathscr{M}}\right). \tag{11.29}$$

Taking into account (11.28) we obtain that (11.29) holds for any $i \leq n+1$.

Induction step. Let the lemma hold for the k-long upper segment of the deductive cut of any $\mathscr{M}(0, m, \mathbf{D})$-inference. Suppose that $\mathrm{Len}\left(\mathrm{Hi}^{\mathrm{Ded}}(\vec{\mathfrak{a}})\right) = k+1$. Then $\mathrm{Ded}(\vec{\mathfrak{a}}) = \langle \vec{\mathfrak{c}}, \vec{\mathfrak{d}} \bullet \mathfrak{e} \rangle$ for some strings $\vec{\mathfrak{c}}$ and $\vec{\mathfrak{d}}$ and for the record $\mathfrak{e}$. Moreover, $\mathfrak{e} \in \mathrm{DASet}(\vec{\mathfrak{a}})$. Then $\vec{\mathfrak{c}} + \vec{\mathfrak{d}}$ is an $\mathscr{M}(0, m, \mathbf{D})$-inference with the deductive cut $\langle \vec{\mathfrak{c}}, \vec{\mathfrak{d}} \rangle$. But $\mathrm{Len}(\vec{\mathfrak{d}}) = k$. Then by the inductive assumption and Corollary 10.2.20 for any $i \leq m$, $\mathrm{Dom}_{\mathscr{M}}(\mathrm{P}_{i}^{\mathscr{M}}, \vec{\mathfrak{a}}) = \mathrm{Dom}_{\mathscr{M}}(\mathrm{P}_{i}^{\mathscr{M}})$. □

Corollary 11.3.5. Let L be an FPJ logic, σ be a signature, $\mathscr{M} \in \mathrm{MC}_{\sigma}(L)$ and $\vec{\mathfrak{a}}$ be an $\mathscr{M}(0, \mathrm{Q}_{\mathscr{M}}, \mathbf{D})$-inference. Then for any $p \in \mathrm{Pr}_{\sigma}(\mathsf{I})$, $\mathrm{Dom}_{\mathscr{M}}(p, \vec{\mathfrak{a}}) = \mathrm{Dom}_{\mathscr{M}}(p)$.

Lemma 11.3.6. Let $\mathscr{M} \in \mathrm{MC}_{\sigma}(L)$ and $\vec{\mathfrak{a}}$ be an $\mathscr{M}(0, \mathrm{Q}_{\mathscr{M}}, \mathbf{D})$-inference. Then

$$\mathrm{JA}_{\sigma}(\vec{\mathfrak{a}}, \mathscr{M}) = \mathrm{JA}_{\sigma}(\mathscr{M}).$$

Proof. By Corollary 11.3.5 for any $p \in \mathrm{Pr}_{\sigma}(\mathsf{I})$ it holds that

$$\mathrm{Dom}_{\mathscr{M}}(p, \vec{\mathfrak{a}}) = \mathrm{Dom}_{\mathscr{M}}(p). \tag{11.30}$$

Now let us prove that

$$\mathrm{JA}_{\sigma}(\vec{\mathfrak{a}}, M) = \mathrm{JA}_{\sigma}(\mathscr{M}). \tag{11.31}$$

By Lemma 11.3.3, $\mathrm{JA}_{\sigma}(\vec{\mathfrak{a}}, \mathscr{M}) \subseteq \mathrm{JA}_{\sigma}(\mathscr{M})$. For any $p \in \mathrm{Pr}_{\sigma}(\mathsf{I})$ it follows from (11.30) that for each $\vec{\mathfrak{a}} \in \mathrm{Dom}_{\mathscr{M}}(p)$ there exists an $\varepsilon \in \mathscr{V}(L, \mathsf{I})$ such that

$$\mathrm{J}_{\varepsilon}\, p(\vec{a}) \in \mathrm{JA}_{\sigma}(\vec{\mathfrak{a}}, \mathscr{M}) \subseteq \mathrm{JA}_{\sigma}(\mathscr{M}).$$

Let $\mathrm{J}_{\delta}\, p(\vec{a}) \in \mathrm{JA}_{\sigma}(\mathscr{M})$. Suppose that $\mathrm{J}_{\delta}\, \mathrm{P}_{i}^{\mathscr{M}}(\vec{\mathfrak{a}}) \notin \mathrm{JA}_{\sigma}(\vec{\mathfrak{a}}, \mathscr{M})$. Then there are $\varepsilon, \delta \in \mathscr{V}(L, \mathsf{I})$ $(\varepsilon \neq \delta)$ such that $\mathrm{J}_{\varepsilon}\, p(\vec{a}) \in \mathrm{JA}_{\sigma}(\mathscr{M})$ and $\mathrm{J}_{\delta}\, \mathrm{P}_{i}^{\mathscr{M}}(\vec{a}) \in \mathrm{JA}_{\sigma}(\mathscr{M})$. This contradicts the fact that $\mathrm{Descr}_{\mathscr{M}}$ is a state description over the set of constants

$\text{Univ}_{\mathcal{M}}$ (see Definitions 10.1.10 and 10.2.6). Consequently $\mathrm{J}_\delta \mathrm{P}_i^{\mathcal{M}}(\vec{a}) \in \mathrm{JA}_\sigma(\vec{\mathfrak{a}}, \mathcal{M})$. Therefore (11.31) also holds. □

Corollary 11.3.7. Let L be an FPJ logic, σ be a signature, $\mathcal{M} \in \mathrm{MC}_\sigma(L)$ and $\vec{\mathfrak{a}}$ be an $\mathcal{M}(1,0,\mathbf{D})$-inference. Then

$$\mathrm{JA}_\sigma(\vec{\mathfrak{a}}, \mathcal{M}) = \mathrm{JA}_\sigma(\mathcal{M}).$$

Proof. By trivial induction on the length of the upper segment of the deductive cut of the $\mathcal{M}(1,0,\mathbf{D})$-inference $\vec{\mathfrak{a}}$. In the basis Lemma 11.3.6 is used. □

Lemma 11.3.8. Let L be an FPJ logic, $\mathcal{M} \in \mathrm{MC}_\sigma(L)$ and $\vec{\mathfrak{a}}$ be an $\mathcal{M}(s,m,d)$-inference, where $s > 0$. Then $\mathrm{JA}_\sigma(\vec{\mathfrak{a}}, \mathcal{M})$ is an $\mathrm{IS}_\sigma\mathrm{PD}(\text{Univ}_{\mathcal{M}}, L)$. I.e. for any $p \in \mathrm{Pr}_\sigma(\mathsf{I})$ and $\vec{\mathfrak{a}} \in \mathrm{Dom}_{\mathcal{M}}(p)$ there exists a unique $\varepsilon \in \mathcal{V}(L,\mathsf{I})$ such that $\mathrm{J}_\varepsilon\, p(\vec{a}) \in \mathrm{JA}_\sigma(\vec{\mathfrak{a}}, \mathcal{M})$ (see 10.1.8).

Proof. By induction on the length of the upper segment of the $(1,0)$-cut of the string $\vec{\mathfrak{a}}$, i.e. on the length of $\mathrm{Hi}(\vec{\mathfrak{a}},1,0)$.

BASIS. Let $\mathrm{Hi}(\vec{\mathfrak{a}},1,0) = \mathbf{e}$. But then $\vec{\mathfrak{a}}$ is an M(1, 0, **D**)-inference, hence by Corollary 11.3.7, $\mathrm{JA}_\sigma(\vec{\mathfrak{a}}, \mathcal{M}) = \mathrm{JA}_\sigma(\mathcal{M})$, but by Remark 10.1.12 (ii), $\mathrm{JA}_\sigma(\mathcal{M})$ is an $\mathrm{IS}_\sigma\mathrm{PD}(\text{Univ}_{\mathcal{M}}, L)$ (see Definition 10.1.8).

INDUCTION STEP. Let the statement hold for the n-long upper segment of the $(1,0)$-cut of $\vec{\mathfrak{a}}$. Suppose that $\mathrm{Len}(\mathrm{Hi}(\vec{\mathfrak{a}},1,0)) = n+1$. Then $\vec{\mathfrak{a}} = \vec{\mathfrak{b}} \bullet \mathfrak{c}$, where $\mathfrak{c}$ is added by the induction step of Definition 10.2.31. Supppose that $\mathfrak{c}$ is added to $\vec{\mathfrak{b}}$ without using any modification rule. Then $\vec{\mathfrak{b}}$ is an $\mathcal{M}$-inference (with stage more than 0) with an n-long upper segment of the $(1,0)$-cut. Then by the inductive assumption $\mathrm{JA}_\sigma\left(\vec{\mathfrak{b}}, \mathcal{M}\right)$ is an $\mathrm{IS}_\sigma\mathrm{PD}(\text{Univ}_{\mathcal{M}}, L)$. However in all these cases clearly $\mathrm{JA}_\sigma(\vec{\mathfrak{a}}, \mathcal{M}) = \mathrm{JA}_\sigma\left(\vec{\mathfrak{b}}, \mathcal{M}\right)$.

If $\mathfrak{c}$ is added by the modification rule $\mathrm{MR}_{\mathcal{M}}[\varepsilon, p]$, then clearly

$$\mathrm{JA}_\sigma(\vec{\mathfrak{a}}, \mathcal{M}) = (\mathrm{JA}_\sigma(\vec{\mathfrak{c}}, \mathcal{M}) \setminus \{\mathrm{J}_\tau\, p(\vec{a})\}) \cup \{\mathrm{J}_\varepsilon\, p(\vec{a})\}, \qquad (11.32)$$

where $\vec{\mathfrak{c}}$ is an $\mathcal{M}$-inference, $\vec{\mathfrak{b}} = \mathrm{Suspend}(\vec{\mathfrak{c}}, \mathrm{J}_\tau\, p(\vec{a}))$, $p \in \mathrm{Pr}_\sigma(\mathsf{I})$, $\vec{\mathfrak{a}} \in \mathrm{Dom}_{\mathcal{M}}(p)$, $\varepsilon \in \mathcal{V}(L,\mathsf{I})$ with $\mathrm{Len}(\mathrm{Hi}(\vec{\mathfrak{c}},1,0)) = n$. Then, by the inductive assumption, the set $\mathrm{JA}_\sigma(\vec{\mathfrak{c}}, \mathcal{M})$ is an $\mathrm{IS}_\sigma\mathrm{PD}(\text{Univ}_{\mathcal{M}}, L)$. Besides that, $\mathrm{J}_\tau\, p(\vec{\mathfrak{a}}) \in \mathrm{JA}_\sigma(\vec{\mathfrak{c}}, \mathcal{M})$ (otherwise we would not be able to apply the rule $\mathrm{MR}_{\mathcal{M}}[\varepsilon, p]$). It is easy to see that in this case $\mathrm{JA}_\sigma(\vec{\mathfrak{a}}, \mathcal{M})$ is also an $\mathrm{IS}_\sigma\mathrm{PD}(\text{Univ}_{\mathcal{M}}, L)$. □

Lemma 11.3.9. Let L be an FPJ logic, σ be a signature, $\mathcal{M} \in \mathrm{MC}_\sigma(L)$ and $\vec{\mathfrak{a}}$ be an $\mathcal{M}(s,m,d)$-inference, where $s > 0$. Then for any $p \in \mathrm{Pr}_\sigma(\mathsf{I})$, $\mathrm{Dom}_{\mathcal{M}}(p, \vec{\mathfrak{a}}) = \mathrm{Dom}_{\mathcal{M}}(p)$.

Proof. This lemma directly follows from Lemma 11.3.8 and the corresponding definitions (10.1.8 and 10.2.17). □

Corollary 11.3.10. Let L be an FPJ logic, σ be a signature, $\mathscr{M} \in \mathrm{MC}_\sigma(L)$ and $\vec{\mathfrak{a}}$ be an $\mathscr{M}(s,m,d)$-inference, where $s > 0$. Then the set

$$\mathrm{Descr}_{\mathscr{M}}\left[\mathrm{JA}_\sigma(\mathscr{M}) \leftarrow \mathrm{JA}_\sigma(\vec{\mathfrak{a}},\mathscr{M})\right]$$

is a state description of signature σ over $\mathrm{Univ}_{\mathscr{M}}$.

Proof. By Lemma 11.3.8, $\mathrm{JA}_\sigma(\vec{\mathfrak{a}},\mathscr{M})$ is an $\mathrm{IS}_\sigma\mathrm{PD}$ over $\mathrm{Univ}_{\mathscr{M}}$. Then by Remark 10.1.15, $\mathrm{Descr}_{\mathscr{M}}\left[\mathrm{JA}_\sigma(\mathscr{M}) \leftarrow \mathrm{JA}_\sigma(\vec{\mathfrak{a}},\mathscr{M})\right]$ is a state description of signature σ over $\mathrm{Univ}_{\mathscr{M}}$. □

Lemma 11.3.11. Let L be an FPJ logic, σ be a signature, $\mathscr{M} \in \mathrm{MC}_\sigma(L)$ and $\vec{\mathfrak{a}} = \vec{\mathfrak{b}} \bullet \mathfrak{c}$ be an $\mathscr{M}(s,m,d)$-inference, where $\mathfrak{c}$ is deductively grounded. Then $\mathrm{JA}_\sigma\left(\vec{\mathfrak{b}} \bullet \mathfrak{c},\mathscr{M}\right) = \mathrm{JA}_\sigma\left(\vec{\mathfrak{b}},\mathscr{M}\right)$.

Proof. It is evident (see Definition 10.2.14). □

We continue investigating the properties of deductive cuts. The next lemma will show that the pure deductive tail of inference does not affect the J-atomic formulas that represent the fact base in the inference. Intuitively, the statement seems to be trivial. Its strict proof can be done by induction on the length of the upper segment of the deductive inference in the modification calculi.

Lemma 11.3.12. Let L be an FPJ logic, σ be a signature, $\mathscr{M} \in \mathrm{MC}_\sigma(L)$ and $\vec{\mathfrak{a}}$ be an $\mathscr{M}(s,m,d)$-inference, where $s > 0$. Then

$$\mathrm{JA}_\sigma(\vec{\mathfrak{a}},\mathscr{M}) = \mathrm{JA}_\sigma\left(\mathrm{Lo}^{\mathrm{Ded}}(\vec{\mathfrak{a}}),M\right).$$

Proof. By induction on the length of $\mathrm{Hi}^{\mathrm{Ded}}(\vec{\mathfrak{a}})$.

BASIS. Let $\mathrm{Len}\left(\mathrm{Hi}^{\mathrm{Ded}}(\vec{\mathfrak{a}})\right) = 0$. Then $\vec{\mathfrak{a}} = \mathrm{Lo}^{\mathrm{Ded}}(\vec{\mathfrak{a}})$ and the statement holds.

INDUCTION STEP. Suppose that the statement holds under the length of the upper segment of deductive cut equal to n. Let $\mathrm{Len}\left(\mathrm{Hi}^{\mathrm{Ded}}(\vec{\mathfrak{a}})\right) = n+1$. Then $\vec{\mathfrak{a}} = \vec{\mathfrak{b}} \bullet \mathfrak{c}$, where the record $\mathfrak{c}$ is deductively grounded. Note that in this case it is necessary that $d = \mathbf{D}$. Then $\vec{\mathfrak{b}}$ is also an $\mathscr{M}(s,m,d)$-inference.

Clearly,

$$\mathrm{Len}\left(\mathrm{Hi}^{\mathrm{Ded}}\left(\vec{\mathfrak{b}}\right)\right) = n, \quad \mathrm{Lo}^{\mathrm{Ded}}\left(\vec{\mathfrak{b}}\right) = \mathrm{Lo}^{\mathrm{Ded}}(\vec{\mathfrak{a}}).$$

By the inductive assumption

$$\mathrm{JA}_\sigma\left(\vec{\mathfrak{b}},\mathscr{M}\right) = \mathrm{JA}_\sigma\left(\mathrm{Lo}^{\mathrm{Ded}}\left(\vec{\mathfrak{b}}\right),\mathscr{M}\right).$$

But

$$\mathrm{JA}_\sigma(\vec{\mathfrak{a}},\mathscr{M}) = \mathrm{JA}_\sigma\left(\vec{\mathfrak{b}},\mathscr{M}\right),$$

since $\vec{\mathfrak{a}} = \vec{\mathfrak{b}} \bullet \mathfrak{c}$, where $\mathfrak{c}$ is deductively grounded (see Lemma 11.3.11). □

We continue to build the notion set for the inferences in the modification calculi and introduce the notion of *extended description of internal syntactical predicates* for the inferences in the modification calculi. This notion is required for the definition of adequate semantics for the modification calculi. To introduce such semantics we will have *to find a syntactic L-structure corresponding to each inference*.

As we know, any syntactic L-structure is entirely defined by a state description. We could obtain a state description for each inference in an arbitrary modification calculus $\mathscr{M}$ by replacing the set of J-atomic formulas contained in the initial description $\mathrm{Descr}_{\mathscr{M}}$ by the set of non-deductively grounded J-atomic formulas contained in this inference.

We can always do this for the inferences whose stage is greater than 0. However, the inference of stage 0 may not yet contain all the necessary J-atomic formulas. Thus the replacement of J-atomic formulas that we mentioned above may not lead to the expected result. I.e. the result of such a substitution may turn out to be a set of external formulas, the set of which does not describe a state.

However, arbitrary inference of stage 0 in a fixed modification calculus $\mathscr{M}$ cannot contain any non-deductively grounded J-atomic formula except for those of $\mathrm{Descr}_{\mathscr{M}}$. But when stage 0 of an inference is completed, all non-deductively grounded J-atomic formulas of $\mathrm{Descr}_{\mathscr{M}}$ will be within the inference.

An extended description of syntactic predicates contains by definition all J-atomic formulas from $\mathrm{Descr}_{\mathscr{M}}$ for any $\mathscr{M}$-inference of stage 0. Subsequently this allows one to correspond a certain syntactic L-structure to each $\mathscr{M}$-inference, even to the $\mathscr{M}$-inference that does not contain all the necessary J-atomic formulas.

Definition 11.3.13. Let L be an FPJ logic, σ be a signature, $\mathscr{M} \in \mathrm{MC}_\sigma(L)$ and $\vec{\mathfrak{a}}$ be an $\mathscr{M}(s,m,d)$-inference. Let by definition

$$\mathrm{JA}^+_\sigma(\vec{\mathfrak{a}},\mathscr{M}) = \begin{cases} \mathrm{JA}_\sigma(\mathscr{M}) & s=0, \\ \mathrm{JA}_\sigma(\vec{\mathfrak{a}},\mathscr{M}) & s>0. \end{cases}$$

The set $\mathrm{JA}^+_\sigma(\vec{\mathfrak{a}},\mathscr{M})$ is called an *extended description of syntactic internal predicate from* $\vec{\mathfrak{a}}$.

Proposition 11.3.14. Let L be an FPJ logic, σ be a signature, $\mathscr{M} \in \mathrm{MC}_\sigma(L)$ and $\vec{\mathfrak{a}}$ be an $\mathscr{M}(s,m,d)$-inference. Then the set $\mathrm{JA}^+_\sigma(\vec{\mathfrak{a}},\mathscr{M})$ is well defined.

Proof. We must prove that $\mathrm{JA}^+_\sigma(\vec{\mathfrak{a}},\mathscr{M})$ does not depend on the choice of s. There is only one case where this dependence is possible. Namely, this is the case where $\vec{\mathfrak{a}}$ is simultaneously an $\mathscr{M}(0,\mathrm{Q}_{\mathscr{M}},\mathbf{D})$-inference and an $\mathscr{M}(1,0,\mathbf{D})$-inference. However, in this case, by Corollary 11.3.5 and Corollary 11.3.7, $\mathrm{JA}^+_\sigma(\vec{\mathfrak{a}},\mathscr{M}) = \mathrm{JA}_\sigma(\mathscr{M})$. □

For the construction of canonical models of theories in modification calculi we require further notions. Canonical models of theories in modification calculi are important in our approach. Further on these theories will be called stage cognition theories (SCT). A set of external L-formulas Γ, the inferences from which will be discussed, will play the role of non-logical axioms for SCT. But an inference in a modification calculus may be interpreted as a reasoning process that can lead not only to conclusions but also to modification of premises. To each of these inferences its own version of the initial set Γ corresponds. (Γ is the set of non-logical axioms or hypotheses.) The set Γ is a metaset of a state description from the modification calculus. Obviously, each inference in a modification calculus corresponds to its own version of this state description.

Definition 11.3.15. Let L be an FPJ logic, σ be a signature, $\mathscr{M} \in \mathrm{MC}_\sigma(L)$ and $\vec{\mathfrak{a}}$ be an $\mathscr{M}(s,m,d)$-inference. Let by definition

$$\mathrm{CSD}(\vec{\mathfrak{a}}) = \mathrm{Descr}_{\mathscr{M}}\left[\mathrm{JA}_\sigma(M) \leftarrow \mathrm{JA}_\sigma^+(\vec{\mathfrak{a}},\mathscr{M})\right].$$

Obviously $\mathrm{CSD}(\vec{\mathfrak{a}}) = \mathrm{Descr}_{\mathscr{M}}$ for $s=0$ (see 11.3.13). Due to this fact and Corollary 11.3.10, $\mathrm{CSD}(\vec{\mathfrak{a}})$ is a state description of signature σ over $\mathrm{Univ}_{\mathscr{M}}$. The set $\mathrm{CSD}(\vec{\mathfrak{a}})$ is called the *current state description of* $\vec{\mathfrak{a}}$.

Definition 11.3.16. Let L be an FPJ logic, σ be a signature, $\mathscr{M} \in \mathrm{MC}_\sigma(L)$, $\Gamma \in \mathrm{AFS}(\mathscr{M})$ and $\vec{\mathfrak{a}}$ be an $\mathscr{M}(s,m,d)$-inference from Γ. Let by definition

$$\mathrm{Cur}(\Gamma,\vec{\mathfrak{a}}) = \Gamma \cup \mathrm{CSD}(\vec{\mathfrak{a}}).$$

The set $\mathrm{Cur}(\Gamma,\vec{\mathfrak{a}})$ will be called the *current version of* Γ w.r.t. $\vec{\mathfrak{a}}$. Obviously $\mathrm{Cur}(\Gamma,\vec{\mathfrak{a}}) = \Gamma$ for $s=0$.

Remark 11.3.17. Let L be an FPJ logic, σ be a signature, $\mathscr{M} \in \mathrm{MC}_\sigma(L)$, $\Gamma \in \mathrm{AFS}(\mathscr{M})$ and $\vec{\mathfrak{a}}$ be an $\mathscr{M}(s,m,d)$-inference Γ. Then the following statements hold:

(i) $\mathrm{JA}_\sigma(\vec{\mathfrak{a}},\mathscr{M}) \subseteq \mathrm{JA}_\sigma^+(\vec{\mathfrak{a}},\mathscr{M})$,
(ii) $\mathrm{Descr}_{\mathscr{M}}[\mathrm{JA}_\sigma(\mathscr{M}) \leftarrow \mathrm{JA}_\sigma(\vec{\mathfrak{a}},M)] \subseteq \mathrm{CSD}(\vec{\mathfrak{a}})$,
(iii) $\Gamma \cup \mathrm{Descr}_{\mathscr{M}}[\mathrm{JA}_\sigma(\mathscr{M}) \leftarrow \mathrm{JA}_\sigma(\vec{\mathfrak{a}},M)] \subseteq \mathrm{Cur}(\Gamma,\vec{\mathfrak{a}})$.

11.4 Deductive Correctness

In this section we discuss the connection between inferences in modification calculi and the derivability in L^1. First we introduce the notion of record strings deductively correct w.r.t. a set of external L-formulas. To transform record strings that are deductively correct w.r.t. the set Γ into the traditional proof from Γ in the L^1calculus it is sufficient to construct a string that consists only of the formulas that occur in the records. Below such a string of formulas will be called a formula extract of the initial record string.

Definition 11.4.1. Let L be an FPJ logic, $\vec{\mathfrak{a}}$ be a record string and Γ be a set of external L-formulas. We will say that *a string* $\vec{\mathfrak{a}}$ *is deductively correct w.r.t.* Γ if for each $k \in \mathrm{AIS}(\vec{\mathfrak{a}})$ at least one of the following holds:

(i) $\mathscr{F}(\vec{\mathfrak{a}}[k]) \in \Gamma$,

(ii) $\mathscr{F}(\vec{\mathfrak{a}}[k])$ is an axiom of L^1,
(iii) There exist $i, j \in \mathrm{AIS}(\vec{\mathfrak{a}})$ such that $i < k$, $j < k$ and

$$\mathscr{F}(\vec{\mathfrak{a}}[j]) = (\mathscr{F}(\vec{\mathfrak{a}}[i]) \to \mathscr{F}(\vec{\mathfrak{a}}[k])),$$

(iv) There exists $i \in \mathrm{AIS}(\vec{\mathfrak{a}})$ such that $i < k$ and $\mathscr{F}(\vec{\mathfrak{a}}[k]) = \forall x \mathscr{F}(\vec{\mathfrak{a}}[i])$.

Remark 11.4.2. Let L be an FPJ logic, $\vec{\mathfrak{a}}$ and $\vec{\mathfrak{b}}$ be record strings, $\vec{\mathfrak{c}} = \vec{\mathfrak{a}} + \vec{\mathfrak{b}}$ and Γ be a set of external L-formulas. Then the following statements hold:

(i) If $\vec{\mathfrak{c}}$ is deductively correct w.r.t. Γ, then the string $\vec{\mathfrak{a}}$ (initial segment of the string $\vec{\mathfrak{c}}$) is also deductively correct w.r.t. Γ;
(ii) If $\vec{\mathfrak{a}}$ and $\vec{\mathfrak{b}}$ are deductively correct w.r.t. Γ, then the string $\vec{\mathfrak{c}}$ (concatenation of $\vec{\mathfrak{a}}$ and $\vec{\mathfrak{b}}$) will also be deductively correct w.r.t. Γ.

Proof. Clear. □

Remark 11.4.3. Let L be an FPJ logic, the record string $\vec{\mathfrak{c}}$ be deductively correct w.r.t. the set of external L-formulas Γ and $\mathfrak{a}$ be a record. Then the string $\vec{\mathfrak{c}} \bullet \mathfrak{a}$ is deductively correct w.r.t. the set $\Gamma \cup \{\mathscr{F}(\mathfrak{a})\}$.

Proof. It is obvious. □

Definition 11.4.4. Let L be an FPJ logic, $\vec{\mathfrak{a}}$ be a record string. A formula string $\mathscr{F}^*\left(\vec{\mathfrak{b}}\right)$, where $\vec{\mathfrak{b}} = \vec{\mathfrak{a}}(\mathrm{AIS}(\vec{\mathfrak{a}}))$, will be called a formula extract of $\vec{\mathfrak{a}}$. The *formula extract of* $\vec{\mathfrak{a}}$ will be denoted $\mathrm{FExtr}(\vec{\mathfrak{a}})$. The function FExtr may be defined by induction on the length of $\vec{\mathfrak{a}}$ or (more easily) on record strings.

BASIS. $\mathrm{FExtr}(\mathbf{e}) = \mathbf{e}$.

INDUCTION STEP.

$$\mathrm{FExtr}(\vec{\mathfrak{a}} \bullet \mathfrak{a}) = \begin{cases} \mathrm{FExtr}(\vec{\mathfrak{a}}) \bullet \mathscr{F}(\mathfrak{a}) & \text{if } \mathrm{DNum}(\mathfrak{a}) = 0, \\ \mathrm{FExtr}(\vec{\mathfrak{a}}) & \text{otherwise.} \end{cases}$$

Lemma 11.4.5. Let L be an FPJ logic, $\vec{\mathfrak{a}}$ be a record string deductively correct w.r.t. the set of external L-formulas Γ. Then $\mathrm{FExtr}(\vec{\mathfrak{a}})$ is an inference from Γ in L^1.

Proof. It immediately follows from Definition 11.4.1. □

Lemma 11.4.6. Let L be an FPJ logic, σ be a signature, $\mathscr{M} \in \mathrm{MC}_\sigma(L)$, $\Gamma \in \mathrm{AFS}(\mathscr{M})$ and $\vec{\mathfrak{a}}$ be an $\mathscr{M}(0,m,d)$-inference from Γ. Then $\mathrm{FExtr}(\vec{\mathfrak{a}})$ is an inference from $\Gamma \cup \mathrm{Descr}_{\mathscr{M}}$ in L^1.

Proof. Taking into account Lemma 11.4.5, it is enough to show that $\vec{\mathfrak{a}}$ is deductively correct w.r.t. Γ. This can be proved by a simple induction on the length of $\mathscr{M}(0,m,d)$-inference. While constructing such an inference we can add to it only records with formulas that:

- Either belong to $\Gamma \cup \mathrm{Descr}_{\mathscr{M}}$,

- Or is an axiom of axiom of L^1,
- Or are obtained by MP or Gen rules from formulas that are contained in the previous records of the inference.

Below we establish a few results that are necessary to prove the main theorem of this section. In this theorem for each modification calculus M, $\Gamma \in \mathrm{AFS}(\mathscr{M})$ and $\mathscr{M}$-inference $\vec{\mathfrak{a}}$ from Γ, we find a set of external St-formulas Ξ such that $\mathrm{FExtr}(\vec{a})$ is an inference from Ξ in the St^1calculus. According to Lemma 11.4.5 it is enough that $\vec{\mathfrak{a}}$ be deductively correct w.r.t. Ξ. □

Lemma 11.4.7. Let L be an FPJ logic, $\vec{\mathfrak{a}}$ be a record string deductively correct w.r.t. a set of external L-formulas Γ and Θ be an external L-formula, $m \in \omega$, $m > 0$. Then the string $\vec{\mathfrak{c}} = \mathrm{Suspend}(\vec{\mathfrak{a}}, \Theta, m)$ is deductively correct w.r.t. $\Gamma \setminus \{\Theta\}$.

Proof. By induction on the length of $\vec{\mathfrak{a}}$.

BASIS. For the empty string the statement clearly holds.

INDUCTION STEP. Suppose the lemma holds for n-long strings. Let the length of $\vec{\mathfrak{a}}$ be equal to $n+1$. Then the string $\vec{\mathfrak{a}}$ can be represented in the form $\vec{\mathfrak{b}} \bullet \mathfrak{a}$, where $\vec{\mathfrak{b}}$ is the n-long initial segment of $\vec{\mathfrak{a}}$ and $\mathfrak{a}$ is the last element of $\vec{\mathfrak{a}}$.

By Definition 10.2.24 in this case

$$\vec{\mathfrak{c}} = \mathrm{Suspend}(\vec{\mathfrak{a}}, \Theta, m) = \mathrm{Suspend}\left(\vec{\mathfrak{b}}, \Theta, m\right) \bullet \mathfrak{b},$$

where

$$\mathfrak{b} = \begin{cases} \mathfrak{a} & \mathfrak{a} \text{ is deductively } \Theta\text{-free in } \vec{\mathfrak{a}}, \\ \mathrm{Deactivate}(a, m) & \mathfrak{a} \text{ deductively depends on } \Theta \text{ in } \vec{\mathfrak{a}}. \end{cases}$$

By Remark 11.4.2(i), $\vec{\mathfrak{b}}$ is deductively correct w.r.t. Γ. Then by the inductive assumption $\vec{\mathfrak{d}} = \mathrm{Suspend}\left(\vec{\mathfrak{b}}, \Theta, m\right)$ is deductively correct w.r.t. $\Gamma \setminus \{\Theta\}$.

Suppose that $k \in \mathrm{AIS}(\vec{\mathfrak{c}})$, where $\vec{\mathfrak{c}} = \mathrm{Suspend}(\vec{\mathfrak{a}}, \Theta, m)$. Then:

- Either $k \in \mathrm{AIS}(\vec{\mathfrak{d}})$, where $\vec{\mathfrak{d}} = \mathrm{Suspend}\left(\vec{\mathfrak{b}}, \Theta, m\right)$,
- Or $k = n+1$, $\mathfrak{b} = \mathfrak{a}$ and $\mathfrak{a}$ is deductively Θ-free in the string $\vec{\mathfrak{a}}$.

We have to show that for any k at least one of the following conditions holds:

(a) $\mathscr{F}(\vec{\mathfrak{c}}[k]) \in \Gamma \setminus \{\Theta\}$,
(b) $\mathscr{F}(\vec{\mathfrak{c}}[k])$ is an axiom of the calculus L,
(c) There exist $i, j \in \mathrm{AIS}(\vec{\mathfrak{c}})$ such that $i < k$, $j < k$ and

$$\mathscr{F}(\vec{\mathfrak{c}}[j]) = (\mathscr{F}(\vec{\mathfrak{c}}[i]) \to \mathscr{F}(\vec{\mathfrak{c}}[k])),$$

(d) There exists $i \in \mathrm{AIS}(\vec{\mathfrak{c}})$ such that $i < k$ and $\mathscr{F}(\vec{\mathfrak{c}}[k]) = \forall x \mathscr{F}(\vec{\mathfrak{c}}[i])$.

If $k \in \mathrm{AIS}(\vec{\mathfrak{d}})$, where $\vec{\mathfrak{d}} = \mathrm{Suspend}\left(\vec{b}, \Theta, m\right)$, then at least one of these conditions holds since by the inductive assumption $\vec{\mathfrak{d}}$ is deductively correct w.r.t. $\Gamma \setminus \{\Theta\}$

Let $k = n + 1$ and $\mathfrak{a}$ be deductively Θ-free in the string $\vec{\mathfrak{a}}$. Then

$$\mathscr{F}(\vec{\mathfrak{c}}[k]) = \mathscr{F}(\vec{\mathfrak{a}}[k]) = \mathscr{F}(a) \neq \Theta.$$

Recall that $k \in \mathrm{AIS}(\vec{\mathfrak{c}}) \subseteq \mathrm{AIS}(\vec{\mathfrak{a}})$ and $\vec{\mathfrak{a}}$ is deductively correct w.r.t. Γ. Then for k at least one of the conditions (i)–(iv) of Definition 11.4.1 holds. Condition (i) clearly implies (a) and (ii) implies (b). If condition (iii) holds, then condition (c) also holds. Otherwise $\mathfrak{a}$ will deductively depend on Θ in $\vec{\mathfrak{a}}$. Similarly, (iv) implies (b). □

Lemma 11.4.8. Let L be an FPJ logic, σ be a signature, $\mathscr{M} \in \mathrm{MC}_\sigma(L)$, $\Gamma \in \mathrm{AFS}(\mathscr{M})$ and $\vec{\mathfrak{a}}$ be an $\mathscr{M}(s,m,d)$-inference from Γ. Then $\vec{\mathfrak{a}}$ is deductively correct w.r.t. the set

$$\Gamma \cup \mathrm{Descr}_{\mathscr{M}}[\mathrm{JA}_\sigma(\mathscr{M}) \leftarrow \mathrm{JA}_\sigma(\vec{\mathfrak{a}}, \mathscr{M})]$$

(the notation was introduced in 10.1.13) .

Proof. By induction on the length of the string $\vec{\mathfrak{a}}$.

BASIS. The lemma clearly holds for the empty string.

INDUCTION STEP. Suppose that the lemma holds for the n-long strings. Let the length of the string $\vec{\mathfrak{a}}$ be equal to $n+1$. Then the string $\vec{\mathfrak{a}}$ can be represented in the form $\vec{\mathfrak{b}} \bullet \mathfrak{a}$, where $\vec{\mathfrak{b}}$ is the n-long initial segment of the string $\vec{\mathfrak{a}}$ and $\mathfrak{a}$ is the last element of $\vec{\mathfrak{a}}$. The string $\vec{\mathfrak{b}}$ is not necessarily an $\mathscr{M}$-inference.

Let us consider the possible cases. In each possible case our argumentation will be based on Definition 10.2.31 and also on Definition 11.4.1.

CASE 1. The record $\mathfrak{a}$ is deductively grounded (cf. 10.2.31 (Ded)), i.e. the type of Reason$(\mathfrak{a})$ is equal to Hyp, LAx, OAx, MP or Gen. Then

$$\mathrm{JA}_\sigma(\vec{\mathfrak{a}}, D) = \mathrm{JA}_\sigma\left(\vec{\mathfrak{b}}, D\right)$$

and $\vec{\mathfrak{b}}$ is an $\mathscr{M}$-inference. By the inductive assumption, $\vec{\mathfrak{b}}$ is deductively correct w.r.t.

$$\Gamma \cup \mathrm{Descr}_{\mathscr{M}}\left[\mathrm{JA}_\sigma(\mathscr{M}) \leftarrow \mathrm{JA}_\sigma\left(\vec{\mathfrak{b}}, \mathscr{M}\right)\right] =$$
$$\Gamma \cup \mathrm{Descr}_{\mathscr{M}}[\mathrm{JA}_\sigma(\mathscr{M}) \leftarrow \mathrm{JA}_\sigma(\vec{\mathfrak{a}}, \mathscr{M})].$$

It is not difficult to show that in this case $\vec{\mathfrak{a}}$ is also deductively correct w.r.t.

$$\Gamma \cup \mathrm{Descr}_{\mathscr{M}}[\mathrm{JA}_\sigma(\mathscr{M}) \leftarrow \mathrm{JA}_\sigma(\vec{\mathfrak{a}}, \mathscr{M})].$$

CASE 2. The record $\mathfrak{a}$ is non-deductively grounded (cf. 10.2.31 (NDed)), i.e. ToR (Reason$(\mathfrak{a})$) is equal to DAx, MR or Conf. Let us consider the subcases.

SUBCASE 2.1. Let

$$\mathrm{ToR}(\mathrm{Reason}(\mathfrak{a})) = \mathrm{Conf}.$$

This is the simplest case. As in Case 1,

$$\mathrm{JA}_\sigma(\vec{\mathfrak{a}},\mathcal{M}) = \mathrm{JA}_\sigma\left(\vec{\mathfrak{b}},\mathcal{M}\right)$$

and $\vec{\mathfrak{b}}$ is an $\mathcal{M}$-inference. Moreover there exists an $i \in \mathrm{AIS}\left(\vec{\mathfrak{b}}\right)$ such that

$$\mathscr{F}(\mathfrak{a}) = \mathscr{F}\left(\vec{\mathfrak{b}}[i]\right).$$

Applying the induction hypothesis we can verify that in this subcase $\vec{\mathfrak{a}}$ is deductively correct w.r.t. $\Gamma\,[\mathrm{JA}_\sigma(\mathcal{M}) \leftarrow \mathrm{JA}_\sigma(\vec{\mathfrak{a}},\mathcal{M})]$.

SUBCASE 2.2. Let $\mathrm{ToR}(\mathrm{Reason}(\mathfrak{a})) = \mathrm{DAx}$. Then

$$\mathrm{JA}_\sigma(\vec{\mathfrak{a}},\mathcal{M}) = \mathrm{JA}_\sigma\left(\vec{\mathfrak{b}},\mathcal{M}\right) \cup \{\mathscr{F}(\mathfrak{a})\}$$

and $\vec{\mathfrak{b}}$ is an $\mathcal{M}$-inference. Then, clearly,

$$\begin{aligned}&\Gamma \cup \mathrm{Descr}_{\mathcal{M}}\,[\mathrm{JA}_\sigma(\mathcal{M}) \leftarrow \mathrm{JA}_\sigma(\vec{\mathfrak{a}},\mathcal{M})] = \\ &\Gamma \cup \mathrm{Descr}_{\mathcal{M}}\left[\mathrm{JA}_\sigma(\mathcal{M}) \leftarrow \mathrm{JA}_\sigma\left(\vec{\mathfrak{b}},\mathcal{M}\right)\right] \cup \{\mathscr{F}(a)\}.\end{aligned}$$

By the inductive assumption and Remark 11.4.3, $\vec{\mathfrak{a}}$ is deductively correct w.r.t. $\Gamma \cup \mathrm{Descr}_{\mathcal{M}}\,[\mathrm{JA}_\sigma(M) \leftarrow \mathrm{JA}_\sigma(\vec{\mathfrak{a}},\mathcal{M})]$.

SUBCASE 2.3. Let $\mathrm{Reason}(a) = \mathrm{MR}_{\mathcal{M}}[p,\varepsilon](i,j)$, where $p \in \mathrm{Pr}_\sigma(\mathsf{I})$, $\varepsilon \in \mathscr{V}(L,\mathsf{I})$, $i \in \mathrm{NAIS}\left(\vec{\mathfrak{b}}\right)$, $j \in \mathrm{AIS}\left(\vec{\mathfrak{b}}\right)$. Then

$$\vec{\mathfrak{b}} = \begin{cases} \vec{\mathfrak{c}} & \text{if } \varepsilon = \tau, \\ \mathrm{Suspend}(\vec{\mathfrak{c}},\mathrm{J}_\tau\, p(\vec{a})) & \text{if } \varepsilon \neq \tau, \end{cases}$$

where $\vec{\mathfrak{a}} \in \mathrm{Dom}_{\mathcal{M}}(p)$ and $\vec{\mathfrak{c}}$ is an n-long $\mathcal{M}$-inference. If $\vec{\mathfrak{b}} = \vec{\mathfrak{c}}$, then our further argumentation will be similar to that in subcase 2.1.

Suppose that $\vec{\mathfrak{b}} = \mathrm{Suspend}(\vec{\mathfrak{c}},\mathrm{J}_\tau\, p(\vec{\mathfrak{a}}))$. By the inductive assumption $\vec{\mathfrak{c}}$ is deductively correct w.r.t. $\Gamma \cup \mathrm{Descr}_{\mathcal{M}}\,[\mathrm{JA}_\sigma(M) \leftarrow \mathrm{JA}_\sigma(\vec{\mathfrak{c}},\mathcal{M})]$. Then by Lemma 11.4.7 the string $\vec{\mathfrak{b}}$ is deductively correct w.r.t. the set

$$\Xi = \Gamma \cup \mathrm{Descr}_{\mathcal{M}}\,[\mathrm{JA}_\sigma(\mathcal{M}) \leftarrow \mathrm{JA}_\sigma(\vec{\mathfrak{c}},\mathcal{M})] \setminus \{\mathrm{J}_\tau\, p(\vec{a})\}.$$

Note that in this subcase $\mathscr{F}(\mathfrak{a}) = \mathrm{J}_\varepsilon\, p(\vec{a})$, therefore

$$\mathrm{JA}_\sigma(\vec{\mathfrak{a}},\mathcal{M}) = (\mathrm{JA}_\sigma(\vec{\mathfrak{c}},\mathcal{M}) \setminus \{\mathrm{J}_\tau\, p(\vec{a})\}) \cup \{\mathrm{J}_\varepsilon\, p(\vec{a})\}.$$

Then it is clear that

$$\Gamma \cup \mathrm{Descr}_{\mathcal{M}}\,[\mathrm{JA}_\sigma(\mathcal{M}) \leftarrow \mathrm{JA}_\sigma(\vec{\mathfrak{a}},\mathcal{M})] = \Xi \cup \mathscr{F}(\mathfrak{a}).$$

By the inductive assumption and Remark 11.4.3, $\vec{\mathfrak{a}}$ is deductively correct w.r.t.

$$\Gamma \cup \mathrm{Descr}_{\mathcal{M}}[\mathrm{JA}_\sigma(\mathcal{M}) \leftarrow \mathrm{JA}_\sigma(\vec{\mathfrak{a}}, \mathcal{M})].$$

□

Now we investigate the connection between inferences in modification calculi and in L^1. The results of this investigation are formulated in the next theorem, which is the main result of this section, and its corrollaries. Further on these results will be often used, especially in the construction of canonical models for theories in modification calculi, i.e. for stage cognition theory.

Theorem 11.4.9. Let L be an FPJ logic, σ be a signature, $\mathcal{M} \in \mathrm{MC}_\sigma(L)$, $\Gamma \in \mathrm{AFS}(\mathcal{M})$, $\vec{\mathfrak{a}}$ be an $\mathcal{M}$-inference from Γ. Then $\mathrm{FExtr}(\vec{\mathfrak{a}})$ is an inference in L^1 from the set of external L-formulas

$$\Gamma \cup \mathrm{Descr}_{\mathcal{M}}[\mathrm{JA}_\sigma(\mathcal{M}) \leftarrow \mathrm{JA}_\sigma(\vec{\mathfrak{a}}, \mathcal{M})].$$

Proof. It follows directly from Lemmas 11.4.8 and 11.4.5. □

Corollary 11.4.10. Let L be an FPJ logic, σ be a signature, $\mathcal{M} \in \mathrm{MC}_\sigma(L)$, $\Gamma \in \mathrm{AFS}(\mathcal{M})$ and $\vec{\mathfrak{a}}$ be an $\mathcal{M}$-inference from Γ. Then $\mathrm{FExtr}(\vec{\mathfrak{a}})$ is an inference in L^1 from $\mathrm{Cur}(\Gamma, \vec{\mathfrak{a}})$.

Proof. The statement follows from 11.4.9, since

$$\Gamma \cup \mathrm{Descr}_{\mathcal{M}}[\mathrm{JA}_\sigma(\mathcal{M}) \leftarrow \mathrm{JA}_\sigma(\vec{\mathfrak{a}}, \mathcal{M})] \subseteq \mathrm{Cur}(\Gamma, \vec{a})$$

(see 11.3.17). □

Corollary 11.4.11. Let L be an FPJ logic, σ be a signature, $\mathcal{M} \in \mathrm{MC}_\sigma(L)$, $\Gamma \in \mathrm{AFS}(\mathcal{M})$ and $\vec{\mathfrak{a}}$ be an $\mathcal{M}$-inference from Γ. Then for any $\Phi \in \mathrm{FSet}(\vec{\mathfrak{a}})$:

$$\Gamma \cup \mathrm{Descr}_{\mathcal{M}}[\mathrm{JA}_\sigma(\mathcal{M}) \leftarrow \mathrm{JA}_\sigma(\vec{\mathfrak{a}}, \mathcal{M})] \vdash_L \Phi.$$

Corollary 11.4.12. Let L be an FPJ logic, σ be a signature, $\mathcal{M} \in \mathrm{MC}_\sigma(L)$, $\Gamma \in \mathrm{AFS}(\mathcal{M})$ and $\vec{\mathfrak{a}}$ be an $\mathcal{M}$-inference from Γ. Then for any $\Phi \in \mathrm{FSet}(\vec{\mathfrak{a}})$

$$\mathrm{Cur}(\Gamma, \vec{\mathfrak{a}}) \vdash_L \Phi.$$

Let us consider two inferences: an $\mathcal{M}$-inference $\vec{\mathfrak{a}}$, where $\mathcal{M}$ is a modification calculus, and an L^1-inference $\vec{\mathfrak{b}}$. Is it possible to add the second inference to the first one so that the result will be an $\mathcal{M}$-inference in a modification calculus? The answer is yes if the first inference is an inference from a certain set of external L-formulas Γ and the second one is from the set $\mathrm{Cur}(\Gamma, \vec{\mathfrak{a}})$. Below we show how to extend an $\mathcal{M}$-inference by using an L^1-inference.

Definition 11.4.13. Let L be an FPJ logic, σ be a signature, $\mathcal{M} \in \mathrm{MC}_\sigma(L)$, $\Gamma \in \mathrm{AFS}(\mathcal{M})$ and $\vec{\mathfrak{a}}$ be an $\mathcal{M}$-inference from Γ. Let Φ be an external L-formula. We say that Φ can be *deductively correctly added* to $\vec{\mathfrak{a}}$ w.r.t Γ if at least one of the following holds:

(i) $\Phi = \mathscr{F}(\vec{\mathfrak{a}}[i])$ for some $i \in \mathrm{AIS}(\vec{\mathfrak{a}})$,
(ii) Φ is an axiom of the calculus L,
(iii) $\Phi \in \Gamma$,
(iv) $\Phi \in \mathrm{Descr}_{\mathscr{M}} \backslash \mathrm{JA}_\sigma(\mathscr{M})$,
(v) There exist $i, j \in \mathrm{AIS}(\vec{\mathfrak{a}})$ such that $\mathscr{F}(\vec{\mathfrak{a}}[j]) = (\mathscr{F}(\vec{\mathfrak{a}}[i]) \to \Phi)$,
(vi) There exists $i \in \mathrm{AIS}(\vec{\mathfrak{a}})$ such that $\Phi = \forall x\, \mathscr{F}(\vec{\mathfrak{a}}[i])$ for some $x \in \mathrm{Var}_\sigma$.

Definition 11.4.14. Define the *operation* of deductively correct extension of an $\mathscr{M}$-inference with external L-formulas. Of course we will add to an inference not an L-formula itself but the record that contains it. This record should also contain a reason for adding the formula to the inference. Unfortunately the conditions (i)–(iv) from the previous definition are not disjoint. Therefore we cannot unambiguously select a reason for adding a formula to the inference. At the same time the value of the discussed operation should be defined unambiguously. In order to do so we establish priorities of the possible reasons and look for minimal indices of the occurrences of the L-formulas in the $\mathscr{M}$-inference.

Let L be an FPJ logic, σ be a signature, $\mathscr{M} \in \mathrm{MC}_\sigma(L)$, $\Gamma \in \mathrm{AFS}(\mathscr{M})$, $s, m \in \omega$ and $\vec{\mathfrak{a}}$ be an $\mathscr{M}(s, m, \mathbf{D})$-inference from Γ. Let Φ be an external L-formula. Let Φ be an external L-formula, which can be added to $\vec{\mathfrak{a}}$ in a deductively correct way. Denote by $\vec{\mathfrak{b}}$ the result of a deductively correct extension of $\vec{\mathfrak{a}}$ with Φ. Then $\vec{\mathfrak{b}}$ can be computed by the following algorithm:

If $\Phi = \mathscr{F}(\vec{\mathfrak{a}}[k])$ *for some* $k \in \mathrm{AIS}(\vec{\mathfrak{a}})$ **Then**
  $\vec{\mathfrak{b}} = \vec{\mathfrak{a}}$
ElseIf Φ *is an axiom of* L **Then**
  $\vec{\mathfrak{b}} = \vec{\mathfrak{a}} \bullet \langle \Phi, \mathrm{LAx}, s, m, 0 \rangle$
ElseIf $\Phi \in \Gamma$ **Then**
  $\vec{\mathfrak{b}} = \vec{\mathfrak{a}} \bullet \langle \Phi, \mathrm{Hyp}, s, m, 0 \rangle$
ElseIf $\Phi \in \mathrm{Descr}_{\mathscr{M}} \backslash \mathrm{JA}_\sigma(\mathscr{M})$ **Then**
  $\vec{\mathfrak{b}} = \vec{\mathfrak{a}} \bullet \langle \Phi, \mathrm{OAx}, s, m, 0 \rangle$
ElseIf $\mathscr{F}(\vec{\mathfrak{a}}[l]) = (\mathscr{F}(\vec{\mathfrak{a}}[k]) \to \Phi)$ *for some* $k, l \in \mathrm{AIS}(\vec{\mathfrak{a}})$ **Then**
  $\langle i, j \rangle = \min\{\langle k, l \rangle \mid \mathscr{F}(\vec{\mathfrak{a}}[l]) = (\mathscr{F}(\vec{\mathfrak{a}}[k]) \to \Phi)\}$
  $\vec{\mathfrak{b}} = \vec{\mathfrak{a}} \bullet \langle \Phi, \mathrm{MP}(i, j), s, m, 0 \rangle$
ElseIf $\Phi = \forall x\, \mathscr{F}(\vec{\mathfrak{a}}[k])$ *for some* $k \in \mathrm{AIS}(\vec{\mathfrak{a}})$ **Then**
  $i = \min\{k \mid \Phi = \forall x\, \mathscr{F}(\vec{\mathfrak{a}}[k])\}$
  $\vec{\mathfrak{b}} = \vec{\mathfrak{a}} \bullet \langle \Phi, \mathrm{Gen}(i), s, m, 0 \rangle$
End If

Let $\vec{\mathfrak{a}} \oplus \Phi$ denote the result of deductively correct addition of the L-formula Φ to the $\mathscr{M}$-inference $\vec{\mathfrak{a}}$.

Lemma 11.4.15. Let L be an FPJ logic, σ be a signature, $\mathscr{M} \in \mathrm{MC}_\sigma(L)$ and $\vec{\mathfrak{a}}$ be an $\mathscr{M}(s, m, \mathbf{D})$-inference. Let Φ be an external L-formula, which can be deductively correctly added to $\vec{\mathfrak{a}}$. Then the record string $\vec{\mathfrak{a}} \oplus \Phi$ is the $\mathscr{M}(s, m, \mathbf{D})$-inference with $\mathrm{JA}_\sigma(\vec{\mathfrak{a}}, \mathscr{M}) = \mathrm{JA}_\sigma(\vec{\mathfrak{a}} \oplus \Phi, M)$.

Proof. Since Φ can be deductively correctly added to $\vec{\mathfrak{a}}$, then for Φ at least one of the conditions (i)–(v) of Definition 11.4.13 holds. Then by Definition 11.4.14, $\vec{\mathfrak{a}} \oplus \Phi$ is either equal to $\vec{\mathfrak{a}}$, or obtained by adding a *deductively grounded* record $\mathfrak{c}$ to the string $\vec{\mathfrak{a}}$, where $\mathscr{F}(\mathfrak{c}) = \Phi$. It is easy to see that $\vec{\mathfrak{a}} \oplus \Phi$ is an inference. Since $\mathfrak{c}$ is deductively grounded,

$$\mathrm{JA}_\sigma(\vec{\mathfrak{a}}, \mathscr{M}) = \mathrm{JA}_\sigma(\vec{\mathfrak{a}} \oplus \Phi, M)$$

(see Definition 10.2.14). □

Lemma 11.4.16. Let L be an FPJ logic, σ be a signature, $\mathscr{M} \in \mathrm{MC}_\sigma(L)$, $\Gamma \in \mathrm{AFS}(\mathscr{M})$ and $\vec{\mathfrak{a}}$ be an $\mathscr{M}(s,m,\mathbf{D})$-inference from Γ. Let $\vec{\Psi}$ be an inference from $\mathrm{Cur}(\Gamma, \vec{\mathfrak{a}})$ in L^1. Then there exists an $\mathscr{M}(s,m,\mathbf{D})$-inference $\vec{\mathfrak{b}}$ such that:

(i) $\mathrm{JA}_\sigma(\vec{\mathfrak{a}}, \mathscr{M}) = \mathrm{JA}_\sigma\left(\vec{\mathfrak{b}}, \mathscr{M}\right)$,

(ii) For any $i \in \mathrm{IS}\left(\vec{\Psi}\right)$ there exists $k \in \mathrm{AIS}\left(\vec{\mathfrak{b}}\right)$ such that $\vec{\Psi}[i] = \mathscr{F}\left(\vec{\mathfrak{b}}[k]\right)$.

Proof. By induction on the length of the inference $\vec{\Psi}$.

Basis. If $\vec{\Psi} = \mathbf{e}$, then $\vec{\mathfrak{b}} = \vec{\mathfrak{a}}$.

Induction step. Suppose that the statement holds for the inferences in L^1, the length of which is not greater than n. Let the length of the inference $\vec{\Psi}$ be equal to $n+1$. Then $\vec{\Psi} = \vec{\Phi} + \Theta$, where $\vec{\Phi}$ is an inference in L^1 of length n, and Θ is an external L-formula.

By the inductive assumption there exists an $\mathscr{M}(s,m,\mathbf{D})$-inference $\vec{\mathfrak{c}}$ such that:

(iii) $\mathrm{JA}_\sigma(\vec{\mathfrak{a}}, \mathscr{M}) = \mathrm{JA}_\sigma(\vec{\mathfrak{c}}, \mathscr{M})$,

(iv) For any $i \in \mathrm{IS}\left(\vec{\Phi}\right)$ there exists $k \in \mathrm{AIS}(\vec{\mathfrak{c}})$ such that $\vec{\Phi}[i] = \mathscr{F}(\vec{\mathfrak{c}}[k])$.

Consider the formula Θ, which is the last one in $\vec{\Psi}$. The following cases are possible:

(a) Θ is an axiom of L^1,

(b) $\Theta \in \mathrm{Cur}(\Gamma, \vec{\mathfrak{a}})$,

(c) There exist $i, j \in \mathrm{IS}\left(\vec{\Phi}\right)$ such that $\vec{\Phi}[j] = \left(\vec{\Phi}[j] \to \Theta\right)$,

(d) There exists $i \in \mathrm{IS}\left(\vec{\Phi}\right)$ such that $\Theta = \forall x \vec{\Phi}[i]$ for some $x \in Var$.

By the induction hypothesis (see (iii) and (iv)) it is simple to verify that in each of the above cases Θ can be added to $\vec{\mathfrak{c}}$ in a deductively correct way.

Let us consider in more detail the case (b). By (iii), $\Theta \in \mathrm{Cur}(\Gamma, \vec{\mathfrak{c}})$.

Consider the subcases:

- Let $\Theta \in \Gamma \cup (\mathrm{Descr}_{\mathscr{M}} \setminus \mathrm{JA}_\sigma(\mathscr{M}))$. Then by Definition 11.4.13(iii), (iv) Θ can be added in a deductively correct way to $\vec{\mathfrak{c}}$.
- Let $\Theta \in \mathrm{JA}_\sigma(\mathscr{M})$. Then $\Theta \in \mathrm{JA}_\sigma(\mathscr{M}) \cap \mathrm{JA}^+_\sigma(\vec{\mathfrak{c}}, \mathscr{M})$. But in this case $\Theta = \mathscr{F}(\vec{\mathfrak{c}}[i])$ for some $i \in \mathrm{AIS}(\vec{\mathfrak{c}})$. Therefore, by Definition 11.4.13(i), Θ can be added in a deductively correct way to $\vec{\mathfrak{c}}$.

Let $\vec{\mathfrak{b}} = \vec{\mathfrak{c}} \oplus \Theta$. By Lemma 11.4.15, $\vec{\mathfrak{b}}$ is an $\mathscr{M}(s,m,\mathbf{D})$-inference and

$$\mathrm{JA}^{+}_{\sigma}(\vec{\mathfrak{a}},\mathscr{M}) = \mathrm{JA}^{+}_{\sigma}(\vec{\mathfrak{c}},\mathscr{M}) = \mathrm{JA}^{+}_{\sigma}\left(\vec{\mathfrak{b}},\mathscr{M}\right).$$

Therefore (i) holds. It is easy to prove that (ii) also holds. □

Corollary 11.4.17. Let L be an FPJ logic, σ be a signature, $\mathscr{M} \in \mathrm{MC}_{\sigma}(L)$, $\Gamma \in \mathrm{AFS}(\mathscr{M})$ and $\vec{\mathfrak{a}}$ be an $\mathscr{M}(s,m,\mathbf{D})$-inference from the set of hypotheses Γ. Let

$$\mathrm{Cur}(\Gamma,\vec{\mathfrak{a}}) \vdash_L \Phi.$$

Then there exists an $\mathscr{M}(s,m,\mathbf{D})$-inference $\vec{\mathfrak{b}}$ such that

(i) $\mathrm{JA}^{+}_{\sigma}(\vec{\mathfrak{a}},\mathscr{M}) = \mathrm{JA}^{+}_{\sigma}\left(\vec{\mathfrak{b}},\mathscr{M}\right)$,
(ii) $\Phi \in \mathrm{FSet}\left(\vec{\mathfrak{b}}\right)$.

Chapter 12
Semantics

In the present chapter appropriate semantics will be developed for the modification calculi as a special type of inference. This semantics will correspond to the intuitive meaning of discrete cognitive processes, sectioned into stages and modules. To each pair $\langle$stage, module$\rangle$ some knowledge state corresponds. Syntactically, a set of external L-formulas corresponds to a knowledge state that earlier we have called a state description. Semantically, a knowledge state has a corresponding L-structure, a model of the description of this state.

What corresponds to the cognitive process itself? Syntactically, it is an inference in the modification calculus that should correspond to a cognitive process. Semantically, it is a sequence of models of the state description, i.e. a certain sequence of L-structures.

12.1 Sequences of L-Structures

Here we start to describe the properties of the sequences of L-structures, where L is an FPJ logic, that may pretend to be the models of the cognitive processes.

Definition 12.1.1. Let L be an FPJ logic, U be a U-frame, $p \in \mathrm{Pr}_U(L)(\mathsf{I})$. The set

$$\{\vec{x} \in \mathrm{Dom}_U(p) \mid p(\vec{x}) \in \mathscr{D}^{\tau}(L)\}$$

is called the *certainty domain* of p in U (written: $\mathrm{Def}_U(p)$).

The complement of the certainty domain of p, i.e. the set

$$\{\vec{x} \in \mathrm{Dom}_U(p) \mid p(\vec{x}) = \tau\},$$

is called its *uncertainty domain* (written: $\mathrm{NDef}_U(p)$).

Definition 12.1.2. Let L be an FPJ logic, U be a U-frame, $p, q \in \mathrm{Pr}_U(L)(\mathsf{I})$, where

$$\mathrm{Dom}_U(p) = \mathrm{Dom}_U(q).$$

O. Anshakov, T. Gergely, *Cognitive Reasoning*, Cognitive Technologies,
DOI 10.1007/978-3-540-68875-4_12, © Springer-Verlag Berlin Heidelberg 2010

We say that q is a *certainification* of p (written: $p \preceq q$ or $q \succeq p$) if $p(\vec{x}) = q(\vec{x})$, for every $\vec{x} \in \mathrm{Def}_U(p)$.

Let us give an informal explanation for the term *certainification*. Suppose $p \preceq q$. Then p and q have the same domain. We represent L-predicates p and q as tables (matrices) filled with truth values from $\mathscr{V}(L, \mathsf{I})$. We call two cells of the tables (matrices) p and q *corresponding* if their indices are the same tuple of arguments. (If a tuple $\vec{x}$ belongs to the common domain of p and q, it can be considered as a string of indices of the corresponding cells.) These cells contain the values $p(\vec{x})$ and $q(\vec{x})$.

If a cell of the table representing p contains a *certain* value (different from τ), then the corresponding cell of the table representing q contains *exactly the same value* (see Definition 12.1.2) .

If a cell of the table for p contains τ (*uncertain*), then the corresponding cell of the table for q may contain any truth value from $\mathscr{V}(L, \mathsf{I}) = \mathscr{V}(L, \mathsf{I})$.

Consequently, the internal L-predicate q can be obtained from p by *substituting certain values for some uncertain values*. Thus, it is quite natural to call q a certainification of p.

Proposition 12.1.3. Let L be an FPJ logic, p and q be two internal L-predicates such that$p \preceq q$. Then $\mathrm{Def}_U(p) \subseteq \mathrm{Def}_U(q)$ (and therefore $\mathrm{NDef}_U(p) \supseteq \mathrm{NDef}_U(q)$).

Proof. It immediately follows from Definition 12.1.2. □

Proposition 12.1.4. Let L be an FPJ logic, U be a U-frame, $\vec{a} \in \mathsf{IS}^*$. Then, the relation $\preceq$ is a nonstrict partial order on the set $\mathrm{Pr}_U(L)(\vec{a}, \mathsf{I})$, i.e. any $p, q, r \in \mathrm{Pr}_U(L)(\vec{a}, \mathsf{I})$ satisfy the following:

(i) $p \preceq p$ (reflexivity),
(ii) $p \preceq q$ and $q \preceq p$ implies $p = q$ (antisymmetry),
(iii) $p \preceq q$ and $q \preceq r$ implies $p \preceq r$ (transitivity).

Proof. Statements (i) and (iii) are obvious. Let us consider (ii) in more detail.

Let $p \preceq q$ and $q \preceq p$. We must show that $p(\vec{x}) = q(\vec{x})$ for any tuple $\vec{x} \in \mathrm{Dom}_U(\vec{a})$. By Proposition 12.1.3, we obtain the equalities

$$\mathrm{Def}_U(p) = \mathrm{Def}_U(q) \text{ and } \mathrm{NDef}_U(p) = \mathrm{NDef}_U(q).$$

Then

$$\mathrm{Dom}_U(\vec{a}) = \mathrm{Def}_U(p) \cup \mathrm{NDef}_U(p).$$

By Definition 12.1.2 for any tuple $\vec{x} \in \mathrm{Def}_U(p)$ we have $p(\vec{x}) = q(\vec{x})$. Moreover for any tuple $\vec{y} \in \mathrm{NDef}_U(p) = \mathrm{NDef}_U(q)$ we have $p(\vec{y}) = q(\vec{y}) = \tau$. □

Definition 12.1.5. Let L be an FPJ logic, $\mathfrak{J}$ and $\mathfrak{K}$ be two L-structures of the same signature σ over the same U-frame. We say that $\mathfrak{K}$ is a *certainification* of $\mathfrak{J}$ (written: $\mathfrak{J} \preceq \mathfrak{K}$ or $\mathfrak{K} \succeq \mathfrak{J}$) if the following conditions are satisfied:

(i) $f^{\mathfrak{J}} = f^{\mathfrak{K}}$ for any function symbol f of signature σ,
(ii) $q^{\mathfrak{J}} = q^{\mathfrak{K}}$ for any external predicate symbol q of signature σ,

(iii) $p^{\mathfrak{J}} \preceq p^{\mathfrak{K}}$ for any internal predicate symbol p of signature σ.

Proposition 12.1.6. Let L be an FPJ logic, σ be a signature and U be a U-frame. Then, the relation $\preceq$ is a nonstrict partial order on the set $\mathbf{Struct}_\sigma(L,U)$, i.e. for all $\mathfrak{J},\mathfrak{K},\mathfrak{L} \in \mathbf{Struct}_\sigma(L,U)$ the following statements hold:

(i) $\mathfrak{J} \preceq \mathfrak{J}$	(reflexivity),
(ii) $\mathfrak{J} \preceq \mathfrak{K}$ and $\mathfrak{K} \preceq \mathfrak{J}$ implies $\mathfrak{J} = \mathfrak{K}$	(antisymmetry),
(iii) $\mathfrak{J} \preceq \mathfrak{K}$ and $\mathfrak{K} \preceq \mathfrak{L}$ implies $\mathfrak{J} \preceq \mathfrak{L}$	(transitivity).

Proof. It is an immediate consequence of Proposition 12.1.4. □

Definition 12.1.7. Let L be an FPJ logic, $\langle \mathfrak{J}_n \mid n \in \omega \rangle$ be a sequence of L-structures of signature σ over the same U-frame. We call this sequence *monotone* if $k \le m$ implies $\mathfrak{J}_k \preceq \mathfrak{J}_m$ for all $k, m \in \omega$.

Remark 12.1.8. Sequences of L-structures, where L is an FPJ logic, can be interpreted in the context of procedural semantics, where the term *procedural semantics* refers to interpreting a sequence of L-structures as a *procedure*. That is, in this context there is a procedure that transforms every element of the sequence into the next element, i.e. the i-th element of the sequence will be transformed into the $(i+1)$-th element. If repeated application of this procedure produces a *monotone* sequence of L-structures, the procedure may be regarded as a *certanification* procedure for L-structures.

The term *monotone sequence* owes its origin to the analogy with *logical monotonicity*, i.e. the principle that forbids rejection of previously accepted facts and knowledge.

Proposition 12.1.9. Let L be an FPJ logic, $\langle \mathfrak{J}_n \mid n \in \omega \rangle$ be a monotone sequence of finite L-structures of signature σ with a finite set of internal predicate symbols. Assume that the set IS is also finite. Then, there is an $m \in \omega$ such that $\mathfrak{J}_k = \mathfrak{J}_m$ for any $k > m$.

Proof. Suppose the converse. Then, the sequence $\langle \mathfrak{J}_n \mid n \in \omega \rangle$ contains an *infinite* subsequence of *different* L-structures. Every L-structure differs from the previous structure in at least one L-predicate. Moreover the uncertainty domain of this L-predicate is strictly smaller than the domain of the corresponding predicate from the previous structure. For every L-structure from the sequence $\langle \mathfrak{J}_n \mid n \in \omega \rangle$, the number of internal L-predicates is finite, and the uncertainty domain of each of these L-predicates is also finite (since the U-frame is finite). Therefore, it is clear that there is an $m \in \omega$ such that the uncertainty domain of every internal predicate of $\mathfrak{J}_m$ is empty. No further reduction of uncertainty domains is possible, which results in a contradiction. □

12.2 Structure Generators

Here we will define the notion of a structure generator, which can be regarded as a generator of models for the modification calculi. As we have already mentioned, sequences of L-structures will be considered as such models.

However, before defining structure generators we have to define some other important notions. The first of them is the generator rule system, which is an analogue to the modification rule system. Only after this will we be able to define structure generators.

Definition 12.2.1. Let L be an FPJ logic, σ be a signature, $p \in \mathrm{Pr}_\sigma(\mathsf{I})$, $\vec{x} \in \mathrm{Var}^\diamondsuit \cap \mathrm{Dom}_\sigma(p)$, $\varepsilon \in \mathscr{V}(L,\mathsf{I})$. Then the ordered pair

$$G = \langle \Phi; \mathrm{J}_\varepsilon\, p(\vec{x}) \rangle ,$$

where Φ is an external L-formula such that $\mathrm{FVar}(\Phi) \subseteq \mathrm{Set}(\vec{x})$, is called a *generating* ε*-rule for the predicate symbol p*.

The formulas Φ and $\mathrm{J}_\varepsilon\, p(\vec{x})$ are called the *generating condition* and the *result* of this rule, respectively.

The rule G is written as either

$$\Phi \Vdash\!\vdash \mathrm{J}_\varepsilon\, p(\vec{x})$$

or

$$\frac{\Phi}{\mathrm{J}_\varepsilon\, p(\vec{x})}.$$

Notation 12.2.2. Let σ be a signature and L be one of the logics considered in this book. Denote by $\mathrm{FS}_\sigma(L,\mathsf{s})$ the collection of all L-formula sets of signature σ and of sort s. For example by $\mathrm{FS}_\sigma(L,\mathsf{E})$ we denote the collection of all sets of external L-formulas of signature σ. I.e., $\Gamma \in \mathrm{FS}_\sigma(L,\mathsf{E})$ is equivalent to "Γ is a set of external L-formulas of signature σ".

Definition 12.2.3. Let L be an FPJ logic, σ be a signature, $\Gamma \in \mathrm{FS}_\sigma(L,\mathsf{E})$, $p \in \mathrm{Pr}_\sigma(\mathsf{I})$,

$$\vec{x} \in \mathrm{Var}^\diamondsuit \cap \mathrm{Dom}_\sigma(p),$$

for any $\varepsilon \in \mathscr{V}(L,\mathsf{I})$ and Φ_ε be an external L-formula such that $\mathrm{FVar}(\Phi_\varepsilon) \subseteq \mathrm{Set}(\vec{x})$. Then any four-element set

$$\{\Phi_\varepsilon \Vdash\!\vdash \mathrm{J}_\varepsilon\, p(\vec{x}) \mid \varepsilon \in \mathscr{V}(L,\mathsf{I})\}$$

is called a *complete generating set* (CGS) for p w.r.t. Γ in L, if it meets the following conditions:

(i) $\Gamma \vdash_L \Phi_{+1} \vee \Phi_{-1} \vee \Phi_0 \vee \Phi_\tau$,
(ii) $\Gamma \vdash_L \Phi_\varepsilon \to \neg\Phi_\delta$ for any $\varepsilon, \delta \in \mathscr{V}(L,\mathsf{I})$ such that $\varepsilon \neq \delta$,

(iii) $\Gamma \vdash_L \mathrm{J}_\varepsilon\, p(\vec{x}) \to \Phi_\varepsilon$ for any $\varepsilon \in \mathscr{D}^\tau(L)$.

Remark 12.2.4. Let L be an FPJ logic, σ be a signature, $\Xi, \Upsilon \in \mathrm{FS}_\sigma(L, \mathsf{E})$, $\Xi \subseteq \Upsilon$, $p \in \mathrm{Pr}_\sigma(\mathsf{I})$ and M be a complete generating set (CGS) for p w.r.t. Ξ in L. Then M is also a complete generating set for p w.r.t. Υ in L.

Definition 12.2.5. Let L be an FPJ logic, σ be a signature, $\Gamma \in \mathrm{FS}_\sigma(L, \mathsf{E})$. The collection of sets

$$R = \{S_p \mid p \in \mathrm{Pr}_\sigma(\mathsf{I})\}$$

is called a *generating rule system of signature* σ w.r.t. Γ in L, if for each $p \in \mathrm{Pr}_\sigma(\mathsf{I})\, S_p$ is a CGS for p w.r.t. Γ in L.

The set of all generating rule systems of signadture σ w.r.t. Γ in L is denoted by $\mathrm{GRS}_\sigma(L, \Gamma)$

We say that R is a *generating rule system of signature* σ in L if there exists a $\Gamma \in \mathrm{FS}_\sigma(L, \mathsf{E})$ such that $R \in \mathrm{GRS}_\sigma(L, \Gamma)$.

The set of all generating rule systems of signadture σ in L is denoted by $\mathrm{GRS}_\sigma(L)$

Definition 12.2.6. Let L be an FPJ logic, σ be a signature, $\Gamma \in \mathrm{FS}_\sigma(L, \mathsf{E})$ and $R \in \mathrm{GRS}_\sigma(L, \Gamma)$. Clearly for any $p \in \mathrm{Pr}_\sigma(\mathsf{I})$ and $\varepsilon \in \mathscr{V}(L, \mathsf{I})$ there exists a unique modification ε-rule for the predicate symbol p, which belongs to an element of this set. The rule, its generating condition and its result are denoted by $R[p, \varepsilon]$, $\mathrm{GC}[R, p, \varepsilon]$ and $\mathrm{Res}[R, p, \varepsilon]$, respectively. The set $\{R[p, \varepsilon] \mid \varepsilon \in \mathscr{V}(L, \mathsf{I})\}$ is called the *p-projection of* R and is denoted by $R[p]$.

Generating rules resemble modification rules. They also allow for changing the truth value of internal predicates if the latter is equal to τ. Condition (iii) of Definition 12.2.3 guarantees "restricted monotony". Certain values of the internal predicate cannot be modified by the application of generating rules.

On the basis of generating rules we can define certain analogues of modification calculi that can be called generating calculi. However, the development of such a technique is not the subject of this book. We think though that the construction of a calculus on the basis of generating rules is a very interesting problem. Note that in such a calculus there is no need for using the justification rule (see 12.2.3 (iii)). In general terms, generating rules themselves seem simpler than modification rules, since they have only one antecedent. Note, that Definition 12.2.3 without condition (iii) brings us to the construction of an entirely non-monotone calculus. The investigation of such calculi is also a very interesting problem but it is out of the scope of this book.

For several reasons we prefer to formulate a calculus on the basis of modification rules and not on the basis of generating rules. First, specific modification rules will be written in a shorter and more transparent form than the corresponding generating rules. The condition of the generation itself looks more complicated than the major premise of the corresponding modification rule. Second, the modification rules (for known cases) are more suitable for computer realisation with the help of logic programming languages than the corresponding generating rules.

In this book the generating rules play a very important role, not from the syntactic point of view but from the semantic one. It is generating rules that are used below to define the semantics of the modification calculi. In order to show that certain semantics is adequate we will need to show that continuing a pure $\mathscr{M}$-inference in the usual way (by the use of modification rules) we obtain those and only those J-atomic formulas that can be obtained by continuing the inference by applying generating rules. A more precise formulation of this statement will be given below.

Lemma 12.2.7. Let L be an FPJ logic, σ be a signature, $\Gamma \in \mathrm{FS}_\sigma(L,\mathsf{E})$,$R \in \mathrm{GRS}_\sigma(L,\Gamma)$ and $\mathfrak{J}$ be an L-structure of signature σ, $p \in \mathrm{Pr}_\sigma(\mathsf{I})$, $\vec{a} \in \mathrm{Dom}\left(p^{\mathfrak{J}}\right)$. Assume that $\mathfrak{J} \vDash_L \Gamma$. Then there exists a unique $\varepsilon \in \mathscr{V}(L,\mathsf{I})$ such that $\mathfrak{J} \vDash_L \mathrm{GC}[R,p,\varepsilon](\vec{a})$.

Proof. The lemma directly follows from Definitions 12.2.3, 12.2.5 and 12.2.6. □

Here we define the main notion of this paragraph, namely, the notion of structure generator. Structure generators can be considered as semantic counterparts of modification calculus. It is with the help of structure generators that modification calculi are generated.

Definition 12.2.8. Let L be an FPJ logic, σ be a signature, $\Gamma \in \mathrm{FS}_\sigma(L,\mathsf{E})$, $R \in \mathrm{GRS}_\sigma(L,\Gamma)$, $p \in \mathrm{Pr}_\sigma(\mathsf{I})$ and $\mathfrak{J}$ be an L-structure of signature σ. Assume that $\mathfrak{J} \vDash_L \Gamma$. Suppose $\mathfrak{K}$ is an L-structure of signature σ such that:

(i) $\mathrm{Frame}_{\mathfrak{K}} = \mathrm{Frame}_{\mathfrak{J}}$,
(ii) $f^{\mathfrak{K}} = f^{\mathfrak{J}}$ for any $f \in \mathrm{Fu}_\sigma$,
(iii) $q^{\mathfrak{K}} = q^{\mathfrak{J}}$ for any $q \in \mathrm{Pr}_\sigma(\mathsf{E})$,
(iv) $r^{\mathfrak{K}} = r^{\mathfrak{J}}$ for any $r \in \mathrm{Pr}_\sigma(\mathsf{I})$ such that $r \neq p$,
(v) For any $\vec{a} \in \mathrm{Dom}\left(p^{\mathfrak{J}}\right) = \mathrm{Dom}\left(p^{\mathfrak{K}}\right)$ and any $\varepsilon \in \mathscr{V}(L,\mathsf{I})$,
$p^{\mathfrak{K}}(\vec{a}) = \varepsilon$ iff $\mathfrak{J} \vDash_L \mathrm{GC}[R,p,\varepsilon](\vec{a})$.

It is not difficult to show that there exists a unique L-structure $\mathfrak{K}$ satisfying the conditions (i)–(v) (see Lemma 12.2.7). This $\mathfrak{K}$ is called the *R-successor* of $\mathfrak{J}$ w.r.t. p and is denoted by $\mathrm{Succ}(R,\mathfrak{J},p)$.

Definition 12.2.9. Let L be an FPJ logic, σ be a signature, $\Gamma \in \mathrm{FS}_\sigma(L,\mathsf{E})$ and $R \in \mathrm{GRS}_\sigma(L,\Gamma)$, $p \in \mathrm{Pr}_\sigma(\mathsf{I})$. We say that R is *Γ-robust* if for any L-structure $\mathfrak{J}$ of signature σ and for any $p \in \mathrm{Pr}_\sigma(\mathsf{I})$,

$$\mathfrak{J} \vDash_L \Gamma \quad \text{implies} \quad \mathrm{Succ}(R,\mathfrak{J},p) \vDash_L \Gamma.$$

By definition let

$$\mathrm{RGRS}_\sigma(L,\Gamma) = \{R \in \mathrm{GRS}_\sigma(L,\Gamma) \mid R \text{ is } \Gamma\text{-robust}\}.$$

Definition 12.2.10. Let L be an FPJ logic, σ be a signature, $\Gamma \in \mathrm{FS}_\sigma(L,\mathsf{E})$, $\mathfrak{J}$ be an L-structure of signature σ, $R \in \mathrm{GRS}_\sigma(L,\Gamma)$ and O be a map of $[1\mathinner{..}Q]$ to the set $\mathrm{Pr}_\sigma(\mathsf{I})$, where $Q \geq 1$. Then the ordered triple

$$\mathfrak{R} = \langle \mathfrak{J}, R, O \rangle$$

is called a *structure generator* of signature σ w.r.t. Γ in L if the following conditions hold:

(i) $\mathfrak{J} \models_L \Gamma$,
(ii) R is Γ-robust.

The L-structure $\mathfrak{J}$ is called *the origin* of $\mathfrak{R}$ and denoted by $\overset{\circ}{\mathfrak{R}}$. $R \in \mathrm{GRS}_\sigma(L, \Gamma)$ is called the *extender* of $\mathfrak{R}$ and is denoted by $\mathfrak{R}^{\triangleright}$, the mapping O is called the *ordering function* of $\mathfrak{R}$ and is denoted by $\mathrm{Ord}_{\mathfrak{R}}$.

The length of the string formed by $\mathrm{Ord}_{\mathfrak{R}}$ is denoted by $\mathrm{Q}_{\mathfrak{R}}$, so $\mathrm{Dom}(\mathrm{Ord}_{\mathfrak{R}}) = [1..\mathrm{Q}_{\mathfrak{R}}]$.

$\mathrm{P}^{\mathfrak{R}}_m$ denotes the predicate symbol $p \in \mathrm{Pr}_\sigma(\mathsf{I})$ such that $p = \mathrm{Ord}_{\mathfrak{R}}(m)$.

$\mathrm{GC}[\mathfrak{R}, p, \varepsilon]$ denotes $\mathrm{GC}[\mathfrak{R}^{\triangleright}, p, \varepsilon]$.

The class of all structure generators of signature *sigma* w.r.t. Γ in L is denoted $\mathrm{SG}_\sigma(L, \Gamma)$.

By definition let

$$\mathrm{SG}_\sigma(L) = \{\mathfrak{R} \in \mathrm{SG}_\sigma(L, \Gamma) \mid \Gamma \in \mathrm{FS}_\sigma(L, \mathsf{E})\}.$$

Below we consider L-structure sequences which are obtained by generators. We will see that such sequences can be considered as models of theories in the modification calculi.

Definition 12.2.11. Let L be an FPJ logic, $\Gamma \in \mathrm{FS}_\sigma(L, \mathsf{E})$, $\mathfrak{R} \in \mathrm{SG}_\sigma(L, \Gamma)$. Define by induction the *sequence of L-structures generated by* $\mathfrak{R}$ (written: $\overrightarrow{\mathfrak{R}}$). This sequence is indexed by the set $\omega \times [0..\mathrm{Q}_{\mathfrak{R}}]$.

BASIS. $\overrightarrow{\mathfrak{R}}[0, m] = \overset{\circ}{\mathfrak{R}} \quad (0 \leq m \leq \mathrm{Q}_{\mathfrak{R}})$,
INDUCTION STEP. Let $s > 0$. Then

(i) $\overrightarrow{\mathfrak{R}}[s, 0] = \overrightarrow{\mathfrak{R}}[s-1..\mathrm{Q}_{\mathfrak{R}}]$,
(ii) $\overrightarrow{\mathfrak{R}}[s, m] = \mathrm{Succ}\left(\mathfrak{R}^{\triangleright}, \overrightarrow{\mathfrak{R}}[s, m-1], \mathrm{P}^{\mathfrak{R}}_m\right) \quad (m > 0)$.

Lemma 12.2.12. Let L be an FPJ logic, $\Gamma \in \mathrm{FS}_\sigma(L, \mathsf{E})$, $\mathfrak{R} \in \mathrm{SG}_\sigma(L, \Gamma)$. Then the sequence $\overrightarrow{\mathfrak{R}}$ (see Definitions 12.2.11) is well defined and for any $s \in \omega, m \in [0..\mathrm{Q}_{\mathfrak{R}}]$,

$$\overrightarrow{\mathfrak{R}}[s, m] \models_L \Gamma.$$

Proof. This lemma can be proved by straightforward induction over s and m (see Definitions 12.2.10, 12.2.8 and 12.2.9). □

Lemma 12.2.13. Let L be an FPJ logic, $\Gamma \in \mathrm{FS}_\sigma(L, \mathsf{E})$, $\mathfrak{R} \in \mathrm{SG}_\sigma(L, \Gamma)$. Then for any $s \in \omega, m \in [0..\mathrm{Q}_{\mathfrak{R}}]$ the following statements hold:

(i) $\mathrm{Frame}_{\overrightarrow{\mathfrak{R}}[s,m]} = \mathrm{Frame}_{\overset{\circ}{\mathfrak{R}}}$,

(ii) $\mathrm{Dom}_{\overrightarrow{\mathfrak{R}}[s,m]}(\mathsf{s}) = \mathrm{Dom}_{\mathring{\mathfrak{R}}}(\mathsf{s})$ for any sort s,

(iii) $f^{\overrightarrow{\mathfrak{R}}[s,m]} = f^{\mathring{\mathfrak{R}}}$ for any $f \in \mathrm{Fu}_\sigma$,

(iv) $q^{\overrightarrow{\mathfrak{R}}[s,m]} = q^{\mathring{\mathfrak{R}}}$ for any $q \in \mathrm{Pr}_\sigma(\mathsf{E})$.

Proof. (i) follows from Definition 12.2.8(i).
(ii) follows from (i).
(iii) and (iv) can be proved by a trivial induction on s and m. □

Corollary 12.2.14. Let L be an FPJ logic, $\Gamma \in \mathrm{FS}_\sigma(L,\mathsf{E})$, $\mathfrak{R} \in \mathrm{SG}_\sigma(L,\Gamma)$. Then for any $s \in \omega$, $s > 0$, $m \in [1\,..\,\mathrm{Q}_{\mathfrak{R}}]$ the following statements hold:

(i) $p^{\overrightarrow{\mathfrak{R}}[s,m]} = p^{\overrightarrow{\mathfrak{R}}[s,m-1]}$ for any $p \in \mathrm{Pr}_\sigma(\mathsf{I})$ such that $p \neq \mathrm{P}^{\mathfrak{R}}_m$;

(ii) $\left(\mathrm{P}^{\mathfrak{R}}_m\right)^{\overrightarrow{\mathfrak{R}}[s,m]}(\vec{a}) = \varepsilon$ iff $\overrightarrow{\mathfrak{R}}[s,m-1] \vDash_L \mathrm{GC}\left[\mathfrak{R}^{\triangleright}, \mathrm{P}^{\mathfrak{R}}_m, \varepsilon\right](\vec{a})$ $(\varepsilon \in \mathscr{V}(L,\mathsf{I}))$;

(iii) $\left(\mathrm{P}^{\mathfrak{R}}_m\right)^{\overrightarrow{\mathfrak{R}}[s,m-1]}(\vec{a}) = \varepsilon$ implies $\left(\mathrm{P}^{\mathfrak{R}}_m\right)^{\overrightarrow{\mathfrak{R}}[s,m]}(\vec{a}) = \varepsilon$ $(\varepsilon \in \mathscr{D}^{\tau}(L))$.

Proof. (a) follows from Definitions 12.2.11 and 12.2.8(iv). (b) follows from Definitions 12.2.11 and 12.2.8(v).

Let us prove (c). Suppose $\left(\mathrm{P}^{\mathfrak{R}}_m\right)^{\overrightarrow{\mathfrak{R}}[s,m-1]}(\vec{a}) = \varepsilon \in \mathscr{D}^{\tau}(L)$. By Lemma 12.2.12, $\overrightarrow{\mathfrak{R}}[s,m-1] \vDash_L \Gamma$. Then by Definition 12.2.3 (iii),

$$\overrightarrow{\mathfrak{R}}[s,m-1] \vDash_L \mathrm{GC}\left[\mathfrak{R}^{\triangleright}, \mathrm{P}^{\mathfrak{R}}_m, \varepsilon\right](\vec{a});$$

hence by (b) we get $\left(\mathrm{P}^{\mathfrak{R}}_m\right)^{\overrightarrow{\mathfrak{R}}[s,m]}(\vec{a}) = \varepsilon$. □

Now we will establish a few basic properties of L-structure sequences generated by a structure generator. All the properties will be in the form of equivalence. Note that the forecoming lemma will be often used further on.

Lemma 12.2.15. Let L be an FPJ logic, $\Gamma \in \mathrm{FS}_\sigma(L,\mathsf{E})$, $\mathfrak{R} \in \mathrm{SG}_\sigma(L,\Gamma)$. Then for any $s \in \omega, m \in [0\,..\,\mathrm{Q}_{\mathfrak{R}}]$ and any valuation v in $\mathrm{Frame}_{\mathring{\mathfrak{R}}}$, the following statements hold:

(i) $f^{\overrightarrow{\mathfrak{R}}[s,m]} = f^{\mathring{\mathfrak{R}}}$ for any $f \in \mathrm{Fu}_\sigma$,

(ii) $q^{\overrightarrow{\mathfrak{R}}[s,m]} = q^{\mathring{\mathfrak{R}}}$ for any $q \in \mathrm{Pr}_\sigma(\mathsf{E})$,

(iii) $t^{\overrightarrow{\mathfrak{R}}[s,m],v} = t^{\mathring{\mathfrak{R}},v}$ for any term t of signature σ,

(iv) $(t_1 = t_2)^{\overrightarrow{\mathfrak{R}}[s,m],v} = (t_1 = t_2)^{\mathring{\mathfrak{R}},v}$ for any terms t_1 and t_2 of signature σ such that $\mathrm{Sort}_\sigma(t_1) = \mathrm{Sort}_\sigma(t_2)$,

(v) $q\left(\vec{t}\right)^{\overrightarrow{\mathfrak{R}}[s,m],v} = q\left(\vec{t}\right)^{\mathring{\mathfrak{R}},v}$ for any $q \in \mathrm{Pr}_\sigma(\mathsf{E}), \vec{t} \in \mathrm{Dom}_\sigma(q)$,

(vi) $p\left(\vec{t}\right)^{\overrightarrow{\mathfrak{R}}[0,m],v} = p\left(\vec{t}\right)^{\mathring{\mathfrak{R}},v}$ for any $p \in \mathrm{Pr}_\sigma(\mathsf{I}), \vec{t} \in \mathrm{Dom}_\sigma(p)$,

(vii) $\left(\mathrm{P}^{\mathfrak{R}}_m\right)^{\overrightarrow{\mathfrak{R}}[s,m]} = \left(\mathrm{P}^{\mathfrak{R}}_m\right)^{\overrightarrow{\mathfrak{R}}[s,k]}$ for any $s,k \in \omega$ $(s > 0, 0 \leq m \leq k \leq \mathrm{Q}_{\mathfrak{R}})$,

(viii) $\left(\mathrm{P}_m^{\mathfrak{R}}\right)^{\overrightarrow{\mathfrak{R}}[s,k]} = \left(\mathrm{P}_m^{\mathfrak{R}}\right)^{\overrightarrow{\mathfrak{R}}[s-1,m]}$ for any $s,k \in \omega$ $(s>0, 0 \le k < m \le \mathrm{Q}_{\mathfrak{R}})$,

(ix) $\left(\mathrm{P}_m^{\mathfrak{R}}(\vec{t})\right)^{\overrightarrow{\mathfrak{R}}[s,m],v} = \left(\mathrm{P}_m^{\mathfrak{R}}(\vec{t})\right)^{\overrightarrow{\mathfrak{R}}[s,k],v}$ for any $s,k \in \omega$ $(s>0, 0 \le m \le k \le \mathrm{Q}_{\mathcal{M}})$ and for any $\vec{t} \in \mathrm{Dom}_\sigma\left(\mathrm{P}_m^{\mathfrak{R}}\right)$,

(x) $\left(\mathrm{P}_m^{\mathfrak{R}}(\vec{t})\right)^{\overrightarrow{\mathfrak{R}}[s,k],v} = \left(\mathrm{P}_m^{\mathfrak{R}}(\vec{t})\right)^{\overrightarrow{\mathfrak{R}}[s-1,m],v}$ for any $s,k \in \omega$ $(s>0, 0 \le k < m \le \mathrm{Q}_{\mathcal{M}})$ and for any $\vec{t} \in \mathrm{Dom}_\sigma\left(\mathrm{P}_m^{\mathfrak{R}}\right)$.

Proof. (i) and (ii) can be proved by a trivial induction on s and m by the use of Definition 12.2.11.

(iii) can be proved by induction over the terms of signature σ by using (i).

(iv) follows from (iii).

(v) follows from (ii) and (iii).

(vi) follows from 12.2.11 (basis of the induction).

(vii) can be proved by an easy induction on k. For $k = m$, it is obvious, that

$$\left(\mathrm{P}_m^{\mathfrak{R}}\right)^{\overrightarrow{\mathfrak{R}}[s,m]} = \left(\mathrm{P}_m^{\mathfrak{R}}\right)^{\overrightarrow{\mathfrak{R}}[s,k]}.$$

Let $\left(\mathrm{P}_m^{\mathfrak{R}}\right)^{\overrightarrow{\mathfrak{R}}[s,m]} = \left(\mathrm{P}_m^{\mathfrak{R}}\right)^{\overrightarrow{\mathfrak{R}}[s,n]}$ hold. The case $m = n+1$ is trivial. Suppose $m \ne n+1$. In this case by we obtain:

$$\begin{aligned} \left(\mathrm{P}_m^{\mathfrak{R}}\right)^{\overrightarrow{\mathfrak{R}}[s,n+1]} &= \left(\mathrm{P}_m^{\mathfrak{R}}\right)^{\mathrm{Succ}\left(\mathfrak{R}^{\rhd},\overrightarrow{\mathfrak{R}}[s,n],\mathrm{P}_{n+1}^{\mathfrak{R}}\right)} && \text{(by 12.2.11 induction step (ii))} \\ &= \left(\mathrm{P}_m^{\mathfrak{R}}\right)^{\overrightarrow{\mathfrak{R}}[s,n]} && \text{(by 12.2.8 (iv) since } m \ne n+1) \\ &= \left(\mathrm{P}_m^{\mathfrak{R}}\right)^{\overrightarrow{\mathfrak{R}}[s,m]} && \text{(by the induction hypothesis)} \end{aligned}$$

(viii) can be proved by an easy induction on k. For $k = 0$ we have:

$$\begin{aligned} \left(\mathrm{P}_m^{\mathfrak{R}}\right)^{\overrightarrow{\mathfrak{R}}[s,0]} &= \left(\mathrm{P}_m^{\mathfrak{R}}\right)^{\overrightarrow{\mathfrak{R}}[s-1,\mathrm{Q}_{\mathfrak{R}}]} && \text{(by 12.2.11 induction step (i))} \\ &= \left(\mathrm{P}_m^{\mathfrak{R}}\right)^{\overrightarrow{\mathfrak{R}}[s-1,m]} && \text{(by (vii))} \end{aligned}$$

Suppose $n+1 < m$ and $\left(\mathrm{P}_m^{\mathfrak{R}}\right)^{\overrightarrow{\mathfrak{R}}[s,n],v} = \left(\mathrm{P}_m^{\mathfrak{R}}\right)^{\overrightarrow{\mathfrak{R}}[s-1,m],v}$ holds. Then similarly to the proof of (vii) we obtain: $\left(\mathrm{P}_m^{\mathfrak{R}}\right)^{\overrightarrow{\mathfrak{R}}[s,n+1],v} = \left(\mathrm{P}_m^{\mathfrak{R}}\right)^{\overrightarrow{\mathfrak{R}}[s-1,m],v}$.

(ix) follows from (vii) and (iii).

(x) follows from (viii) and (iii). □

Now we will show that a sequence of L-structures generated by a structure generator is *monotonous*, i.e. for this sequence, certainification of matrices (that correspond to internal predicates) does take place, i.e. this sequence is a real result of certainification of matrices corresponding to internal predicates.

Lemma 12.2.16. Let L be an FPJ logic, $\Gamma \in \mathrm{FS}_\sigma(L, \mathsf{E})$, $\mathfrak{R} \in \mathrm{SG}_\sigma(L, \Gamma)$. Let $s \in \omega$, $m, k \in [0..\mathrm{Q}_{\mathfrak{R}}]$, where $m \leq k$. Then $\overrightarrow{\mathfrak{R}}[s,m] \preceq \overrightarrow{\mathfrak{R}}[s,k]$.

Proof. By Definition 12.2.11,

$$\mathrm{Frame}_{\overrightarrow{\mathfrak{R}}[s,m]} = \mathrm{Frame}_{\overrightarrow{\mathfrak{R}}[s,k]} = \mathrm{Frame}_{\overset{\circ}{\mathfrak{R}}}.$$

By Lemma 12.2.15(i), $f^{\overrightarrow{\mathfrak{R}}[s,m]} = f^{\overrightarrow{\mathfrak{R}}[l,k]}$ for any $f \in \mathrm{Fu}_\sigma$. By Lemma 12.2.15 (ii), for any $q \in \mathrm{Pr}_\sigma(\mathsf{E})$, $q^{\overrightarrow{\mathfrak{R}}[s,m]} = q^{\overrightarrow{\mathfrak{R}}[l,k]}$.

By Definition 12.1.5 in order to prove

$$\overrightarrow{\mathfrak{R}}[s,m] \preceq \overrightarrow{\mathfrak{R}}[s,k]$$

it is enough to show that

$$p^{\overrightarrow{\mathfrak{R}}[s,m]} \preceq p^{\overrightarrow{\mathfrak{R}}[l,k]}$$

for any $p \in \mathrm{Pr}_\sigma(\mathsf{I})$. That is we have to check that for any $\vec{a} \in \mathrm{Dom}\left(p^{\mathfrak{J}}\right)$,

$$p^{\overrightarrow{\mathfrak{R}}[s,m]}(\vec{a}) \neq \tau \quad \text{implies} \quad p^{\overrightarrow{\mathfrak{R}}[s,m]}(\vec{a}) = p^{\overrightarrow{\mathfrak{R}}[l,k]}(\vec{a}).$$

Let $p = \mathrm{P}_i^{\mathfrak{R}}$, $\vec{a} \in \mathrm{Dom}\left(p^{\mathfrak{J}}\right)$, $\left(\mathrm{P}_i^{\mathfrak{R}}\right)^{\overrightarrow{\mathfrak{R}}[s,m],v}(\vec{a}) \neq \tau$. Consider the possible cases.

CASE 1. Let $i \leq m$. Then $i \leq k$ and

$$\begin{aligned}
\left(\mathrm{P}_i^{\mathfrak{R}}\right)^{\overrightarrow{\mathfrak{R}}[s,m],v}(\vec{a}) &= \left(\mathrm{P}_i^{\mathfrak{R}}\right)^{\overrightarrow{\mathfrak{R}}[s,i],v}(\vec{a}) && \text{(by 12.2.16 (vii))}\\
&= \left(\mathrm{P}_i^{\mathfrak{R}}\right)^{\overrightarrow{\mathfrak{R}}[s,k],v}(\vec{a}) && \text{(by 12.2.16 (vii))}
\end{aligned}$$

CASE 2. Let $k < i$. Then $m < i$ and

$$\begin{aligned}
\left(\mathrm{P}_i^{\mathfrak{R}}\right)^{\overrightarrow{\mathfrak{R}}[s,m],v}(\vec{a}) &= \left(\mathrm{P}_i^{\mathfrak{R}}\right)^{\overrightarrow{\mathfrak{R}}[s-1,i],v}(\vec{a}) && \text{(by 12.2.16 (viii))}\\
&= \left(\mathrm{P}_i^{\mathfrak{R}}\right)^{\overrightarrow{\mathfrak{R}}[s,k],v}(\vec{a}) && \text{(by 12.2.16 (viii))}
\end{aligned}$$

CASE 3. Let $m < i \leq k$.

$$
\begin{aligned}
\left(\mathrm{P}_i^{\mathfrak{R}}\right)^{\vec{\mathfrak{R}}[s,m],v}(\vec{a}) &= \left(\mathrm{P}_i^{\mathfrak{R}}\right)^{\vec{\mathfrak{R}}[s-1,i],v}(\vec{a}) && \text{(by 12.2.16 (viii))} \\
&= \left(\mathrm{P}_i^{\mathfrak{R}}\right)^{\vec{\mathfrak{R}}[s,i-1],v}(\vec{a}) && \text{(by 12.2.16 (viii))} \\
&= \left(\mathrm{P}_i^{\mathfrak{R}}\right)^{\vec{\mathfrak{R}}[s,i],v}(\vec{a}) && \text{(by 12.2.14 (c)).}
\end{aligned}
$$

Note that the last equality holds since $\left(\mathrm{P}_i^{\mathfrak{R}}\right)^{\vec{\mathfrak{R}}[s,m],v}(\vec{a}) \neq \tau$. □

Theorem 12.2.17. Let L be an FPJ logic, $\Gamma \in \mathrm{FS}_\sigma(L,\mathsf{E})$, $\mathfrak{R} \in \mathrm{SG}_\sigma(L,\Gamma)$. A natural lexicographic order can be defined on the set $\{\langle s,m\rangle \mid s \in \omega, 0 \leq m \leq \mathrm{Q}_{\mathfrak{R}}\}$. Then the sequence defined in Definition 12.2.11 is monotone w.r.t. this order, i.e. $\langle s,m\rangle \leq \langle l,k\rangle$ implies $\vec{\mathfrak{R}}[s,m] \preceq \vec{\mathfrak{R}}[l,k]$ for any $s,l \in \omega, m,k \in [0..\mathrm{Q}_{\mathfrak{R}}]$.

Proof. Let $\langle s,m\rangle \leq \langle l,k\rangle$. We prove the statement by induction on l.
BASIS. Let $l = s$. Then, $m \leq k$. By Lemma 12.2.16,

$$\vec{\mathfrak{R}}[s,m] \preceq \vec{\mathfrak{R}}[l,k].$$

INDUCTION STEP. Suppose

$$\vec{\mathfrak{R}}[s,m] \preceq \vec{\mathfrak{R}}[n,k]$$

and let us show that

$$\vec{\mathfrak{R}}[s,m] \preceq \vec{\mathfrak{R}}[n+1,k].$$

Indeed,

$$
\begin{aligned}
\vec{\mathfrak{R}}[s,m] &\preceq \vec{\mathfrak{R}}[n,k] && \text{(by the induction hypothesis)} \\
&\preceq \vec{\mathfrak{R}}[n,\mathrm{Q}_{\mathfrak{R}}] && \text{(by 12.2.16)} \\
&\preceq \vec{\mathfrak{R}}[n+1,0] && \text{(by 12.2.11 induction step (i))} \\
&\preceq \vec{\mathfrak{R}}[n+1,k] && \text{(by 12.2.16)}
\end{aligned}
$$

□

Now we analyse the stabilisation of an L-structure sequence generated by a structure generator. By stabilisation we mean a situation where the elements of the sequence are repeated from a certain element on.

Lemma 12.2.18. Let L be an FPJ logic, $\Gamma \in \mathrm{FS}_\sigma(L,\mathsf{E})$, $\mathfrak{R} \in \mathrm{SG}_\sigma(L,\Gamma)$. Let $s \in \omega\,(s > 0)$ be such that

$$\left(\mathrm{P}_m^{\mathfrak{N}}\right)^{\vec{\mathfrak{N}}[s,m]} = \left(\mathrm{P}_m^{\mathfrak{N}}\right)^{\vec{\mathfrak{N}}[s-1,m]} \tag{12.1}$$

holds for any $m \in [1\,..\,\mathrm{Q}_{\mathfrak{N}}]$. Then

$$\vec{\mathfrak{N}}[s,m] = \vec{\mathfrak{N}}[s,0] \tag{12.2}$$

holds for any $m \in [1\,..\,\mathrm{Q}_{\mathfrak{N}}]$.

Proof. By induction on m.

BASIS. Obviously, the equality (12.2) holds for $m = 0$.

INDUCTION STEP. Let the equality (12.2) hold for $m = k$. Let us prove it for $m = k+1$. By Definition 12.2.11 and the inductive assumption:

(a) $f^{\vec{\mathfrak{N}}[s,k+1]} = f^{\vec{\mathfrak{N}}[s,k]} = f^{\vec{\mathfrak{N}}[s,0]}$ for any $f \in \mathrm{Fu}_\sigma$,
(b) $q^{\vec{\mathfrak{N}}[s,k+1]} = q^{\vec{\mathfrak{N}}[s,k]} = q^{\vec{\mathfrak{N}}[s,0]}$ for any $q \in \mathrm{Pr}_\sigma(\mathsf{E})$,
(c) $p^{\vec{\mathfrak{N}}[s,k+1]} = p^{\vec{\mathfrak{N}}[s,k]} = p^{\vec{\mathfrak{N}}[s,0]}$ for any $p \in \mathrm{Pr}_\sigma(\mathsf{I})$ such that $p \neq \mathrm{P}_{k+1}^{\mathfrak{N}}$.

Let us prove that

(d) $\left(\mathrm{P}_{k+1}^{\mathfrak{N}}\right)^{\vec{\mathfrak{N}}[s,k+1]} = \left(\mathrm{P}_{k+1}^{\mathfrak{N}}\right)^{\vec{\mathfrak{N}}[s,0]}$.

Indeed,

$$\begin{aligned}\left(\mathrm{P}_{k+1}^{\mathfrak{N}}\right)^{\vec{\mathfrak{N}}[s,k+1]} &= \left(\mathrm{P}_{k+1}^{\mathfrak{N}}\right)^{\vec{\mathfrak{N}}[s-1,k+1]} && \text{(by (12.1))}\\ &= \left(\mathrm{P}_{k+1}^{\mathfrak{N}}\right)^{\vec{\mathfrak{N}}[s,0]} && \text{(by 12.2.16 (viii))}\end{aligned}$$

(a)–(d) imply $\vec{\mathfrak{N}}[s,k+1] = \vec{\mathfrak{N}}[s,0]$. □

Lemma 12.2.19 (Stabilisation Lemma). Let L be an FPJ logic, $\Gamma \in \mathrm{FS}_\sigma(L,\mathsf{E})$, $\mathfrak{N} \in \mathrm{SG}_\sigma(L,\Gamma)$. Let $s \in \omega$, $s > 0$ and

$$\left(\mathrm{P}_m^{\mathfrak{N}}\right)^{\vec{\mathfrak{N}}[s,m]} = \left(\mathrm{P}_m^{\mathfrak{N}}\right)^{\vec{\mathfrak{N}}[s-1,m]} \tag{12.3}$$

be true for any $m \in [0\,..\,\mathrm{Q}_{\mathfrak{N}}]$. Then

$$\vec{\mathfrak{N}}[n,m] = \vec{\mathfrak{N}}[s,0] \tag{12.4}$$

holds for any $n \geq s$ and any $m \in [0\,..\,\mathrm{Q}_{\mathfrak{N}}]$.

Proof. By induction on n.

BASIS. By Lemma 12.2.18 the statement holds for $n = s$.

INDUCTION STEP. Suppose that (12.2) holds for all $m \in [0..Q_{\mathfrak{R}}]$ and $n = l$. Let us show that the equality (12.2) holds also for $n = l+1$ (and for all $m \in [0..Q_{\mathfrak{R}}]$). By induction on m.

Basis. Let $m = 0$. Then

$$\begin{aligned} \overrightarrow{\mathfrak{R}}[l+1,0] &= \overrightarrow{\mathfrak{R}}[l,Q_{\mathfrak{R}}] && \text{(by 12.2.11 induction step (i))} \\ &= \overrightarrow{\mathfrak{R}}[s,0] && \text{(by the induction hypothesis)} \end{aligned}$$

Induction step. Let the equality (12.2) hold for $m = k$. Let us prove it for $m = k+1$. By the inductive assumption,

$$\overrightarrow{\mathfrak{R}}[l+1,k] = \overrightarrow{\mathfrak{R}}[s,0]\,. \tag{12.5}$$

By Definition 12.2.11, $\overrightarrow{\mathfrak{R}}[l+1,k+1]$ may differ from $\overrightarrow{\mathfrak{R}}[l+1,k]$ only in the interpretation of the predicate symbol $P^{\mathfrak{R}}_{k+1}$. Thus it is enough to show that

$$\left(P^{\mathfrak{R}}_{k+1}\right)^{\overrightarrow{\mathfrak{R}}[l+1,k+1]} = \left(P^{\mathfrak{R}}_{k+1}\right)^{\overrightarrow{\mathfrak{R}}[s,0]}. \tag{12.6}$$

Assume the converse. Let

$$\left(P^{\mathfrak{R}}_{k+1}\right)^{\overrightarrow{\mathfrak{R}}[l+1,k+1]} \neq \left(P^{\mathfrak{R}}_{k+1}\right)^{\overrightarrow{\mathfrak{R}}[s,0]}.$$

Hence, by taking into account (12.5), we obtain

$$\left(P^{\mathfrak{R}}_{k+1}\right)^{\overrightarrow{\mathfrak{R}}[l+1,k+1]} \neq \left(P^{\mathfrak{R}}_{k+1}\right)^{\overrightarrow{\mathfrak{R}}[l+1,k]}.$$

Therefore for some $\vec{a} \in \mathrm{Dom}\left(\left(P^{\mathfrak{R}}_{k+1}\right)^{\mathfrak{I}}\right)$,

$$\left(P^{\mathfrak{R}}_{k+1}\right)^{\overrightarrow{\mathfrak{R}}[l+1,k]}(\vec{a}) = \tau, \tag{12.7}$$

but

$$\left(P^{\mathfrak{R}}_{k+1}\right)^{\overrightarrow{\mathfrak{R}}[l+1,k+1]}(\vec{a}) = \varepsilon \neq \tau.$$

By Definitions 12.2.11 (induction step (ii)) and 12.2.8 (v),

$$\overrightarrow{\mathfrak{R}}[l+1,k] \models_L \mathrm{GC}\left[R, P^{\mathfrak{R}}_{k+1}, \varepsilon\right](\vec{a})\,. \tag{12.8}$$

By using Lemma 12.2.18 we obtain

$$\overrightarrow{\mathfrak{R}}[s,k] = \overrightarrow{\mathfrak{R}}[s,0]\,. \tag{12.9}$$

Then (12.5), (12.7) and (12.9) imply that

$$\left(\mathrm{P}_{k+1}^{\mathfrak{R}}\right)^{\overrightarrow{\mathfrak{R}}[s,k]}(\vec{a}) = \tau. \tag{12.10}$$

Then from (12.5), (12.8) and (12.9) we obtain

$$\overrightarrow{\mathfrak{R}}[s,k] \vDash_L \mathrm{GC}\left[R, \mathrm{P}_{k+1}^{\mathfrak{R}}, \varepsilon\right](\vec{a}). \tag{12.11}$$

From (12.10) and (12.11) by Definition 12.2.11 (induction step (ii)), we obtain

$$\left(\mathrm{P}_{k+1}^{\mathfrak{R}}\right)^{\overrightarrow{\mathfrak{R}}[s,k+1]}(\vec{a}) = \varepsilon \neq \tau. \tag{12.12}$$

However, (12.12), (12.10) and (12.9) imply that $\overrightarrow{\mathfrak{R}}[s,k+1] \neq \overrightarrow{\mathfrak{R}}[s,0]$, which is impossible because of Lemma 12.2.18. Then (12.6) holds, which was to be proved.

□

Chapter 13
Iterative Representation of Structure Generators

13.1 Immersions and Snaps

In this chapter we establish a relationship between validities in L-structures and I-structures, where L is an FPJ logic and I is its iterative version w.r.t. some value $\tau \in \mathscr{V}(L, \mathsf{I})$, i.e. $I = \mathrm{I}^{\tau}(L)$. I-structures represent the history of the cognitive process. The cognitive process itself is simulated by an inference in a modification calculus.

A generator is the semantic analogue of a modification calculus. We will show that for each generator there exists a unique I-structure which, in a certain sense, is an extension of the origin of this generator.

We begin with a number of definitions and statements that connect formulas and truth values in L and I logics. Such a connection may be defined if we additionally suppose that we know: (i) the stage and the module of an inference in a certain modification calculus and (ii) the ordering over the internal predicate symbols.

Agreement 13.1.1. Here and hereafter we denote the value "uncertain" from the set $\mathscr{V}(L, \mathsf{I})$, where L is an arbitrary FPJ logic, by τ.

Definition 13.1.2. Let L be an FPJ logic, σ be a signature. Suppose O is a map of $[1..Q]$ to $\mathrm{Pr}_{\sigma}(\mathsf{I})$, where $Q \geq 1$, i.e. the map O forms a finite sequence (string) of internal predicate symbols of signature σ.

For the sake of uniformity we will write P_i^O instead of $O(i)$, i.e. by P_i^O we denote the predicate symbol $p \in \mathrm{Pr}_{\sigma}(\mathsf{I})$ that is the i-th member of the sequence formed by the map O. The length of this sequence (which is equal to Q from the segment $[1..Q]$) is denoted by Q_O.

For any $p \in \mathrm{Pr}_{\sigma}(\mathsf{I})$, let

$$\mathrm{R}^O(p) = \begin{cases} \max\{i \mid p = \mathrm{P}_i^O\} & \text{if } p \in \mathrm{Range}(O), \\ \mathrm{Q}_O & \text{otherwise.} \end{cases}$$

The number $\mathrm{R}^O(p)$ is called the *upper rank* of p w.r.t. O.

Suppose $s \in \omega$, $0 \leq m \leq \mathrm{Q}_O$.

O. Anshakov, T. Gergely, *Cognitive Reasoning*, Cognitive Technologies,
DOI 10.1007/978-3-540-68875-4_13, © Springer-Verlag Berlin Heidelberg 2010

Let us inductively define the $\langle O,s,m\rangle$-*immersion* of external L-formulas (written: $\Phi^{\langle O,s,m\rangle}$) . By definition put:

(i) $\mathrm{J}_{\varepsilon}\, p\,(\vec{t})^{\langle O,s,m\rangle} = \begin{cases} \mathrm{J}_{(\varepsilon,s)}\, p\,(\vec{t}) & \text{if } s=0 \text{ or } \mathrm{R}^{O}(p) \leq m, \\ \mathrm{J}_{(\varepsilon,s-1)}\, p\,(\vec{t}) & \text{if } s>0 \text{ and } \mathrm{R}^{O}(p) > m, \end{cases}$
where $p \in \mathrm{Pr}_{\sigma}(\mathsf{I})$, $\varepsilon \in \mathscr{V}(L,\mathsf{I})$, $\vec{t} \in \mathrm{Dom}_{\sigma}(p)$;

(ii) $q\,(\vec{t})^{\langle O,s,m\rangle} = q\,(\vec{t})$,
where $q \in \mathrm{Pr}_{\sigma}(\mathsf{E})$, $\vec{t} \in \mathrm{Dom}_{\sigma}(q)$;

(iii) $(t_1 = t_2)^{\langle O,s,m\rangle} = (t_1 = t_2)$,
where t_1 and t_2 are terms such that $\mathrm{Sort}_{\sigma}(t_1) = \mathrm{Sort}_{\sigma}(t_2)$;

(iv) $(\Phi \rightarrow \Psi)^{\langle O,s,m\rangle} = \left(\Phi^{\langle O,s,m\rangle} \rightarrow \Psi^{\langle O,s,m\rangle}\right)$,
where Φ and Ψ are external formulas of the logic L;

(v) $(\neg\Phi)^{\langle O,s,m\rangle} = \left(\neg\Phi^{\langle O,s,m\rangle}\right)$,
where Φ is an external formula of the logic L;

(vi) $(\forall x\, \Phi)^{\langle O,s,m\rangle} = \left(\forall x\, \Phi^{\langle O,s,m\rangle}\right)$,
where Φ is an external formula of the logic L, x is an individual variable.

Lemma 13.1.3. Let L be an FPJ logic, σ be a signature. Suppose

$$O\colon\ [1\,..\,\mathrm{Q}_O] \rightarrow \mathrm{Pr}_{\sigma}(\mathsf{I})\,.$$

Then for any external L-formula Φ and for any $s > 0$:

$$\Phi^{\langle O,s,0\rangle} = \Phi^{\langle O,s-1,\mathrm{Q}_O\rangle}.$$

Proof. By induction over external L-formulas.

BASIS. By Definition 13.1.2(i),

$$\begin{aligned} \mathrm{J}_{\varepsilon}\, p\,(\vec{t})^{\langle O,s,0\rangle} &= \mathrm{J}_{(\varepsilon,s-1)}\, p\,(\vec{t}) \\ &= \mathrm{J}_{\varepsilon}\, p\,(\vec{t})^{\langle O,s-1,\mathrm{Q}_O\rangle}, \end{aligned}$$

where $p \in \mathrm{Pr}_{\sigma}(\mathsf{I})$, $\varepsilon \in \mathscr{V}(L,\mathsf{I})$, $\vec{t} \in \mathrm{Dom}_{\sigma}(p)$.

By Definition 13.1.2(ii),

$$\begin{aligned} q\,(\vec{t})^{\langle O,s,0\rangle} &= q\,(\vec{t}) \\ &= q\,(\vec{t})^{\langle O,s-1,\mathrm{Q}_O\rangle}, \end{aligned}$$

where $q \in \mathrm{Pr}_{\sigma}(\mathsf{E})$, $\vec{t} \in \mathrm{Dom}_{\sigma}(q)$. Similarly, by 13.1.2(iii), we get

$$(t_1 = t_2)^{\langle O,s,0\rangle} = (t_1 = t_2)^{\langle O,s-1,\mathrm{Q}_O\rangle},$$

where t_1 and t_2 are terms such that $\mathrm{Sort}_{\sigma}(t_1) = \mathrm{Sort}_{\sigma}(t_2)$.

INDUCTION STEP. It is trivial. □

Definition 13.1.4. Let L be an FPJ logic, $I = \mathrm{I}^{\tau}(L)$, $\alpha \in \mathscr{V}(I, \mathsf{I})$, $s \in \omega$. By Lemma 7.4.11, there exists a unique $\varepsilon \in \mathscr{V}(L, \mathsf{I})$ such that $\alpha \in \mathrm{Ch}(\varepsilon, s)$. This ε is called the *s-snap of the* truth value α and denoted by $\mathrm{Snap}(\alpha, s)$.

Proposition 13.1.5. Let L be an FPJ logic, $I = \mathrm{I}^{\tau}(L)$. Then the following statements hold:

(i) $\mathrm{Snap}(\alpha, s) = \varepsilon$ iff $\alpha \in \mathrm{Ch}(\varepsilon, s)$,
where $\alpha \in \mathscr{V}(I, \mathsf{I})$, $\varepsilon \in \mathscr{V}(L, \mathsf{I})$, $s \in \omega$;

(ii) $\mathrm{Snap}(\langle \varepsilon, n \rangle, s) = \begin{cases} \varepsilon & n \le s, \\ \tau & n > s, \end{cases}$
where $\varepsilon \in \mathscr{D}^{\tau}(L)$, $s, n \in \omega$;

(iii) $\alpha = \langle \varepsilon, s \rangle$ iff $\mathrm{Snap}(\alpha, s) = \varepsilon$ and $\mathrm{Snap}(\alpha, k) = \tau$ for any $k < s$ ($\varepsilon \in \mathscr{D}^{\tau}(L)$, $s, k \in \omega$, $\alpha \in \mathscr{V}(I, \mathsf{I})$);

(iv) $\alpha = \tau$ iff $\mathrm{Snap}(\alpha, s) = \tau$ for any $s \in \omega$ ($\alpha \in \mathscr{V}(I, \mathsf{I})$).

Proof. It is obvious. See Definition 13.1.4 and Definition 7.4.10. □

Below we investigate the relation of validities in L- and I-structures. We introduce a new notion of cut. We have earlier defined the $\langle O, s, m \rangle$-cut as a syntactic function that maps a set of L-formulas into a set of I-formulas. Now we introduce the $[O, s, m]$-*snap* as a semantic function that maps a class of I-structures of a certain signature into a class of L-structures of the same signature.

Definition 13.1.6. Let L be an FPJ logic, $I = \mathrm{I}^{\tau}(L)$, σ be a signature,

$$O\colon\ [1\,..\,\mathrm{Q}_O] \to \mathrm{Pr}_{\sigma}(\mathsf{I}),$$

$s \in \omega$, $0 \le m \le \mathrm{Q}_O$. $p \in \mathrm{Pr}_{\sigma}(\mathsf{I})$. Let $\mathfrak{K}$ be an I-structure of signature σ. An L-structure $\mathfrak{J}$ is called the $[O, s, m]$-*snap* of the I-structure $\mathfrak{K}$ if it meets the following conditions:

(i) $\mathrm{Frame}_{\mathfrak{J}} = \mathrm{Frame}_{\mathfrak{K}}$,
(ii) $f^{\mathfrak{J}} = f^{\mathfrak{K}}$ for any $f \in \mathrm{Fu}_{\sigma}$,
(iii) $q^{\mathfrak{J}} = q^{\mathfrak{K}}$ for any $q \in \mathrm{Pr}_{\sigma}(\mathsf{E})$,
(iv) $p^{\mathfrak{J}}(\vec{a}) = \begin{cases} \mathrm{Snap}\left(p^{\mathfrak{K}}(\vec{a}), s\right) & s = 0 \text{ or } \mathrm{R}^{O}(p) \le m, \\ \mathrm{Snap}\left(p^{\mathfrak{K}}(\vec{a}), s-1\right) & s > 0 \text{ and } \mathrm{R}^{O}(p) > m, \end{cases}$
where $\vec{a} \in \mathrm{Dom}\left(p^{\mathfrak{J}}\right)$.

$\mathfrak{K}[O, s, m]$ will denote the $[O, s, m]$-snap of the I-structure $\mathfrak{K}$.

Lemma 13.1.7. Let L be an FPJ logic, $I = \mathrm{I}^{\tau}(L)$, $\mathfrak{K}$ be an I-structure of signature σ,

$$O\colon\ [1\,..\,\mathrm{Q}_O] \to \mathrm{Pr}_{\sigma}(\mathsf{I}),$$

$s \in \omega$, $0 \le m \le \mathrm{Q}_O$, $\mathfrak{J} = \mathfrak{K}[O, s, m]$ and v be a valuation in $\mathrm{Frame}_{\mathfrak{J}} = \mathrm{Frame}_{\mathfrak{K}}$. Then the following statements hold:

(i) $t^{\mathfrak{J},v} = t^{\mathfrak{K},v}$ for any term t of signature σ,

(ii) $(t_1 = t_2)^{\mathfrak{J},v} = (t_1 = t_2)^{\mathfrak{K},v}$ for any terms t_1 and t_2 of signature σ such that $\mathrm{Sort}_\sigma(t_1) = \mathrm{Sort}_\sigma(t_2)$,

(iii) $q(\vec{t})^{\mathfrak{J},v} = q(\vec{t})^{\mathfrak{K},v}$ for any $q \in \mathrm{Pr}_\sigma(\mathsf{E})$, $\vec{t} \in \mathrm{Dom}_\sigma(q)$,

(iv) $\left(\mathrm{J}_\varepsilon\, p(\vec{t})\right)^{\mathfrak{J},v} = \left(\mathrm{J}_{(\varepsilon,s)}\, p(\vec{t})\right)^{\mathfrak{K},v}$ for any $\varepsilon \in \mathscr{V}(L,\mathsf{I})$, $p \in \mathrm{Pr}_\sigma(\mathsf{I})$, $\vec{t} \in \mathrm{Dom}_\sigma(p)$, if $s = 0$ or $\mathrm{R}^O(p) \le m$,

(v) $\left(\mathrm{J}_\varepsilon\, p(\vec{t})\right)^{\mathfrak{J},v} = \left(\mathrm{J}_{(\varepsilon,s-1)}\, p(\vec{t})\right)^{\mathfrak{K},v}$ for any $\varepsilon \in \mathscr{V}(L,\mathsf{I})$, $p \in \mathrm{Pr}_\sigma(\mathsf{I})$, $\vec{t} \in \mathrm{Dom}_\sigma(p)$, if $s > 0$ and $\mathrm{R}^O(p) > m$.

Proof. (i) can be proved by induction on the terms of signature σ by Definition 13.1.6 (ii).

(ii) follows from (i).

(iii) follows from (i) and Definition 13.1.6(iii).

(iv) can be proved as follows:

$$
\begin{aligned}
\left(\mathrm{J}_\varepsilon\, p(\vec{t})\right)^{\mathfrak{J},v} = \mathbf{t} \quad &\text{iff} \quad p^{\mathfrak{J}}\left(\vec{t}^{(\mathfrak{J},v)*}\right) = \varepsilon && \text{(by 7.4.9)}\\
&\text{iff} \quad \mathrm{Snap}\left(p^{\mathfrak{K}}\left(\vec{t}^{(\mathfrak{K},v)*}\right), s\right) = \varepsilon && \text{(by 13.1.6 (iv))}\\
&\text{iff} \quad p^{\mathfrak{K}}\left(\vec{t}^{(\mathfrak{K},v)*}\right) \in \mathrm{Ch}(\varepsilon, s) && \text{(by 13.1.5 (i))}\\
&\text{iff} \quad \left(\mathrm{J}_{(\varepsilon,s)}\, p(\vec{t})\right)^{\mathfrak{K},v} = \mathbf{t} && \text{(by 7.4.2)}
\end{aligned}
$$

(v) can be proved similarly to (iv). □

Lemma 13.1.8. Let L be an FPJ logic, $I = \mathrm{I}^\tau(L)$, σ be a signature,

$$O\colon\ [1\,..\,\mathrm{Q}_O] \to \mathrm{Pr}_\sigma(\mathsf{I}),$$

$s \in \omega$, $0 \le m \le \mathrm{Q}_O$. Let $\mathfrak{K}$ be an I-structure of signature σ. Then for any valuation v in $\mathrm{Frame}_{\mathfrak{K}}$ and any external L-formula Φ of signature σ,

$$\left(\Phi^{\langle O,s,m\rangle}\right)^{\mathfrak{K},v} = \Phi^{\mathfrak{K}[O,s,m],v}.$$

Proof. Let $\mathfrak{J} = \mathfrak{K}[O,s,m]$. By induction on external L-formulas we prove that $\left(\Phi^{\langle O,s,m\rangle}\right)^{\mathfrak{K},v} = \Phi^{\mathfrak{J},v}$.

BASIS. Consider the possible cases. It can be proved by sraightforward case analysis.

CASE 1. $\Phi = (t_1 = t_2)$, where t_1 and t_2 are terms such that $\mathrm{Sort}_\sigma(t_1) = \mathrm{Sort}_\sigma(t_2)$. Then

$$\left((t_1 = t_2)^{\langle O,s,m\rangle}\right)^{\mathfrak{K},v} = (t_1 = t_2)^{\mathfrak{K},v} \qquad \text{(by 13.1.2 (iii))}$$
$$= (t_1 = t_2)^{\mathfrak{J},v} \qquad \text{(by 13.1.7 (ii))}$$

CASE 2. $\Phi = q(\vec{t})$, where $q \in \mathrm{Pr}_\sigma(\mathsf{E})$, $\vec{t} \in \mathrm{Dom}_\sigma(q)$. Then

$$\left(q(\vec{t})^{\langle O,s,m\rangle}\right)^{\mathfrak{K},v} = q(\vec{t})^{\mathfrak{K},v} \qquad \text{(by 13.1.2 (ii))}$$
$$= q(\vec{t})^{\mathfrak{J},v} \qquad \text{(by 13.1.7 (iii))}$$

CASE 3. $\Phi = \mathrm{J}_\varepsilon\, p(\vec{t})$, where $\varepsilon \in \mathscr{V}(L,\mathsf{I})$, $p \in \mathrm{Pr}_\sigma(\mathsf{I})$, $\vec{t} \in \mathrm{Dom}_\sigma(p)$. Consider the subcases.

SUBCASE 3.1. Let $s = 0$ or $\mathrm{R}^O(p) \le m$. Then

$$\left(\mathrm{J}_\varepsilon\, p(\vec{t})^{\langle O,s,m\rangle}\right)^{\mathfrak{K},v} = \left(\mathrm{J}_{(\varepsilon,s)}\, p(\vec{t})\right)^{\mathfrak{K},v} \qquad \text{(by 13.1.2 (i))}$$
$$= \left(\mathrm{J}_\varepsilon\, p(\vec{t})\right)^{\mathfrak{J},v} \qquad \text{(by 13.1.7 (iv))}$$

SUBCASE 3.2. Let $s > 0$ and $\mathrm{R}^O(p) > m$. Then

$$\left(\mathrm{J}_\varepsilon\, p(\vec{t})^{\langle O,s,m\rangle}\right)^{\mathfrak{K},v} = \left(\mathrm{J}_{(\varepsilon,s-1)}\, p(\vec{t})\right)^{\mathfrak{K},v} \qquad \text{(by 13.1.2 (i))}$$
$$= \left(\mathrm{J}_\varepsilon\, p(\vec{t})\right)^{\mathfrak{J},v} \qquad \text{(by 13.1.7 (v))}$$

INDUCTION STEP. It is trivial. See Definition 13.1.2 (iv)–(vi). □

13.2 Implementations and Extensions

Below the notion of iterative formula initialisation is introduced. Thus we define a method of transformation of L-formulas into I-formulas such that the result corresponds to the initial state of the cognitive process.

Lemma 13.2.1. Let L be an FPJ logic, $I = \mathrm{I}^\tau(L)$, σ be a signature,

$$O\colon\ [1\,..\,\mathrm{Q}_O] \to \mathrm{Pr}_\sigma(\mathsf{I}),$$

$s \in \omega$, $0 \le m \le \mathrm{Q}_O$. Let $\mathfrak{K}$ be an I-structure of signature σ, v is a valuation in $\mathrm{Frame}_{\mathfrak{K}}$. Then

(i) $\left(\mathrm{P}_m^O\right)^{\mathfrak{K}[O,s,m]} = \left(\mathrm{P}_m^O\right)^{\mathfrak{K}[O,s,k]}$ for any $s,k \in \omega\,(s > 0, 0 \le m \le k \le \mathrm{Q}_O)$,
(ii) $\left(\mathrm{P}_m^O\right)^{\mathfrak{K}[O,s,k]} = \left(\mathrm{P}_m^O\right)^{\mathfrak{K}[O,s-1,m]}$ for any $s,k \in \omega\,(s > 0, 0 \le k < m \le \mathrm{Q}_O)$.

Proof. Let $\vec{a} \in \mathrm{Dom}_\sigma\left(\left(\mathrm{P}_m^O\right)^{\mathfrak{K}}\right)$. Suppose $s > 0$, $0 \le m \le k \le \mathrm{Q}_O$. Then by Definition 13.1.6,

$$\begin{aligned}\left(\mathrm{P}_m^O\right)^{\mathfrak{K}[O,s,k]}(\vec{a}) &= \mathrm{Snap}\left(\left(\mathrm{P}_m^O\right)^{\mathfrak{K}}(\vec{a}), s\right)\\ &= \left(\mathrm{P}_m^O\right)^{\mathfrak{K}[O,s,m]}(\vec{a}).\end{aligned}$$

Thus (i) is proved.

Suppose $s > 0$, $0 \le k < m \le \mathrm{Q}_O$. Then by Definition 13.1.6,

$$\begin{aligned}\left(\mathrm{P}_m^O\right)^{\mathfrak{K}[O,s,k]}(\vec{a}) &= \mathrm{Snap}\left(\left(\mathrm{P}_m^O\right)^{\mathfrak{K}}(\vec{a}), s-1\right)\\ &= \left(\mathrm{P}_m^O\right)^{\mathfrak{K}[O,s-1,m]}(\vec{a}).\end{aligned}$$

Thus (ii) is proved. □

Now we introduce the notion of *implementation* of structure generators. The implementation of a structure generator is an I-structure which can represent any cognitive process that is defined by the given structure generator.

Definition 13.2.2. Let L be an FPJ logic, $I = \mathrm{I}^\tau(L)$, $\Gamma \in \mathrm{FS}_\sigma(L, \mathsf{E})$,

$$\mathfrak{R} = \langle \mathfrak{J}, R, O\rangle \in \mathrm{SG}_\sigma(L, \Gamma).$$

An I-structure $\mathfrak{K}$ of signature σ is called the *implementation* of $\mathfrak{R}$ (written: $\mathrm{Imp}(\mathfrak{R})$) if it meets the following conditions:

(i) $\mathrm{Frame}_{\mathfrak{K}} = \mathrm{Frame}_{\mathfrak{J}}$,
(ii) $f^{\mathfrak{K}} = f^{\mathfrak{J}}$ for any $f \in \mathrm{Fu}_\sigma$,
(iii) $q^{\mathfrak{K}} = q^{\mathfrak{J}}$ for any $q \in \mathrm{Pr}_\sigma(\mathsf{E})$,
(iv) $$\left(\mathrm{P}_m^{\mathfrak{R}}\right)^{\mathfrak{K}}(\vec{a}) = \begin{cases} \langle \varepsilon, s\rangle & \text{if } \left(\mathrm{P}_m^{\mathfrak{R}}\right)^{\vec{\mathfrak{R}}[s,m]}(\vec{a}) = \varepsilon \ne \tau \text{ and}\\ & \quad\left(\mathrm{P}_m^{\mathfrak{R}}\right)^{\vec{\mathfrak{R}}[l,m]}(\vec{a}) = \tau \text{ for any } l < s,\\ \tau & \text{otherwise,}\end{cases}$$

where $\varepsilon \in \mathscr{D}^\tau(L)$, $\mathrm{P}_m^{\mathfrak{R}} = O(m)$ (see Definition 12.2.10); the sequence $\vec{\mathfrak{R}}$ is defined in Definition 12.2.11, $\vec{a} \in \mathrm{Dom}\left(\left(\mathrm{P}_m^{\mathfrak{R}}\right)^{\mathfrak{J}}\right)$.

Proposition 13.2.3. The implementation of an $\mathfrak{R} \in \mathrm{SG}_\sigma(L)$, where L is an FPJ logic, is well defined.

Proof. Consider the set

$$S = \left\{ l \in \omega \mid \left(\mathrm{P}_k^{\mathfrak{R}}\right)^{\vec{\mathfrak{R}}[l,k]} (\vec{a}) \neq \tau \right\}.$$

If $S = \emptyset$ then $\left(\mathrm{P}_k^{\mathfrak{R}}\right)^{\mathfrak{K}} (\vec{a}) = \tau$. In the converse case let $s = \min S$,

$$\varepsilon = \left(\mathrm{P}_k^{\mathfrak{R}}\right)^{\vec{\mathfrak{R}}[s,k]} (\vec{a}).$$

Since the sequence $\vec{\mathfrak{R}}$ is monotone: $\left(\mathrm{P}_k^{\mathfrak{R}}\right)^{\vec{\mathfrak{R}}[l,k]} (\vec{a}) = \varepsilon$ for any $l \geq s$. Moreover for any $m < s$, $\left(\mathrm{P}_k^{\mathfrak{R}}\right)^{\vec{\mathfrak{R}}[m,k]} (\vec{a}) = \tau$ holds. □

Now we define the extention of L-structures, where L is an FPJ logic. An extention of an L-structure is an I-structure, where $I = \mathrm{I}^{\tau}(L)$. The extention $\mathfrak{K}$ of the L-structure $\mathfrak{J}$ has the following property: if a formula is valid in $\mathfrak{J}$, then its zero cut ($\langle O, 0, m\rangle$-cut) will be valid in $\mathfrak{K}$. That is, the extention seems to include the image of the initial L-structure at the zero stage.

Definition 13.2.4. Let L be an FPJ logic, $I = \mathrm{I}^{\tau}(L)$, $\mathfrak{J}$ be an L-structure of signature σ. An I-structure $\mathfrak{K}$ of signature σ is called an *extention* of $\mathfrak{J}$, if the following conditions hold:

(i) $\mathrm{Frame}_{\mathfrak{K}} = \mathrm{Frame}_{\mathfrak{J}}$,
(ii) $f^{\mathfrak{K}} = f^{\mathfrak{J}}$ for any $f \in \mathrm{Fu}_{\sigma}$,
(iii) $q^{\mathfrak{K}} = q^{\mathfrak{J}}$ for any $q \in \mathrm{Pr}_{\sigma}(\mathsf{E})$,
(iv) $p^{\mathfrak{J}}(\vec{a}) = \varepsilon$ iff $p^{\mathfrak{K}}(\vec{a}) \in \mathrm{Ch}(\varepsilon, 0)$,
for any $p \in \mathrm{Pr}_{\sigma}(\mathsf{I})$, $\vec{a} \in \mathrm{Dom}\left(p^{\mathfrak{J}}\right) = \mathrm{Dom}\left(p^{\mathfrak{K}}\right)$, $\varepsilon \in \mathscr{V}(L, \mathsf{I})$.

Proposition 13.2.5. Let L be an FPJ logic, $I = \mathrm{I}^{\tau}(L)$, $\mathfrak{J}$ be an L-structure of signature σ. An I-structure $\mathfrak{K}$ can be defined as follows:

(i) $\mathrm{Frame}_{\mathfrak{K}} = \mathrm{Frame}_{\mathfrak{J}}$,
(ii) $f^{\mathfrak{K}} = f^{\mathfrak{J}}$ for any $f \in \mathrm{Fu}_{\sigma}$,
(iii) $q^{\mathfrak{K}} = q^{\mathfrak{J}}$ for any $q \in \mathrm{Pr}_{\sigma}(\mathsf{E})$,
(iv) $p^{\mathfrak{K}}(\vec{a}) = \begin{cases} \langle p^{\mathfrak{J}}(\vec{a}), 0\rangle & \text{if } p^{\mathfrak{J}}(\vec{a}) \neq \tau, \\ \tau & \text{if } p^{\mathfrak{J}}(\vec{a}) = \tau, \end{cases}$
for any $p \in \mathrm{Pr}_{\sigma}(\mathsf{I})$, $\vec{a} \in \mathrm{Dom}\left(p^{\mathfrak{J}}\right) = \mathrm{Dom}\left(p^{\mathfrak{K}}\right)$.

It is easy to show that $\mathfrak{K}$ will be an extention of $\mathfrak{J}$ in the sense of Definition 13.2.5. Such an extention is called *trivial.*

Proposition 13.2.6. Let L be an FPJ logic, $I = \mathrm{I}^{\tau}(L)$, $\mathfrak{K}$ be an I-structure of signature σ,

$$O\colon [1..\mathrm{Q}_O] \to \mathrm{Pr}_{\sigma}(\mathsf{I}),$$

$m \in [1..\mathrm{Q}_O]$. Then $\mathfrak{K}$ is a extention of $\mathfrak{K}[O, 0, m]$.

Proof. It is evident that the conditions (i)–(iii) of Definition 13.2.4 hold (see Definition 13.1.6 (i)–(iii)). The condition (iv) of Definition 13.2.4 also holds (see Definition 13.1.6 (iv) and Proposition 13.1.5). □

Remark 13.2.7. Let L be an FPJ logic, $I = \mathrm{I}^{\tau}(L)$, $\mathfrak{J}$, $\mathfrak{K}$ be two L-structures and $\mathfrak{L}$ be an I-structure of signature σ. Suppose $\mathfrak{L}$ is an extention of both $\mathfrak{J}$ and $\mathfrak{K}$. Then $\mathfrak{J} = \mathfrak{K}$.

Proof. It is obvious (see Definition 13.2.4). □

The following lemma shows that the realisation of the generator of a structure is the extention of the origin of this structure.

Lemma 13.2.8. Let L be an FPJ logic, $\mathfrak{R} \in \mathrm{SG}_{\sigma}(L)$, $\mathfrak{J} = \mathring{\mathfrak{R}}$, $\mathfrak{K} = \mathrm{Imp}(\mathfrak{R})$. Then $\mathfrak{K}$ is an extention of $\mathfrak{J}$.

Proof. Obviously, $\mathfrak{K}$ satisfies the conditions (i)–(iii) of Definition 13.2.4. In order to prove validity of the condition (iv) note that (iv) is equivalent to the following statement:

For any $p \in \mathrm{Pr}_{\sigma}(\mathsf{I})$, $\vec{a} \in \mathrm{Dom}\left(p^{\mathfrak{J}}\right) = \mathrm{Dom}\left(p^{\mathfrak{K}}\right)$:

- If $p^{\mathfrak{J}}(\vec{a}) \neq \tau$ then $p^{\mathfrak{K}}(\vec{a}) = \left\langle p^{\mathfrak{J}}(\vec{a}), 0\right\rangle$,
- If $p^{\mathfrak{J}}(\vec{a}) = \tau$ then $p^{\mathfrak{K}}(\vec{a}) \neq \langle \delta, 0\rangle$ for any $\delta \in \mathscr{D}^{\tau}(L)$.

The last statement holds because of Definitions 13.2.2 (iv) and 12.2.11 (the case $s = 0$ is the basis of induction). □

The lemma below connects the value of an internal predicate in the realisation of a structure generator with the value of the corresponding predicate in the element of the sequence induced by this generator. This lemma will be used further on for establishing a relation between the cuts of the realisation of the atomic structure generator and the L-structures, where L is an FPJ logic, that are elements of the sequence induced by this generator.

Lemma 13.2.9. Let L be an FPJ logic, $\Gamma \in \mathrm{FS}_{\sigma}(L, \mathsf{E})$, $\mathfrak{R} \in \mathrm{SG}_{\sigma}(L, \Gamma)$, $\mathfrak{K} = \mathrm{Imp}(\mathfrak{R})$. Then for any $s \in \omega$, $m \in [1\,..\,\mathrm{Q}_{\mathfrak{R}}]$, $\vec{a} \in \mathrm{Dom}\left(\left(\mathrm{P}_m^{\mathfrak{R}}\right)^{\mathring{\mathfrak{R}}}\right)$,

$$\left(\mathrm{P}_m^{\mathfrak{R}}\right)^{\mathfrak{K}}(\vec{a}) \in \mathrm{Ch}(\varepsilon, s) \quad \text{iff} \quad \left(\mathrm{P}_m^{\mathfrak{R}}\right)^{\vec{\mathfrak{R}}[s,m]}(\vec{a}) = \varepsilon.$$

Proof. By induction on s.

Basis. The statement holds for $s = 0$ by Lemma 13.2.8 and Definition 13.2.4 (iv).

Induction step. Let the statement hold for all $s \leq n$. Let us prove it for $s = n + 1$. Consider the possible cases.

Case 1. $\varepsilon = \tau$.

($\Rightarrow$) Let $\left(\mathrm{P}_m^{\mathfrak{N}}\right)^{\mathfrak{K}}(\vec{a}) \in \mathrm{Ch}(\tau, n+1)$. Then $\left(\mathrm{P}_m^{\mathfrak{N}}\right)^{\mathfrak{K}}(\vec{a}) \in \mathrm{Ch}(\tau, l)$ for any $l \leq n$ (see Lemma 7.4.12(ii)). Then by the inductive assumption, $\left(\mathrm{P}_m^{\mathfrak{N}}\right)^{\vec{\mathfrak{N}}[l,m]}(\vec{a}) = \tau$ for any $l \leq n$. Suppose $\left(\mathrm{P}_m^{\mathfrak{N}}\right)^{\vec{\mathfrak{N}}[n+1,m]}(\vec{a}) = \delta \neq \tau$. Then by Definition 13.2.2(iv),

$$\left(\mathrm{P}_m^{\mathfrak{N}}\right)^{\mathfrak{K}}(\vec{a}) = \langle \delta, n+1 \rangle \notin \mathrm{Ch}(\tau, n+1).$$

This is a contradiction. Therefore $\left(\mathrm{P}_m^{\mathfrak{N}}\right)^{\vec{\mathfrak{N}}[n+1,m]}(\vec{a}) = \tau$.

($\Leftarrow$) Let $\left(\mathrm{P}_m^{\mathfrak{N}}\right)^{\vec{\mathfrak{N}}[n+1,m]}(\vec{a}) = \tau$. Since the sequence $\vec{\mathfrak{N}}$ is monotone (see Theorem 12.2.17), $\left(\mathrm{P}_m^{\mathfrak{N}}\right)^{\vec{\mathfrak{N}}[l,m]}(\vec{a}) = \tau$ for any $l \leq n+1$ (see Definitions 12.1.5 and 12.1.2). In this case it is not possible that $\left(\mathrm{P}_m^{\mathfrak{N}}\right)^{\mathfrak{K}}(\vec{a}) = \langle \delta, l \rangle$ for some $\delta \neq \tau$ and $l \leq n+1$ (see 13.2.2(iv)). Then

$$\left(\mathrm{P}_m^{\mathfrak{N}}\right)^{\mathfrak{K}}(\vec{a}) \in \mathrm{Ch}(\tau, n+1).$$

CASE 2. $\varepsilon \neq \tau$.

($\Rightarrow$) Let $\left(\mathrm{P}_m^{\mathfrak{N}}\right)^{\mathfrak{K}}(\vec{a}) \in \mathrm{Ch}(\varepsilon, n+1)$. Then $\left(\mathrm{P}_m^{\mathfrak{N}}\right)^{\mathfrak{K}}(\vec{a}) = \langle \varepsilon, l \rangle$ for some $l \leq n+1$. If $\left(\mathrm{P}_m^{\mathfrak{N}}\right)^{\mathfrak{K}}(\vec{a}) = \langle \varepsilon, n+1 \rangle$ then by 13.2.2(iv), $\left(\mathrm{P}_m^{\mathfrak{N}}\right)^{\vec{\mathfrak{N}}[n+1,m]}(\vec{a}) = \varepsilon$. Let $\left(\mathrm{P}_m^{\mathfrak{N}}\right)^{\mathfrak{K}}(\vec{a}) = \langle \varepsilon, l \rangle$ for some $l \leq n$. Then $\left(\mathrm{P}_m^{\mathfrak{N}}\right)^{\mathfrak{K}}(\vec{a}) \in \mathrm{Ch}(\varepsilon, l)$. By the inductive assumption, $\left(\mathrm{P}_m^{\mathfrak{N}}\right)^{\vec{\mathfrak{N}}[l,m]}(\vec{a}) = \varepsilon$. Since the sequence $\vec{\mathfrak{N}}$ is monotone (see Theorem 12.2.17), $\left(\mathrm{P}_m^{\mathfrak{N}}\right)^{\vec{\mathfrak{N}}[n+1,m]}(\vec{a}) = \varepsilon$.

($\Leftarrow$) Let $\left(\mathrm{P}_m^{\mathfrak{N}}\right)^{\vec{\mathfrak{N}}[n+1,m]}(\vec{a}) = \varepsilon$. Let

$$l = \min\left\{ i \in \omega \mid \left(\mathrm{P}_m^{\mathfrak{N}}\right)^{\vec{\mathfrak{N}}[i,m]}(\vec{a}) = \varepsilon \right\}.$$

Then $l \leq n+1$. Since the sequence $\vec{\mathfrak{N}}$ is monotone, $\left(\mathrm{P}_m^{\mathfrak{N}}\right)^{\vec{\mathfrak{N}}[k,m]}(\vec{a}) = \tau$ for any $k < l$. In this case by 13.2.2(iv), $\left(\mathrm{P}_m^{\mathfrak{N}}\right)^{\mathfrak{K}}(\vec{a}) = \langle \varepsilon, l \rangle$. Then $\left(\mathrm{P}_m^{\mathfrak{N}}\right)^{\mathfrak{K}}(\vec{a}) \in \mathrm{Ch}(\varepsilon, n+1)$, as was to be proved. □

Lemma 13.2.10. Let L be an FPJ logic, $\Gamma \in \mathrm{FS}_\sigma(L, \mathsf{E})$, $\mathfrak{N} \in \mathrm{SG}_\sigma(L, \Gamma)$, $\mathfrak{K} = \mathrm{Imp}(\mathfrak{N})$. Then for any $s \in \omega$, $m \in [1..\mathrm{Q}_{\mathfrak{N}}]$,

$$\mathfrak{K}[O, s, m] = \vec{\mathfrak{N}}[s, m].$$

Proof. By 12.2.11, 13.1.6 and 13.2.2 we obtain that, for any $s \in \omega$, $m \in [0..\mathrm{Q}_{/\!/}]$,

(i) $\mathrm{Frame}_{\mathfrak{K}[O,s,m]} = \mathrm{Frame}_{\mathfrak{K}} = \mathrm{Frame}_{\overset{\circ}{\mathfrak{N}}} = \mathrm{Frame}_{\vec{\mathfrak{N}}[s,m]}$,

(ii) $f^{\mathfrak{K}[O,s,m]} = f^{\mathfrak{K}} = f^{\mathring{\mathfrak{R}}} = f^{\overrightarrow{\mathfrak{R}}[s,m]}$ for any $f \in \mathrm{Fu}_\sigma$,
(iii) $q^{\mathfrak{K}[O,s,m]} = q^{\mathfrak{K}} = q^{\mathring{\mathfrak{R}}} = q^{\overrightarrow{\mathfrak{R}}[s,m]}$ for any $q \in \mathrm{Pr}_\sigma(\mathsf{E})$.

Therefore, in order to prove the lemma it is enough to show that

(i) $\left(\mathrm{P}_i^{\mathfrak{R}}\right)^{\mathfrak{K}[O,s,m]} = \left(\mathrm{P}_i^{\mathfrak{R}}\right)^{\overrightarrow{\mathfrak{R}}[s,m]}$ for any $i \in [1..\mathrm{Q}_{\mathscr{M}}]$.

Let $\vec{a} \in \mathrm{Dom}\left(\left(\mathrm{P}_i^{\mathfrak{R}}\right)^{\mathring{\mathfrak{R}}}\right)$. Consider the possible cases.

CASE 1. $i \leq m$.

$$\begin{aligned}
\left(\mathrm{P}_i^{\mathfrak{R}}\right)^{\mathfrak{K}[O,s,m]}(\vec{a}) = \varepsilon \quad &\text{iff} \quad \mathrm{Snap}\left(\left(\mathrm{P}_i^{\mathfrak{R}}\right)^{\mathfrak{K}}(\vec{a}), s\right) = \varepsilon && \text{(by 13.1.6 (iv))}\\
&\text{iff} \quad \left(\mathrm{P}_i^{\mathfrak{R}}\right)^{\mathfrak{K}}(\vec{a}) \in \mathrm{Ch}(\varepsilon, s) && \text{(by 13.1.5 (i))}\\
&\text{iff} \quad \left(\mathrm{P}_i^{\mathfrak{R}}\right)^{\overrightarrow{\mathfrak{R}}[s,i]}(\vec{a}) = \varepsilon && \text{(by 13.2.9)}\\
&\text{iff} \quad \left(\mathrm{P}_i^{\mathfrak{R}}\right)^{\overrightarrow{\mathfrak{R}}[s,m]}(\vec{a}) = \varepsilon && \text{(by 12.2.15 (vii))}
\end{aligned}$$

CASE 2. $i > m$.

$$\begin{aligned}
\left(\mathrm{P}_i^{\mathfrak{R}}\right)^{\mathfrak{K}[O,s,m]}(\vec{a}) = \varepsilon \quad &\text{iff} \quad \mathrm{Snap}\left(\left(\mathrm{P}_i^{\mathfrak{R}}\right)^{\mathfrak{K}}(\vec{a}), s-1\right) = \varepsilon && \text{(by 13.1.6 (iv))}\\
&\text{iff} \quad \left(\mathrm{P}_i^{\mathfrak{R}}\right)^{\mathfrak{K}}(\vec{a}) \in \mathrm{Ch}(\varepsilon, s-1) && \text{(by 13.1.5 (i))}\\
&\text{iff} \quad \left(\mathrm{P}_i^{\mathfrak{R}}\right)^{\overrightarrow{\mathfrak{R}}[s-1,i]}(\vec{a}) = \varepsilon && \text{(by 13.2.9)}\\
&\text{iff} \quad \left(\mathrm{P}_i^{\mathfrak{R}}\right)^{\overrightarrow{\mathfrak{R}}[s,m]}(\vec{a}) = \varepsilon && \text{(by 12.2.15 (viii))}
\end{aligned}$$

The lemma is proved. □

Lemma 13.2.11. Let L be an FPJ logic, $\Gamma \in \mathrm{FS}_\sigma(L, \mathsf{E})$, $\mathfrak{R} \in \mathrm{SG}_\sigma(L, \Gamma)$, $\mathfrak{K} = \mathrm{Imp}(\mathfrak{R})$, Φ be an arbitrary external L-formula and v be an arbitrary valuation in $\mathrm{Frame}_{\mathfrak{K}}$. Suppose $s \in \omega$, $0 \leq m \leq \mathrm{Q}_{\mathfrak{R}}$. Then

$$\left(\Phi^{\langle O,s,m\rangle}\right)^{\mathfrak{K},v} = \Phi^{\overrightarrow{\mathfrak{R}}[s,m],v}.$$

Proof. The lemma follows directly from 13.1.8 and 13.2.10. □

Definition 13.2.12. Let L be an FPJ logic, $I = \mathrm{I}^\tau(L)$, σ be a signature. Let us define the map

$$\text{Init}\colon \mathscr{F}_\sigma(L,\mathsf{E}) \to \mathscr{F}_\sigma(I,\mathsf{E})$$

by induction as follows: (Φ^{Init} denotes the image of the formula Φ in the map Init)

(i) $\left(\mathrm{J}_\varepsilon p(\vec{t})\right)^{\text{Init}} = \mathrm{J}_{(\varepsilon,0)}\, p(\vec{t})$,
where $\varepsilon \in \mathscr{V}(L,\mathsf{I})$, $p \in \mathrm{Pr}_\sigma(\mathsf{I})$, $\vec{t} \in \mathrm{Dom}_\sigma\left(\mathrm{P}_i^O\right)$;
(ii) $q(\vec{t})^{\text{Init}} = q(\vec{t})$,
where $q \in \mathrm{Pr}_\sigma(\mathsf{E})$, $\vec{t} \in \mathrm{Dom}_\sigma(q)$;
(iii) $(t_1 = t_2)^{\text{Init}} = (t_1 = t_2)$,
where t_1 and t_2 are terms such that $\mathrm{Sort}_\sigma(t_1) = \mathrm{Sort}_\sigma(t_2)$;
(iv) $(\Phi \to \Psi)^{\text{Init}} = \left(\Phi^{\text{Init}} \to \Psi^{\text{Init}}\right)$,
where Φ and Ψ are external formula of the logic L;
(v) $(\neg\Phi)^{\text{Init}} = \left(\neg\Phi^{\text{Init}}\right)$,
where Φ is an external formula of the logic L;
(vi) $(\forall x\Phi)^{\text{Init}} = \left(\forall x\Phi^{\text{Init}}\right)$,
where Φ is an external formula of the logic L, x is an individual variable.

Φ^{Init} is called the *iterative initialisation of* Φ.

If Γ is a set of external L-formulas, then the set $\left\{\Phi^{\text{Init}} \mid \Phi \in \Gamma\right\}$ is called *iterative initialisation of* Γ and is denoted by Γ^{Init}.

Remark 13.2.13. Let L be an FPJ logic, σ be a signature,

$$O\colon [1\,..\,\mathrm{Q}_O] \to \mathrm{Pr}_\sigma(\mathsf{I}).$$

By Definition 13.1.2, it is easy to prove that for any external L-formula Φ of signature σ

$$\Phi^{\text{Init}} = \Phi^{\langle O,0,m\rangle},$$

for any $m \in [0\,..\,\mathrm{Q}_O]$.

We establish in the following two statements the connection between the validity of a set of formulas in an L-structure, where L is an FPJ logic, and the validity of the iterative initialisation of this set in the I-structure, which is the extention of the given L-structure.

Lemma 13.2.14. Let L be an FPJ logic, $I = \mathrm{I}^\tau(L)$, σ be a signature,

$$O\colon [1\,..\,\mathrm{Q}_O] \to \mathrm{Pr}_\sigma(\mathsf{I}).$$

Let $\mathfrak{K}$ be an I-structure of signature σ, Γ be a set of external L-formulas of signature σ, such that $\mathfrak{K} \models_I \Gamma^{\text{Init}}$. Then

$$\mathfrak{K}[O,0,m] \models_L \Gamma$$

for any $m \in [0\,..\,\mathrm{Q}_O]$.

Proof. Let v be an arbitrary valuation in $\text{Frame}_{\mathfrak{K}}$, $\Phi \in \Gamma$. Then

$$\begin{aligned}
\Phi^{\mathfrak{K}[O,0,m],v} &= \left(\Phi^{\langle O,0,m\rangle}\right)^{\mathfrak{K},v} && \text{(by 13.1.8)}\\
&= \left(\Phi^{\mathrm{Init}}\right)^{\mathfrak{K},v} && \text{(by 13.2.13)}\\
&= \mathbf{t} && \text{(by hypothesis)}
\end{aligned}$$

□

Below we define the cut of the modification rules, which is a function providing each modification rule with a certain formula of the first-order I logic

Corollary 13.2.15. Let L be an FPJ logic, $I = \mathrm{I}^{\tau}(L)$, σ be a signature,

$$O\colon\ [1\,..\,\mathrm{Q}_O] \to \mathrm{Pr}_{\sigma}(\mathsf{I})\,.$$

Let $\mathfrak{K}$ be an I-structure of signature σ, Γ be a set of external L-formulas of signature σ, such that $\mathfrak{K} \models_I \Gamma^{\mathrm{Init}}$. Then $\mathfrak{K}$ is an extention of some model of the set Γ.

Proof. By 13.2.6, $\mathfrak{K}$ is a extention of $\mathfrak{K}[O,0,m]$. □

13.3 Iterative Images

Definition 13.3.1. Let L be an FPJ logic, $I = \mathrm{I}^{\tau}(L)$, σ be a signature with a finite set of internal predicate symbols,

$$O\colon\ [1\,..\,\mathrm{Q}_O] \to \mathrm{Pr}_{\sigma}(\mathsf{I})\,,$$

$s \in \omega$, $s > 0$. Let

$$\Phi \Vdash \mathrm{J}_{\varepsilon}\, p(\vec{x})$$

be a genetating rule (see 10.2.1).

Then the external I-formula

$$\Phi^{\langle O,s,m-1\rangle} \to \mathrm{J}_{\varepsilon}\, p(\vec{x})^{\langle O,s,m\rangle}\,,$$

where $p = O(m)$, i.e. the formula

$$\Phi^{\langle O,s,m-1\rangle} \to \mathrm{J}_{(\varepsilon,s)}\, p(\vec{x})$$

is called *the* $\langle O,\, s\rangle$*-immersion* of this rule.

The $\langle O,\, s\rangle$-immersion of a generating rule r is denoted by $r^{\langle O,s\rangle}$.

To go on along the way of developing the method of embedding structure generators into I-structures we will now define the iterative image of the generating rule system.

Definition 13.3.2. Assume that L is an FPJ logic and $I = \mathrm{I}^{\tau}(L)$. Let $\Gamma \in \mathrm{FS}_{\sigma}(L, \mathsf{E})$, $R \in \mathrm{GRS}_{\sigma}(L, \Gamma)$, $O\colon [1\,..\,\mathrm{Q}_O] \to \mathrm{Pr}_{\sigma}(\mathsf{I})$. Then we call the set

$$\left\{R[p,\varepsilon]^{\langle O,s\rangle} \mid p \in \mathrm{Pr}_{\sigma}(\mathsf{I}), \varepsilon \in \mathscr{V}(L,\mathsf{I}), s \in \omega, s > 0\right\}$$

the iterative image (I-image) of R w.r.t. O and denote it by $\mathrm{I}^{O}(R)$.

The following lemma states that the iterative image of modification rule system is valid in the realisation of the structure generator that includes it.

Lemma 13.3.3. Let L be an FPJ logic, $I = \mathrm{I}^{\tau}(L)$, $\Gamma \in \mathrm{FS}_{\sigma}(L, \mathsf{E})$, $\mathfrak{R} \in \mathrm{SG}_{\sigma}(L, \Gamma)$, $R = \mathfrak{R}^{\triangleright}$, $O = \mathrm{Ord}_{\mathfrak{R}}$, $\mathfrak{K} = \mathrm{Imp}(\mathfrak{R})$. Then $\mathfrak{K} \vDash_I \mathrm{I}^{O}(R)$, i.e.

$$\left(R\left[\mathrm{P}_{m}^{\mathfrak{R}}, \varepsilon\right]^{\langle O,s\rangle}\right)^{\mathfrak{K},v} = \mathbf{t}$$

for any valuation v in $\mathrm{Frame}_{\mathfrak{K}}$.

Proof. Suppose $s \in \omega$, $s > 0$, $m \in [1\,..\,\mathrm{Q}_{\mathfrak{R}}]$, $\varepsilon \in \mathscr{V}(L, \mathsf{I})$. We need to prove that

$$\left(R\left[\mathrm{P}_{m}^{\mathfrak{R}}, \varepsilon\right]^{\langle O,s\rangle}\right)^{\mathfrak{K},v} = \mathbf{t}$$

for any valuation v in $\mathrm{Frame}_{\mathfrak{K}}$.

Let v be an arbitrary valuation in $\mathrm{Frame}_{\mathfrak{K}}$. By 13.3.1, $R\left[\mathrm{P}_{m}^{\mathfrak{R}}, \varepsilon\right]^{\langle O,s\rangle}$ is the formula

$$\Phi^{\langle O,s,m-1\rangle} \to \left(\mathrm{J}_{\varepsilon}\,\mathrm{P}_{m}^{\mathfrak{R}}(\vec{x})\right)^{\langle O,s,m\rangle}, \tag{13.1}$$

where $\vec{x} \in \mathrm{Var}^{\diamondsuit} \cap \mathrm{Dom}_{\sigma}(p)$. Let

$$\left(\Phi^{\langle O,s,m-1\rangle}\right)^{\mathfrak{K},v} = \mathbf{t}. \tag{13.2}$$

Then by Lemma 13.2.11,

$$\Phi^{\overrightarrow{\mathfrak{R}}[s,m-1],v} = \mathbf{t}. \tag{13.3}$$

Let $\vec{a} = v^{*}(\vec{x})$. Then (12.5) may be written as

$$\overrightarrow{\mathfrak{R}}[s,m-1] \vDash_L \Phi(\vec{a}). \tag{13.4}$$

The formula Φ is the generating condition of the rule $R\left[\mathrm{P}_{m}^{\mathfrak{R}}, \varepsilon\right]$, which is denoted by $\mathrm{GC}\left[R, \mathrm{P}_{m}^{\mathfrak{R}}, \varepsilon\right]$ (see Definition 12.2.1). Then (12.6) could be rewritten as

$$\overrightarrow{\mathfrak{R}}[s,m-1] \vDash_L \mathrm{GC}\left[R, \mathrm{P}_{m}^{\mathfrak{R}}, \varepsilon\right](\vec{a}). \tag{13.5}$$

From (12.7) by Definitions 12.2.11 (induction step (ii)) and 12.2.8 (v) we obtain

$$\left(\mathrm{P}_m^{\mathfrak{R}}\right)^{\vec{\mathfrak{R}}[s,m]}(\vec{a}) = \varepsilon.$$

Then

$$\left(\mathrm{J}_\varepsilon \mathrm{P}_m^{\mathfrak{R}}(\vec{x})\right)^{\vec{\mathfrak{R}}[s,m],v} = \mathbf{t},$$

hence by Lemma 13.2.11 we obtain

$$\left(\left(\mathrm{J}_\varepsilon \mathrm{P}_m^{\mathfrak{R}}(\vec{x})\right)^{\langle O,s,m\rangle}\right)^{\mathfrak{K},v} = \mathbf{t}. \tag{13.6}$$

Thus it was shown that (12.2) implies (12.8), that is, (12.1) is I-true in $\mathfrak{K}$, as was to be proved. □

Below we introduce the so-called stabilisation condition (axiom scheme) at stage s. Following this we prove that the realisation of any generator structure is a model of any axiom obtained by this scheme.

Definition 13.3.4. Let L be an FPJ logic, $I = \mathrm{I}^\tau(L)$, σ be a signature with a finite set of internal predicates, $p \in \mathrm{Pr}_\sigma(\mathsf{I})$. Then the external I-formula

$$\bigwedge_{p\in\mathrm{Pr}_\sigma(\mathsf{I})} \forall\vec{x}\left(\mathrm{J}_{(\tau,s-1)}\, p(\vec{x}) \to \mathrm{J}_{(\tau,s)}\, p(\vec{x})\right) \to \bigwedge_{p\in\mathrm{Pr}_\sigma(\mathsf{I})} \forall\vec{x}\left(\mathrm{J}_{(\tau,s-1)}\, p(\vec{x}) \to \mathrm{J}_{(\tau,n)}\, p(\vec{x})\right),$$

where $n > s > 0$, is called *the (s,n)-stabilisation condition of signature* σ (*stabilisation condition at stage s w.r.t. n for the signature* σ) and is denoted by $\mathrm{Stable}_\sigma(s,n)$.

Lemma 13.3.5. Let L be an FPJ logic, $I = \mathrm{I}^\tau(L)$, $\Gamma \in \mathrm{FS}_\sigma(L,\mathsf{E})$, $\mathfrak{R} \in \mathrm{SG}_\sigma(L,\Gamma)$, $O = \mathrm{Ord}_{\mathfrak{R}}$, $\mathfrak{K} = \mathrm{Imp}(\mathfrak{R})$, $s,n \in \omega$, $n > s > 0$. Then $\mathfrak{K} \vDash_I \mathrm{Stable}_\sigma(s,n)$.

Proof. $\mathrm{Stable}_\sigma(s,n)$ can be written as follows

$$\bigwedge_{m=1}^{Q_{\mathfrak{R}}} \forall\vec{x}\left(\mathrm{J}_{(\tau,s-1)}\mathrm{P}_m^{\mathfrak{R}}(\vec{x}) \to \mathrm{J}_{(\tau,s)}\mathrm{P}_m^{\mathfrak{R}}(\vec{x})\right) \to \bigwedge_{m=1}^{Q_{\mathfrak{R}}} \forall\vec{x}\left(\mathrm{J}_{(\tau,s-1)}\mathrm{P}_m^{\mathfrak{R}}(\vec{x}) \to \mathrm{J}_{(\tau,n)}\mathrm{P}_m^{\mathfrak{R}}(\vec{x})\right).$$

By Definition 13.1.2(i) this formula is equivalent to the formula

$$\bigwedge_{m=1}^{Q_{\mathfrak{R}}} \forall\vec{x}\left(\left(\mathrm{J}_\tau\mathrm{P}_m^{\mathfrak{R}}(\vec{x})\right)^{\langle O,s-1,m\rangle} \to \left(\mathrm{J}_\tau\mathrm{P}_m^{\mathfrak{R}}(\vec{x})\right)^{\langle O,s,m\rangle}\right) \to$$
$$\to \bigwedge_{m=1}^{Q_{\mathfrak{R}}} \forall\vec{x}\left(\left(\mathrm{J}_\tau\mathrm{P}_m^{\mathfrak{R}}(\vec{x})\right)^{\langle O,s-1,m\rangle} \to \left(\mathrm{J}_\tau\mathrm{P}_m^{\mathfrak{R}}(\vec{x})\right)^{\langle O,n,m\rangle}\right).$$

Let v be an arbitrary valuation in $\mathrm{Frame}_{\mathfrak{K}}$. Suppose

$$\left(\bigwedge_{m=1}^{Q_{\mathfrak{R}}} \forall\vec{x}\left(\left(J_\tau P_m^{\mathfrak{R}}(\vec{x})\right)^{\langle O,s-1,m\rangle} \to \left(J_\tau P_m^{\mathfrak{R}}(\vec{x})\right)^{\langle O,s,m\rangle}\right)\right)^{\mathfrak{K},v} = \mathbf{t}. \tag{13.7}$$

Then for any $m \in [1 .. Q_{\mathfrak{R}}]$,

$$\left(\left(J_\tau P_m^{\mathfrak{R}}(\vec{x})\right)^{\langle O,s-1,m\rangle}\right)^{\mathfrak{K},v} = \mathbf{t} \quad \text{implies} \quad \left(\left(J_\tau P_m^{\mathfrak{R}}(\vec{x})\right)^{\langle O,s,m\rangle}\right)^{\mathfrak{K},v} = \mathbf{t}. \tag{13.8}$$

By Lemma 13.2.11, (13.8) is equivalent to

$$\left(J_\tau P_m^{\mathfrak{R}}(\vec{x})\right)^{\vec{\mathfrak{R}}[s-1,m],v} = \mathbf{t} \quad \text{implies} \quad \left(J_\tau P_m^{\mathfrak{R}}(\vec{x})\right)^{\vec{\mathfrak{R}}[s,m],v} = \mathbf{t}. \tag{13.9}$$

Since v is an arbitrary valuation we obtain that for any $m \in [1..Q_{\mathfrak{R}}]$ and $\vec{a} \in \mathrm{Dom}\left(\left(P_m^{\mathfrak{R}}\right)^{\mathfrak{K}}\right)$,

$$\left(P_m^{\mathfrak{R}}\right)^{\vec{\mathfrak{R}}[s-1,m]}(\vec{a}) = \tau \quad \text{implies} \quad \left(P_m^{\mathfrak{R}}\right)^{\vec{\mathfrak{R}}[s,m]}(\vec{a}) = \tau. \tag{13.10}$$

However, since the sequence $\vec{\mathfrak{R}}$ is monotone (see 12.2.17) we obtain

$$\left(P_m^{\mathfrak{R}}\right)^{\vec{\mathfrak{R}}[s-1,m]} = \left(P_m^{\mathfrak{R}}\right)^{\vec{\mathfrak{R}}[s,m]}. \tag{13.11}$$

Hence by Stabilisation Lemma 12.2.19 we obtain, that for any $n \geq s$ and any $m \in [0 .. Q_{\mathfrak{R}}]$ it holds that:

$$\vec{\mathfrak{R}}[n,m] = \vec{\mathfrak{R}}[s,0]. \tag{13.12}$$

Then,

$$\begin{aligned}
\left(P_m^{\mathfrak{R}}\right)^{\vec{\mathfrak{R}}[n,m]} &= \left(P_m^{\mathfrak{R}}\right)^{\vec{\mathfrak{R}}[s,0]} && \text{(by (13.12))} \\
&= \left(P_m^{\mathfrak{R}}\right)^{\vec{\mathfrak{R}}[s-1,m]} && \text{(by 12.2.16 (viii))}
\end{aligned}$$

Consequently

$$\left(P_m^{\mathfrak{R}}\right)^{\vec{\mathfrak{R}}[s-1,m]} = \left(P_m^{\mathfrak{R}}\right)^{\vec{\mathfrak{R}}[n,m]}$$

for any $n \geq s$ and any $m \in [0 .. Q_{\mathfrak{R}}]$. Then by using 13.2.11 it is easy to show that

$$\left(\bigwedge_{m=1}^{Q_{\mathfrak{R}}} \forall\vec{x}\left(\left(J_\tau P_m^{\mathfrak{R}}(\vec{x})\right)^{\langle O,s-1,m\rangle} \to \left(J_\tau P_m^{\mathfrak{R}}(\vec{x})\right)^{\langle O,n,m\rangle}\right)\right)^{\mathfrak{K},v} = \mathbf{t}. \tag{13.13}$$

Thus (13.7) implies (13.13) for any valuation v in $\mathrm{Frame}_{\mathfrak{K}}$. Then $\mathfrak{K} \vDash_I \mathrm{Stable}_\sigma(s,n)$, as was to be proved. □

Lemma 13.3.6. Let L be an FPJ logic, I be its iterative version w.r.t. $\tau \in \mathscr{V}(L,\mathsf{I})$, $\Gamma \in \mathrm{FS}_\sigma(L,\mathsf{E})$, $R \in \mathrm{RGRS}_\sigma(L,\Gamma)$,

$$O\colon\ [1\,..\,\mathrm{Q}_O] \to \mathrm{Pr}_\sigma(\mathsf{I}),$$

$\mathfrak{K}$ be an I-structure. Suppose

$$\mathfrak{K} \vDash_I \Gamma^{\mathrm{Init}} \cup \mathrm{I}^O(R),$$

$\mathfrak{J} = \mathfrak{K}[O,0,0]$. Then $\mathfrak{R} = \langle \mathfrak{J}, R, O\rangle \in \mathrm{SG}_\sigma(L,\Gamma)$ and for any $s \in \omega$, $m \in [1\,..\,\mathrm{Q}_O]$,

$$\vec{\mathfrak{R}}[s,m] = \mathfrak{K}[O,s,m].$$

Proof. First note that $\mathfrak{J} \vDash_L \Gamma$(see Lemma 13.2.14). By the assumption,

$$R \in \mathrm{RGRS}_\sigma(L,\Gamma).$$

Therefore

$$\mathfrak{R} = \langle \mathfrak{J}, R, O\rangle \in \mathrm{SG}_\sigma(L,\Gamma)$$

(see Definition 12.2.10). Let us prove by induction on s and m that, for any $s \in \omega$, $m \in [1\,..\,\mathrm{Q}_O]$,

$$\vec{\mathfrak{R}}[s,m] = \mathfrak{K}[O,s,m].$$

BASIS. For each $m \in [1\,..\,\mathrm{Q}_O]$, $\vec{\mathfrak{R}}[s,0] = \mathfrak{J} = \mathfrak{K}[O,s,m]$ (see Definition 12.2.11 and Lemma 13.2.10).

INDUCTION STEP. Suppose that $\vec{\mathfrak{R}}[n,m] = \mathfrak{K}[O,n,m]$ for any $m \in [1\,..\,\mathrm{Q}_O]$. Let us prove that for any $m \in [1\,..\,\mathrm{Q}_O]$, $\vec{\mathfrak{R}}[n+1,m] = \mathfrak{K}[O,n+1,m]$. Continue our proof by induction on m.

Basis.

$$\begin{aligned}
\vec{\mathfrak{R}}[n+1,0] &= \vec{\mathfrak{R}}[n,\mathrm{Q}_O] && \text{(by 12.2.11, induction step)}\\
&= \mathfrak{K}[O,n,\mathrm{Q}_O] && \text{(by the induction hypothesis)}\\
&= \mathfrak{K}[O,n+1,0] && \text{(by 13.2.1(ii))}
\end{aligned}$$

Induction step. Suppose that, for any $j \le k$, $\vec{\mathfrak{R}}[n+1,j] = \mathfrak{K}[O,n+1,j]$. Let us prove that $\vec{\mathfrak{R}}[n+1,k+1] = \mathfrak{K}[O,n+1,k+1]$. By Definition 13.1.6 and Lemma 12.2.13,

(i) $\mathrm{Frame}_{\vec{\mathfrak{R}}[n+1,k+1]} = \mathrm{Frame}_{\mathfrak{J}} = \mathrm{Frame}_{\mathfrak{K}} = \mathrm{Frame}_{\mathfrak{K}[O,n+1,k+1]}$,

(ii) $f^{\vec{\mathfrak{R}}[n+1,k+1]} = f^{\mathfrak{J}} = f^{\mathfrak{K}} = f^{\mathfrak{K}[O,n+1,k+1]}$ for any $f \in \mathrm{Fu}_\sigma$,

(iii) $q^{\overrightarrow{\mathfrak{R}}[n+1,k+1]} = q^{\mathfrak{J}} = q^{\mathfrak{K}} = q^{\mathfrak{K}[O,n+1,k+1]}$ for any $q \in \mathrm{Pr}_\sigma(\mathsf{E})$.

To finish the proof we must show that

(i) $p^{\overrightarrow{\mathfrak{R}}[n+1,k+1]} = p^{\mathfrak{K}[O,n+1,k+1]}$ for any $p \in \mathrm{Pr}_\sigma(\mathsf{I})$.

Let us prove that

$$\left(\mathrm{P}_i^O\right)^{\overrightarrow{\mathfrak{R}}[n+1,k+1]}(\vec{a}) = \left(\mathrm{P}_i^O\right)^{\mathfrak{K}[O,n+1,k+1]}(\vec{a})$$

for any $i \in [1\,..\,\mathrm{Q}_O]$, $\vec{a} \in \mathrm{Dom}\left(\left(\mathrm{P}_i^O\right)^{\mathfrak{J}}\right)$. Consider the possible cases

CASE 1. Let $i \leq k$. Then

$$\begin{aligned}
\left(\mathrm{P}_i^O\right)^{\overrightarrow{\mathfrak{R}}[n+1,k+1]}(\vec{a}) &= \left(\mathrm{P}_i^O\right)^{\overrightarrow{\mathfrak{R}}[n+1,i]}(\vec{a}) && \text{(by 12.2.15 (vii))}\\
&= \left(\mathrm{P}_i^O\right)^{\mathfrak{K}[O,n+1,i]}(\vec{a}) && \text{(by the inductive assumption)}\\
&= \left(\mathrm{P}_i^O\right)^{\mathfrak{K}[O,n+1,k+1]}(\vec{a}) && \text{(by 13.2.1 (i))}
\end{aligned}$$

CASE 2. Let $k+1 < i$. Then

$$\begin{aligned}
\left(\mathrm{P}_i^O\right)^{\overrightarrow{\mathfrak{R}}[n+1,k+1]}(\vec{a}) &= \left(\mathrm{P}_i^O\right)^{\overrightarrow{\mathfrak{R}}[n,i]}(\vec{a}) && \text{(by 12.2.15 (viii))}\\
&= \left(\mathrm{P}_i^O\right)^{\mathfrak{K}[O,n,i]}(\vec{a}) && \text{(by the inductive assumption)}\\
&= \left(\mathrm{P}_i^O\right)^{\mathfrak{K}[O,n+1,k+1]}(\vec{a}) && \text{(by 13.2.1 (ii))}
\end{aligned}$$

CASE 3. Let $i = k+1$. By Definition 12.2.11,

$$\overrightarrow{\mathfrak{R}}[n+1,k+1] = \mathrm{Succ}\left(R, \overrightarrow{\mathfrak{R}}[n+1,k], \mathrm{P}_{k+1}^O\right). \tag{13.14}$$

Let $\varepsilon \in \mathscr{V}(L, \mathsf{I})$,

$$\begin{aligned}
&\left(\mathrm{P}_{k+1}^O\right)^{\overrightarrow{\mathfrak{R}}[n+1,k+1]}(\vec{a}) = \varepsilon\\
&\quad\text{iff}\quad \overrightarrow{\mathfrak{R}}[n+1,k] \vDash_L \mathrm{GC}\left[R, \mathrm{P}_{k+1}^O, \varepsilon\right](\vec{a}) && \text{(by (12.1) and 12.2.8(v))}\\
&\quad\text{iff}\quad \mathfrak{K}[O,n+1,k] \vDash_L \mathrm{GC}\left[R, \mathrm{P}_{k+1}^O, \varepsilon\right](\vec{a}) && \text{(by the inductive assumption)}\\
&\quad\text{iff}\quad \mathfrak{K} \vDash_I \mathrm{GC}\left[R, \mathrm{P}_{k+1}^O, \varepsilon\right](\vec{a})^{\langle O,n+1,k\rangle} && \text{(by 13.1.8)}\\
&\quad\text{iff}\quad \mathfrak{K} \vDash_I \mathrm{J}_{(\varepsilon,k+1)}\mathrm{P}_{k+1}^O(\vec{a}) && \text{(by 13.3.1 since } \mathfrak{K} \vDash_I \mathrm{I}^O(R)\text{)}\\
&\quad\text{iff}\quad \mathfrak{K}[O,n+1,k+1] \vDash_L \mathrm{J}_\varepsilon \mathrm{P}_{k+1}^O(\vec{a}) && \text{(by 13.1.7(iv))}\\
&\quad\text{iff}\quad \left(\mathrm{P}_{k+1}^O\right)^{\mathfrak{K}[O,n+1,k+1]}(\vec{a}) = \varepsilon && \text{(by 7.4.9)}
\end{aligned}$$

□

Lemma 13.3.7. Let L be an FPJ logic, I be its iterative version w.r.t. $\tau \in \mathscr{V}(L,\mathsf{I})$, $\Gamma \in \mathrm{FS}_\sigma(L,\mathsf{E})$, $R \in \mathrm{RGRS}_\sigma(L,\Gamma)$, $O\colon [1\,..\,\mathrm{Q}_O] \to \mathrm{Pr}_\sigma(\mathsf{I})$ and $\mathfrak{K}$ be an I-structure of signature σ. Suppose

$$\mathfrak{K} \vDash_I \Gamma^{\mathrm{Init}} \cup \mathrm{I}^O(R).$$

Then for any $s \in \omega$, $s > 0$, $m \in [1\,..\,\mathrm{Q}_O]$, $\vec{a} \in \mathrm{Dom}\left(\left(\mathrm{P}^O_m\right)^{\mathfrak{K}}\right)$ the following conditions hold:

$$\mathfrak{K} \vDash_I \bigvee_{\varepsilon \in \mathscr{V}(L,\mathsf{I})} \left(\mathrm{GC}\left[R,\mathrm{P}^O_m,\varepsilon\right]\right)^{\langle O,s,m-1\rangle}(\vec{a}), \tag{13.15}$$

$$\mathfrak{K} \vDash_I \bigwedge_{\substack{\varepsilon,\delta \in \mathscr{V}(L,\mathsf{I})\\ \varepsilon \neq \delta}} \left(\mathrm{GC}\left[R,\mathrm{P}^O_m,\varepsilon\right] \to \neg\mathrm{GC}\left[R,\mathrm{P}^O_m,\delta\right]\right)^{\langle O,s,m-1\rangle}(\vec{a}). \tag{13.16}$$

Proof. The condition (13.15), according to Lemma 13.1.8, is equivalent to

$$\mathfrak{K}[O,s,m-1] \vDash_L \bigvee_{\varepsilon \in \mathscr{V}(L,\mathsf{I})} \left(\mathrm{GC}\left[R,\mathrm{P}^O_m,\varepsilon\right]\right)(\vec{a}), \tag{13.17}$$

The condition (13.16) is equivalent to

$$\mathfrak{K}[O,s,m-1] \vDash_L \bigwedge_{\substack{\varepsilon,\delta \in \mathscr{V}(L,\mathsf{I})\\ \varepsilon \neq \delta}} \left(\mathrm{GC}\left[R,\mathrm{P}^O_m,\varepsilon\right] \to \neg\mathrm{GC}\left[R,\mathrm{P}^O_m,\delta\right]\right)(\vec{a}). \tag{13.18}$$

Let $\mathfrak{J} = \mathfrak{K}[O,0,0]$. Then by Lemma 13.3.6, the triple

$$\mathfrak{R} = \langle \mathfrak{J}, R, O\rangle \in \mathrm{SG}_\sigma(L,\Gamma)$$

is such that for any $s \in \omega$, $m \in [1\,..\,\mathrm{Q}_O]$,

$$\vec{\mathfrak{R}}[s,m] = \mathfrak{K}[O,s,m].$$

Then conditions (13.17) and (13.18) are equivalent to

$$\vec{\mathfrak{R}}[s,m-1] \vDash_L \bigvee_{\varepsilon \in \mathscr{V}(L,\mathsf{I})} \left(\mathrm{GC}\left[R,\mathrm{P}^O_m,\varepsilon\right]\right)(\vec{a}) \tag{13.19}$$

and

$$\vec{\mathfrak{R}}[s,m-1] \vDash_L \bigwedge_{\substack{\varepsilon,\delta \in \mathscr{V}(L,\mathsf{I})\\ \varepsilon \neq \delta}} \left(\mathrm{GC}\left[R,\mathrm{P}^O_m,\varepsilon\right] \to \neg\mathrm{GC}\left[R,\mathrm{P}^O_m,\delta\right]\right)(\vec{a}), \tag{13.20}$$

respectively. Conditions (13.19) and (13.20) hold due to Lemma 12.2.12 and Definitions 12.2.6 and 12.2.3 (i) and (ii). □

Definition 13.3.8. Let L be an FPJ logic, $I = \mathrm{I}^\tau(L)$ and the external I-formula Φ be of the form $\Theta \to \Xi$. Then the formula $\Xi \to \Theta$ is called the *conversion* of Φ (written:

$\mathrm{Conv}(\Phi)$). If Γ is a set of I-formulas of the form $\Theta \to \Xi$, then by definition put

$$\mathrm{Conv}(\Gamma) = \{\mathrm{Conv}(\Phi) \mid \Phi \in \Gamma\}.$$

Theorem 13.3.9 (Conversion Theorem). Let L be an FPJ logic, $I = \mathrm{I}^{\tau}(L)$, $\Gamma \in \mathrm{FS}_{\sigma}(L, \mathsf{E})$, $R \in \mathrm{RGRS}_{\sigma}(L, \Gamma)$ and $O\colon [1\,..\,\mathrm{Q}_O] \to \mathrm{Pr}_{\sigma}(\mathsf{I})$. Let $\mathfrak{K}$ be an I-structure of signature σ. Suppose

$$\mathfrak{K} \models_I \Gamma^{\mathrm{Init}} \cup \mathrm{I}^O(R).$$

Then

$$\mathfrak{K} \models_I \mathrm{Conv}\left(\mathrm{I}^O(R)\right).$$

Proof. Let $s \in \omega$, $s > 0$, $m \in [1\,..\,\mathrm{Q}_O]$, $\varepsilon \in \mathscr{V}(L, \mathsf{I})$. By Definition 13.3.1, the conversion of $R\left[\mathrm{P}_m^O, \varepsilon\right]^{\langle O,s\rangle}$ is the formula

$$\left(\mathrm{J}_{\varepsilon}\, \mathrm{P}_m^O(\vec{x})\right)^{\langle O,s,m\rangle} \to \mathrm{GC}\left[R, \mathrm{P}_m^O, \varepsilon\right]^{\langle O,s,m-1\rangle}, \qquad (13.21)$$

where $\vec{x} \in \mathrm{Var}^{\diamond} \cap \mathrm{Dom}_{\sigma}(p)$. Let v be an arbitrary valuation in $\mathrm{Frame}_{\mathfrak{K}}$ and let $\vec{a} = v^*(\vec{x})$. Suppose

$$\left(\left(\mathrm{J}_{\varepsilon}\, \mathrm{P}_m^O(\vec{x})\right)^{\langle O,s,m\rangle}\right)^{\mathfrak{K},v} = \mathbf{t}. \qquad (13.22)$$

Hence, taking into account Definition 13.1.2, we obtain

$$\left(\mathrm{J}_{(\varepsilon,s)}\, \mathrm{P}_m^O(\vec{x})\right)^{\mathfrak{K},v} = \mathbf{t}. \qquad (13.23)$$

We must prove that

$$\left(\mathrm{GC}\left[R, \mathrm{P}_m^O, \varepsilon\right]^{\langle O,s,m-1\rangle}\right)^{\mathfrak{K},v} = \mathbf{t}. \qquad (13.24)$$

Suppose the converse. Then by Lemma 13.3.7 we obtain

$$\left(\mathrm{GC}\left[R, \mathrm{P}_m^O, \delta\right]^{\langle O,s,m-1\rangle}\right)^{\mathfrak{K},v} = \mathbf{t} \qquad (13.25)$$

for some $\delta \neq \varepsilon$. By the assumption, $\mathfrak{K} \models \Pi^O(R)$. Then from (13.25) it follows

$$\left(\mathrm{J}_{(\delta,s)}\, \mathrm{P}_m^O(\vec{t})\right)^{\mathfrak{K},v} = \mathbf{t},$$

where $\delta \neq \varepsilon$, which contradicts (13.23) proved earlier. Therefore our supposition was not true. Thus, (13.22) implies (13.24). Because of the arbitrary choice of s, m, ε and v we have $\mathfrak{K} \models_I \mathrm{Conv}\left(\mathrm{I}^O(R)\right)$, which was to be proved. □

We need the next lemma to prove the theorem on the uniqueness of the extention.

Lemma 13.3.10. Let L be an FPJ logic, $I = \mathrm{I}^{\tau}(L)$, $\mathfrak{K}$ and $\mathfrak{L}$ be two I-structures such that:

(i) $\mathrm{Frame}_{\mathfrak{K}} = \mathrm{Frame}_{\mathfrak{L}}$,
(ii) $f^{\mathfrak{K}} = f^{\mathfrak{L}}$ for any $f \in \mathrm{Fu}_{\sigma}$,

(iii) $q^{\mathfrak{K}} = q^{\mathfrak{L}}$ for any $q \in \mathrm{Pr}_\sigma(\mathsf{E})$,
(iv) $\left(\mathrm{J}_{(\varepsilon,s)}\, p(\vec{x})\right)^{\mathfrak{K},v} = \left(\mathrm{J}_{(\varepsilon,s)}\, p(\vec{x})\right)^{\mathfrak{L},v}$ for any $p \in \mathrm{Pr}_\sigma(\mathsf{I})$, $\vec{x} \in \mathrm{Var}^{\Diamond} \cap \mathrm{Dom}_\sigma(p)$, $\varepsilon \in \mathscr{V}(L,\mathsf{I})$, $s \in \omega$ and for any valuation v in $\mathrm{Frame}_{\mathfrak{K}}$.

Then $\mathfrak{K} = \mathfrak{L}$.

Proof. It is enough to show that $p^{\mathfrak{K}} = p^{\mathfrak{L}}$ for any $p \in \mathrm{Pr}_\sigma(\mathsf{I})$. Suppose the converse. Then there exist $p \in \mathrm{Pr}\mathit{edS}_\sigma(\mathsf{I})$ and $\vec{a} \in \mathrm{Dom}\left(p^{\mathfrak{K}}\right) = \mathrm{Dom}\left(p^{\mathfrak{L}}\right)$ such that $p^{\mathfrak{K}}(\vec{a}) \neq p^{\mathfrak{L}}(\vec{a})$. Then the following cases are possible:

(i) $p^{\mathfrak{K}}(\vec{a}) = \langle \varepsilon, s\rangle, p^{\mathfrak{L}}(\vec{a}) = \langle \delta, l\rangle$ for some $s,l \in \omega$ and $\varepsilon, \delta \in \mathscr{D}^{\tau}(L)$ such that $\varepsilon \neq \delta$;
(ii) $p^{\mathfrak{K}}(\vec{a}) = \langle \varepsilon, s\rangle, p^{\mathfrak{L}}(\vec{a}) = \langle \varepsilon, l\rangle$ for an $\varepsilon \in \mathscr{D}^{\tau}(L)$ and some $s,l \in \omega$ such that $s \neq l$;
(iii) $p^{\mathfrak{K}}(\vec{a}) = \langle \varepsilon, s\rangle$ for some $\varepsilon \in \mathscr{D}^{\tau}(L)$, $s \in \omega$, $p^{\mathfrak{L}}(\vec{a}) = \tau$;
(iv) $p^{\mathfrak{L}}(\vec{a}) = \langle \varepsilon, s\rangle$ for some $\varepsilon \in \mathscr{D}^{\tau}(L)$, $s \in \omega$, $p^{\mathfrak{K}}(\vec{a}) = \tau$.

Let us take a valuation v in $\mathrm{Frame}_{\mathfrak{K}}$ such that $v^*(\vec{x}) = \vec{a}$ and consider each of the above enumerated cases. By direct checking we can verify that for each case:

$$\left(\mathrm{J}_{(\varepsilon,s)}\, p(\vec{x})\right)^{\mathfrak{K},v} \neq \left(\mathrm{J}_{(\varepsilon,s)}\, p(\vec{x})\right)^{\mathfrak{L},v}.$$

□

The next theorem states that there exists not more than one extention of a fixed L-structure that respects the fixed modification rule system, i.e. in which the iterative images of this system are valid. A corollary of this theorem is that the realisation of an atomic structure generator is a unique extension of its origin.

This theorem is obviously the main theorem of this section since it is an analogue of the theorem of uniqueness of model in the JSM theory [Anshakov et al, 1993]. The property of uniqueness is especially important from the point of view of constructivity.

Theorem 13.3.11 (Theorem on the uniqueness of extension). Let σ be a signature, L be an FPJ logic, $I = \mathrm{I}^{\tau}(L)$, $\Gamma \in \mathrm{FS}_\sigma(L,\mathsf{E})$, $R \in \mathrm{RGRS}_\sigma(L,\Gamma)$ and

$$O\colon\ [1\,..\,\mathrm{Q}_O] \to \mathrm{Pr}_\sigma(\mathsf{I}).$$

Let $\mathfrak{J}$ be an L-structure of signature σ and $\mathfrak{K}$ and $\mathfrak{L}$ be two I-structures such that:

(a) $\mathfrak{K}$ and $\mathfrak{L}$ are extensions of $\mathfrak{J}$,
(b) $\mathfrak{K} \models_I \Gamma^{\mathrm{Init}} \cup \mathrm{I}^{O}(R), \mathfrak{L} \models_I \Gamma^{\mathrm{Init}} \cup \mathrm{I}^{O}(R)$.

Then $\mathfrak{K} = \mathfrak{L}$.

Proof. Due to (a) the following statements hold:

(i) $\mathrm{Frame}_{\mathfrak{K}} = \mathrm{Frame}_{\mathfrak{L}} = \mathrm{Frame}_{\mathfrak{J}}$,
(ii) $f^{\mathfrak{K}} = f^{\mathfrak{L}} = f^{\mathfrak{J}}$ for any $f \in \mathrm{Fu}_\sigma$,

(iii) $q^{\mathfrak{K}} = q^{\mathfrak{L}} = q^{\mathfrak{I}}$ for any $q \in \mathrm{Pr}_\sigma(\mathsf{E})$.

Due to Lemma 13.3.10 in order to prove $\mathfrak{K} = \mathfrak{L}$ it is enough to show that:

(iv) $\left(\mathrm{J}_{(\varepsilon,s)}\, p(\vec{x})\right)^{\mathfrak{K},v} = \left(\mathrm{J}_{(\varepsilon,s)}\, p(\vec{x})\right)^{\mathfrak{L},v}$ for any $p \in \mathrm{Pr}_\sigma(\mathsf{I})$, $\vec{x} \in \mathrm{Var}^{\diamondsuit} \cap \mathrm{Dom}_\sigma(p)$, $\varepsilon \in \mathscr{V}(L,\mathsf{I})$, $s \in \omega$ and arbitrary valuation v in $\mathrm{Frame}_{\mathfrak{K}}$.

In order to verify the validity of (iv) we prove the fulfillment of the following two conditions by induction on s and m:

(A) $\left(\mathrm{J}_{(\varepsilon,s)}\, \mathrm{P}_i^O(\vec{x})\right)^{\mathfrak{K},v} = \left(\mathrm{J}_{(\varepsilon,s)}\, \mathrm{P}_i^O(\vec{x})\right)^{\mathfrak{L},v}$ for all $\vec{x} \in \mathrm{Var}^{\diamondsuit} \cap \mathrm{Dom}_\sigma(p)$, $\varepsilon \in \mathscr{V}(L,\mathsf{I})$, $i \le m \in [0..\mathrm{Q}_O]$ and for any valuation v in $\mathrm{Frame}_{\mathfrak{K}}$;

(B) $\left(\Phi^{\langle O,s,m\rangle}\right)^{\mathfrak{K},v} = \left(\Phi^{\langle O,s,m\rangle}\right)^{\mathfrak{L},v}$ for any external L-formula Φ, for any $m \in [0..\mathrm{Q}_O]$ and for any valuation v in $\mathrm{Frame}_{\mathfrak{K}}$;

However, first we note that by (ii)

$$t^{\mathfrak{K},v} = t^{\mathfrak{L},v} = t^{\mathfrak{I},v}$$

for any term t of signature σ and for any valuation v in $\mathrm{Frame}_{\mathfrak{K}}$. Then by Definition 13.1.2, we obtain, that for any $s \in \omega$, $m \in [0..\mathrm{Q}_O]$:

(1*) $\left((t_1 = t_2)^{\langle O,s,m\rangle}\right)^{\mathfrak{K},v} = \left((t_1 = t_2)^{\langle O,s,m\rangle}\right)^{\mathfrak{L},v}$ for any terms t_1 and t_2 such that $\mathrm{Sort}_\sigma(t_1) = \mathrm{Sort}_\sigma(t_2)$,

(2*) $\left(q(\vec{t})^{\langle O,s,m\rangle}\right)^{\mathfrak{K},v} = \left(q(\vec{t})^{\langle O,s,m\rangle}\right)^{\mathfrak{L},v}$ for any $q \in \mathrm{Pr}_\sigma(\mathsf{E})$, $\vec{t} \in \mathrm{Dom}_\sigma(q)$.

If we proved (B) by induction over the external formulas then to conclude the basis of induction it would be sufficient to establish that for any $s \in \omega$ and $m \in [0..\mathrm{Q}_O]$:

(3*) $\left(\left(\mathrm{J}_\varepsilon\, p(\vec{t})\right)^{\langle O,s,m\rangle}\right)^{\mathfrak{K},v} = \left(\left(\mathrm{J}_\varepsilon\, p(\vec{t})\right)^{\langle O,s,m\rangle}\right)^{\mathfrak{L},v}$ for any $\varepsilon \in \mathscr{V}(L,\mathsf{I})$, $p \in \mathrm{Pr}_\sigma(\mathsf{I})$, $\vec{t} \in \mathrm{Dom}_\sigma(p)$.

However we cannot prove the statement (3*) yet. Let us prove it by induction on s.

BASIS. Let $s = 0$.

(A) Since (a),

$$\begin{aligned} p^{\mathfrak{K}}(\vec{a}) \in \mathrm{Ch}(\varepsilon,0) \quad &\text{iff} \quad p^{\mathfrak{L}}(\vec{a}) \in \mathrm{Ch}(\varepsilon,0) \\ &\text{iff} \quad p^{\mathfrak{I}}(\vec{a}) = \varepsilon \end{aligned}$$

for any $p \in \mathrm{Pr}_\sigma(\mathsf{I})$, $\vec{a} \in \mathrm{Dom}(p)$. Hence it is easy to show that (A) holds for $s = 0$ and any $m \in [0..\mathrm{Q}_O]$.

(B) Induction over external formulas.

Basis. (1*) and (2*) are proved for all s. It is easy to verify that for $s = 0$, (3*) is equvalent to (A) (see Definition 13.1.2). But (A) has been proved above for $s = 0$.

Induction step. It is trivial.

INDUCTION STEP. Let (A) and (B) hold for $s = n$ and for all $m \in [0..\mathrm{Q}_O]$. Let us prove these statements for $s = n+1$ by induction on m.

Basis. $m = 0$.

(A) is trivial since there are no predicate symbols with numbers less than or equal to 0.

(B) By 13.1.3, $\Phi^{\langle O,n+1,0\rangle} = \Phi^{\langle O,n,\mathrm{Q}_{\mathscr{M}}\rangle}$, hence by using the inductive assumption it easy to get $\left(\Phi^{\langle O,n+1,0\rangle}\right)^{\mathfrak{K},v} = \left(\Phi^{\langle O,n+1,0\rangle}\right)^{\mathfrak{L},v}$.

Induction step. Suppose that for $m = k$ (and $s = n+1$) (A) and (B) hold. Let us prove these for $m = k+1$.

(A) We should prove that for all $i \leq k+1$

$$\left(\mathrm{J}_{(\varepsilon,n+1)}\,\mathrm{P}_i^O(\vec{x})\right)^{\mathfrak{K},v} = \left(\mathrm{J}_{(\varepsilon,n+1)}\,\mathrm{P}_i^O(\vec{x})\right)^{\mathfrak{L},v}$$

for any $\vec{x} \in \mathrm{Var}^{\diamond} \cap \mathrm{Dom}_{\sigma}(p)$, $\varepsilon \in \mathscr{V}(L,\mathsf{I})$ and for any valuation v in $\mathrm{Frame}_{\mathfrak{K}}$. If $i \leq k$, then by the inductive assumption this equality holds. Consider the equality

$$\left(\mathrm{J}_{(\varepsilon,n+1)}\,\mathrm{P}_{k+1}^O(\vec{x})\right)^{\mathfrak{K},v} = \left(\mathrm{J}_{(\varepsilon,n+1)}\,\mathrm{P}_{k+1}^O(\vec{x})\right)^{\mathfrak{L},v}. \tag{13.26}$$

Assume the converse. Let for definiteness

$$\left(\mathrm{J}_{(\varepsilon,n+1)}\,\mathrm{P}_{k+1}^O(\vec{x})\right)^{\mathfrak{K},v} = \mathbf{t}, \tag{13.27}$$

$$\left(\mathrm{J}_{(\varepsilon,n+1)}\,\mathrm{P}_{k+1}^O(\vec{x})\right)^{\mathfrak{L},v} = \mathbf{f}. \tag{13.28}$$

By the assumption, $\mathfrak{K} \models_I \Gamma^{\mathrm{Init}} \cup \mathrm{I}^O(R)$. Then by Conversion Theorem (13.3.9),

$$\mathfrak{K} \models_I \mathrm{Conv}\left(\mathrm{I}^O(R)\right).$$

Therefore (13.27) implies

$$\left(\mathrm{GC}\left[R,\mathrm{P}_{k+1}^O,\varepsilon\right]^{\langle O,n+1,k\rangle}\right)^{\mathfrak{K},v} = \mathbf{t}. \tag{13.29}$$

Then by the inductive assumption (for (B))

$$\left(\mathrm{GC}\left[R,\mathrm{P}_{k+1}^O,\varepsilon\right]^{\langle O,n+1,k\rangle}\right)^{\mathfrak{L},v} = \mathbf{t}. \tag{13.30}$$

By the assumption, $\mathfrak{L} \models_I \mathrm{I}^O(R)$. Then (13.30) implies

$$\left(\mathrm{J}_{(\varepsilon,n+1)}\,\mathrm{P}_{k+1}^O(\vec{x})\right)^{\mathfrak{L},v} = \mathbf{t},$$

which contradicts (13.28). Thus our supposition was not true. Therefore (13.26) holds.

We have proved that (A) does hold for $s = n+1$ and $m = k+1$.

We have to prove that for any arbitrary external formula Φ,

$$\left(\Phi^{\langle O,n+1,k+1\rangle}\right)^{\mathfrak{K},v}=\left(\Phi^{\langle O,n+1,k+1\rangle}\right)^{\mathfrak{L},v}.$$

The proof is by induction over external formulas.

Basis. (1^*) and (2^*) hold for $s=n+1$ and $m=k+1$. Let us show that for these s and k (3^*) holds. Recall that by 13.1.2,

$$\left(\mathrm{J}_{\varepsilon}\,\mathrm{P}_i^O(\vec{x})\right)^{\langle O,n+1,k+1\rangle}=\begin{cases}\mathrm{J}_{(\varepsilon,n+1)}\,\mathrm{P}_i^O(\vec{t}) & \text{if } i\le k+1,\\ \mathrm{J}_{(\varepsilon,n)}\,\mathrm{P}_i^O(\vec{t}) & \text{if } i>k+1.\end{cases}$$

Then for proving (3^*) it is enough to show that

$$\left(\mathrm{J}_{(\varepsilon,n+1)}\,\mathrm{P}_i^O(\vec{x})\right)^{\mathfrak{K},v}=\left(\mathrm{J}_{(\varepsilon,n+1)}\,\mathrm{P}_i^O(\vec{x})\right)^{\mathfrak{L},v}\tag{13.31}$$

for all $i\le k+1$, and

$$\left(\mathrm{J}_{(\varepsilon,n)}\,\mathrm{P}_j^O(\vec{x})\right)^{\mathfrak{K},v}=\left(\mathrm{J}_{(\varepsilon,n)}\,\mathrm{P}_j^O(\vec{x})\right)^{\mathfrak{L},v}\tag{13.32}$$

for all $j>k+1$.

By the already proved (A) (13.31) holds for any $i\le k+1$. Note that by Definition 13.1.2,

$$\left(\mathrm{J}_{(\varepsilon,n)}\,\mathrm{P}_j^O(\vec{x})\right)=\left(\mathrm{J}_{\varepsilon}\,\mathrm{P}_j^O(\vec{x})\right)^{\langle O,n,\mathrm{Q}_{\mathscr{M}}\rangle},\tag{13.33}$$

and by the inductive assumption ((B) for $s=n$),

$$\left(\left(\mathrm{J}_{\varepsilon}\,\mathrm{P}_j^O(\vec{x})\right)^{\langle O,n,\mathrm{Q}_O\rangle}\right)^{\mathfrak{K},v}=\left(\left(\mathrm{J}_{\varepsilon}\,\mathrm{P}_j^O(\vec{x})\right)^{\langle O,n,\mathrm{Q}_O\rangle}\right)^{\mathfrak{L},v}.\tag{13.34}$$

Then (13.33) and (13.34) imply (13.32). Thus (3^*) holds for $s=n+1$ and $m=k+1$.

Induction step. It is trivial. □

Corollary 13.3.12. Assume that L is an FPJ logic, $I=\mathrm{I}^{\tau}(L)$ and σ is a signature. Let $\Gamma\in\mathrm{FS}_{\sigma}(L,\mathsf{E})$ and $\mathfrak{R}\in\mathrm{SG}_{\sigma}(L,\Gamma)$. Then $\mathfrak{K}=\mathrm{Imp}(\mathfrak{R})$ is the unique extention of $\overset{\circ}{\mathfrak{R}}$, which is an I-model of the set $\mathrm{I}^{\mathrm{Ord}_{\mathfrak{R}}}(\mathfrak{R}^{\triangleright})$.

Proof. See 13.2.8, 13.3.3 and 13.3.11. □

The next lemma states that the corresponding elements of the sequences of L-structures generated by two generators with isomorphic origins are isomorphic.

Lemma 13.3.13. Let L be an FPJ logic, $I=\mathrm{I}^{\tau}(L)$, σ be a signature, $\Gamma\in\mathrm{FS}_{\sigma}(L,\mathsf{E})$, $\mathfrak{J}$ and $\mathfrak{K}$ be isomorphic L-models of Γ, $R\in\mathrm{RGRS}_{\sigma}(L,\Gamma)$ and

$$O\colon\ [1..\mathrm{Q}_O]\to\mathrm{Pr}_{\sigma}(\mathsf{I}).$$

Suppose $\mathfrak{R}=\langle\mathfrak{J},R,O\rangle$, $\mathfrak{S}=\langle\mathfrak{K},R,O\rangle$. Then for any $s\in\omega$, $m\in[0..\mathrm{Q}_O]$,

$$\overrightarrow{\mathfrak{R}}[s,m]\cong\overrightarrow{\mathfrak{S}}[s,m].$$

Proof. Let

$$\varphi\colon \mathrm{Dom}_{\mathfrak{J}} \to \mathrm{Dom}_{\mathfrak{K}}$$

be an isomorphism. By 12.2.11, for any $s \in \omega$, $m \in [0\,..\,\mathrm{Q}_O]$,

$$\mathrm{Frame}_{\vec{\mathfrak{R}}[s,m]} = \mathrm{Frame}_{\mathfrak{J}}, \qquad \mathrm{Frame}_{\vec{\mathfrak{S}}[s,m]} = \mathrm{Frame}_{\mathfrak{K}},$$

and

$$\mathrm{Dom}_{\vec{\mathfrak{R}}[s,m]} = \mathrm{Dom}_{\mathfrak{J}}, \qquad \mathrm{Dom}_{\vec{\mathfrak{S}}[s,m]} = \mathrm{Dom}_{\mathfrak{K}}.$$

Let us prove by induction on s that for any $m \in [0\,..\,\mathrm{Q}_O]$,

$$\varphi\colon \mathrm{Dom}_{\vec{\mathfrak{R}}[s,m]} \to \mathrm{Dom}_{\vec{\mathfrak{S}}[s,m]}$$

is an isomorphism.

BASIS. $s = 0$. The statement trivially holds since, by 12.2.11, for any $m \in [0\,..\,\mathrm{Q}_O]]$,

$$\vec{\mathfrak{R}}[0,m] = \mathfrak{J}, \qquad \vec{\mathfrak{S}}[0,m] = \mathfrak{K}.$$

INDUCTION STEP. Let φ be an isomorphism of the L-structures $\vec{\mathfrak{R}}[s,m]$ and $\vec{\mathfrak{S}}[s,m]$ for $s = n$ and any $m \in [0\,..\,\mathrm{Q}_O]$. Prove that this also holds for $s = n+1$ by induction on m.

Basis. $m = 0$. By 12.2.11,

$$\vec{\mathfrak{R}}[n+1,0] = \vec{\mathfrak{R}}[n,\mathrm{Q}_O], \qquad \vec{\mathfrak{S}}[n+1,0] = \vec{\mathfrak{S}}[n,\mathrm{Q}_O].$$

By the inductive assumption,

$$\varphi\colon \mathrm{Dom}_{\vec{\mathfrak{R}}[n,\mathrm{Q}_O]} \to \mathrm{Dom}_{\vec{\mathfrak{S}}[n,\mathrm{Q}_O]}$$

is an isomorphism.

Induction step. Suppose

$$\varphi\colon \mathrm{Dom}_{\vec{\mathfrak{R}}[n+1,k]} \to \mathrm{Dom}_{\vec{\mathfrak{S}}[n+1,k]}$$

is an isomorphism. Prove that this map is an isomorphism of the L-structures $\vec{\mathfrak{R}}[n+1,k+1]$ and $\vec{\mathfrak{S}}[n+1,k+1]$.

The operations and external predicates from the L-structures $\vec{\mathfrak{R}}[n+1,k+1]$ and $\vec{\mathfrak{S}}[n+1,k+1]$ are the same as the corresponding operations and external predicates from $\mathfrak{J}\,(\mathfrak{K})$. Therefore, to show that φ is an isomorphism of the L-structures $\vec{\mathfrak{R}}[n+1,k+1]$ and $\vec{\mathfrak{S}}[n+1,k+1]$ it is enough to check the validity of

$$\left(\Phi^{\langle O,n+1,k+1\rangle}\right)^{\mathfrak{K},v} = \left(\Phi^{\langle O,n+1,k+1\rangle}\right)^{\mathfrak{L},v}.$$

The proof is by induction over external formulas.

Basis. (1^*) and (2^*) hold for $s = n+1$ and $m = k+1$. Let us show that for these s and k (3^*) holds. Recall that by 13.1.2,

$$\left(\mathrm{J}_\varepsilon \mathrm{P}_i^O(\vec{x})\right)^{\langle O,n+1,k+1\rangle} = \begin{cases} \mathrm{J}_{(\varepsilon,n+1)} \mathrm{P}_i^O(\vec{t}) & \text{if } i \leq k+1, \\ \mathrm{J}_{(\varepsilon,n)} \mathrm{P}_i^O(\vec{t}) & \text{if } i > k+1. \end{cases}$$

Then for proving (3^*) it is enough to show that

$$\left(\mathrm{J}_{(\varepsilon,n+1)} \mathrm{P}_i^O(\vec{x})\right)^{\mathfrak{K},v} = \left(\mathrm{J}_{(\varepsilon,n+1)} \mathrm{P}_i^O(\vec{x})\right)^{\mathfrak{L},v} \tag{13.31}$$

for all $i \leq k+1$, and

$$\left(\mathrm{J}_{(\varepsilon,n)} \mathrm{P}_j^O(\vec{x})\right)^{\mathfrak{K},v} = \left(\mathrm{J}_{(\varepsilon,n)} \mathrm{P}_j^O(\vec{x})\right)^{\mathfrak{L},v} \tag{13.32}$$

for all $j > k+1$.

By the already proved (A) (13.31) holds for any $i \leq k+1$. Note that by Definition 13.1.2,

$$\left(\mathrm{J}_{(\varepsilon,n)} \mathrm{P}_j^O(\vec{x})\right) = \left(\mathrm{J}_\varepsilon \mathrm{P}_j^O(\vec{x})\right)^{\langle O,n,\mathrm{Q}_{\mathcal{O}}\rangle}, \tag{13.33}$$

and by the inductive assumption ((B) for $s = n$),

$$\left(\left(\mathrm{J}_\varepsilon \mathrm{P}_j^O(\vec{x})\right)^{\langle O,n,\mathrm{Q}_O\rangle}\right)^{\mathfrak{K},v} = \left(\left(\mathrm{J}_\varepsilon \mathrm{P}_j^O(\vec{x})\right)^{\langle O,n,\mathrm{Q}_O\rangle}\right)^{\mathfrak{L},v}. \tag{13.34}$$

Then (13.33) and (13.34) imply (13.32). Thus (3^*) holds for $s = n+1$ and $m = k+1$.

Induction step. It is trivial. □

Corollary 13.3.12. Assume that L is an FPJ logic, $I = \mathrm{I}^\tau(L)$ and σ is a signature. Let $\Gamma \in \mathrm{FS}_\sigma(L, \mathsf{E})$ and $\mathfrak{R} \in \mathrm{SG}_\sigma(L, \Gamma)$. Then $\mathfrak{K} = \mathrm{Imp}(\mathfrak{R})$ is the unique extention of $\mathring{\mathfrak{R}}$, which is an I-model of the set $\mathrm{I}^{\mathrm{Ord}_{\mathfrak{R}}}(\mathfrak{R}^{\triangleright})$.

Proof. See 13.2.8, 13.3.3 and 13.3.11. □

The next lemma states that the corresponding elements of the sequences of L-structures generated by two generators with isomorphic origins are isomorphic.

Lemma 13.3.13. Let L be an FPJ logic, $I = \mathrm{I}^\tau(L)$, σ be a signature, $\Gamma \in \mathrm{FS}_\sigma(L, \mathsf{E})$, $\mathfrak{J}$ and $\mathfrak{K}$ be isomorphic L-models of Γ, $R \in \mathrm{RGRS}_\sigma(L, \Gamma)$ and

$$O\colon\ [1\,..\,\mathrm{Q}_O] \to \mathrm{Pr}_\sigma(\mathsf{I}).$$

Suppose $\mathfrak{R} = \langle \mathfrak{J}, R, O\rangle$, $\mathfrak{S} = \langle \mathfrak{K}, R, O\rangle$. Then for any $s \in \omega$, $m \in [0\,..\,\mathrm{Q}_O]$,

$$\overrightarrow{\mathfrak{R}}[s,m] \cong \overrightarrow{\mathfrak{S}}[s,m].$$

Proof. Let

$$\varphi\colon \mathrm{Dom}_{\mathfrak{J}} \to \mathrm{Dom}_{\mathfrak{K}}$$

be an isomorphism. By 12.2.11, for any $s \in \omega$, $m \in [0\,..\,\mathrm{Q}_O]$,

$$\mathrm{Frame}_{\overrightarrow{\mathfrak{R}}[s,m]} = \mathrm{Frame}_{\mathfrak{J}}, \qquad \mathrm{Frame}_{\overrightarrow{\mathfrak{S}}[s,m]} = \mathrm{Frame}_{\mathfrak{K}},$$

and

$$\mathrm{Dom}_{\overrightarrow{\mathfrak{R}}[s,m]} = \mathrm{Dom}_{\mathfrak{J}}, \qquad \mathrm{Dom}_{\overrightarrow{\mathfrak{S}}[s,m]} = \mathrm{Dom}_{\mathfrak{K}}.$$

Let us prove by induction on s that for any $m \in [0\,..\,\mathrm{Q}_O]$,

$$\varphi\colon \mathrm{Dom}_{\overrightarrow{\mathfrak{R}}[s,m]} \to \mathrm{Dom}_{\overrightarrow{\mathfrak{S}}[s,m]}$$

is an isomorphism.

BASIS. $s = 0$. The statement trivially holds since, by 12.2.11, for any $m \in [0\,..\,\mathrm{Q}_O]]$,

$$\overrightarrow{\mathfrak{R}}[0,m] = \mathfrak{J}, \qquad \overrightarrow{\mathfrak{S}}[0,m] = \mathfrak{K}.$$

INDUCTION STEP. Let φ be an isomorphism of the L-structures $\overrightarrow{\mathfrak{R}}[s,m]$ and $\overrightarrow{\mathfrak{S}}[s,m]$ for $s = n$ and any $m \in [0\,..\,\mathrm{Q}_O]$. Prove that this also holds for $s = n+1$ by induction on m.

Basis. $m = 0$. By 12.2.11,

$$\overrightarrow{\mathfrak{R}}[n+1,0] = \overrightarrow{\mathfrak{R}}[n,\mathrm{Q}_O], \qquad \overrightarrow{\mathfrak{S}}[n+1,0] = \overrightarrow{\mathfrak{S}}[n,\mathrm{Q}_O].$$

By the inductive assumption,

$$\varphi\colon \mathrm{Dom}_{\overrightarrow{\mathfrak{R}}[n,\mathrm{Q}_O]} \to \mathrm{Dom}_{\overrightarrow{\mathfrak{S}}[n,\mathrm{Q}_O]}$$

is an isomorphism.

Induction step. Suppose

$$\varphi\colon \mathrm{Dom}_{\overrightarrow{\mathfrak{R}}[n+1,k]} \to \mathrm{Dom}_{\overrightarrow{\mathfrak{S}}[n+1,k]}$$

is an isomorphism. Prove that this map is an isomorphism of the L-structures $\overrightarrow{\mathfrak{R}}[n+1,k+1]$ and $\overrightarrow{\mathfrak{S}}[n+1,k+1]$.

The operations and external predicates from the L-structures $\overrightarrow{\mathfrak{R}}[n+1,k+1]$ and $\overrightarrow{\mathfrak{S}}[n+1,k+1]$ are the same as the corresponding operations and external predicates from $\mathfrak{J}\,(\mathfrak{K})$. Therefore, to show that φ is an isomorphism of the L-structures $\overrightarrow{\mathfrak{R}}[n+1,k+1]$ and $\overrightarrow{\mathfrak{S}}[n+1,k+1]$ it is enough to check the validity of

$$\varphi\left(\left(\mathrm{P}_i^O\right)^{\vec{\mathfrak{R}}[n+1,k+1]}(\vec{a})\right) = \left(\mathrm{P}_i^O\right)^{\vec{\mathfrak{S}}[n+1,k+1]}(\varphi^*(\vec{a})) \tag{13.35}$$

for any $i \in [0\,..\,\mathrm{Q}_O]$ and $\vec{a} \in \mathrm{Dom}\left(\left(\mathrm{P}_i^O\right)^{\vec{\mathfrak{R}}[n+1,k+1]}\right)$. Consider the possible cases:

CASE 1. $i \neq k+1$. Then by 12.2.11,

$$\left(\mathrm{P}_i^O\right)^{\vec{\mathfrak{R}}[n+1,k+1]} = \left(\mathrm{P}_i^O\right)^{\vec{\mathfrak{R}}[n+1,k]} \quad \text{and} \quad \left(\mathrm{P}_i^O\right)^{\vec{\mathfrak{S}}[n+1,k+1]} = \left(\mathrm{P}_i^O\right)^{\vec{\mathfrak{S}}[n+1,k]}.$$

Then the equality (13.35) holds by the inductive assumption.

CASE 2. $i = k+1$. By the inductive assumption and Corollary 8.4.9

$$\vec{\mathfrak{R}}[n+1,k] \vDash_L \mathrm{GC}\left[R, \mathrm{P}_{k+1}^O, \varepsilon\right](\vec{a})$$
$$\text{iff} \quad \vec{\mathfrak{S}}[n+1,k] \vDash_L \mathrm{GC}\left[R, \mathrm{P}_{k+1}^O, \varepsilon\right](\varphi^*(\vec{a})) \tag{13.36}$$

for any $\varepsilon \in \mathscr{V}(L, \mathsf{I})$. By 12.2.11 and (13.36)

$$\varphi\left(\left(\mathrm{P}_{k+1}^O\right)^{\vec{\mathfrak{R}}[n+1,k+1]}(\vec{a})\right) = \left(\mathrm{P}_{k+1}^O\right)^{\vec{\mathfrak{S}}[n+1,k+1]}(\varphi^*(\vec{a}))$$

as was to be proved. □

We now establish that the implementations of two structure generators with isomorphic origins are isomorphic I-structures.

Lemma 13.3.14. Let L be an FPJ logic, $I = \mathrm{I}^\tau(L)$, σ be a signature, $\Gamma \in \mathrm{FS}_\sigma(L, \mathsf{E})$, $\mathfrak{J}$ and $\mathfrak{K}$ be isomorphic L-models of Γ, $R \in \mathrm{RGRS}_\sigma(L, \Gamma)$ and

$$O\colon\ [1\,..\,\mathrm{Q}_O] \to \mathrm{Pr}_\sigma(\mathsf{I}).$$

Let $\mathfrak{R} = \langle \mathfrak{J}, R, O\rangle$, $\mathfrak{S} = \langle \mathfrak{K}, R, O\rangle$. Then $\mathrm{Imp}(\mathfrak{R}) \cong \mathrm{Imp}(\mathfrak{S})$.

Proof. Let

$$\varphi\colon \mathrm{Dom}_{\mathfrak{J}} \to \mathrm{Dom}_{\mathfrak{K}}$$

is an isomorphism. By 13.2.2(i),

$$\mathrm{Frame}_{\mathrm{Imp}(\mathfrak{R})} = \mathrm{Frame}_{\mathfrak{J}}, \qquad \mathrm{Frame}_{\mathrm{Imp}(\mathfrak{S})} = \mathrm{Frame}_{\mathfrak{K}},$$

hence

$$\mathrm{Dom}_{\mathrm{Imp}(\mathfrak{R})} = \mathrm{Dom}_{\mathfrak{J}}, \qquad \mathrm{Dom}_{\mathrm{Imp}(\mathfrak{S})} = \mathrm{Dom}_{\mathfrak{K}}.$$

Let us show that

$$\varphi\colon \mathrm{Dom}_{\mathrm{Imp}(\mathfrak{R})} \to \mathrm{Dom}_{\mathrm{Imp}(\mathfrak{S})}$$

is an **It**-isomorphism.

The operations and external predicates from $\mathrm{Imp}(\mathfrak{N})$ ($\mathrm{Imp}(\mathfrak{S})$) are the same as the corresponding operations and external predicates from $\mathfrak{J}(\mathfrak{K})$ Therefore, to show that φ is an isomorphism of the I-structures $\mathrm{Imp}(\mathfrak{N})$ and $\mathrm{Imp}(\mathfrak{S})$ it is enough to check that

$$\varphi\left(\left(\mathrm{P}_m^O\right)^{\mathrm{Imp}(\mathfrak{N})}(\vec{a})\right) = \left(\mathrm{P}_m^O\right)^{\mathrm{Imp}(\mathfrak{S})}(\varphi^*(\vec{a})) \tag{13.37}$$

for any $m \in [0..\mathrm{Q}_O]$ and $\vec{a} \in \mathrm{Dom}\left(\left(\mathrm{P}_m^O\right)^{\mathrm{Imp}(\mathfrak{N})}\right)$. In the proof of Lemma 13.3.13 it was established that the map φ considered is an isomorphism of L-structures $\overrightarrow{\mathfrak{N}}[s,m]$ and $\overrightarrow{\mathfrak{S}}[s,m]$ for any $s \in \omega$ and $m \in [0..\mathrm{Q}_{\mathscr{M}}]$.

In this case,

$$\left(\mathrm{P}_m^O\right)^{\overrightarrow{\mathfrak{N}}[s,m]}(\vec{a}) = \varepsilon \quad \text{iff} \quad \left(\mathrm{P}_m^O\right)^{\overrightarrow{\mathfrak{S}}[s,m]}(\varphi^*(\vec{a})) = \varepsilon$$

for any $\vec{a} \in \mathrm{Dom}\left(\left(\mathrm{P}_m^O\right)^{\overrightarrow{\mathfrak{N}}[s,m]}\right) = \mathrm{Dom}\left(\left(\mathrm{P}_m^O\right)^{\mathrm{Imp}(\mathfrak{N})}\right)$ and for any $\varepsilon \in \mathscr{V}(L,\mathsf{I})$. Then by 13.2.2,

$$\left(\mathrm{P}_m^O\right)^{\mathrm{Imp}(\mathfrak{N})}(\vec{a}) = \alpha \quad \text{iff} \quad \left(\mathrm{P}_m^O\right)^{\mathrm{Imp}(\mathfrak{S})}(\varphi^*(\vec{a})) = \alpha$$

for any $\vec{a} \in \mathrm{Dom}\left(\left(\mathrm{P}_m^O\right)^{\mathrm{Imp}(\mathfrak{N})}\right)$ and for any $\alpha \in \mathscr{V}(I,\mathsf{I})$. The last statement is obviously equivalent to (13.37). □

The corollary below states that if the extentions of two isomorphic L-structures preserve the same modification rules then they are isomorphic I-structures.

Corollary 13.3.15. Assume that L is an FPJ logic, $I = \mathrm{I}^{\tau}(L)$ and σ is a signature. Let $\Gamma \in \mathrm{FS}_{\sigma}(L,\mathsf{E})$, $\mathfrak{J}$ and $\mathfrak{K}$ be isomorphic L-models of Γ, $R \in \mathrm{RGRS}_{\sigma}(L,\Gamma)$ and $O\colon [1..\mathrm{Q}_O] \to \mathrm{Pr}_{\sigma}(\mathsf{I})$. Let $\mathfrak{L}$ and $\mathfrak{M}$ be two I-structures such that:

(a) $\mathfrak{L}$ is an extention of $\mathfrak{J}$, $\mathfrak{M}$ is a extention of $\mathfrak{K}$,
(b) $\mathfrak{L} \models t\Gamma^{\mathrm{Init}} \cup \mathrm{I}^O(R)$, $\mathfrak{M} \models_I \Gamma^{\mathrm{Init}} \cup \mathrm{I}^O(R)$.

Then $\mathfrak{L} \cong \mathfrak{M}$.

Proof. Let $\mathfrak{N} = \langle \mathfrak{J}, R, O \rangle$, $\mathfrak{S} = \langle \mathfrak{K}, R, O \rangle$. By Corollary 13.3.12,

$$\mathfrak{L} = \mathrm{Imp}(\mathfrak{N}), \qquad \mathfrak{M} = \mathrm{Imp}(\mathfrak{S}).$$

Then by 13.3.14, $\mathfrak{L} \cong \mathfrak{M}$. □

Lemma 13.3.16. Let L be an FPJ logic, $I = \mathrm{I}^{\tau}(L)$, σ be a signature with a finite set of internal predicate symbols, D be a σ-vocabulary, Δ be a state description of signature σ over D, $\Gamma \in \mathrm{FS}_{\sigma}(L,\mathsf{E})$. Let $R \in \mathrm{RGRS}_{\sigma}(L,\Gamma)$ and

$$O\colon [1..\mathrm{Q}_O] \to \mathrm{Pr}_{\sigma}(\mathsf{I}).$$

Suppose $\mathfrak{L}$ and $\mathfrak{M}$ are two I-structures of signature σ such that:

(a) $\mathfrak{L} \models_I \Delta^{\text{Init}} \cup \Gamma^{\text{Init}} \cup \mathrm{I}^O(R)$,
(b) $\mathfrak{M} \models_I \Delta^{\text{Init}} \cup \Gamma^{\text{Init}} \cup \mathrm{I}^O(R)$.

Then $\mathfrak{L} \cong \mathfrak{M}$.

Proof. By Lemma 13.2.14 there exist two L-structures $\mathfrak{J}$ and $\mathfrak{K}$ such that

$$\mathfrak{J} \models_L \Delta, \qquad \mathfrak{K} \models_L \Delta.$$

and $\mathfrak{L}$ and $\mathfrak{M}$ are extensions of $\mathfrak{J}$ and $\mathfrak{K}$, respectively. By 10.1.1, $\mathfrak{J} \cong \mathfrak{K}$. Hence by 13.3.15, $\mathfrak{L} \cong \mathfrak{M}$. □

Corollary 13.3.17. Let L be an FPJ logic, $I = \mathrm{I}^\tau(L)$, σ be a signature with a finite set of internal predicate symbols, D be a σ-vocabulary, Δ be a state description of signature σ over D, $\Gamma \in \mathrm{FS}_\sigma(L, \mathsf{E})$, $R \in \mathrm{RGRS}_\sigma(L, \Gamma)$ and $O\colon [1\,..\,\mathrm{Q}_O] \to \mathrm{Pr}_\sigma(\mathsf{I})$. Suppose that Ξ is a set of external I-formulas of signature σ, such that

$$\Delta^{\text{Init}} \cup \Gamma^{\text{Init}} \cup \mathrm{I}^O(R) \subseteq \Xi.$$

Then the following statements hold:

(i) Any two I-models of Ξ are isomorphic;
(ii) Let $\mathfrak{K}$ be an I-model of Ξ. Then for any external I-formula Φ of signature σ,

$$\Xi \vdash_I \Phi \quad \text{iff} \quad \mathfrak{K} \models_I \Phi.$$

Proof. (i) Each I-model of the set Ξ at the same time is an I-model of $\Delta^{\text{Init}} \cup \Gamma^{\text{Init}} \cup \mathrm{I}^O(R)$. By Lemma 13.3.16 any I-structures which are I-models of $\Delta^{\text{Init}} \cup \Gamma^{\text{Init}} \cup \mathrm{I}^O(R)$ are isomorphic.

(ii) $(\Rightarrow)\Xi \vdash_I \Phi$ implies $\mathfrak{K} \models_I \Phi$, since $\mathfrak{K}$ is an I-model of Γ.

$(\Leftarrow)$ Let $\mathfrak{K} \models_I \Phi$. Suppose $\mathfrak{L} \models_I \Xi$. Then by 13.3.17, $\mathfrak{L} \cong \mathfrak{K}$. Hence $\mathfrak{K} \models_I \Phi$. Thus, $\Xi \models_I \Phi$. Then by the Completeness Theorem $\Xi \vdash_I \Phi$. □

Chapter 14
Modification Theories

14.1 Validity and Derivability

Now we investigate the connection between derivability in modification calculi and validity in L-structures, where L is an FPJ logic. With few exceptions, in this section we will consider only *pure* inferences in modification calculi, i.e. inferences from own state descriptions of modification calculi. Our tools of investigation will include:

- (From the syntactic point of view) "current state descriptions" that correspond to an inference, and results of resumption operation for certain cuts of this inference
- (From the semantic point of view) syntactic L-structures for these descriptions

In the following few statements we establish a relation between the inferences in modification calculi, derivability in the L^1 calculus from the current state description and validity in the corresponding syntactical L-structures. The results we thus obtain will correspond to our intuition and so are anticipated.

Lemma 14.1.1. Let L be an FPJ logic, σ be a signature, $\mathscr{M} \in \mathrm{MC}_\sigma(L)$, $\Gamma \in \mathrm{AFS}(\mathscr{M})$, $\vec{a}$ be an $\mathscr{M}(s,m,\mathbf{D})$-inference from Γ. Suppose

$$\mathfrak{J} = \mathrm{Sint}_L(\mathrm{CSD}(\vec{\mathfrak{a}}), \mathrm{Univ}_{\mathscr{M}}).$$

Then for any external L-formula Φ of signature σ:

(i) $\mathrm{CSD}(\vec{\mathfrak{a}}) \vdash_L \Phi$ iff $\mathfrak{J} \models_L \Phi$,
(ii) If $\mathrm{Cur}(\Gamma, \vec{\mathfrak{a}})$ is L-consistent then $\mathrm{Cur}(\Gamma, \vec{\mathfrak{a}}) \vdash_L \Phi$ iff $\mathfrak{J} \models_L \Phi$.

Proof. (i) By the use of Propositions 10.1.20 and 10.1.23.
(ii) Let $\mathrm{Cur}(\Gamma, \vec{\mathfrak{a}})$ be L-consistent. Then it has an L-model $\mathfrak{K}$. Clearly,

$$\mathrm{CSD}(\vec{\mathfrak{a}}) \subseteq \mathrm{Cur}(\Gamma, \vec{\mathfrak{a}}).$$

Then $\mathfrak{K}$ is an L-model of $\mathrm{CSD}(\vec{\mathfrak{a}})$. By 10.1.22, $\mathfrak{K} \cong \mathfrak{J}$, whence $\mathfrak{J}$ is also an L-model of $\mathrm{Cur}(\Gamma, \vec{\mathfrak{a}})$. Since any model of $\mathrm{Cur}(\Gamma, \vec{\mathfrak{a}})$ is isomorphic to $\mathfrak{J}$, it is not difficult to

O. Anshakov, T. Gergely, *Cognitive Reasoning*, Cognitive Technologies, DOI 10.1007/978-3-540-68875-4_14, © Springer-Verlag Berlin Heidelberg 2010

prove that

$$\mathrm{Cur}\,(\Gamma,\vec{\mathfrak{a}}) \vdash_L \Phi \quad \text{iff} \quad \mathfrak{J} \models_L \Phi.$$

(For a similar proof see Proposition 10.1.23.) □

Lemma 14.1.2. Let L be an FPJ logic, σ be a signature, $\mathscr{M} \in \mathrm{MC}_\sigma(L)$, $\Gamma \in \mathrm{AFS}\,(\mathscr{M})$ and $\vec{\mathfrak{a}}$ be an $\mathscr{M}(s,m,\mathbf{D})$-inference from Γ. Assume that $\mathrm{Cur}\,(\Gamma,\vec{\mathfrak{a}})$ is L-consistent. Then

$$\mathfrak{J} = \mathrm{Sint}_L\,(\mathrm{CSD}\,(\vec{\mathfrak{a}})\,,\mathrm{Univ}_{\mathscr{M}})$$

is an L-model of $\mathrm{FSet}\,(\vec{\mathfrak{a}})$.

Proof. Let $\Phi \in \mathrm{FSet}\,(\vec{\mathfrak{a}})$. Then Φ is contained in the sequence $\mathrm{FExtr}\,(\vec{\mathfrak{a}})$. By Corollary 11.4.10, $\mathrm{FExtr}\,(\vec{\mathfrak{a}})$ is an L^1-inference from $\mathrm{Cur}\,(\Gamma,\vec{\mathfrak{a}})$. Then $\mathrm{Cur}\,(\Gamma,\vec{\mathfrak{a}}) \vdash_L \Phi$. Consequently, $\mathfrak{J} \models_L \Phi$, since $\mathfrak{J}$ is an L-model of $\mathrm{Cur}\,(\Gamma,\vec{\mathfrak{a}})$ (see Lemma 14.1.1). □

Corollary 14.1.3. Let L be an FPJ logic, σ be a signature, $\mathscr{M} \in \mathrm{MC}_\sigma(L)$ and $\vec{\mathfrak{a}}$ be a pure $\mathscr{M}(s,m,\mathbf{D})$-inference. Then

$$\mathfrak{J} = \mathrm{Sint}_L\,(\mathrm{CSD}\,(\vec{\mathfrak{a}})\,,\mathrm{Univ}_{\mathscr{M}})$$

is an L-model of $\mathrm{FSet}\,(\vec{\mathfrak{a}})$.

Proof. By Definitions 11.3.15 and 11.3.16,

$$\mathrm{Cur}\,(\mathrm{Descr}_{\mathscr{M}},\vec{\mathfrak{a}}) = \mathrm{CSD}\,(\vec{\mathfrak{a}})\,.$$

$\mathrm{CSD}\,(\vec{\mathfrak{a}})$ is consistent, since it has a model $\mathfrak{J}$ (see Lemma 14.1.1). Then by Lemma 14.1.2 we obtain what we set out to prove. □

Below we introduce a notation that will be widely used in the present chapter. Namely, we will define a convenient abbreviation/shortening for the result of applying the resumption operation to the lower segment of the $(k,\,l)$-cut of an inference. Operation Resume returns the inference to the previous state which corresponds to the rank $\langle k,l\rangle$.

We assume that inferences in the modification calculi imitate the cognitive process that passes through a series of stages. Each of these stages corresponds to the completion of plausible reasoning at a certain stage and module. As for the pair of stage and module it is the rank of the inference. Therefore, a certain knowledge state corresponds to each admissible value of the rank. With the help of resumption of the lower segments of an inference we obtain an inference that corresponds to the smaller rank, i.e. to the previous moment of the cognitive process. This way we can study the history of this process.

For each possible value of the rank that is not greater than the rank of the inference in question the result of the above-mentioned resumption will be also an inference. This inference has a current state description and a corresponding syntactic model. The sequence of such syntactic models (sequence of knowledge states) is considered as a model of our inference.

The main task of the present section is to show that precisely the same sequence can be obtained in a different way, too. Namely, we can generate an L-structure, where L is an FPJ logic, by using a certain structure generator. We show that *to any pure inference* in a fixed modification calculus *corresponds one and the same sequence* of L-structures which can be considered not only as a model of a concrete inference but also as a model of the modification calculus itself.

Definition 14.1.4. Let L be an FPJ logic, σ be a signature, $\mathscr{M} \in \mathrm{MC}_\sigma(L)$, $\Gamma \in \mathrm{AFS}(\mathscr{M})$ and $\vec{\mathfrak{a}}$ be an $\mathscr{M}(s,m,\mathbf{D})$- inference from Γ. Let $\langle k,l\rangle \leq \langle s,m\rangle$. By definition put

$$\vec{\mathfrak{a}}_{\overleftarrow{k,l}} = \mathrm{Resume}\left(\mathrm{Lo}\left(\vec{\mathfrak{a}},k,l\right)\right).$$

Proposition 14.1.5. Let L be an FPJ logic, σ be a signature, $\mathscr{M} \in \mathrm{MC}_\sigma(L)$, $\Gamma \in \mathrm{AFS}(\mathscr{M})$ and $\vec{\mathfrak{a}}$ be an $\mathscr{M}(s,m,\mathbf{D})$-inference from Γ, $\langle k,l\rangle \leq \langle s,m\rangle$. Then $\vec{\mathfrak{a}}_{\overleftarrow{k,l}}$ is an $\mathscr{M}(\overline{\Delta}_\sigma(k,l),\mathbf{D})$-inference from Γ. (Here $\mathscr{M}(\overline{\Delta}_\sigma(k,l),\mathbf{D})$ denotes the $\mathscr{M}(k',l',\mathbf{D})$ such that $\langle k',l'\rangle = \overline{\Delta}_\sigma(k,l)$.)

Proof. This statement directly follows from Definition 14.1.4 and the theorem on (k,l)-section (Corollary 11.2.30). □

Lemma 14.1.6. Let L be an FPJ logic, σ be a signature, $\mathscr{M} \in \mathrm{MC}_\sigma(L)$, $\Gamma \in \mathrm{AFS}(\mathscr{M})$ and $\vec{\mathfrak{a}}$ be an $\mathscr{M}(s,m,\mathbf{D})$-inference from Γ. Let $0 \leq k < s$. Then

$$\vec{\mathfrak{a}}_{\overleftarrow{k,\mathrm{Q}_{\mathscr{M}}}} = \vec{\mathfrak{a}}_{\overleftarrow{k+1,0}}.$$

Proof. This lemma straightforwardly follows from Lemma 11.2.21. □

Definition 14.1.7. Let L be an FPJ logic, σ be a signature, $\mathscr{M} \in \mathrm{MC}_\sigma(L)$, $\Gamma \in \mathrm{AFS}(\mathscr{M})$ and $\vec{\mathfrak{a}}$ be an $\mathscr{M}(s,m,\mathbf{D})$-inference from Γ. Let $\langle k,l\rangle \leq \langle s,m\rangle$. Then let by definition:

$$\mathrm{Sint}_L(\vec{\mathfrak{a}},k,l) = \mathrm{Sint}_L\left(\mathrm{CSD}\left(\vec{\mathfrak{a}}_{\overleftarrow{k,l}},\mathrm{Univ}_{\mathscr{M}}\right)\right).$$

Lemma 14.1.8. Let L be an FPJ logic, σ be a signature, $\mathscr{M} \in \mathrm{MC}_\sigma(L)$, $\Gamma \in \mathrm{AFS}(\mathscr{M})$ and $\vec{\mathfrak{a}}$ be an $\mathscr{M}(s,m,\mathbf{D})$-inference from Γ, $\langle k,l\rangle \leq \langle s,m\rangle$. Assume that the set $\mathrm{Cur}\left(\Gamma,\vec{\mathfrak{a}}_{\overleftarrow{k,l}}\right)$ is L-consistent. Then $\mathrm{Sint}_L(\vec{\mathfrak{a}},k,l)$ is an L-model of $\mathrm{FSet}\left(\vec{\mathfrak{a}}_{\overleftarrow{k,l}}\right)$.

Proof. This statement straightforwardly follows from Lemma 14.1.2 and Definition 14.1.7. □

Proposition 14.1.9. Let L be an FPJ logic, σ be a signature, $\mathscr{M} \in \mathrm{MC}_\sigma(L)$ and $\vec{\mathfrak{a}}$ be an $\mathscr{M}(s,m,\mathbf{D})$-inference, where $s > 0$, $m > 0$. Then for any $\vec{\mathfrak{a}} \in \mathrm{Dom}_{\mathscr{M}}\left(\mathrm{P}_m^{\mathscr{M}}\right)$ there exists a unique $\varepsilon \in \mathscr{V}(L,\mathsf{I})$ such that the record $\left\langle \mathrm{J}_\varepsilon \mathrm{P}_m^{\mathscr{M}}(\vec{\mathfrak{a}}),r,s,m,0,0\right\rangle$ occurs in $\vec{\mathfrak{a}}$, where $\mathrm{ToR}(r)$ is equal to either Conf or MP.

Proof. By Definition 10.2.31 one can get into the regime $\mathbf{D}$ if and only if

$$\mathrm{Dom}_{\mathscr{M}}\left(\mathrm{P}_m^{\mathscr{M}},\vec{\mathfrak{a}},s\right) = \mathrm{Dom}_{\mathscr{M}}\left(\mathrm{P}_m^{\mathscr{M}}\right).$$

I.e. at the stage s for all elements of $\mathrm{Dom}_{\mathscr{M}}\left(\mathrm{P}_m^{\mathscr{M}}\right)$ either new values should be obtained or the old values should be confirmed.

The uiqueness follows from the fact that the major premises of the modification rules are disjoint. □

The following definition establishes a relation between modification calculi and structure generators. The meaning of this definition is as follows: structure generators are compatible with modification calculi provided that, from the current state description, the generator conditions can be derived if and only if an inference can be continued so that it provides the results of the generating rule.

Definition 14.1.10. Let L be an FPJ logic, σ be a signature, $\mathscr{M} \in \mathrm{AMC}_\sigma(L)$, $\Gamma \in \mathrm{FS}_\sigma(L, \mathsf{E})$ and $\mathfrak{R} \in \mathrm{SG}_\sigma(L, \Gamma)$. We say that $\mathfrak{R}$ is *$\mathscr{M}$-compatible*, if the following conditions hold:

(i) $\mathring{\mathfrak{R}} = \mathrm{Sint}_L(\mathscr{M})$;
(ii) $\mathrm{Ord}_{\mathfrak{R}} = \mathrm{Ord}_{\mathscr{M}}$;
(iii) For any $s \in \omega$, $m \in [1..\mathrm{Q}_{\mathscr{M}}]$, $\vec{\mathfrak{a}} \in \mathrm{Dom}_{\mathscr{M}}\left(\mathrm{P}_m^{\mathscr{M}}\right)$, $\varepsilon \in \mathscr{V}(L, \mathsf{I})$ and any pure $\mathscr{M}(s, m, \mathbf{D})$-inference $\vec{\mathfrak{a}}$, if

$$\mathrm{CSD}\left(\vec{\mathfrak{a}}_{\overleftarrow{s,m-1}}\right) \vdash_L \Gamma$$

then

$$\mathrm{J}_\varepsilon \, \mathrm{P}_m^{\mathscr{M}}(\vec{\mathfrak{a}}) \in \mathrm{CSD}\left(\vec{\mathfrak{a}}_{\overleftarrow{s,m}}\right) \quad \text{iff} \quad \mathrm{CSD}\left(\vec{\mathfrak{a}}_{\overleftarrow{s,m-1}}\right) \vdash_L \mathrm{GC}\left[\mathfrak{R}^{\triangleright}, \mathrm{P}_m^{\mathscr{M}}, \varepsilon\right](\vec{\mathfrak{a}}).$$

By definition let:

$$\mathrm{SG}_\sigma(L, \Gamma, \mathscr{M}) = \{\mathfrak{R} \in \mathrm{SG}_\sigma(L, \Gamma) \mid \mathfrak{R} \text{ is } \mathscr{M}\text{-compatible}\},$$
$$\mathrm{SG}_\sigma(L, \mathscr{M}) = \{\mathfrak{R} \in \mathrm{SG}_\sigma(L) \mid \mathfrak{R} \text{ is } \mathscr{M}\text{-compatible}\}.$$

The next theorem shows that, with the help of compatible structure generators, we can define a model of the pure inference in a modification calculus (we consider a certain L-structure sequence as a model). Moreover, this theorem shows that this model does not depend on the choice of the inference. It only depends on the modification calculus, i.e. it can be considered the very model of the modification calculus itself.

Theorem 14.1.11. Let L be an FPJ logic, σ be a signature, $\mathscr{M} \in \mathrm{MC}_\sigma(L)$, $\mathfrak{R} \in \mathrm{SG}_\sigma(L, \mathscr{M})$ and $\vec{\mathfrak{a}}$ be an arbitrary pure $\mathscr{M}(\overline{\Delta}_\sigma(s,m), \mathbf{D})$-inference. Then for any pair $\langle k, l\rangle \leq \langle s, m\rangle$,

$$\mathrm{Sint}_L(\vec{\mathfrak{a}}, k, l) = \vec{\mathfrak{R}}[k, l].$$

Proof. By 14.1.5, $\vec{\mathfrak{a}}_{\overleftarrow{k,l}}$ is a pure $\mathscr{M}(\overline{\Delta}_\sigma(k,l), \mathbf{D})$-inference. Continue the proof by induction on k and l.

BASIS. Let $k = 0$. Then

$$\begin{aligned}
\mathrm{Sint}_L(\vec{\mathfrak{a}},0,l) &= \mathrm{Sint}_L\left(\mathrm{CSD}\left(\overleftarrow{\mathfrak{a}_{0,l}}\right),\mathscr{M}\right) && \text{(by 14.1.7)}\\
&= \mathrm{Sint}_L(\mathscr{M}) && \text{(by 11.3.15)}\\
&= \vec{\mathfrak{R}}[0,l] && \text{(by 12.2.11)}
\end{aligned}$$

INDUCTION STEP. Suppose that the theorem holds for $k=n$. Let us prove it for $k=n+1$. The proof is by induction on l.

Basis. Let $l=0$. Then

$$\begin{aligned}
\mathrm{Sint}_L(\vec{\mathfrak{a}},n+1,0) &= \mathrm{Sint}_L\left(\mathrm{CSD}\left(\overleftarrow{\mathfrak{a}_{n+1,0}}\right),\mathscr{M}\right) && \text{(by 14.1.7)}\\
&= \mathrm{Sint}_L\left(\mathrm{CSD}\left(\overleftarrow{\mathfrak{a}_{n,\mathrm{Q}_{\mathscr{M}}}},\mathscr{M}\right)\right) && \text{(by 14.1.6)}\\
&= \mathrm{Sint}_L(\vec{\mathfrak{a}},n,\mathrm{Q}_{\mathscr{M}}) && \text{(by 14.1.7)}\\
&= \vec{\mathfrak{R}}[n,\mathrm{Q}_{\mathscr{M}}] && \text{(by induction hypothesis)}\\
&= \vec{\mathfrak{R}}[n+1,0] && \text{(by 12.2.11)}
\end{aligned}$$

Induction step. Suppose the theorem has already been proved for $l<\mathrm{Q}_{\mathscr{M}}$. Now let us prove it for $l+1$. Let:

(i) $\mathfrak{K}=\vec{\mathfrak{R}}[n+1,l]$,
(ii) $\mathfrak{L}=\vec{\mathfrak{R}}[n+1,l+1]$,
(iii) $\mathfrak{M}=\mathrm{Sint}_L(\vec{\mathfrak{a}},n+1,l+1)$.

By the inductive assumption: $\mathfrak{K}=\mathrm{Sint}_L(\vec{\mathfrak{a}},n+1,l)$. Then by 14.1.7 and (iii) we get:

$$\mathfrak{K}=\mathrm{Sint}_L\left(\mathrm{CSD}\left(\overleftarrow{\mathfrak{a}_{n+1,l}}\right),\mathscr{M}\right) \tag{14.1}$$

$$\mathfrak{M}=\mathrm{Sint}_L\left(\mathrm{CSD}\left(\overleftarrow{\mathfrak{a}_{n+1,l+1}}\right),\mathscr{M}\right). \tag{14.2}$$

By (i) and Lemma 12.2.12, $\mathfrak{K}\vDash_L\Gamma$. Hence by Lemma 14.1.1 and (12.1),

$$\mathrm{CSD}\left(\overleftarrow{\mathfrak{a}_{n+1,l}}\right)\vdash_L\Gamma. \tag{14.3}$$

To complete the induction step we have to prove that $\mathfrak{M}=\mathfrak{L}$. Note that

$$\mathrm{Frame}_{\mathfrak{M}}=\mathrm{Frame}_{\mathfrak{K}}=\mathrm{Frame}_{\mathfrak{L}}=\mathrm{Frame}_{\sigma}(\mathscr{M}).$$

Moreover, it is trivial (see 12.2.11 and the corresponding definitions and lemmas).

(i) $f^{\mathfrak{M}}=f^{\mathfrak{K}}=f^{\mathfrak{L}}$ for any $f\in\mathrm{Fu}_{\sigma}$,
(ii) $q^{\mathfrak{M}}=q^{\mathfrak{K}}=q^{\mathfrak{L}}$ for any $q\in\mathrm{Pr}_{\sigma}(\mathsf{E})$,
(iii) $p^{\mathfrak{M}}=p^{\mathfrak{K}}=p^{\mathfrak{L}}$ for any $p\in\mathrm{Pr}_{\sigma}(\mathsf{I})$ such that $p\neq\mathrm{P}^{\mathfrak{R}}_{l+1}$.

It remains to prove that $\left(\mathrm{P}_{l+1}^{\mathfrak{R}}\right)^{\mathfrak{M}} = \left(\mathrm{P}_{l+1}^{\mathfrak{R}}\right)^{\mathfrak{L}}$. Let $\vec{a} \in \mathrm{Dom}_{\mathscr{M}}\left(\mathrm{P}_{l+1}^{\mathfrak{R}}\right)$, $\varepsilon \in \mathscr{D}^{\tau}(L)$. Then

$$
\begin{aligned}
&\left(\mathrm{P}_{l+1}^{\mathfrak{R}}\right)^{\mathfrak{M}}(\vec{a}) = \varepsilon \\
&\quad \text{iff} \quad \mathrm{J}_{\varepsilon}\,\mathrm{P}_{l+1}^{\mathfrak{R}}(\vec{a}) \in \mathrm{CSD}\left(\vec{\mathfrak{a}}_{\overleftarrow{n+1,l+1}}\right) && \text{(by 10.1.16)} \\
&\quad \text{iff} \quad \mathrm{CSD}\left(\vec{\mathfrak{a}}_{\overleftarrow{n+1,l}}\right) \vdash_L \mathrm{GC}\left[\mathfrak{R}^{\rhd}, \mathrm{P}_{l+1}^{\mathfrak{R}}, \varepsilon\right](\vec{a}) && \text{(by 14.1.10 and (14.3))} \\
&\quad \text{iff} \quad \mathfrak{K} \vDash_L \mathrm{GC}\left[\mathfrak{R}^{\rhd}, \mathrm{P}_{l+1}^{\mathfrak{R}}, \varepsilon\right](\vec{a}) && \text{(by 14.1.1)} \\
&\quad \text{iff} \quad \left(\mathrm{P}_{l+1}^{\mathfrak{R}}\right)^{\mathfrak{L}}(\vec{a}) = \varepsilon. && \text{(by 12.2.14)}
\end{aligned}
$$

The above arguments are enough to claim that $\left(\mathrm{P}_{l+1}^{\mathfrak{R}}\right)^{\mathfrak{M}} = \left(\mathrm{P}_{l+1}^{\mathfrak{R}}\right)^{\mathfrak{L}}$. Then $\mathfrak{M} = \mathfrak{L}$, i.e. $\mathrm{Sint}_L(\vec{\mathfrak{a}}, n+1, l+1) = \overrightarrow{\mathfrak{R}}[n+1, l+1]$. The induction step is concluded. □

The above theorem provides a method of defining the semantics for the modification calculus $\mathscr{M}$ in the case when there exists an $\mathscr{M}$-compatible structure generator. However, we would prefer to have a general, efficient method for transforming modification rules into generating rules. We have no intention to define a universal function that transforms any modification rule into a generating rule, but below we provide conditions that should be satisfied by a function that transforms modification rules into generating rules if it is defined on a set of rules of some MRS.

Obviously by using such a rule transformation function we can present a structure generator $\mathfrak{R}$ that will correspond to each modification calculus M. Particularly interesting is the case when the obtained structure generator $\mathfrak{R}$ is $\mathscr{M}$-compatible . The modification calculus $\mathscr{M}$ will be called *conformable* if to it a corresponding $\mathscr{M}$-compatible structure generator can be found by a rule transformation function that we will call conforming function.

Specific classes of conformable modification calculi will be considered in the next section.

Definition 14.1.12. Let L be an FPJ logic, σ be a signature, $\Gamma \in \mathrm{FS}_{\sigma}(L, \mathsf{E})$ and $R \in \mathrm{MRS}_{\sigma}(L)$. Denote by G_{σ} the set of all generating rule of signature σ. Suppose that g is a map of the set

$$\{R[p, \varepsilon] \mid p \in \mathrm{Pr}_{\sigma}(\mathsf{I}), \varepsilon \in \mathscr{V}(L, \mathsf{I})\}$$

to the set G_{σ} such that the following conditions hold:

(i) g is a computable (recursive) function,
(ii) For any $\varepsilon \in \mathscr{V}(L, \mathsf{I})$, the g-image of a modification ε-rule is a generating ε-rule,
(iii) For any $p \in \mathrm{Pr}_{\sigma}(\mathsf{I})$, $g(R[p])$ is a a complete generating set (CGS) for p w.r.t. Γ.

Then the map g is called a *conforming function for* R w.r.t. Γ.

Proposition 14.1.13. Let L be an FPJ logic, σ be a signature, $\Gamma \in \mathrm{FS}_\sigma(L, \mathsf{E})$, $R \in \mathrm{MRS}_\sigma(L)$ and g be a conforming function for R w.r.t. Γ. Then the collection

$$\{g(R[p]) \mid p \in \mathrm{Pr}_\sigma(\mathsf{I})\}$$

is a generating rule system of signature σ w.r.t. Γ in L.

Proof. By Definition 12.2.5 it is sufficient to prove that for any $p \in \mathrm{Pr}_\sigma(\mathsf{I})$, $g(R[p])$ is a CGS for p w.r.t. Γ. But this is valid by Definition 14.1.12(iii). □

Notation 14.1.14. Let L be an FPJ logic, σ be a signature, $\Gamma \in \mathrm{FS}_\sigma(L, \mathsf{E})$, $R \in \mathrm{MRS}_\sigma(L)$ and g be a conforming function for R w.r.t. Γ. Then the generating rule system

$$\{g(R[p]) \mid p \in \mathrm{Pr}_\sigma(\mathsf{I})\}$$

is denoted by $g(R)$.

For any $p \in \mathrm{Pr}_\sigma(\mathsf{I})$ and $\varepsilon \in \mathscr{V}(L, \mathsf{I})$, the generating ε-rule $g(R[p, \varepsilon])$ is denoted by $g(R)[p, \varepsilon]$.

Similarly for any $p \in \mathrm{Pr}_\sigma(\mathsf{I})$, the set $g(R[p])$ is denoted by $g(R)[p]$.

For any $p \in \mathrm{Pr}_\sigma(\mathsf{I})$ and $\varepsilon \in \mathscr{V}(L, \mathsf{I})$, the generating condition of $g(R)[p, \varepsilon]$ is denoted by $\mathrm{GC}[g(R), p, \varepsilon]$, and the result of $g(R)[p, \varepsilon]$ is denoted by $\mathrm{Res}[g(R), p, \varepsilon]$.

Definition 14.1.15. Let L be an FPJ logic, σ be a signature, $\Gamma \in \mathrm{FS}_\sigma(L, \mathsf{E})$, $R \in \mathrm{MRS}_\sigma(L)$ and g be a conforming function for R w.r.t. Γ. We say that g is *Γ-robust* if $g(R) \in \mathrm{RGRS}_\sigma(L, \Gamma)$.

Notation 14.1.16. Let L be an FPJ logic, σ be a signature, $\Gamma \in \mathrm{FS}_\sigma(L, \mathsf{E})$, $\mathscr{M} \in \mathrm{MC}_\sigma(L)$ and g be a conforming function for $\mathrm{MR}_{\mathscr{M}}$ w.r.t. Γ. Then for any $p \in \mathrm{Pr}_\sigma(\mathsf{I})$ and $\varepsilon \in \mathscr{V}(L, \mathsf{I})$, the generating ε-rule $g(\mathrm{MR}_{\mathscr{M}})[p, \varepsilon]$ is denoted by $g(\mathscr{M})[p, \varepsilon]$.

Similarly for any $p \in \mathrm{Pr}_\sigma(\mathsf{I})$, the set $g(\mathrm{MR}_{\mathscr{M}})[p]$ is denoted by $g(\mathscr{M})[p]$.

For any $p \in \mathrm{Pr}_\sigma(\mathsf{I})$ and $\varepsilon \in \mathscr{V}(L, \mathsf{I})$:

- The generating condition of $g(\mathscr{M})[p, \varepsilon]$ is denoted by $\mathrm{GC}[g(\mathscr{M}), p, \varepsilon]$,
- The result of $g(\mathscr{M})[p, \varepsilon]$ is denoted by $\mathrm{Res}[g(\mathscr{M}), p, \varepsilon]$.

Definition 14.1.17. Let L be an FPJ logic, σ be a signature, $\Gamma \in \mathrm{FS}_\sigma(L, \mathsf{E})$, $\mathscr{M} \in \mathrm{MC}_\sigma(L)$ and g be a Γ-robust conforming function for $\mathrm{MR}_{\mathscr{M}}$, $\mathrm{Sint}_L(\mathscr{M}) \vDash_L \Gamma$. Then the ordered triple

$$\langle \mathrm{Sint}_L(\mathscr{M}), g(\mathrm{MR}_{\mathscr{M}}), \mathrm{Ord}_{\mathscr{M}} \rangle$$

is called the *g-image* of $\mathscr{M}$ and is denoted by $g(\mathscr{M})$.

Proposition 14.1.18. Let L be an FPJ logic, σ be a signature, $\Gamma \in \mathrm{FS}_\sigma(L, \mathsf{E})$, $\mathscr{M} \in \mathrm{MC}_\sigma(L)$,

$$\mathrm{Sint}_L(\mathscr{M}) \vDash_L \Gamma$$

and g be a Γ-robust conforming function for $\mathrm{MR}_{\mathscr{M}}$. Then the triple $g(\mathscr{M})$ is a structure generator of signature σ ($\mathrm{SG}_\sigma()$) w.r.t. Γ.

Proof. $\mathrm{Sint}_L(\mathscr{M})$ is an L-structure (see Definition 10.1.16, Notation 10.1.19 and Definition 10.2.6). By the assumption, $\mathrm{Sint}_L(\mathscr{M}) \models_L \Gamma$. By Proposition 14.1.13 and Definition 14.1.15, $g(\mathrm{MR}_{\mathscr{M}}) \in \mathrm{RGRS}_\sigma(L,\Gamma)$. By Definition 10.2.6, $\mathrm{Ord}_{\mathscr{M}}$ is a bijection of $\mathrm{Pr}_\sigma(\mathsf{I})$ onto $[1..\mathrm{Q}_{\mathscr{M}}]$. Therefore by Definition 12.2.10 the triple $g(\mathscr{M}) \in \mathrm{SG}_\sigma(L,\Gamma)$. □

Definition 14.1.19. Let L be an FPJ logic, σ be a signature, $\Gamma \in \mathrm{FS}_\sigma(L,\mathsf{E})$, $\mathscr{M} \in \mathrm{MC}_\sigma(L)$ and g be a Γ-robust conforming function for $\mathrm{MR}_{\mathscr{M}}$ such that

$$g(\mathscr{M}) \in \mathrm{SG}_\sigma(L,\Gamma,\mathscr{M}).$$

Then g is called *Γ-$\mathscr{M}$-compatible*. We say that g is *$\mathscr{M}$-compatible*, if g is Γ-$\mathscr{M}$-compatible for some Γ.

Definition 14.1.20. Let L be an FPJ logic, σ be a signature, $\Gamma \in \mathrm{FS}_\sigma(L,\mathsf{E})$ and $\mathscr{M} \in \mathrm{MC}_\sigma(L)$. We say that $\mathscr{M}$ is *Γ-conformable* if there exists a Γ-$\mathscr{M}$-compatible conforming function for $\mathrm{MR}_{\mathscr{M}}$. We say that $\mathscr{M}$ is *conformable* if $\mathscr{M}$ is Γ-conformable for some Γ.

By definition let:

$$\mathrm{CMC}_\sigma(L,\Gamma) = \{\mathscr{M} \in \mathrm{MC}_\sigma(L) \mid \mathscr{M} \text{ is } \Gamma\text{-conformable}\},$$
$$\mathrm{CMC}_\sigma(L) = \{\mathscr{M} \in \mathrm{MC}_\sigma(L) \mid \mathscr{M} \text{ is conformable}\}.$$

Proposition 14.1.21. Let L be an FPJ logic, σ be a signature, $\mathscr{M} \in \mathrm{CMC}_\sigma(L)$ and $\vec{\mathfrak{a}}$ be an arbitrary pure $\mathscr{M}(\overline{\Delta}_\sigma(s,m),\mathbf{D})$-inference. Suppose $\Gamma \in \mathrm{FS}_\sigma(L,\mathsf{E})$, g is an Γ-$\mathscr{M}$-compatible conforming function for $\mathrm{MR}_{\mathscr{M}}$. Let $\mathfrak{R} = g(\mathscr{M})$. Then for any pair $\langle k,l\rangle \leq \langle s,m\rangle$,

$$\mathrm{Sint}_L(\vec{\mathfrak{a}},k,l) = \overrightarrow{\mathfrak{R}}[k,l].$$

Proof. This statement directly follows from Definitions 14.1.19 and 14.1.20 and Theorem 14.1.11. □

14.2 Modification Theories

Theorem 14.1.11 states that to any two $\mathscr{M}$-inferences of the same rank corresponds one and the same syntactical L-structure, which is the model of the current state descriptions of both $\mathscr{M}$-inferences. It is easy to show that, in the case of these $\mathscr{M}$-inferences, even the current state descriptions will be identical, though the inferences themselves may differ from each other.

The inference is a way of justifying a certain statement. The so-justified statements are considered valid. It is natural to require that *the validity of a statement should be independent from the way of its justification.* In the first-order theories the fact of provability of a formula does not depend on the mode of its proof. We would like to learn how to construct a theory on the basis of modification calculi that satisfies this requirement. We hope to do this with the help of Theorem 14.1.11.

The current section is directly devoted to the theories built up on the basis of modification calculi. As stated above, a modification calculus can be considered as a special type of inference. A theory over modification calculi can be considered as their extension, much like first-order theories are regarded as extensions of first-order calculi.

However, adding non-logical axioms to a first-order calculus provides a possibility to prove such statements that could be never proved within the first-order calculus itself. The situation with modification calculi is different. Each modification calculus contains a state description that represents the initial situation in such detail that the additional axioms (if they do not contradict this description) do not allow us to obtain any additional consequences.

Then why do we need additional axioms? Because they allow us to obtain consequences faster than without them. This is especially important in the case of computer realisation of an inference in the modification structures. Additional axioms may contain general statements valid for all elements of a given sort. We could prove these general statements by using state descriptions. However, to achieve this we have to prove special cases of *each general statement for each element of the necessary sort s* and then, in order to get a general statement, apply the following formula from the state description

$$\forall x \left(\bigvee_{a \in \mathrm{Dom}_{\sigma}(s)} x = a \right).$$

Note that the domains of all sorts in the state descriptions are finite but not necessarily small.

Using plausible rules in the modification calculi makes the case even more intriguing. Plausible rules do not guarantee consistency. Moreover, even if the initial state description was consistent and the additional axioms did not contradict it, the inference in the modification calculi might result in such a stage where contradictions may appear. However, if we do not continue the inference (do not use contradictional statements and neglect them) then contradiction may disappear in subsequent stages.

Such a situation is close to the usual research practice in those areas which cannot be called “exact science”. In these disciplines a noticed contradiction will not stop the research process. The contradictory results are not used for a while and, sooner or later, a moment comes when the contradiction will be overcome.

It must also be noted that an inference (including an inference in the modification calculi) can be considered as a means of justifying a certain statement. It can also be considered as a path to this statement. In traditional first-order theories no such path can lead us to a dead end, i.e. any inference can be extended as far as the statement we need. The situation also differs in the case of inferences in modification calculus, which is due to unreliable rules of plausible inference.

Some inferences in modification calculi may turn out to be dead ends. Therefore, we have to narrow down the set of inferences in order to connect derivability with appropriate semantics. Below the notion of $\mathscr{M}$-correct inference will be defined. Informally, $\mathscr{M}$-correct inferences are those paths which do not end up in a dead end.

Definition 14.2.1. Let L be an FPJ logic, σ be a signature, $\mathscr{M} \in \mathrm{MC}_\sigma(L)$, $\Gamma \in \mathrm{AFS}(\mathscr{M})$ and $\vec{\mathfrak{a}}$ be an $\mathscr{M}(s,m,\mathbf{D})$-inference from Γ. $\vec{\mathfrak{a}}$ is called *$\mathscr{M}$-correct* if there exists a pure $\mathscr{M}(s,m,\mathbf{D})$-inference $\vec{\mathfrak{b}}$ such that $\mathrm{JA}_\sigma{}^+(\vec{\mathfrak{a}},\mathscr{M}) = \mathrm{JA}_\sigma{}^+\left(\vec{\mathfrak{b}},\mathscr{M}\right)$.

Lemma 14.2.2. Let L be an FPJ logic, σ be a signature, $\mathscr{M} \in \mathrm{MC}_\sigma(L)$, $\Gamma \in \mathrm{AFS}(\mathscr{M})$, $\vec{\mathfrak{a}}$ be an $\mathscr{M}$-correct $\mathscr{M}(s,m,\mathbf{D})$-inference from Γ and $\vec{\mathfrak{b}}$ be a pure $\mathscr{M}(s,m,\mathbf{D})$-inference such that

$$\mathrm{JA}_\sigma{}^+(\vec{\mathfrak{a}},\mathscr{M}) = \mathrm{JA}_\sigma{}^+\left(\vec{\mathfrak{b}},\mathscr{M}\right).$$

Then $\mathrm{CSD}(\vec{\mathfrak{a}}) = \mathrm{CSD}\left(\vec{\mathfrak{b}}\right)$.

Proof. It is obvious (see Definition 11.3.15). □

Corollary 14.2.3. Let L be an FPJ logic, σ be a signature, $\mathscr{M} \in \mathrm{CMC}_\sigma(L)$, $\vec{\mathfrak{a}}$ be an $\mathscr{M}$-correct $\mathscr{M}(s,m,\mathbf{D})$-inference from Γ such that $\Gamma \in \mathrm{AFS}(\mathscr{M})$, g be an $\mathscr{M}$-compatible conforming function for $\mathrm{MR}_{\mathscr{M}}$ and $\mathfrak{R} = g(\mathscr{M})$. Assume that $\mathrm{Cur}(\Gamma,\vec{\mathfrak{a}})$ is L-consistent. Then $\overrightarrow{\mathfrak{R}}[s,m]$ is an L-model of the set of formulas $\mathrm{FSet}(\vec{\mathfrak{a}})$.

Proof. Obviously, $\vec{\mathfrak{a}}_{\overleftarrow{s,m}} = \vec{\mathfrak{a}}$. Then by Lemma 14.1.8, $\mathrm{Sint}_L(\vec{\mathfrak{a}},s,m)$ is an L-model of the set of formulas $\mathrm{FSet}(\vec{\mathfrak{a}})$. Show that

$$\mathrm{Sint}_L(\vec{\mathfrak{a}},s,m) = \overrightarrow{\mathfrak{R}}[s,m].$$

Let $\vec{\mathfrak{b}}$ be a pure $\mathscr{M}(s,m,\mathbf{D})$-inference such that $\mathrm{JA}_\sigma{}^+(\vec{\mathfrak{a}},\mathscr{M}) = \mathrm{JA}_\sigma{}^+\left(\vec{\mathfrak{b}},\mathscr{M}\right)$. Then

$$\begin{aligned}
\mathrm{Sint}_L(\vec{\mathfrak{a}},s,m) &= \mathrm{Sint}_L(\mathrm{CSD}(\vec{\mathfrak{a}}),\mathscr{M}) && \text{(by 14.1.7)}\\
&= \mathrm{Sint}_L\left(\mathrm{CSD}\left(\vec{\mathfrak{b}}\right),\mathscr{M}\right) && \text{(by 14.1.11)}\\
&= \overrightarrow{\mathfrak{R}}[s,m] && \text{(by 14.1.21)}
\end{aligned}$$

□

We have already, albeit non-formally, used the notion of derivability in modification calculi. It was supposed that if there is an inference then there should also be derivability. However, until now we have not defined what derivability is, so now we intend to do so. Since we have $\mathscr{M}$-inference on the one hand and $\mathscr{M}$-correct $\mathscr{M}$-inference on the other, we need to introduce two notions of derivability that correspond to these two types of inference. We will also study the relationship between these two types of derivability. Besides that we are going to introduce the usual notions connected to derivability, first of all the notions of consistency and inconsistency.

Definition 14.2.4. Let L be an FPJ logic, σ be a signature, $\mathscr{M} \in \mathrm{MC}_\sigma(L)$, $\Gamma \in \mathrm{AFS}(\mathscr{M})$, $\langle s,m\rangle \in \omega \times [0..\mathrm{Q}_{\mathscr{M}}]$. An external L-formula Φ is called:

- *$\mathscr{M}(s,m)$-derivable from Γ* (written: $\Gamma \vdash^{\mathscr{M}}_{s,m} \Phi$), if there exists an $\mathscr{M}(s,m,\mathbf{D})$-inference $\vec{\mathfrak{a}}$ from Γ such that $\Phi \in \mathrm{FSet}(\vec{\mathfrak{a}})$ and
- *Strictly $\mathscr{M}(s,m)$-derivable from Γ* (written: $\Gamma \Vdash^{\mathscr{M}}_{s,m} \Phi$), if there exists an $\mathscr{M}$-correct $\mathscr{M}(s,m,\mathbf{D})$-inference $\vec{\mathfrak{a}}$ from Γ such that $\Phi \in \mathrm{FSet}(\vec{\mathfrak{a}})$.

Proposition 14.2.5. Let L be an FPJ logic, σ be a signature, $\mathscr{M} \in \mathrm{MC}_\sigma(L)$, $\Gamma \in \mathrm{AFS}(\mathscr{M})$, $\langle s,m\rangle \in \omega \times [0..\mathrm{Q}_{\mathscr{M}}]$ and Φ be an external L-formula. Then

$$\Gamma \Vdash^{\mathscr{M}}_{s,m} \Phi \quad \text{implies} \quad \Gamma \vdash^{\mathscr{M}}_{s,m} \Phi.$$

Proof. It is obvious (see Definition 14.2.1). □

Lemma 14.2.6. Let L be an FPJ logic, σ be a signature, $\mathscr{M} \in \mathrm{MC}_\sigma(L)$, $\Gamma \in \mathrm{AFS}(\mathscr{M})$ and $\langle s,m\rangle \in \omega \times [0..\mathrm{Q}_{\mathscr{M}}]$ and Φ be an external L-formula. Then:

(i) If $\Gamma \vdash^{\mathscr{M}}_{s,m} \Phi$ then there exists an $\mathscr{M}(s,m,\mathbf{D})$-inference $\vec{\mathfrak{a}}$ from Γ such that $\mathrm{Cur}(\Gamma,\vec{\mathfrak{a}}) \vdash^{\Phi}_{L}$,
(ii) If $\Gamma \Vdash^{\mathscr{M}}_{s,m} \Phi$ then there exists an $\mathscr{M}$-correct $\mathscr{M}(s,m,\mathbf{D})$-inference $\vec{\mathfrak{a}}$ from Γ such that $\mathrm{Cur}(\Gamma,\vec{\mathfrak{a}}) \vdash^{\Phi}_{L}$.

Proof. (i) By Definition 14.2.4 there exists an $\mathscr{M}(s,m,\mathbf{D})$-inference $\vec{\mathfrak{a}}$ from Γ such that $\Phi \in \mathrm{FSet}(\vec{\mathfrak{a}})$. Then by 11.4.12, $\mathrm{Cur}(\Gamma,\vec{\mathfrak{a}}) \vdash^{\Phi}_{L}$, which was required.
(ii) It is similar. □

Lemma 14.2.7. Let L be an FPJ logic, σ be a signature, $\mathscr{M} \in \mathrm{MC}_\sigma(L)$, $\Gamma \in \mathrm{AFS}(\mathscr{M})$ and $\vec{\mathfrak{a}}$ be an $\mathscr{M}(s,m,\mathbf{D})$-inference from Γ. Suppose $\mathrm{Cur}(\Gamma,\vec{\mathfrak{a}}) \vdash_L \Phi$. Then:

(i) $\Gamma \vdash^{\mathscr{M}}_{s,m} \Phi$;
(ii) $\Gamma \Vdash^{\mathscr{M}}_{s,m} \Phi$, if $\vec{\mathfrak{a}}$ is $\mathscr{M}$-correct.

Proof. By 11.4.17, there exists an $\mathscr{M}(s,m,\mathbf{D})$-inference $\vec{\mathfrak{b}}$ from Γ such that

$$\mathrm{JA}_\sigma{}^{+}(\vec{\mathfrak{a}},\mathscr{M}) = \mathrm{JA}_\sigma{}^{+}\left(\vec{\mathfrak{b}},M\right), \tag{14.4}$$

$$\Phi \in \mathrm{FSet}\left(\vec{\mathfrak{b}}\right).$$

If $\vec{\mathfrak{a}}$ is an $\mathscr{M}$-correct $\mathscr{M}(s,m,\mathbf{D})$-inference from Γ then there exists an $\mathscr{M}(s,m,\mathbf{D})$-inference $\vec{\mathfrak{c}}$ from $\mathrm{Descr}_{\mathscr{M}}$ such that

$$\mathrm{JA}_\sigma{}^{+}(\vec{\mathfrak{a}},\mathscr{M}) = \mathrm{JA}_\sigma{}^{+}(\vec{\mathfrak{c}},\mathscr{M}). \tag{14.5}$$

Then by 11.4.17, there exists an $\mathscr{M}(s,m,\mathbf{D})$-inference $\vec{\mathfrak{d}}$ from $\Delta = \mathrm{Descr}_{\mathscr{M}}$ such that

$$\mathrm{JA}_\sigma{}^{+}(\vec{\mathfrak{c}},\mathscr{M}) = \mathrm{JA}_\sigma{}^{+}(\vec{\mathfrak{d}},\mathscr{M}). \tag{14.6}$$

It follows from (14.4)–(14.6) that

$$\mathrm{JA}_\sigma{}^+\left(\vec{\mathfrak{b}},\mathscr{M}\right)=\mathrm{JA}_\sigma{}^+\left(\vec{\mathfrak{d}},\mathscr{M}\right).$$

This means that the $\mathscr{M}(s,m,\mathbf{D})$-inference $\vec{\mathfrak{b}}$ is $\mathscr{M}$-correct.

Then by Definition 14.2.4, $\Gamma\vdash^{\mathscr{M}}_{s,m}\Phi$. If $\vec{\mathfrak{a}}$ is $\mathscr{M}$-correct, then $\Gamma\Vdash^{\mathscr{M}}_{s,m}\Phi$. □

Definition 14.2.8. Let L be an FPJ logic, σ be a signature, $\mathscr{M}\in\mathrm{MC}_\sigma(L)$, $\Gamma\in\mathrm{AFS}(\mathscr{M})$, s be a stage index and m be a module index. The set Γ is called $\mathscr{M}(s,m)$-*inconsistent*, if there exists an external L-formula Φ such that $\Gamma\vdash^{\mathscr{M}}_{s,m}\Phi$ and $\Gamma\vdash^{\mathscr{M}}_{s,m}\neg\Phi$. Otherwise the set Γ is called $\mathscr{M}(s,m)$-*consistent.*

The set Γ is called *strictly* $\mathscr{M}(s,m)$-*inconsistent*, if there exists an external L-formula Φ such that $\Gamma\Vdash^{\mathscr{M}}_{s,m}\Phi$ and $\Gamma\Vdash^{\mathscr{M}}_{s,m}\neg\Phi$. Otherwise the set Γ is called *strictly* $\mathscr{M}(s,m)$-*consistent.*

Proposition 14.2.9. Let L be an FPJ logic, σ be a signature, $\mathscr{M}\in\mathrm{MC}_\sigma(L)$, $\langle s,m\rangle\in\omega\times[0..\mathrm{Q}_{\mathscr{M}}]$, $\Gamma\in\mathrm{AFS}(\mathscr{M})$. Then:

(i) If Γ is strictly $\mathscr{M}(s,m)$-inconsistent then Γ is $\mathscr{M}(s,m)$-inconsistent;
(ii) If Γ is $\mathscr{M}(s,m)$-consistent, then Γ is strictly $\mathscr{M}(s,m)$-consistent.

Proof. It is evident. □

Definition 14.2.10. Let L be an FPJ logic, σ be a signature, $\mathscr{M}\in\mathrm{MC}_\sigma(L)$, $\Gamma\in\mathrm{AFS}(\mathscr{M})$. The set Γ is called *absolutely* $\mathscr{M}$-*consistent*, if it is $\mathscr{M}(s,m)$-*consistent* for any s and m.

The set Γ is called *strictly absolutely* $\mathscr{M}$-*consistent*, if it is *strictly* $\mathscr{M}(s,m)$-*consistent* for any s and m.

In the proofs of the above statements we did not use any additional restrictions on the modification calculi. As for the following two statements they are valid only for the *comformable* modification calculi.

Lemma 14.2.11. Let L be an FPJ logic, σ be a signature, $\mathscr{M}\in\mathrm{CMC}_\sigma(L)$. Then the set $\mathrm{Descr}_{\mathscr{M}}$ is absolutely $\mathscr{M}$-consistent.

Proof. Suppose the converse. Then for some s and m there exist $\mathscr{M}(s,m,\mathbf{D})$-inferences $\vec{\mathfrak{a}}$ and $\vec{\mathfrak{b}}$ such that $\Phi\in\mathrm{FSet}(\vec{\mathfrak{a}})$ and $\neg\Phi\in\mathrm{FSet}\left(\vec{\mathfrak{b}}\right)$. By Lemma 14.1.8, $\mathrm{Sint}_L(\vec{\mathfrak{a}},s,m)$ is an L-model of the set of formulas $\mathrm{FSet}(\vec{\mathfrak{a}}_{s,m}\leftarrow)=\mathrm{FSet}(\vec{\mathfrak{a}})$. Similarly, $\mathrm{Sint}_L\left(\vec{\mathfrak{b}},s,m\right)$ is an L-model of the set $\mathrm{FSet}\left(\vec{\mathfrak{b}}\right)$. Let g be an $\mathscr{M}$-compatible conforming function for $\mathrm{MR}_{\mathscr{M}}$. Then by Proposition 14.1.21,

$$\mathrm{Sint}_L(\vec{\mathfrak{a}},s,m)=\vec{\mathfrak{R}}[s,m]=\mathrm{Sint}_L\left(\vec{\mathfrak{b}},s,m\right),$$

where $\mathfrak{R}=g(\mathscr{M})$. Therefore there exists an L-structure which is an L-model of both set $\mathrm{FSet}(\vec{\mathfrak{a}})$ and set $\mathrm{FSet}\left(\vec{\mathfrak{b}}\right)$. Then simultaneously it is an L-model of both formulas Φ and $\neg\Phi$, which is impossible. □

Proposition 14.2.12. Let L be an FPJ logic, σ be a signature, $\mathscr{M} \in \mathrm{CMC}_\sigma(L)$, $\Gamma \in \mathrm{AFS}(\mathscr{M})$, g be an $\mathscr{M}$-compatible conforming function for $\mathrm{MR}_{\mathscr{M}}$ and $\mathfrak{R} = g(\mathscr{M})$. Suppose that Γ is strictly $\mathscr{M}(s,m)$-consistent. Let $\vec{\mathfrak{a}}$ be an arbitrary $\mathscr{M}$-correct $\mathscr{M}(s,m,\mathbf{D})$-inference from Γ. Then the L-structure $\overrightarrow{\mathfrak{R}}[s,m]$ is an L-model of $\mathrm{FSet}(\vec{\mathfrak{a}})$.

Proof. Prove that $\mathrm{Cur}(\Gamma,\vec{\mathfrak{a}})$ is L-consistent. Suppose the opposite. Then there exists an L-formula Φ of signature σ such that

$$\mathrm{Cur}(\Gamma,\vec{\mathfrak{a}}) \vdash_L \Phi \quad \text{and} \quad \mathrm{Cur}(\Gamma,\vec{\mathfrak{a}}) \vdash_L \neg\Phi.$$

Then by Lemma 14.2.7, $\Gamma \Vdash^{\mathscr{M}}_{s,m} \Phi$ and $\Gamma \Vdash^{\mathscr{M}}_{s,m} \neg\Phi$. Hence $\mathrm{Cur}(\Gamma,\vec{\mathfrak{a}})$ is strictly $\mathscr{M}(s,m)$-inconsistent. Thus our assumption is not true. Hence $\mathrm{Cur}(\Gamma,\vec{\mathfrak{a}})$ is L-consistent. Then by Corollary 14.2.3, $\overrightarrow{\mathfrak{R}}[s,m]$ is an L-model of the set of formulas $\mathrm{FSet}(\vec{\mathfrak{a}})$. □

It is now time to define the staged cognitive theory (SCT), which is the main notion of this present section. We hope that the staged cognitive theory has the same relation to modification calculi as the first-order theory has to the first-order predicate calculus. The SCT can be considered as a system of conventions considered in the cognitive processes. This system of conventions may define the structure, the order and the way of application of the plausible reasoning rules as well as those initial conditions that should be satisfied by all objects of the subject domain. These conditions can be considered as preliminary knowledge.

Below we define some notions related to SCT, in particular the notions of derivability and consistency, and prove a few easy statements about SCT.

Definition 14.2.13. Let L be an FPJ logic, σ be a signature, $\mathscr{M} \in \mathrm{MC}_\sigma(L)$, $\Gamma \in \mathrm{AFS}(\mathscr{M})$. Then the ordered pair $Q = \langle \Gamma, \mathscr{M} \rangle$ is said to be a *stage cognition theory of signature* σ in L.

The set of all state cognition theory of signature σ in L is denoted by $\mathrm{SCT}_\sigma(L)$

Definition 14.2.14. Let L be an FPJ logic, σ be a signature,

$$Q = \langle \Gamma, \mathscr{M} \rangle \in \mathrm{SCT}_\sigma(L).$$

The set Γ is called the *set of non-logical axioms* of Q and is denoted by Ax_Q.

The modification calculus $\mathscr{M}$ is said to be the *modification calculus* of Q and is denoted by MC_Q.

The system of modification rules $\mathrm{MR}_{\mathscr{M}}$ is called the *modification rule system* of Q and is denoted by MR_Q.

The set $\mathrm{Univ}_{\mathscr{M}}$ is called the *universe* of Q and is denoted by Univ_Q.

The set $\mathrm{Descr}_{\mathscr{M}}$ is called the *initial state description* of Q and is denoted by Descr_Q.

The mapping $\mathrm{Ord}_{\mathscr{M}}$ is called the *order mapping* of Q and is denoted by Ord_Q.

The number $\mathrm{Q}_{\mathscr{M}}$ is called the *most index of an active internal predicate* of Q and is denoted by Q_Q.

The internal predicate simbol $\mathrm{P}_i^{\mathscr{M}}$ is denoted by P_i^Q.

Definition 14.2.15. Let L be an FPJ logic and σ be a signature. A $Q \in \mathrm{SCT}_\sigma(L)$ is called *conformable* if $\mathrm{MC}_Q \in \mathrm{CMC}_\sigma(L)$.

By definition let:

$$\mathrm{CSCT}_\sigma(L) = \{Q \in \mathrm{SCT}_\sigma(L) \mid Q \text{ is conformable}\}.$$

Definition 14.2.16. Let L be an FPJ logic, σ be a signature,

$$Q = \langle \Gamma, \mathscr{M} \rangle \in \mathrm{SCT}_\sigma(L),$$

Φ be an external L-formula, $s, m \in \omega$ and $m \leq \mathrm{Q}_Q$. Φ is called *provable in Q at stage s in module m ($Q(s,m)$-provable*; (written: $\Vdash^Q_{s,m} \Phi$) if $\Gamma \Vdash^{\mathscr{M}}_{s,m} \Phi$, i.e. Φ is strictly $\mathscr{M}(s,m)$-derivable from Ax_Q.

Definition 14.2.17. Let L be an FPJ logic, σ be a signature,

$$Q = \langle \Gamma, \mathscr{M} \rangle \in \mathrm{SCT}_\sigma(L),$$

$s, m \in \omega$, $m \leq \mathrm{Q}_Q$. Q is said to be *inconsistent at stage s in module m* ((s,m)-inconsistent) if the set Ax_Q is $\mathscr{M}(s,m)$-inconsistent, i.e. there exists an external L-formula Φ such that $\Vdash^Q_{s,m} \Phi$ and $\Vdash^Q_{s,m} \neg\Phi$. In the opposite case Q is said to be *consistent at stage s in module m ((s,m)-consistent)*.

Proposition 14.2.18. Let L be an FPJ logic, σ be a signature,

$$Q = \langle \Gamma, \mathscr{M} \rangle \in \mathrm{SCT}_\sigma(L),$$

$s, m \in \omega$, $m \leq \mathrm{Q}_Q$. Then:

- Q is (s,m)-consistent if the set Ax_Q is strictly $\mathscr{M}(s,m)$-consistent,
- Q is (s,m)-inconsistent if the set Ax_Q is strictly $\mathscr{M}(s,m)$-inconsistent.

Proof. See Definitions 14.2.17, 14.2.16 and 14.2.8. □

Definition 14.2.19. Let L be an FPJ logic and σ be a signature. A $Q \in \mathrm{SCT}_\sigma(L)$ is said to be *absolutely consistent* if it is (s,m)-consistent for any s and m ($s, m \in \omega, m \leq \mathrm{Q}_Q$).

Proposition 14.2.20. Let L be an FPJ logic and σ be a signature. A $Q \in \mathrm{SCT}_\sigma(L)$ is absolutely consistent if and only if the set Ax_Q is absolutely $\mathscr{M}$-consistent, where $\mathscr{M} = \mathrm{MC}_Q$.

Proof. See Definitions 14.2.10 and 14.2.19 and Proposition 14.2.18. □

Definition 14.2.21. Let L be an FPJ logic, σ be a signature, $Q \in \mathrm{SCT}_\sigma(L)$. Suppose

$$\xi = \langle \mathfrak{J}_{nk} \mid n \in \omega, 0 \leq k \leq \mathrm{Q}_Q \rangle$$

is a sequence of L-structures with double indexing over the same U-frame.

(i) The sequence ξ is called an (s,m)*-model* of Q (where $s,m \in \omega$, $m \leq \mathrm{Q}_Q$) if, for any external L-formula Φ of signature σ,

$$\Vdash^{Q}_{s,m} \Phi \quad \text{implies} \quad \mathfrak{J}_{sm} \vDash_L \Phi.$$

(ii) The sequence ξ is called an *absolute model* of Q if for any external L-formula Φ, stage s and module m $(0 \leq m \leq \mathrm{Q}_Q)$,

$$\Vdash^{Q}_{s,m} \Phi \quad \text{implies} \quad \mathfrak{J}_{sm} \vDash_L \Phi.$$

Remark 14.2.22. Let L be an FPJ logic and σ be a signature. A sequence of L-structures

$$\xi = \langle \mathfrak{J}_{nk} \mid n \in \omega, 0 \leq k \leq \mathrm{Q}_Q \rangle$$

is an absolute model of a $Q \in \mathrm{SCT}_\sigma(L)$ if and only if ξ is an (s,m)-model of Q for any $s,m \in \omega\,(m \leq \mathrm{Q}_Q)$.

Proposition 14.2.23. Let L be an FPJ logic, σ be a signature, $Q \in \mathrm{SCT}_\sigma(L)$, $s,m \in \omega$ $(m \leq \mathrm{Q}_Q)$. If Q has an (s,m)-model then it is (s,m)-consistent.

Proof. Suppose the converse. Let $\xi = \langle \mathfrak{J}_{nk} \mid n \in \omega, 0 \leq k \leq K \rangle$ be a sequence of L-structures with double indexing over the same U-frame. Let ξ be an (s,m)-model of the (s,m)-inconsistent Q. Then for some external L-formula Φ,

$$\Vdash^{Q}_{s,m} \Phi \quad \text{and} \quad \Vdash^{Q}_{s,m} \neg\Phi.$$

Then by Definition 14.2.21, $\mathfrak{J}_{sm} \vDash_L \Phi$ and $\mathfrak{J}_{sm} \vDash_L \neg\Phi$, which is impossible. □

Proposition 14.2.24. Let L be an FPJ logic, σ be a signature, $Q \in \mathrm{SCT}_\sigma(L)$. If Q has an absolute model than it is absolutely consistent.

Proof. Suppose that ξ is an absolute model of Q. Then ξ is an (s,m)-model of Q for any $s,m \in \omega$ $(m \leq \mathrm{Q}_Q)$ (see Definition 14.2.21). In this case by Proposition 14.2.23, Q is (s,m)-consistent for any $s,m \in \omega$ $(m \leq \mathrm{Q}_Q)$. Then by Definition 14.2.19, Q is absolutely consistent. □

The above statements are valid for any SCT. In contrast to this some of the following statements will be valid only for conformable SCT. We will prove them by the use of Proposition 14.1.21, which is valid only for conformable modification calculi or for their derivatives.

The following statement can be considered as a correctness theorem for SCT. It states that a formula which is provable in a consistent theory will be valid in the standard model. Since all models for a modification calculus are isomorphic, this statement fully corresponds to the usual perception of a correctness theorem. Moreover, we have two versions of the correctness theorem: one is for (s,m)-consistent SCT and the other one is for absolutely consistent SCT.

Proposition 14.2.25. Let L be an FPJ logic, σ be a signature, $s,m \in \omega$, $m \leq \mathrm{Q}_Q$,

$$Q = \langle \Gamma, \mathscr{M} \rangle \in \mathrm{CSCT}_\sigma(L)$$

be (s,m)-consistent. Suppose g is an $\mathscr{M}$-compatible conforming function for $\mathrm{MR}_{\mathscr{M}}$, $\mathfrak{R} = g(\mathscr{M})$. Then the sequence of L-structures $\overrightarrow{\mathfrak{R}}$ is an (s,m)-model of Q.

Proof. Let $\Vdash^{Q}_{s,m} \Phi$. Then $\Gamma \vdash^{\mathscr{M}}_{s,m} \Phi$ (see Definition 14.2.16). Therefore there exists an $\mathscr{M}$-correct $\mathscr{M}(s,m,\mathbf{D})$-inference $\vec{\mathfrak{a}}$ from Γ such that $\Phi \in \mathrm{FSet}(\vec{\mathfrak{a}})$ (see Definition 14.2.4). By Proposition 14.2.18, the set Γ is strictly $\mathscr{M}(s,m)$-consistent. Then by Proposition 14.2.12 the L-structure $\overrightarrow{\mathfrak{R}}[s,m]$ is an L-model of the set $\mathrm{FSet}(\vec{\mathfrak{a}})$. Therefore, $\overrightarrow{\mathfrak{R}}[s,m] \vDash_L \Phi$. Hence by Definition 14.2.21, the sequence of L-structures $\overrightarrow{\mathfrak{R}}$ is an (s,m)-model of Q. □

Proposition 14.2.26. Let L be an FPJ logic, σ be a signature, $Q \in \mathrm{CSCT}_\sigma(L)$ be absolutely consistent, $\mathscr{M} = \mathrm{MC}_Q$. Suppose g is an $\mathscr{M}$-compatible conforming function for $\mathrm{MR}_{\mathscr{M}}$, $\mathfrak{R} = g(\mathscr{M})$. Then the sequence of L-structures $\overrightarrow{\mathfrak{R}}$ is an absolute model of the theory Q.

Proof. By Definition 14.2.19, Q is (s,m)-consistent for any s and m ($s,m \in \omega$, $m \leq \mathrm{Q}_Q$). Then by Proposition 14.2.25, $\overrightarrow{\mathfrak{R}}$ is an $(s,\ m)$-model of Q for any s and m. By Remark 14.2.22, $\overrightarrow{\mathfrak{R}}$ is an absolute model of Q. □

All models of a modification calculus are isomorphic, and the validity in one model is equivalent to the validity in all models. Therefore, the following statement can be considered as a completeness theorem for SCT.

Proposition 14.2.27. Let L be an FPJ logic, σ be a signature,

$$Q = \langle \Gamma, \mathscr{M} \rangle \in \mathrm{CSCT}_\sigma(L),$$

$s,m \in \omega$, $m \leq \mathrm{Q}_Q$, Suppose g is an $\mathscr{M}$-compatible conforming function for $\mathrm{MR}_{\mathscr{M}}$, $\mathfrak{R} = g(\mathscr{M})$. Then for any external L-formula Φ,

$$\overrightarrow{\mathfrak{R}}[s,m] \vDash_L \Phi \quad \text{implies} \quad \Vdash^{Q}_{s,m} \Phi.$$

Proof. Let $\overrightarrow{\mathfrak{R}}[s,m] \vDash_L \Phi$. Let $\vec{\mathfrak{a}}$ be a pure $\mathscr{M}(s,m,\mathbf{D})$-inference. Then $\vec{\mathfrak{a}} = \vec{\mathfrak{a}}_{\overleftarrow{s,m}}$. By Theorem 14.1.11

$$\overrightarrow{\mathfrak{R}}[s,m] = \mathrm{Sint}_L(\vec{\mathfrak{a}}, s, m).$$

Therefore, $\mathrm{Sint}_L(\vec{\mathfrak{a}}, s, m) \vDash_L \Phi$. But

$$\begin{aligned} \mathrm{Sint}_L(\vec{\mathfrak{a}}, s, m) &= \mathrm{Sint}_L\left(\mathrm{CSD}\left(\vec{\mathfrak{a}}_{\overleftarrow{s,m}}\right), D\right) && \text{(by 14.1.7)} \\ &= \mathrm{Sint}_L(\mathrm{CSD}(\vec{\mathfrak{a}}), D) && \text{(since } \vec{\mathfrak{a}}_{\overleftarrow{s,m}} = \vec{\mathfrak{a}}\text{)} \end{aligned}$$

Then by Remark 14.1.1, $\mathrm{CSD}(\vec{\mathfrak{a}}) \vdash_L \Phi$. In this case by Corollary 11.4.17, there exists a pure $\mathscr{M}(s,m,\mathbf{D})$-inference $\vec{\mathfrak{b}}$ such that $\Phi \in \mathrm{FSet}\left(\vec{\mathfrak{b}}\right)$. However since

$\mathrm{Descr}_{\mathscr{M}} \subseteq \Gamma$, then each pure $\mathscr{M}(s,m,\mathbf{D})$-inference is obviously an $\mathscr{M}$-correct $\mathscr{M}(s,m,\mathbf{D})$-inference from Γ. Then $\Gamma \Vdash^{\mathscr{M}}_{s,m} \Phi$. Hence $\Vdash^{Q}_{s,m} \Phi$, as we wished to prove. □

Above we have repeatedly spoken about the so called temporal contradiction. Now we will consider some general arguments on this question, and following that an example will be provided.

This example is a continuation of the already considered examples with Mick the robot, which use the simplest FPJ logic and modification calculi in adapting to the varying conditions of the environment.

It is obvious that from the formal point of view the conclusion in modification calculus is non-monotonic because the structure of such an inference allows us not only to add to it the conclusions of the rules, but also to modify some of the premises. From the point of view of formal semantics this circumstance is connected to the possibility to revise the truth values of the internal atomic formulas.

However, from the point of view of substance non-monotonicity is reduced to the replacement of the truth value "uncertain" (that corresponds to "ignorance") by one of the certainty truth values which will not be revised any further. I.e. the valuation of a statement declared once as true cannot be turned to doubted and then believed to be false within the paradigm accepted in this book.

We assume that the further development of the inference technique proposed in this book permits to build calculi completely non-monotonic from both formal and substantive points of view. However, already there are formal tools with the help of which we can demonstrate various effects of non-monotonicity (though through some artificial examples).

Possibility of appearance, and then disappearance of contradiction is an expected consequence of the non-monotonic nature of a modification inference. Modification theories can be temporarily inconsistent. For example, a modification theory Q can be $(s,m-1)$-consistent, (s,m)-inconsistent, but $(s,m+1)$-consistent, i.e. the contradiction not present at the initial stage can appear and then disappear.

From the informal point of view it is possible to explain the phenomenon of temporal contradiction with (i) the combination of applications of both plausible and reliable (deductive) rules in a modification inference, (ii) the discrete nature of modification inference.

Only deductive rules cannot result contradiction if the initial data (the initial state description) are consistent. Also within the limits of our paradigm plausible rules themselves cannot result contradiction as we never r entirely rely on plausible rules. With their help we obtain statements with truth values always with a certain amount of uncertainty. Contradiction arises when we are equally confident in some statement and in its negation.

If we combine applications of plausible and deductive (reliable) rules, then the occurrence of contradictions is not unexpected because of considering the conclusions of plausible rules (which actually are only hypotheses) as premises of deductive rules (i.e. in the same manner as the strictly proved statements).

It is possible to illustrate the influence of the discrete nature of a modification inference on contradictions with the following example. Let the modification theory

Q contain a non-logical axiom of the form $\Phi \to \Psi$, where Φ and Ψ are J-atomic formulas. In the initial state Φ and Ψ are false, so the axiom $\Phi \to \Psi$ is true. By means of combination of plausible and reliable rules we can obtain such a situation where Φ (the premise of the considered axiom) already becomes true, Ψ (the consequence) will still remain false.

Due to the discrete nature of inference we have no possibility to establish the validity of both formulas in one module. This effect of delay results the occurrence of a contradiction. A further construction of inference may yield in avoiding this contradiction.

Example 14.2.28. This example will develop the example from Sect. 5.2 (the story about Mick the robot). The various aspects of the example from Sect. 5.2 were stated formally in Examples 8.4.7, 10.1.11, 10.1.17, 10.2.7 and 10.2.35.

Take the modification calculus $\mathscr{M}$ from Example 10.2.7. Let

$$\Gamma = \{!\mathrm{F}(\mathrm{D7},\mathrm{PC},\mathrm{S}) \wedge !\mathrm{A}(\mathrm{D7},\mathrm{PC}) \to !\mathrm{A}(\mathrm{D7},\mathrm{S})\}, \qquad Q = \langle \Gamma, \mathscr{M} \rangle .$$

In Table 14.1 we wrote the proof in Q ($\mathscr{M}$-inference of Γ) that illustrates the phenomenon of temporal contradiction. The non-logical axiom of the modification theory Q is in Row 10 of Table 14.1. In explicit form the contradiction appears in Stage 1 and Modul 2 (see row 18). This contradiction disappears at Modul 3 of Stage 1 after deactivation of row 8 and row 18 (row 18 deductively depends on row 8).

It easy to prove that Q is (s,m)-consistent for $s = 0$ and $m = 0,1,2,3$ and for $s = 1$ and $m = 0,1,3$, but Q is $(1,2)$-inconsistent.

The single non-logical axiom of Q seems to be valid from the viewpoint of common sense. The informal meaning of this axiom is as follows:

Let:

- On day D7 within a short time interval after the event PC there occurs the event S;
- On D7 there occurs PC.

Then:

- On D7 there occurs S".

Note that at the initial stage (Stage 0) the discussed axiom is also valid from the viewpoint of formal semantics. Indeed, the initial **Si**-structure is defined by the initial state desctiption, in particular, by the axiom $?\mathrm{F}(\mathrm{D7},\mathrm{PC},\mathrm{S})$, which is in Row 7 of Table 14.1. Then the atomic formula $\mathrm{F}(\mathrm{D7},\mathrm{PC},\mathrm{S})$ takes the value '**?**'. In this case the antecedent $!\mathrm{F}(\mathrm{D7},\mathrm{PC},\mathrm{S})$ of the axiom from Row 10 of Table 14.1 is false. Therefore, this axiom is true in the initial **Si**-structure.

However, the further reasoning leads to the modification of the formula from Row 7. This formula is replaced with the formula $!\mathrm{F}(\mathrm{D7},\mathrm{PC},\mathrm{S})$ from Row 16. Formally, we add Row 16 to the inference from Table 14.1 and deactivate the formula

Table 14.1 Temporal contradiction

#	Formula	Reason	Stage	Module	DN
1	$\mathrm{D1} \neq \mathrm{D4}$	OAx	0	0	
2	$\mathrm{D2} \neq \mathrm{D5}$	OAx	0	0	
3	$?\mathrm{C}(\mathrm{PC},\mathrm{S})$	DAx	0	1	13
4	$?\mathrm{C}(\mathrm{LM},\mathrm{RT})$	DAx	0	1	14
5	$!\mathrm{F}(\mathrm{D1},\mathrm{PC},\mathrm{S}), !\mathrm{F}(\mathrm{D4},\mathrm{PC},\mathrm{S})$	DAx	0	2	
6	$!\mathrm{F}(\mathrm{D2},\mathrm{LM},\mathrm{RT}), !\mathrm{F}(\mathrm{D5},\mathrm{LM},\mathrm{RT})$	DAx	0	2	
7	$?\mathrm{F}(\mathrm{D7},\mathrm{PC},\mathrm{S})$	DAx	0	2	16
8	$!\mathrm{A}(\mathrm{D7},\mathrm{PC})$	DAx	0	3	
9	$?\mathrm{A}(\mathrm{D7},\mathrm{S})$	DAx	0	3	19
10	$!\mathrm{F}(\mathrm{D7},\mathrm{PC},\mathrm{S}) \wedge !\mathrm{A}(\mathrm{D7},\mathrm{PC}) \rightarrow !\mathrm{A}(\mathrm{D7},\mathrm{S})$	Q Ax	0	3	
11	$\exists d_1 \exists d_2 (d_1 \neq d_2 \wedge !\mathrm{F}(d_1,\mathrm{PC},\mathrm{S}) \wedge !\mathrm{F}(d_2,\mathrm{PC},\mathrm{S}))$	Ded (1,5)	1	0	
12	$\exists d_1 \exists d_2 (d_1 \neq d_2 \wedge !\mathrm{F}(d_1,\mathrm{LM},\mathrm{RT}) \wedge !\mathrm{F}(d_2,\mathrm{LM},\mathrm{RT}))$	Ded (2,6)	1	0	
13	$!\mathrm{C}(\mathrm{PC},\mathrm{S})$	MR [C,!] (3,11)	1	1	
14	$!\mathrm{C}(\mathrm{LM},\mathrm{RT})$	MR [C,!] (4,12)	1	1	
15	$!\mathrm{A}(\mathrm{D7},\mathrm{PC}) \wedge !\mathrm{C}(\mathrm{PC},\mathrm{S})$	Ded (8,13)	1	1	
16	$!\mathrm{F}(\mathrm{D7},\mathrm{PC},\mathrm{S})$	MR [F,!] (7,15)	1	2	
17	$!\mathrm{A}(\mathrm{D7},\mathrm{S})$	Ded (8,10,16)	1	2	
18	$?\mathrm{A}(\mathrm{D7},\mathrm{S}) \wedge !\mathrm{A}(\mathrm{D7},\mathrm{S})$	Ded (9,17)	1	2	19
19	$!\mathrm{A}(\mathrm{D7},\mathrm{S})$	MR [A,!] (9,15,)	1	3	

from Row 7. Then we can apply the deductive rules to the axiom from Row 10, get the consequence of this axiom and place this consequence in Row 17.

We immediately get the contradiction. This contradiction is expressed explicitly in Row 18. The formula from Row 18 asserts that the internal atomic formula $\mathrm{A}(\mathrm{D7},\mathrm{S})$ is simultaneously "certain" (**!**) and "uncertain" (**?**), which is impossible.

But further we apply the modification rule to the Row 9. We replace the formula $?\mathrm{A}(\mathrm{D7},\mathrm{S})$ from Row 9 with the formula $!\mathrm{A}(\mathrm{D7},\mathrm{S})$ from Row 19. Formally we add Row 19 to the discussed inference and deactivate all the formulas that deductively depend on the formula from Row 9.

Note that as a result of this modification we deactivate all the formulas that generate the contradiction. This restores the consistency.

Let us stress that Rows 17 and 19 contain the identical formulas, but the addition of Row !7 generates the contradiction whereas the addition of Row 19 eliminates this contradiction and restores the consistency.

To explain this circumstance, note that we obtain Row 17 by the deductive rules; therefore we can only add this row but we cannot modify anything in our inference when we add this row. Thereagainst Row 19 is obtained by the modification rule and we add this row and deactivate some above rows at the same time. Namely, this deactivation restores the consistency.

Note that we get the contradiction just with the deductive rules. This can be explained by the fact that in modification inferences the deductive rules are applied in combination with plausible rules. When we apply a deductive rule, we expect that all the premises of this rule are reliable, but some of these premises can be merely plausible. (They can be obtained by plausible rules.)

This circumstance can lead to contradictions. But these contradictions are not fatal. They can be eliminated with further reasoning.

The phenomenon of temporal contradiction can also be explained by the discrete nature of modification inference. The contradiction can appear when one already has modified the antecedent (but not yet consequence) of some implication. Therefore there may be a stage and module of an inference, where the discussed implication can be false. In this case one can obtain the ordinary contradiction in the form of $\varphi \wedge \neg\varphi$.

At the next stage or module one can modify the consequence and deactivate records with false formulas. Thus the contradiction disappears.

Let us now study how SCT and I-theories are related, where I is an iterative version of an FPJ logic. The first-order iterative calculus is an advanced tool for representing the history of cognitive reasoning. It is more powerful than the modification calculi, because it allows proper work with infinite universes of various sorts as well.

One of our main objectives is to establish the connection between derivability in SCT and derivability in a certain iterative theory, which will be called the iterative image of SCT. This connection will be established in the last theorem of this chapter.

Definition 14.2.29. Let L be an FPJ logic, σ be a signature with a finite set of internal predicate symbols; $O\colon\ [1\,..\,\mathrm{Q}_O] \to \mathrm{Pr}_\sigma(\mathsf{I})$; $S \subseteq \omega \times [1\,..\,\mathrm{Q}_O]$. Let Γ be a set of external L-formulas. The set

$$\left\{\Phi^{\langle O,s,m\rangle} \mid \Phi \in \Gamma, \langle s,m\rangle \in S\right\}$$

is called the *$\langle O, S\rangle$-immersion* of Γ and is denoted by $\Gamma^{\langle O,S\rangle}$.

Definition 14.2.30. Let L be an FPJ logic, σ be a signature, $Q \in \mathrm{SCT}_\sigma(L)$. The set

$$\{\langle s,m\rangle \in \omega \times [1\,..\,\mathrm{Q}_Q] \mid Q \text{ is } (s,m)\text{-consistent}\}$$

is called the *consistency domain* of Q and is denoted by DoC_Q.

Definition 14.2.31. Assume that L is an FPJ logic and σ is a signature. Let

$$Q = \langle \Gamma, \mathscr{M}\rangle \in \mathrm{CSCT}_\sigma(L) \quad \text{and} \quad \Xi \in \mathrm{FS}_\sigma(L,\mathsf{E})\,.$$

Suppose g is an Ξ-robust $\mathscr{M}$-compatible conforming function for $\mathrm{MR}_{\mathscr{M}}$. The set of external I-formulas

$$(\Gamma \backslash \mathrm{JA}_\sigma(\mathscr{M}))^{\langle \mathrm{Ord}_{.\mathscr{M}}, \mathrm{DoC}_Q \rangle} \cup \Delta^{\mathrm{Init}} \cup \Xi^{\mathrm{Init}} \cup \mathrm{I}^{\mathrm{Ord}_{.\mathscr{M}}}(g(\mathrm{MR}_{.\mathscr{M}}))$$

is called the *iterative g-image* (*I-g-image*) of Q and is denoted by $\mathrm{I}^g(Q)$.

Lemma 14.2.32. Let L be an FPJ logic, σ be a signature, $Q = \langle \Gamma, \mathscr{M} \rangle \in \mathrm{SCT}_\sigma(L)$. Then $\Vdash^Q_{s,m} \Phi$ holds for any $\Phi \in \Gamma \backslash \mathrm{JA}_\sigma(\mathscr{M})$ and any $s \in \omega$, $m \in [1 \,..\, \mathrm{Q}_{.\mathscr{M}}]$.

Proof. It is obvious because we can add a record with any formula from the set $\Gamma \backslash \mathrm{JA}_\sigma(\mathscr{M})$ to an $\mathscr{M}$-inference at each stage and in each module (see 10.2.30 and 10.2.31). □

Theorem 14.2.33. Assume that L is an FPJ logic and σ is a signature. Let

$$Q = \langle \Gamma, \mathscr{M} \rangle \in \mathrm{CSCT}_\sigma(L) \quad \text{and} \quad \Xi \in \mathrm{FS}_\sigma(L, \mathsf{E}).$$

Suppose $\mathrm{Sint}_L(\mathscr{M}) \vDash_L \Xi$, g is a Ξ-robust $\mathscr{M}$-compatible conforming function for $\mathrm{MR}_{.\mathscr{M}}$, $\mathfrak{R} = g(\mathscr{M})$, $\mathfrak{K} = \mathrm{Imp}(\mathfrak{R})$. Then $\mathfrak{K}$ is an I-model of the set $\mathrm{I}^g(Q)$.

Proof. According to 14.2.31 to prove this theorem it is enough to show the validity of the following statements:

(i) $\mathfrak{K} \vDash_I (\Gamma \backslash \mathrm{JA}_\sigma(\mathscr{M}))^{\langle \mathrm{Ord}_{.\mathscr{M}}, \mathrm{DoC}_Q \rangle}$;
(ii) $\mathfrak{K} \vDash_I \mathrm{Descr}_{.\mathscr{M}}{}^{\mathrm{Init}}$;
(iii) $\mathfrak{K} \vDash_I \Xi^{\mathrm{Init}}$;
(iv) $\mathfrak{K} \vDash_I \mathrm{I}^{\mathrm{Ord}_{.\mathscr{M}}}(g(\mathrm{MR}_{.\mathscr{M}}))$.

Let us prove these statements.

(i) Suppose

$$\langle s, m \rangle \in \mathrm{DoC}_Q. \tag{14.7}$$

Then

$$\begin{aligned} \Phi \in \Gamma \backslash \mathrm{JA}_\sigma(\mathscr{M}) \quad &\text{implies} \quad \Vdash^Q_{s,m} \Phi && \text{(by 14.2.32)} \\ &\text{implies} \quad \vec{\mathfrak{R}}[s,m] \vDash_L \Phi && \text{(by (14.7) and 14.2.25)} \\ &\text{implies} \quad \mathfrak{K} \vDash_I \Phi^{\langle \mathrm{Ord}_{.\mathscr{M}}, s, m \rangle} && \text{(by 13.2.11)} \end{aligned}$$

Due to the arbitrariness of $\langle s, m \rangle \in \mathrm{DoC}_Q$ and $\Phi \in \Gamma \backslash \mathrm{JA}_\sigma(\mathscr{M})$, (i) holds.

(ii) By 10.1.6,

$$\mathrm{Sint}_L(\mathscr{M}) \vDash_L \mathrm{Descr}_{.\mathscr{M}}, \tag{14.8}$$

but by 12.2.11 (basis of induction),

$$\mathrm{Sint}_L(\mathscr{M}) = \vec{\mathfrak{R}}[0, m] \tag{14.9}$$

for any $m \in [1 \,..\, \mathrm{Q}_{.\mathscr{M}}]$.

Let v be an arbitrary valuation in $\mathrm{Frame}_{\mathfrak{K}}$, $m \in [1 \,..\, \mathrm{Q}_{.\mathscr{M}}]$. Then for any $\Phi \in \mathrm{Descr}_{.\mathscr{M}}$,

$$\left(\Phi^{\text{Init}}\right)^{\mathfrak{K},v} = \left(\Phi^{\langle \text{Ord}_{\mathscr{M}},0,m\rangle}\right)^{\mathfrak{K},v} \qquad \text{(by 13.2.13)}$$
$$= \Phi^{\overrightarrow{\mathfrak{R}}[0,m],v} \qquad \text{(by 13.2.11)}$$
$$= \Phi^{\text{Sint}_L(\mathscr{M}),v} \qquad \text{(by (14.9))}$$
$$= \mathbf{t} \qquad \text{(by (14.8))}$$

By the arbitrariness of Φ and v, the statement (ii) holds.
(iii) This statement can be proved similarly.
(iv) By 13.3.3, (iv) holds. □

Corollary 14.2.34. Let L be an FPJ logic, σ be a signature,

$$Q = \langle \Gamma, \mathscr{M} \rangle \in \text{CSCT}_\sigma(L)$$

and g be an $\mathscr{M}$-compatible conforming function for $\text{MR}_{\mathscr{M}}$. Then the set $\text{I}^g(Q)$ is I-consistent.

Theorem 14.2.35. Let L be an FPJ logic, σ be a signature,

$$Q = \langle \Gamma, \mathscr{M} \rangle \in \text{CSCT}_\sigma(L)$$

be an (s,m)-consistent and g be an $\mathscr{M}$-compatible conforming function for $\text{MR}_{\mathscr{M}}$. Then for any external L-formula Φ of signature σ,

$$\Vdash^Q_{s,m} \Phi \quad \text{iff} \quad \text{I}^g(Q) \vdash_I \Phi^{\langle \text{Ord}_{\mathscr{M}},s,m\rangle}.$$

Proof. Let $\mathfrak{R} = g(\mathscr{M})$, $\mathfrak{K} = \text{Imp}(\mathfrak{R})$ and Φ be an arbitrary external L-formula of signature σ. Then

$$\Vdash^Q_{s,m} \Phi \quad \text{iff} \quad \overrightarrow{\mathfrak{R}}[s,m] \vDash_L \Phi \qquad \text{(by 14.2.25 and 14.2.27)}$$
$$\text{iff} \quad \mathfrak{K} \vDash_I \Phi^{\langle \text{Ord}_{\mathscr{M}},s,m\rangle} \qquad \text{(by 13.2.11)}$$
$$\text{iff} \quad \text{I}^g(Q) \vdash_I \Phi^{\langle \text{Ord}_{\mathscr{M}},s,m\rangle} \qquad \text{(by 13.3.17 and 14.2.31)}$$

□

Chapter 15
Conformability

The imitation of cognitive reasoning will be based on the technique of modification calculi. In this section two classes of modification calculi will be introduced differing in special syntactical restrictions. We will study the calculi of these classes and show that any calculus of these classes is conformable and, consequently, correct and complete w.r.t. the semantics defined in the previous section.

15.1 Locality

We now define some notions that, despite their syntactical character, will play an important role in the definition of adequate semantics for the modification calculi. Among others the notion of *locality* will be defined and the basic properties of *local formulas* will be established. The notion of locality will be necessary to define some additional restrictions concerning the modification rule systems. It will be possible to define adequate semantics only for those modification calculi whose MRSs meet these restrictions.

Definition 15.1.1. Let L be a PJ logic, σ be a signature, $p \in \mathrm{Pr}_\sigma, \vec{t} \in \mathrm{Dom}_\sigma(p)$ and Φ be an external L-formula of signature σ. Φ is said to be *local* w.r.t. the atomic formula $p(\vec{t})$, if the following two conditions hold:

(i) Every occurrence of the symbol p in the L-formula Φ is a part of the occurrence of the atomic subformula $p(\vec{t})$ in this formula,
(ii) No occurrence of the atomic subformula $p(\vec{t})$ of Φ is in the scope of quantifiers over the variables that occur in this formula.

In order to catch the intuitive meaning of locality let us represent the predicate corresponding to the symbol p as a matrix, the cells of which are indexed by the arguments of this predicate (see Section 5.2 and Example 8.4.7). The content of each cell of the matrix represents a value of the predicate p. Locality of a formula means that its value depends only on one cell of this matrix. Namely it depends on the cell indexed by the tuple of values $\vec{t}$.

O. Anshakov, T. Gergely, *Cognitive Reasoning*, Cognitive Technologies,
DOI 10.1007/978-3-540-68875-4_15, © Springer-Verlag Berlin Heidelberg 2010

Now we give a recursive version of the definition of local formulas.

Definition 15.1.2. Let L be a PJ logic, σ be a signature, $p \in \mathrm{Pr}_\sigma, \vec{t} \in \mathrm{Dom}_\sigma(p)$. L-formulas of signature σ local w.r.t. atomic formula $p(\vec{t})$ can be recursively defined in the following way:

BASIS. The following L-formulas of signature σ are local w.r.t. $p(\vec{t})$:

(i) Atomic L-formulas of signature σ of the form $s_1 = s_2$, where $\mathrm{Sort}_\sigma(s_1) = \mathrm{Sort}_\sigma(s_2)$
(ii) Atomic L-formulas of signature σ of the form $q(\vec{s})$, where $q \in \mathrm{Pr}_\sigma$, $q \neq p$, $\vec{s} \in \mathrm{Dom}_\sigma(q)$
(iii) J-atomic L-formulas of signature σ of the form $\mathrm{J}_\alpha q(\vec{s})$, where $q \in \mathrm{Pr}_\sigma(\mathsf{I})$, $q \neq p$, $\vec{s} \in \mathrm{Dom}_\sigma(q)$, $\alpha \in \mathrm{JI}(L)$
(iv) Atomic formula $p(\vec{t})$
(v) J-atomic L-formulas of the form $\mathrm{J}_\alpha p(\vec{t})$, where $p \in \mathrm{Pr}_\sigma(\mathsf{I})$ and $\alpha \in \mathrm{JI}(L)$

INDUCTION STEP. Let Φ and Ψ be external L-formulas local w.r.t. $p(\vec{t})$, then $\neg\Phi$, $\Phi \to \Psi$ and $\forall x\Phi$, where x does not occur in $p(\vec{t})$, are also local w.r.t. $p(\vec{t})$.

Lemma 15.1.3. Let L be a classical or PJ logic and σ be a signature. Assume that all individual variables with free occurrence in the formula Φ are contained in the string $\vec{x}$, i.e. $\mathrm{FVar}(\Phi) \subseteq \mathrm{Set}(\vec{x})$. Let $\mathfrak{J}$ be an L-structure of signature σ and let v and u be two valuations of individual variables such that $v^*(\vec{x}) = u^*(\vec{x})$. Then

$$\Phi^{\mathfrak{J},v} = \Phi^{\mathfrak{J},u}.$$

Proof. It is similar to the proof of the corresponding statement of the first-order classical logic [Mendelson, 1997]. □

Notation 15.1.4. Let σ be a signature, L be a classical or PJ logic and Φ be an L-formula of signature σ such that $\mathrm{FVar}(\Phi) \subseteq \mathrm{Set}(\vec{x})$. Let $\mathfrak{J}$ be an L-structure of signature σ, $\vec{a} \in \mathrm{Dom}_{\mathfrak{J}}{}^*(\vec{x})$. The value of the formula Φ in the L-structure $\mathfrak{J}$ for any valuation v such that $v^*(\vec{x}) = \vec{a}$ is denoted by $\Phi^{\mathfrak{J}}(\vec{a})$. This notation is well-defined because of Lemma 15.1.3.

$\mathfrak{J} \models_L \Phi(\vec{a})$ will denote the fact that $\Phi^{\mathfrak{J},v} = \mathbf{t}$ for any valuation v of the individual variables such that $v^*(\vec{x}) = \vec{a}$.

This notation is also well defined because of Lemma 15.1.3.

Earlier we have mentioned that the intuitive meaning of locality is the following: an external formula local w.r.t. an internal atomic formula $p(\vec{a})$ does not depend on the cells of the matrix representing predicate p except for the cell indexed by $\vec{a}$. In the next lemma we show that such an independence does exist from the point of view of formalised semantics.

Lemma 15.1.5. Let L be an FPJ logic, σ be a signature, D be a σ-vocabulary, $\Delta \in \mathbf{SD}_\sigma(L,D)$, $p \in \mathrm{Pr}_\sigma(\mathsf{I})$ and $\vec{a} \in \mathrm{Dom}_\sigma(D,p)$. Assume:

(i) $\vec{b} \in \mathrm{Dom}_\sigma(D,p)$, $\vec{b} \neq \vec{a}$ and $\mathrm{J}_\varepsilon p\left(\vec{b}\right) \in \Delta$;

(ii) $\mathfrak{J} = \mathrm{Sint}_L(\Delta, D)$;
(iii) $\mathfrak{K} = \mathrm{Sint}_L\left(\Delta\left[\mathrm{J}_\varepsilon p\left(\vec{b}\right) \leftarrow J_\delta p\left(\vec{b}\right)\right], D\right)$, where $\delta \neq \varepsilon$.

Let Φ be a closed external L-formula, which is local w.r.t. the atomic formula $p(\vec{a})$. Then

$$\mathfrak{J} \models_L \Phi \quad \text{iff} \quad \mathfrak{K} \models_L \Phi.$$

Proof. Syntactical L-structures $\mathfrak{J}$ and $\mathfrak{K}$ only differ slightly. Both have the same U-frame $\mathrm{Frame}_\sigma(D)$. Creating these L-structures according to Definition 10.1.16 we can check that:

(a) $f^{\mathfrak{J}} = f^{\mathfrak{K}}$ for any $f \in \mathrm{Fu}_\sigma$;
(b) $q^{\mathfrak{J}} = q^{\mathfrak{K}}$ for any $q \in \mathrm{Pr}_\sigma(\mathsf{E})$;
(c) $r^{\mathfrak{J}} = r^{\mathfrak{K}}$ for any $r \in \mathrm{Pr}_\sigma(\mathsf{I})$ such that $r \neq p$;
(d) $p^{\mathfrak{K}}(\vec{c}) = \begin{cases} p^{\mathfrak{J}}(\vec{c}) & \text{if } \vec{c} \neq \vec{b}, \\ \delta & \text{if } \vec{c} = \vec{b}. \end{cases}$

Particularly

$$p^{\mathfrak{K}}(\vec{a}) = p^{\mathfrak{J}}(\vec{a}). \tag{15.1}$$

From (a) we can obtain by a trivial induction on terms that $t^{\mathfrak{J}} = t^{\mathfrak{K}}$ for any closed term t of signature σ. We continue the proof by induction on the height of the external closed L-formulas local w.r.t. $p(\vec{a})$.

BASIS. Clearly,

$$\mathfrak{J} \models_L \varphi \quad \text{iff} \quad \mathfrak{K} \models_L \varphi$$

in the case when φ is an external atomic formula or when φ is a J-atomic formula that contains an internal predicate symbol different from p.

Now let us consider the case when the J-atomic formula contains the internal predicate symbol p. According to the assumption, this formula is local w.r.t. $p(\vec{a})$. Therefore it is the formula $\mathrm{J}_\kappa p(\vec{a})$ for some $\kappa \in \mathscr{V}(L, \mathsf{I})$. Now by (15.1),

$$\mathfrak{J} \models_L \mathrm{J}_\kappa p(\vec{a}) \quad \text{iff} \quad \mathfrak{K} \models_L \mathrm{J}_\kappa p(\vec{a}).$$

INDUCTION STEP. Assume that the statement is true for all formulas of height less than or equal to n. Let the height of the formula Φ be equal to $n+1$. Let us consider the possible cases.

CASE 1. Φ is $\neg\Psi$. The height of the formula Ψ is equal to n. If Φ is local w.r.t. $p(\vec{a})$, then Ψ will be such too. Then by the induction hypothesis

$$\mathfrak{J} \models_L \Psi \quad \text{iff} \quad \mathfrak{K} \models_L \Psi,$$

whence it clearly follows that

$$\mathfrak{J} \models_L \neg\Psi \quad \text{iff} \quad \mathfrak{K} \models_L \neg\Psi.$$

CASE 2. Φ is $\Psi \to \Theta$. This case is similar to the previous one.

CASE 3. Φ is $\forall x \Psi(x)$. Let Sort $(x) = \mathsf{s} \in \mathsf{IS}$. It is easy to see that $\mathfrak{J} \models_L \forall x \Psi(x)$ if and only if $\mathfrak{J} \models_L \Psi(c)$ for any $c \in \mathrm{Dom}_\sigma(D, \mathsf{s})$. (The same holds for the L-structure $\mathfrak{K}$.)

If Φ is local w.r.t $p(\vec{a})$, then for any $c \in \mathrm{Dom}_\sigma(D, \mathsf{s})$ the formula $\Psi(c)$ will also be local w.r.t $p(\vec{a})$. Locality of a formula requires that each appearence of the predicate symbol p in this formula may be only *inside the atomic formula* $p(\vec{a})$. Hence atomic subformulas containing p in $\forall x \Psi(x)$ cannot contain variables.

The height of the formula Ψ is n. Then, by the induction assumption,

$$\mathfrak{J} \models_L \Psi(c) \quad \text{iff} \quad \mathfrak{K} \models_L \Psi(c),$$

for any $c \in \mathrm{Dom}_\sigma(D, \mathsf{s})$. Then, clearly,

$$\mathfrak{J} \models_L \forall x \Psi(x) \quad \text{iff} \quad \mathfrak{K} \models_L \forall x \Psi(x).$$

□

The next lemma emphasises the syntactic aspect of the fact that any external L-formula local w.r.t. $p(\vec{a})$ is independent from any cell of the matrix representing predicate p except for the cell represented by the formula $p(\vec{a})$.

Lemma 15.1.6. Let L be an FPJ logic, σ be a signature, D be a σ-vocabulary, $\Delta \in \mathbf{SD}_\sigma(L, D)$, $p \in \mathrm{Pr}_\sigma(\mathsf{I})$ and $\vec{a} \in \mathrm{Dom}_\sigma(D, p)$. Assume:

(i) $\vec{b} \in \mathrm{Dom}_\sigma(D, p)$, $\vec{b} \neq \vec{a}$ and $\mathrm{J}_\varepsilon\, p\left(\vec{b}\right) \in \Delta$;
(ii) $\mathfrak{J} = \mathrm{Sint}_L(\Delta, D)$;
(iii) Φ is a closed external L-formula local w.r.t. atomic formula $p(\vec{a})$;
(iv) $\mathfrak{J} \models_L \Phi$.

Then $\Delta \setminus \left\{\mathrm{J}_\varepsilon\, p\left(\vec{b}\right)\right\} \vdash_L \Phi$.

Proof. Let $\delta \in \mathscr{V}(L, \mathsf{I})$. Introduce the following notations

$$\Delta^\delta = \Delta\left[\mathrm{J}_\varepsilon\, p\left(\vec{b}\right) \leftarrow J_\delta p\left(\vec{b}\right)\right],$$

$$\mathfrak{K}^\delta = \mathrm{Sint}_L\left(\Delta^\delta, D\right).$$

By the use of these notations, $\Delta = \Delta^\varepsilon$, $\mathfrak{J} = \mathfrak{K}^\varepsilon$. Let

$$\Gamma = \Delta \setminus \left\{\mathrm{J}_\varepsilon\, p\left(\vec{b}\right)\right\}.$$

Then

$$\Delta^\delta = \Gamma \cup \left\{J_\delta p\left(\vec{b}\right)\right\} \tag{15.2}$$

for any $\delta \in \mathscr{V}(L, \mathsf{I})$.

By (iv) and the previous lemma,

$$\mathfrak{K}^{\delta} \models_L \Phi \tag{15.3}$$

for any $\delta \in \mathscr{V}(L, \mathsf{I})$.

For any $\delta \in \mathscr{V}(L, \mathsf{I})$, $\mathfrak{K}^{\delta}$ is a syntactical L-structure for Δ^{δ} over D. Then by (15.3) and the assumption of 10.1.23 we get

$$\Delta^{\delta} \vdash_L \Phi,$$

for any $\delta \in \mathscr{V}(L, \mathsf{I})$. Taking into account (15.2) for any for any $\delta \in \mathscr{V}(L, \mathsf{I})$ we obtain the following statement:

$$\Gamma \cup \left\{ \mathrm{J}_{\delta}\, p\left(\vec{b}\right) \right\} \vdash_L \Phi.$$

Then by the Deduction Theorem for the calculus L^1 for any $\delta \in \mathscr{V}(L, \mathsf{I})$ we get:

$$\Gamma \vdash_L \mathrm{J}_{\delta}\, p\left(\vec{b}\right) \to \Phi.$$

Now by the use of the well-known admissible inference rule which occurs in both the classical logic and logic L, we get

$$\Gamma \vdash_L \bigvee_{\delta \in \mathscr{V}(L, \mathsf{I})} \mathrm{J}_{\delta}\, p\left(\vec{b}\right) \to \Phi.$$

But the formula

$$\bigvee_{\delta \in \mathscr{V}(L, \mathsf{I})} \mathrm{J}_{\delta}\, p\left(\vec{b}\right)$$

is an axiom obtained by the schema (B1) of the calculus L^1. Then by *modus ponens* we get $\Gamma \vdash_L \Phi$, as was to be proved. □

The following theorem will finalise our statement according to which an external L-formula local w.r.t. $p(\vec{a})$ is independent from any cell of the matrix representing an internal predicate symbol p except for the cell corresponding to the formula $p(\vec{a})$.

Theorem 15.1.7. Let L be an FPJ logic, σ be a signature, D be a σ-vocabulary, $\Delta \in \mathbf{SD}_{\sigma}(L, D)$, $p(\vec{a}) \in \mathrm{AF}_{\sigma}{}^{D}(\mathsf{I})$. Assume:

(i) Σ is the set of all formulas of the type $\mathrm{J}_{\varepsilon}\, p\left(\vec{b}\right) \in \Delta$ such that $\vec{b} \in \mathrm{Dom}_{\sigma}(D, p)$ and $\vec{b} \neq \vec{a}$;
(ii) $\mathfrak{J} = \mathrm{Sint}_L(\Delta, D)$;
(iii) Φ is a closed external L-formula local w.r.t. atomic formula $p(\vec{a})$;
(iv) $\mathfrak{J} \models_L \Phi$.

Then $\Delta \backslash \Sigma \vdash_L \Phi$.

Proof. By simple induction on the cardinality of the set Σ by using the previous lemma. □

Next we introduce the notion of pure atomic formula that will be used just as the notion of locality to formulate additional restrictions for the modification rule system (MRS).

Definition 15.1.8. Let σ be a signature, $p \in \mathrm{Pr}_\sigma$, $\vec{x} \in \mathrm{Var}^{\diamondsuit} \cap \mathrm{Dom}_\sigma(p)$. Then $p(\vec{x})$ is called a *pure* atomic formula of signature σ . I.e., a *pure* atomic formula does not contain any term except individual variables and each variable occurs at most once in it.

Proposition 15.1.9. Let L be a PJ logic, σ be a signature and D be a finite σ-vocabulary, $\Delta \in \mathbf{SD}_\sigma(L,D)$, $p(\vec{x})$ is a pure atomic formula of signature σ, $\vec{a} \in \mathrm{Dom}_\sigma(D,p)$. Assume that the L-formula $\Phi(\vec{x})$ is local w.r.t. the pure atomic formula $p(\vec{x})$. Then $\Phi(\vec{a})$ is local w.r.t. the atomic formula $p(\vec{a})$.

Proof. The statement can be proved by a simple induction on L-formulas local w.r.t. $p(\vec{x})$. □

Below are two important statements and their consequences that connect derivability of local formulas from two current state descriptions. It follows from these consequences that any external closed formula Φ, local w.r.t. the J-atomic formula φ, is derivable from $\mathrm{CSD}(\vec{\mathfrak{a}})$ if and only if it is derivable from $\mathrm{CSD}\left(\vec{\mathfrak{a}}_{\overleftarrow{s,m-1}}\right)$ in the case when φ does not occur in the record of the inference $\vec{a}$ of rank $\langle s,m\rangle$. Informally speaking, if φ does not appear in an inference of rank $\langle s,m\rangle$, then all closed external formulas local w.r.t. the formula φ which can be obtained by extending this inference through deductive tools can also be obtained with the same tools from the inference of the previous rank. The converse also holds.

These statements further permit proofs by induction on indices of modules for an inference in a modification calculus. These statements provide us with the induction step: the possibility to pass from the module of index m to the module of index $m-1$ and then the induction hypotheses can be applied.

Lemma 15.1.10. Let L be an FPJ logic, σ be a signature, $\mathscr{M} \in \mathrm{MC}_\sigma(L)$ and $\vec{\mathfrak{a}}$ be a pure $\mathscr{M}(s,k,d)$-inference, where $s>0$, $m>0$, $m-1 \le k \le m$. Let $p \in \mathrm{Pr}_\sigma(\mathsf{I})$, $\vec{a} \in \mathrm{Dom}_{\mathscr{M}}(p)$, $\mathrm{Ord}_{\mathscr{M}}(p)=m$. Suppose:

(i) Φ is an external L-formula local w.r.t. $p(\vec{a})$,
(ii) For any $\varepsilon \in \mathscr{V}(L,\mathsf{I})$, if $\mathrm{J}_\varepsilon\, p(\vec{a}) \in \mathrm{FSet}(\vec{\mathfrak{a}})$ then $\mathrm{J}_\varepsilon\, p(\vec{a}) \in \mathrm{FSet}\left(\vec{\mathfrak{a}}_{\overleftarrow{s,m-1}}\right)$,
(iii) $\mathrm{CSD}(\vec{\mathfrak{a}}) \vdash_L \Phi$.

Then $\mathrm{CSD}\left(\vec{\mathfrak{a}}_{\overleftarrow{s,m-1}}\right) \vdash_L \Phi$.

Proof. The case when $k=m-1$ is trivial (in this case $\vec{\mathfrak{a}}=\vec{\mathfrak{a}}_{\overleftarrow{s,m-1}}$).

Suppose $k=m$. Let

$$\Xi = \mathrm{CSD}(\vec{\mathfrak{a}}) \setminus \left(\mathrm{JA}_\sigma(\vec{\mathfrak{a}},p,\mathscr{M}) \setminus \{\mathrm{J}_\delta\, p(\vec{a})\}\right), \tag{15.4}$$

where $\mathrm{J}_\delta\, p(\vec{a}) \in \mathrm{FSet}(\vec{\mathfrak{a}})$. Then Ξ does not contain any formula of the form $\mathrm{J}_\varepsilon\, p\left(\vec{b}\right)$ except for the formula $\mathrm{J}_\delta\, p(\vec{a})$. But this formula was supplied to the $\mathscr{M}$-inference $\vec{\mathfrak{a}}$ from the previous $(m-1)$-th module because of (ii). Since $\mathrm{Ord}_{\mathscr{M}}(p) = m$, $\mathrm{CSD}\left(\vec{\mathfrak{a}}_{\overleftarrow{s,m-1}}\right)$ may differ from $\mathrm{CSD}(\vec{\mathfrak{a}}))$ only in formulas from $\mathrm{JA}_\sigma(\vec{\mathfrak{a}}, p, \mathscr{M})$. Then:

$$\Xi \subseteq \mathrm{CSD}\left(\vec{\mathfrak{a}}_{\overleftarrow{s,m-1}}\right). \tag{15.5}$$

By Theorem 15.1.7, (iii) and (15.4) imply

$$\Xi \vdash_L \Phi. \tag{15.6}$$

From (15.5) and (15.6) we get $\mathrm{CSD}\left(\vec{\mathfrak{a}}_{\overleftarrow{s,m-1}}\right) \vdash_L \Phi$, as was to be proved. □

Lemma 15.1.11. Let L be an FPJ logic, σ be a signature, $\mathscr{M} \in \mathrm{MC}_\sigma(L)$ and $\vec{\mathfrak{a}}$ be a pure $\mathscr{M}(s,k,d)$-inference, where $s > 0$, $m > 0$, $m-1 \le k \le m$. Let $p \in \mathrm{Pr}_\sigma(\mathsf{I})$, $\vec{a} \in \mathrm{Dom}_{\mathscr{M}}(p)$, $\mathrm{Ord}_{\mathscr{M}}(p) = m$. Suppose:

(i) Φ is a closed external L-formula local w.r.t. $p(\vec{a})$,
(ii) For any $\varepsilon \in \mathscr{V}(L,\mathsf{I})$, if $\mathrm{J}_\varepsilon\, p(\vec{a}) \in \mathrm{FSet}(\vec{\mathfrak{a}})$ then $\mathrm{J}_\varepsilon\, p(\vec{a}) \in \mathrm{FSet}\left(\vec{\mathfrak{a}}_{\overleftarrow{s,m-1}}\right)$,
(iii) $\mathrm{CSD}\left(\vec{\mathfrak{a}}_{\overleftarrow{s,m-1}}\right) \vdash_L \Phi$.

Then $\mathrm{CSD}(\vec{\mathfrak{a}}) \vdash_L \Phi$.

Proof. The case of $k = m-1$ is trivial (in this case $\vec{\mathfrak{a}} = \vec{\mathfrak{a}}_{\overleftarrow{s,m-1}}$).

Let $k = m$. Suppose $\mathrm{CSD}(\vec{\mathfrak{a}}) \nvdash_L \Phi$. Then by Lemma 14.1.1,

$$\mathrm{Sint}_L(\mathrm{CSD}(\vec{\mathfrak{a}}), M) \nvDash_L \Phi.$$

Therefore, since Φ is closed

$$\mathrm{Sint}_L(\mathrm{CSD}(\vec{\mathfrak{a}}), M) \vDash_L \neg\Phi,$$

whence by Lemma 14.1.1 we get

$$\mathrm{CSD}(\vec{\mathfrak{a}}) \vdash_L \neg\Phi. \tag{15.7}$$

But by 15.1.2, $\neg\Phi$ is also local w.r.t. $p(\vec{a})$. Then by Corollary 15.1.12, (15.7) implies $\mathrm{CSD}\left(\vec{\mathfrak{a}}_{\overleftarrow{s,m-1}}\right) \vdash_L \neg\Phi$, which contradicts the assumption. □

Corollary 15.1.12. Let L be an FPJ logic, σ be a signature, $\mathscr{M} \in \mathrm{MC}_\sigma(L)$ and $\vec{\mathfrak{a}}$ be a pure $\mathscr{M}(s,k,d)$-inference, where $s > 0$, $m > 0$, $m-1 \le k \le m$. Let $p \in \mathrm{Pr}_\sigma(\mathsf{I})$, $\vec{a} \in \mathrm{Dom}_{\mathscr{M}}(p)$, $\mathrm{Ord}_{\mathscr{M}}(p) = m$. Suppose:

(i) Φ is an external L-formula local w.r.t. $p(\vec{a})$,
(ii) For any $\varepsilon \in \mathscr{V}(L,\mathsf{I})$, $\vec{\mathfrak{a}}$ contains no record $\mathfrak{b}$ such that

$$\mathscr{F}(\mathfrak{b}) = \mathrm{J}_\varepsilon\, p(\vec{a}) \quad \text{and} \quad \mathrm{Rank}(\mathfrak{b}) = \langle s,m\rangle,$$

(iii) $\mathrm{CSD}(\vec{\mathfrak{a}}) \vdash_L \Phi$.

Then $\mathrm{CSD}\left(\vec{\mathfrak{a}}_{\overleftarrow{s,m-1}}\right) \vdash_L \Phi$.

Proof. This statement follows directly from Lemma 15.1.10. □

Corollary 15.1.13. Let L be an FPJ logic, σ be a signature, $\mathscr{M} \in \mathrm{MC}_\sigma(L)$ and $\vec{\mathfrak{a}}$ be a pure $\mathscr{M}(s,k,d)$-inference, where $s > 0$, $m > 0$, $m-1 \le k \le m$. Let $p \in \mathrm{Pr}_\sigma(\mathsf{I})$, $\vec{a} \in \mathrm{Dom}_{\mathscr{M}}(p)$, $\mathrm{Ord}_{\mathscr{M}}(p) = m$. Suppose:

(i) Φ is a closed external L-formula local w.r.t. $p(\vec{a})$,
(ii) for any $\varepsilon \in \mathscr{V}(L,\mathsf{I})$, $\vec{\mathfrak{a}}$ contains no record $\mathfrak{b}$ such that

$$\mathscr{F}(\mathfrak{b}) = \mathrm{J}_\varepsilon\, p\left(\vec{\mathfrak{b}}\right) \quad \text{and} \quad \mathrm{Rank}(\mathfrak{b}) = \langle s,m\rangle,$$

(iii) $\mathrm{CSD}\left(\vec{\mathfrak{a}}_{\overleftarrow{s,m-1}}\right) \vdash_L \Phi$.

Then $\mathrm{CSD}(\vec{\mathfrak{a}}) \vdash_L \Phi$.

Proof. This statement follows directly from Lemma 15.1.11. □

Note that Lemmas 15.1.10 and 15.1.11 are really stronger than the corresponding consequences (15.1.12 and 15.1.13). However, the consequences describe the most important particular case that, further on, will be most frequently used.

The consequences fully describe the situation when the formula $\mathrm{J}_\varepsilon\, p(\vec{a})$ is included in the inference within the module m by the use of the modification rule, where $\varepsilon \ne \tau$. But in the case when $\mathrm{J}_\varepsilon\, p(\vec{a})$ is added by the confirmation rule or when $\varepsilon = \tau$, condition (ii) of Corollaries 15.1.12 and 15.1.13 is too strict. It is necessary to weaken this condition and thus to strengthen the corresponding statements. It is Lemmas 15.1.12 and 15.1.13 that are stronger statements. There are situations, though not many, when in order to obtain an important result one of these lemmas is to be applied. At the same time the application of the consequences does not lead to the required consequences.

The next statement shows that a few conditions connected with pure inferences in modification calculi are equivalent for the closed J-atomic formulas. Particularly interesting and useful is the case when the validity of such a formula in a syntactic L-structure that corresponds to some inference leads to the necessary occurrence of this formula in this inference.

Lemma 15.1.14. Let L be an FPJ logic, σ be a signature, $\mathscr{M} \in \mathrm{MC}_\sigma(L)$, $p \in \mathrm{Pr}_\sigma(\mathsf{I})$, $\vec{a} \in \mathrm{Dom}_{\mathscr{M}}(p)$, $\varepsilon \in \mathscr{V}(L,\mathsf{I})$. Let $\vec{\mathfrak{a}}$ be a pure $\mathscr{M}(s)$-inference, where $s > 0$. Then the following conditions are equivalent:

(i) $\mathrm{J}_\varepsilon\, p(\vec{a}) \in \mathrm{CSD}(\vec{\mathfrak{a}})$;
(ii) $\mathrm{J}_\varepsilon\, p(\vec{a}) \in \mathrm{FSet}(\vec{\mathfrak{a}})$;
(iii) $\mathrm{CSD}(\vec{\mathfrak{a}}) \vdash_L \mathrm{J}_\varepsilon\, p(\vec{a})$;
(iv) $\mathrm{Sint}_L(\mathrm{CSD}(\vec{\mathfrak{a}}), \mathrm{Univ}_{\mathscr{M}}) \models_L \mathrm{J}_\varepsilon\, p(\vec{a})$.

Proof. Obviously, (i) implies (ii) and by Corollary 11.4.12, (ii) implies (iii). Since $\mathrm{Sint}_L(\mathrm{CSD}(\vec{\mathfrak{a}}), D)$ is a model of $\mathrm{CSD}(\vec{\mathfrak{a}})$, (iii) implies (iv). Let (iv) be true. Let us show that (i) holds. Suppose the converse. Then since $\mathrm{CSD}(\vec{\mathfrak{a}})$ is a state description of signature σ over D (see Definition 11.3.15), there exists a $\delta \in \mathscr{V}(L, \mathsf{I})\,(\delta \neq \varepsilon)$ such that $\mathrm{J}_\delta\, p(\vec{a}) \in \mathrm{CSD}(\vec{\mathfrak{a}})$. Then

$$\mathrm{Sint}_L(\mathrm{CSD}(\vec{\mathfrak{a}}), \mathrm{Univ}_{\mathscr{M}}) \vDash_L \mathrm{J}_\delta\, p(\vec{a}),$$

which contradicts (iv). □

Lemma 15.1.15. Let L be an FPJ logic, σ be a signature, $\mathscr{M} \in \mathrm{MC}_\sigma(L)$ and $\vec{\mathfrak{a}} = \vec{\mathfrak{b}} \bullet \mathfrak{c}$ be a pure $\mathscr{M}$-inference. Then for any $p \in \mathrm{Pr}_\sigma(\mathsf{I})$, $\vec{a} \in \mathrm{Dom}_{\mathscr{M}}(p)$, $\varepsilon \in \mathscr{D}^\tau(L)$,

$$\mathrm{J}_\varepsilon\, p(\vec{a}) \in \mathrm{CSD}\left(\mathrm{Resume}\left(\vec{\mathfrak{b}}\right)\right) \quad \text{implies} \quad \mathrm{J}_\varepsilon\, p(\vec{a}) \in \mathrm{CSD}(\vec{\mathfrak{a}}).$$

Proof. By an easy analysis of the possible cases of adding an element into an $\mathscr{M}$-inference (see Definition 10.2.31). □

Lemma 15.1.16. Let L be an FPJ logic, σ be a signature, $\mathscr{M} \in \mathrm{MC}_\sigma(L)$ and $\vec{\mathfrak{a}}$ be a pure $\mathscr{M}$-inference. Suppose that $\vec{\mathfrak{a}} = \vec{\mathfrak{b}} + \vec{\mathfrak{c}}$ ($\vec{\mathfrak{b}}$ is the initial segment of $\vec{\mathfrak{a}}$, $\left\langle \vec{\mathfrak{b}}, \vec{\mathfrak{c}} \right\rangle$ is a cut of $\vec{\mathfrak{a}}$). Then for any $p \in \mathrm{Pr}_\sigma(\mathsf{I})$, $\vec{a} \in \mathrm{Dom}_{\mathscr{M}}(p)$, $\varepsilon \in \mathscr{D}^\tau(L)$,

$$\mathrm{J}_\varepsilon\, p(\vec{a}) \in \mathrm{CSD}\left(\mathrm{Resume}\left(\vec{\mathfrak{b}}\right)\right) \quad \text{implies} \quad \mathrm{J}_\varepsilon\, p(\vec{a}) \in \mathrm{CSD}(\vec{\mathfrak{a}}). \tag{15.8}$$

Proof. This evident statement is easy to prove by induction on the length of the upper segment of the cut $S = \left\langle \vec{\mathfrak{b}}, \vec{\mathfrak{c}} \right\rangle$.

BASIS. Let $\mathrm{Len}(\vec{\mathfrak{c}}) = 0$. Then $\vec{\mathfrak{b}} = \vec{\mathfrak{a}}$. In this case (15.8) is trivial.

INDUCTION STEP. Suppose that the statement holds for the n-long upper segment of the considered cut. Then for some $\vec{\mathfrak{d}}$ and $\mathfrak{f}$, $\vec{\mathfrak{c}} = \vec{\mathfrak{d}} \bullet \mathfrak{f}$. In this case, $\vec{\mathfrak{a}} = \left(\vec{\mathfrak{b}} + \vec{\mathfrak{d}}\right) \bullet \mathfrak{f}$. Let $\vec{\mathfrak{g}} = \mathrm{Resume}\left(\vec{\mathfrak{b}} + \vec{\mathfrak{d}}\right)$, $k = \mathrm{Len}(\vec{\mathfrak{g}})$, $\vec{\mathfrak{h}} = \mathrm{Resume}\left(\vec{\mathfrak{b}}, k\right)$ and $\vec{i} = \mathrm{Resume}\left(\vec{\mathfrak{d}}, k\right)$. Then $\left\langle \vec{\mathfrak{h}}, \vec{\mathfrak{i}} \right\rangle$ is a cut of $\vec{\mathfrak{g}}$. The length of the upper segment of this cut is equal to n. Note that

$$\mathrm{Resume}\left(\vec{\mathfrak{h}}\right) = \mathrm{Resume}\left(\mathrm{Resume}\left(\vec{\mathfrak{b}}, k\right)\right) = \mathrm{Resume}\left(\vec{\mathfrak{b}}\right),$$

since $k = \mathrm{Len}(\vec{\mathfrak{g}}) \leq \mathrm{Len}\left(\vec{\mathfrak{b}}\right)$ (see Lemma 11.1.14). Then by the induction hypothesis

$$\mathrm{J}_\varepsilon\, p(\vec{a}) \in \mathrm{CSD}\left(\mathrm{Resume}\left(\vec{\mathfrak{b}}\right)\right) \quad \text{implies} \quad \mathrm{J}_\varepsilon\, p(\vec{a}) \in \mathrm{CSD}(\vec{\mathfrak{g}})). \tag{15.9}$$

By Lemma 15.1.15,

$$\mathrm{J}_\varepsilon\, p(\vec{a}) \in \mathrm{CSD}(\vec{\mathfrak{g}}) \quad \text{implies} \quad \mathrm{J}_\varepsilon\, p(\vec{a}) \in \mathrm{CSD}(\vec{\mathfrak{a}}). \tag{15.10}$$

(15.9) and (15.10) imply (15.8). □

Corollary 15.1.17. Let L be an FPJ logic, σ be a signature, $\mathscr{M} \in \mathrm{MC}_\sigma(L)$ and $\vec{\mathfrak{a}}$ be a pure $\mathscr{M}$-inference. $s \in \omega$, $m \in [0..\mathrm{Q}_{\mathscr{M}}]$. Then for any $p \in \mathrm{Pr}_\sigma(\mathsf{I})$, $\vec{a} \in \mathrm{Dom}_{\mathscr{M}}(p)$, $\varepsilon \in \mathscr{D}^\tau(L)$,

$$\mathrm{J}_\varepsilon\, p(\vec{a}) \in \mathrm{CSD}\left(\vec{\mathfrak{a}}_{\overleftarrow{s,m}}\right) \quad \text{implies} \quad \mathrm{J}_\varepsilon\, p(\vec{a}) \in \mathrm{CSD}(\vec{\mathfrak{a}}).$$

Proof. By Definition 14.1.4, $\vec{\mathfrak{a}}_{\overleftarrow{s,m}}$ is the result of application of the operation Resume () to the upper segment of the (s, m)-cut of $\vec{\mathfrak{a}}$. □

15.2 Atomic Sorts

It is time to apply the notion of locality to the formulation of syntactic requirements for the modification calculi which permit the definition of the classes of conformable modification calculi.

Definition 15.2.1. Let L be an FPJ logic, σ be a signature, $p \in \mathrm{Pr}_\sigma(\mathsf{I})$. $\vec{x} \in \mathrm{Var}^\diamond \cap \mathrm{Dom}_\sigma(p)$. A finite set $\{\Phi_\varepsilon \mid \varepsilon \in \mathscr{V}(L,\mathsf{I})\}$ of external L-formulas is called a *rule-lifting or rule-founding system (RL-system) for the formula $p(\vec{x})$* (and for the predicate symbol p) if it meets the following conditions:

(i) $\mathrm{FVar}(\Phi_\varepsilon) \subseteq \mathrm{Set}(\vec{x})$ for any $\varepsilon \in \mathscr{V}(L,\mathsf{I})$,
(ii) For any $\varepsilon \in \mathscr{V}(L,\mathsf{I})$, Φ_ε is local w.r.t. pure atomic formula $p(\vec{x})$,
(iii) $\Phi_\tau = \bigwedge\limits_{\delta \in \mathscr{D}^\tau(L)} \neg\Phi_\delta$,
(iv) $\vdash_L \Phi_\varepsilon \to \neg\Phi_\delta$ for any $\varepsilon, \delta \in \mathscr{D}^\tau(L)$ such that $\varepsilon \neq \delta$.

Definition 15.2.2. Let L be an FPJ logic, σ be a signature, $p \in \mathrm{Pr}_\sigma(\mathsf{I})$. $\vec{x} \in \mathrm{Var}^\diamond \cap \mathrm{Dom}_\sigma(p)$. Let $M = \{\Phi_\varepsilon \mid \varepsilon \in \mathscr{V}(L,\mathsf{I})\}$ be a rule-lifting system for $p(\vec{x})$. Let

$$S = \{\mathrm{J}_\tau\, p(\vec{x}), \Psi_\varepsilon \Vdash \mathrm{J}_\varepsilon\, p(\vec{x}) \mid \varepsilon \in \mathscr{V}(L,\mathsf{I})\}$$

be a complete modification set (CMC) for p. Then we say that S is *atomically connected (A-connected)* with M (M is atomically connected with S, M and S are atomically connected), if $\Psi_\varepsilon = \Phi_\varepsilon$ for any $\varepsilon \in \mathscr{V}(L,\mathsf{I})$.

It is evident that any CMC S A-connected with some RL-system M is disjont. This remark implies the fact that the following two notions are well defined.

Let $R \in \mathrm{MRS}_\sigma(L)$, $p \in \mathrm{Pr}_\sigma(\mathsf{I})$. We say that p is *atomic in R* if there exists an RL-system M A-connected with $R[p]$. By definition put

$$\mathrm{A}[R] = \{p \in \mathrm{Pr}_\sigma(\mathsf{I}) \mid p \text{ is atomic in } R\}.$$

Let $\mathscr{M} \in \mathrm{MC}_\sigma(L)$, $p \in \mathrm{Pr}_\sigma(\mathsf{I})$. We say that p is *atomic in $\mathscr{M}$* if p is atomic in $\mathrm{MR}_{\mathscr{M}}$. By definition put

$$\mathrm{A}[\mathscr{M}] = \mathrm{A}[\mathrm{MR}_{\mathscr{M}}].$$

An $R \in \mathrm{MRS}_\sigma(L)$ is called *atomic* if any $p \in \mathrm{Pr}_\sigma(\mathsf{I})$ is atomic in R, i.e. $\mathrm{A}[R] = \mathrm{Pr}_\sigma(\mathsf{I})$.

By definition let:

$$\mathrm{AMRS}_\sigma(L) = \{R \in \mathrm{MRS}_\sigma(L) \mid R \text{ is atomic}\}.$$

An $\mathscr{M} \in \mathrm{MC}_\sigma(L)$ is called *atomic* if $\mathrm{MR}_{\mathscr{M}} \in \mathrm{AMRS}_\sigma(L)$, i.e. $\mathrm{A}[\mathscr{M}] = \mathrm{Pr}_\sigma(\mathsf{I})$. By definition let:

$$\mathrm{AMC}_\sigma(L) = \{\mathscr{M} \in \mathrm{MC}_\sigma(L) \mid \mathscr{M} \text{ is atomic}\}$$

Definition 15.2.3. Let L be an FPJ logic, σ be a signature, $R \in \mathrm{MRS}_\sigma(L)$, $p \in \mathrm{Pr}_\sigma(\mathsf{I})$ and

$$M = \{\Phi_\varepsilon \mid \varepsilon \in \mathscr{V}(L,\mathsf{I})\}$$

be an RL-sytem for p such that the set of rules

$$S = \{R[p,\varepsilon] \mid \varepsilon \in \mathscr{V}(L,\mathsf{I})\}$$

is A-connected with M. Then for any $\varepsilon \in \mathscr{V}(L,\mathsf{I})$ the formula Φ_ε is called the *kernel* of the rule $R[p,\varepsilon]$ and is denoted by $\mathrm{Ker}[R,p,\varepsilon]$.

Let $\mathscr{M} \in \mathrm{MC}_\sigma(L)$, $p \in \mathrm{A}[\mathscr{M}]$. Then by definition put

$$\mathrm{Ker}[M,p,\varepsilon] = \mathrm{Ker}[\mathrm{MR}_{\mathscr{M}},p,\varepsilon].$$

Obviously, for any $R \in \mathrm{AMRS}_\sigma(L)$ the equality $\mathrm{Ker}[R,p,\varepsilon] = LP[R,p,\varepsilon]$ will be always valid. Thus the notion of a modification rule kernel is unnecessary. However, later on during the study of so-called set sorts we will consider the cases when the correspondence between the major premise and the rule kernel is more sophisticated. Nevertheless, a rule kernel will be local w.r.t. its conclusion and the statement formulated below will be valid.

Lemma 15.2.4. Let L be an FPJ logic, σ be a signature, $\mathfrak{J}$ be an L-structure of signature σ, $p \in \mathrm{Pr}_\sigma(\mathsf{I})$, $\vec{a} \in \mathrm{Dom}\left(p^{\mathfrak{J}}\right)$ and $R \in \mathrm{MRS}_\sigma(L)$. Then there exists a unique $\varepsilon \in \mathscr{V}(L,\mathsf{I})$ such that

$$\mathfrak{J} \models_L \mathrm{Ker}[R,p,\varepsilon](\vec{a}).$$

Proof. The lemma follows directly from Definitions 15.2.1 and 15.2.3. □

Now we intend to introduce the generation rules for the atomic case and then we make a detailed analysis of the first large class of modification calculi for which an adequate semantics can be defined. This is the class of atomic modification calculi. First we define the comforming function which transfers modification rules into generation rules.

Definition 15.2.5. Let L be an FPJ logic, σ be a signature and $R \in \mathrm{MRS}_\sigma(L)$, $p \in \mathrm{A}[R]$, $\vec{x} \in \mathrm{Var}^{\diamond} \cap \mathrm{Dom}_\sigma(p)$, $\varepsilon \in \mathscr{V}(L,\mathsf{I})$. Then the generating ε-rule

$$\Phi \Vdash \mathrm{J}_{\varepsilon}\, p(\vec{x}),$$

where

$$\Phi = \begin{cases} \mathrm{J}_{\varepsilon}\, p(\vec{x}) \vee (\mathrm{J}_{\tau}\, p(\vec{x}) \wedge \mathrm{Ker}[R,p,\varepsilon](\vec{x})) & \text{if } \varepsilon \in \mathscr{D}^{\tau}(L), \\ \mathrm{J}_{\tau}\, p(\vec{x}) \wedge \mathrm{Ker}[R,p,\tau](\vec{x}) & \text{if } \varepsilon = \tau, \end{cases}$$

is called the *atomic generating image* (AG-image) of the modification rule $R[p,\varepsilon]$ (written: $\mathrm{AG}[R,p,\varepsilon]$). The formula Φ (the generating condition of $\mathrm{AG}[R,p,\varepsilon]$) is denoted by $\mathrm{GC}[\mathrm{AG}(R),p,\varepsilon]$.

Definition 15.2.6. Let σ be a signature and $R \in \mathrm{MRS}_{\sigma}(L)$, $p \in \mathrm{A}[R]$. Then the set

$$\{\mathrm{AG}[R,p,\varepsilon] \mid \varepsilon \in \mathscr{V}(L,\mathsf{I})\}$$

is called the *atomic generating image* (AG-image) of $R[p]$ and is denoted by $\mathrm{AG}[R,p]$.

Proposition 15.2.7. Let L be an FPJ logic, σ be a signature and $R \in \mathrm{MRS}_{\sigma}(L)$, $p \in \mathrm{A}[R]$. Then the set $\mathrm{AG}[R,p]$ is a complete generating set (CGS) for p.

Proof. We must establish the validity of the conditions (i), (ii) and (iii) from Definition 12.2.3.

By axiom schemata (B1) and (B2) from 8.5.2 we obtain

$$\vdash_L \bigvee_{\varepsilon \in \mathscr{V}(L,\mathsf{I})} \mathrm{J}_{\varepsilon}\, p(\vec{x}) \tag{15.11}$$

and

$$\vdash_L \mathrm{J}_{\varepsilon}\, p(\vec{x}) \rightarrow \neg \mathrm{J}_{\delta}\, p(\vec{x}) \tag{15.12}$$

for any $\varepsilon, \delta \in \mathscr{V}(L,\mathsf{I})$ such that $\varepsilon \neq \delta$.

By the use of Lemma 15.2.4 and the Completeness Theorem (8.5.30) it is not difficult to show that

$$\vdash_L \bigvee_{\varepsilon \in \mathscr{V}(L,\mathsf{I})} \mathrm{Ker}[R,p,\varepsilon](\vec{x}) \tag{15.13}$$

and

$$\vdash_L \mathrm{Ker}[R,p,\varepsilon](\vec{x}) \rightarrow \neg \mathrm{Ker}[R,p,\delta](\vec{x}) \tag{15.14}$$

for any $\varepsilon, \delta \in \mathscr{V}(L,\mathsf{I})$ such that $\varepsilon \neq \delta$.

From (15.11), (15.12), (15.13), (15.14) and Definition 15.2.5 we can get

$$\vdash_L \bigvee_{\varepsilon \in \mathscr{V}(L,\mathsf{I})} \mathrm{GC}[\mathrm{AG}(R),p,\varepsilon](\vec{x}) \tag{15.15}$$

and

$$\vdash_L \mathrm{GC}[\mathrm{AG}(R),p,\varepsilon](\vec{x}) \rightarrow \neg \mathrm{GC}[\mathrm{AG}(R),p,\delta](\vec{x}) \tag{15.16}$$

for any $\varepsilon, \delta \in \mathscr{V}(L,\mathsf{I})$ such that $\varepsilon \neq \delta$.

Expressions (15.15) and (15.16) imply the conditions (i) and (ii) of Definition 12.2.3, respectively.

Recall that

$$\mathrm{GC}\left[\mathrm{AG}(R), p, \varepsilon\right](\vec{x}) = \mathrm{J}_{\varepsilon}\, p(\vec{x}) \vee (\mathrm{J}_{\tau}\, p(\vec{x}) \wedge \mathrm{Ker}[R, p, \varepsilon](\vec{x})) \tag{15.17}$$

for each $\varepsilon \in \mathscr{D}^{\tau}(L)$ (see Definition 15.2.5). It is evident that (15.17) implies (iii) of Definition 12.2.3. □

Definition 15.2.8. Let L be an FPJ logic, σ be a signature and $R \in \mathrm{AMRS}_{\sigma}(L)$. Denote by $\mathscr{G}_{\sigma}$ the set of all generating rule of signature σ. Denote by AG^{R} the map of the set

$$\{R[p,\varepsilon] \mid p \in \mathrm{Pr}_{\sigma}(\mathsf{I}), \varepsilon \in \mathscr{V}(L,\mathsf{I})\}$$

to the set $\mathscr{G}_{\sigma}$ such that for any $p \in \mathrm{Pr}_{\sigma}(\mathsf{I})$, $\varepsilon \in \mathscr{V}(L,\mathsf{I})$

$$\mathrm{AG}^{R}(R[p,\varepsilon]) = \mathrm{AG}[R,p,\varepsilon]$$

(see Definition 15.2.5). The map AG^{R} is called the *atomic generating function for* R.

Since $\mathrm{Pr}_{\sigma}(\mathsf{I}) = \mathrm{A}[R]$ for any $R \in \mathrm{AMRS}_{\sigma}(L)$, the map AG^{R} is well defined.

Proposition 15.2.9. Let L be an FPJ logic, σ be a signature and $R \in \mathrm{AMRS}_{\sigma}(L)$. Then the map AG^{R} is a conforming function for R w.r.t. $\emptyset$.

Proof. We must check the validity of the conditions (i), (ii) and (iii) from Definition 14.1.12.

It is evident that (i) and (ii) are valid (see Definition 15.2.5).

To prove (iii) we must show that for any $p \in \mathrm{Pr}_{\sigma}(\mathsf{I})$, $\mathrm{AG}^{R}(R[p])$ is a complete generating set (CGS) for p. Obviously $\mathrm{AG}^{R}(R[p]) = \mathrm{AG}[R,p]$ (see Definitions 15.2.6 and 15.2.8). Then (iii) follows from Proposition 15.2.7. □

Definition 15.2.10. Let L be an FPJ logic, σ be a signature and $\mathscr{M} \in \mathrm{AMC}_{\sigma}(L)$. By definition put $\mathrm{AG}^{\mathscr{M}} = \mathrm{AG}^{\mathrm{MR}_{\mathscr{M}}}$. The map $\mathrm{AG}^{\mathscr{M}}$ is called the *atomic generating function for* $\mathscr{M}$.

Now we can find the AG^{R}-image $\mathrm{AG}^{R}(R)$ by the use of AG for each atomic modification rule system R, which, in accordance with Proposition 14.1.13, is a generating rule system (GRS) w.r.t. the empty set of formulas. The emptiness of this set means that in the atomic case we do not have to impose any additional restrictions on the rules. Also in accordance with Definition 14.1.17 we can define the $\mathrm{AG}^{\mathscr{M}}$-image of any atomic modification calculus M. This $\mathrm{AG}^{\mathscr{M}}$-image, by Proposition 14.1.18, is a generator structure. However, we cannot yet state that the $\mathrm{AG}^{\mathscr{M}}$-image of the modification calculus $\mathscr{M}$ is an $\mathscr{M}$-compatible structure generator.

Below we will show that for each atomic modification calculus $\mathscr{M}$ the function $\mathrm{AG}^{\mathscr{M}}$ is $\emptyset$-$\mathscr{M}$-compatible. Thus it will be proved that any atomic modification calculus is $\emptyset$-comformable. The next lemma sets the stage for the proof of the statement in question. Note that in this lemma it is essential to apply the locality of the formulas. Attention is to be paid to that, in the case of atomic modification calculi, we will have comformability and compatibility w.r.t. the empty set of formulas. These points make atomic rules convenient and easy to use.

Lemma 15.2.11. Let L be an FPJ logic, σ be a signature, $\mathscr{M} \in \mathrm{MC}_\sigma(L)$ and $\vec{\mathfrak{a}}$ be a pure $\mathscr{M}(s,m,\mathbf{D})$-inference, where $s > 0$, $m > 0$, $\mathrm{P}_m^{\mathscr{M}} \in \mathrm{A}[R]$. Then for any $\varepsilon \in \mathscr{D}^\tau(L)$, $\vec{a} \in \mathrm{Dom}_{\mathscr{M}}(\mathrm{P}_m^{\mathscr{M}})$,

$$\mathrm{J}_\varepsilon \mathrm{P}_m^{\mathscr{M}}(\vec{a}) \in \mathrm{CSD}(\vec{\mathfrak{a}})$$

if and only if at least one of the following conditions hold:

(i) $\mathrm{J}_\varepsilon \mathrm{P}_m^{\mathscr{M}}(\vec{a}) \in \mathrm{CSD}\left(\vec{\mathfrak{a}}_{\overleftarrow{s,m-1}}\right)$;
(ii) $\mathrm{J}_\tau \mathrm{P}_m^{\mathscr{M}}(\vec{a}) \in \mathrm{CSD}\left(\vec{\mathfrak{a}}_{\overleftarrow{s,m-1}}\right)$ and $\mathrm{CSD}\left(\vec{\mathfrak{a}}_{\overleftarrow{s,m-1}}\right) \vdash_L \mathrm{Ker}\left[R,\mathrm{P}_m^{\mathscr{M}},\varepsilon\right](\vec{a})$.

Proof. ($\Rightarrow$) Suppose

$$\mathrm{J}_\varepsilon \mathrm{P}_m^{\mathscr{M}}(\vec{a}) \in \mathrm{CSD}(\vec{\mathfrak{a}}).$$

Let $\mathrm{J}_\delta \mathrm{P}_m^{\mathscr{M}}(\vec{a}) \in \mathrm{CSD}\left(\vec{\mathfrak{a}}_{\overleftarrow{s,m-1}}\right)$. By Corollary 15.1.17 this is possible only in two cases: $\delta = \varepsilon$ and $\delta = \tau$. Consider the possible cases.

CASE 1. If $\delta = \varepsilon$, then (i) holds.

CASE 2. Let $\delta = \tau$. Then $\mathrm{J}_\varepsilon \mathrm{P}_m^{\mathscr{M}}(\vec{a})$ was added to the $\mathscr{M}$-inference by the use of the rule $R\left[\mathrm{P}_m^{\mathscr{M}},\varepsilon\right]$. Since $\mathrm{P}_m^{\mathscr{M}} \in \mathrm{A}[R]$,

$$LP\left[R,\mathrm{P}_m^{\mathscr{M}},\varepsilon\right](\vec{a}) = \mathrm{Ker}\left[R,\mathrm{P}_m^{\mathscr{M}},\varepsilon\right](\vec{a}).$$

Thus $\mathrm{Ker}\left[R,\mathrm{P}_m^{\mathscr{M}},\varepsilon\right](\vec{a})$ appears in $\vec{\mathfrak{a}}$ before the application of the corresponding nondeductive rule to the formula $\mathrm{P}_m^{\mathscr{M}}(\vec{a})$. Let $\vec{\mathfrak{b}}$ be the initial segment of $\vec{\mathfrak{a}}$ ending with the formula $\mathrm{Ker}\left[R,\mathrm{P}_m^{\mathscr{M}},\varepsilon\right](\vec{a})$, $\vec{\mathfrak{c}} = Resume\left(\vec{\mathfrak{b}}\right)$. Then by Corollary 11.4.12, $\mathrm{CSD}(\vec{\mathfrak{a}}) \vdash_L \mathrm{Ker}\left[R,\mathrm{P}_m^{\mathscr{M}},\varepsilon\right](\vec{a})$. But $\mathrm{Ker}\left[R,\mathrm{P}_m^{\mathscr{M}},\varepsilon\right](\vec{a})$ is local w.r.t. $\mathrm{P}_m^{\mathscr{M}}(\vec{a})$. Then by Corollary 15.1.12,

$$\mathrm{CSD}\left(\vec{\mathfrak{a}}_{\overleftarrow{s,m-1}}\right) \vdash_L \mathrm{Ker}\left[R,\mathrm{P}_m^{\mathscr{M}},\varepsilon\right](\vec{a}),$$

as was to be proved.

($\Leftarrow$) Suppose that either (i) or (ii) holds.

Consider the possible cases.

CASE 1. Let (i) hold. Then by Corollary 15.1.17, $\mathrm{J}_\varepsilon \mathrm{P}_m^{\mathscr{M}}(\vec{a}) \in \mathrm{CSD}(\vec{\mathfrak{a}})$, as was to be proved.

CASE 2. Let (ii) hold. Suppose $\mathrm{J}_\varepsilon \mathrm{P}_m^{\mathscr{M}}(\vec{a}) \notin \mathrm{CSD}(\vec{\mathfrak{a}})$. Then $\mathrm{J}_\delta \mathrm{P}_m^{\mathscr{M}}(\vec{a}) \in \mathrm{CSD}(\vec{\mathfrak{a}})$ for some $\delta \neq \varepsilon$. By (ii), $\mathrm{J}_\tau \mathrm{P}_m^{\mathscr{M}}(\vec{a}) \in \mathrm{CSD}\left(\vec{\mathfrak{a}}_{\overleftarrow{s,m-1}}\right)$. Therefore, $\mathrm{J}_\delta \mathrm{P}_m^{\mathscr{M}}(\vec{a})$ could appear in the $\mathscr{M}$-inference $\vec{\mathfrak{a}}$ only by the rule $R\left[\mathrm{P}_m^{\mathscr{M}},\delta\right]$. Thus $\vec{\mathfrak{a}}$ contains the premise of this rule: the formula $\mathrm{Ker}\left[R,\mathrm{P}_m^{\mathscr{M}},\delta\right](\vec{a})$. Moreover, this premise appears in $\vec{\mathfrak{a}}$ before the record with the formula $\mathrm{J}_\delta \mathrm{P}_m^{\mathscr{M}}(\vec{a})$ of rank $\langle s,m\rangle$. But $\mathrm{Ker}\left[R,\mathrm{P}_m^{\mathscr{M}},\delta\right](\vec{a})$ is local w.r.t. $\mathrm{P}_m^{\mathscr{M}}(\vec{a})$. Then by Corollary 15.1.12,

$$\mathrm{CSD}\left(\vec{\mathfrak{a}}_{\overleftarrow{s,m-1}}\right) \vdash_L \mathrm{Ker}\left[R, \mathrm{P}_m^{\cdot\mathscr{M}}, \delta\right](\vec{a}). \tag{15.18}$$

But (ii) and (15.18) cannot hold simultaneously since for $\varepsilon \neq \delta$ the formulas

$$\mathrm{Ker}\left[R, \mathrm{P}_m^{\cdot\mathscr{M}}, \varepsilon\right](\vec{a}) \quad \text{and} \quad \mathrm{Ker}\left[R, \mathrm{P}_m^{\cdot\mathscr{M}}, \delta\right](\vec{a})$$

are disjoint and $\mathrm{CSD}\left(\vec{\mathfrak{a}}_{\overleftarrow{s,m-1}}\right)$ is consistent. Thus our supposition does not hold. Therefore, $\mathrm{J}_\varepsilon \mathrm{P}_m^{\cdot\mathscr{M}}(\vec{a}) \in \mathrm{CSD}(\vec{\mathfrak{a}})$, as was to be proved. □

Proposition 15.2.12. Let L be an FPJ logic, σ be a signature and $\mathscr{M} \in \mathrm{AMC}_\sigma(L)$. Then

$$\mathrm{AG}^{\cdot\mathscr{M}}(\mathscr{M}) \in \mathrm{SG}_\sigma(L, \mathscr{M}).$$

Proof. By Propositions 14.1.18 and 15.2.9, $\mathrm{AG}^{\cdot\mathscr{M}}(\mathscr{M}) \in \mathrm{SG}_\sigma(L)$. We must establish the validity of the conditions (i), (ii) and (iii) from Definition 14.1.10. By Definition 14.1.17,

$$\mathrm{AG}^{\cdot\mathscr{M}}(\mathscr{M}) = \left\langle \mathrm{Sint}_L(\mathscr{M}), \mathrm{AG}^{\cdot\mathscr{M}}(\mathrm{MR}_{\cdot\mathscr{M}}), \mathrm{Ord}_{\cdot\mathscr{M}} \right\rangle.$$

Then (i) and (ii) from Definition 14.1.10 are valid. By Lemma 15.2.11 and Definitions 15.2.5 and 15.2.8, we get that for any $\varepsilon \in \mathscr{D}^\tau(L)$, $\vec{a} \in \mathrm{Dom}_{\cdot\mathscr{M}}\left(\mathrm{P}_m^{\cdot\mathscr{M}}\right)$,

$$\mathrm{J}_\varepsilon \mathrm{P}_m^{\cdot\mathscr{M}}(\vec{a}) \in \mathrm{CSD}(\vec{\mathfrak{a}}) \quad \text{iff} \quad \mathrm{CSD}\left(\vec{\mathfrak{a}}_{\overleftarrow{s,m-1}}\right) \vdash_L \mathrm{GC}\left[\mathrm{AG}^{\cdot\mathscr{M}}(\mathrm{MR}_{\cdot\mathscr{M}}), \mathrm{P}_m^{\cdot\mathscr{M}}, \varepsilon\right](\vec{a}).$$

Then also

$$\mathrm{J}_\tau \mathrm{P}_m^{\cdot\mathscr{M}}(\vec{a}) \in \mathrm{CSD}(\vec{\mathfrak{a}}) \quad \text{iff} \quad \mathrm{CSD}\left(\vec{\mathfrak{a}}_{\overleftarrow{s,m-1}}\right) \vdash_L \mathrm{GC}\left[\mathrm{AG}^{\cdot\mathscr{M}}(\mathrm{MR}_{\cdot\mathscr{M}}), \mathrm{P}_m^{\cdot\mathscr{M}}, \tau\right](\vec{a}),$$

since there exists a unique $\delta \in \mathscr{V}(L, \mathsf{I})$ such that $\mathrm{J}_\delta \mathrm{P}_m^{\cdot\mathscr{M}}(\vec{a}) \in \mathrm{CSD}(\vec{\mathfrak{a}})$, and there exists a unique $\delta \in \mathscr{V}(L, \mathsf{I})$ such that

$$\mathrm{CSD}\left(\vec{\mathfrak{a}}_{\overleftarrow{s,m-1}}\right) \vdash_L \mathrm{GC}\left[\mathrm{AG}^{\cdot\mathscr{M}}(\mathrm{MR}_{\cdot\mathscr{M}}), \mathrm{P}_m^{\cdot\mathscr{M}}, \delta\right](\vec{a}).$$

Therefore (iii) is also valid. □

Corollary 15.2.13. Let L be an FPJ logic, σ be a signature and $\mathscr{M} \in \mathrm{AMC}_\sigma(L)$. Then the following conditions hold:

(i) $\mathrm{AG}^{\cdot\mathscr{M}}$ is an $\emptyset$-$\mathscr{M}$-compatible conforming function;
(ii) $\mathscr{M}$ is $\emptyset$-conformable.

Proof. (i) By Proposition 15.2.9, $\mathrm{AG}^{\cdot\mathscr{M}}$ is a conforming function for $\mathrm{MR}_{\cdot\mathscr{M}}$ w.r.t. $\emptyset$. It is evident that any conforming function w.r.t. $\emptyset$ is $\emptyset$-robust (see Definitions 12.2.9 and 14.1.15). Therefore, by Proposition 15.2.12, $\mathrm{AG}^{\cdot\mathscr{M}}$ is $\emptyset$-$\mathscr{M}$-compatible (see Definition 14.1.19).

(ii) follows from (i) (see Definition 14.1.20). □

Thus we have seen that each atomic modification calculus is $\emptyset$-comformable, i.e. for such modification calculus an adequate semantics can be constructed with the help of the tools proposed in the previous section.

Part IV
Handling Complex Structures

Chapter 16
Introductory Explanation

16.1 Atomic Sorts

In the previous chapter the first results connecting syntactic and semantic constructs were obtained under the assumption that the modification calculi possess the property of *conformability*. This property means that there exists a *generating rule system* which corresponds to the modification rule system of a modification calculus in a special, strictly defined meaning. To define the iterative image of the modification theory we need to assume the same condition. Similarly to modification rules, generating rules are syntactic constructs but they are used in a semantic context.

The notions of conformability and generating rules need to be introduced because we wish to propose such a modification calculus that permits to work with *set sort* objects. A set sort means such a sort, the objects of which can be interpreted as sets. On the domain of such a sort there should be defined a binary operation, which plays the role of the union operation, and a constant that stands for the empty set.

However, even the existence of the analogues of set-theoretic operations satisfying the axioms of Boolean algebra is not sufficient to consider a sort to be a set sort. The set sort objects should behave like sets and be arguments of internal predicates. We shall revert to this subject later on. For the moment we consider the sorts that are in some sense converse to set sorts and which are called atomic sorts.

Although the definition of atomic sorts is simpler than the definition of set sorts, it requires preliminary introduction of some auxiliary notions. Therefore, for the time being, instead of giving the definition of atomic sorts we will give their intuitive explanation.

Objects of an atomic sort considered as arguments of internal predicates behave as isolated and *independent from each other* individuals. This independence can be formally expressed in the form of the locality requirement formulated in the Sect. 15.1. Informally this means that the value of a predicate for an object of atomic sort does not at all depend on the value of the same predicate for/of any other object of the same sort.

O. Anshakov, T. Gergely, *Cognitive Reasoning*, Cognitive Technologies,
DOI 10.1007/978-3-540-68875-4_16, © Springer-Verlag Berlin Heidelberg 2010

In addition to the locality requirement, the *exhaustiveness requirement* stems from the idea of independence. Indeed, suppose that a modification rule is used, and as a result the value of an internal predicate should change. The value of a predicate for an atomic object will not depend on the value of this predicate for any other objects of the same sort. Thus it should be immediately decided which value should be assigned to the predicate without waiting for the information about the value of this predicate for other objects of the same sort. In fact, regardless of values that this predicate can obtain for other objects of the same sort, nothing will change for the given predicate and object. Therefore, the value to be assigned to the predicate should be known precisely and immediately. For this the major premise of at least one modification rule from the complete modification system (CMS) for any string of arguments should be true for the considered predicate.[1] This is exactly what is meant by *exhaustiveness*.

Thus if we deal only with objects of atomic sort then the modification rules will satisfy the locality condition and CMS will satisfy the exhaustiveness condition. In this case modification rules can be very easily transformed into generating rules; so easily that separate introduction of the generating rules is not necessary. The way of generating the sequences of L-structures as models of modification theories may be defined directly by modification rules

The necessity to separate the generating rules for the sequences of L-structures from the modification rules emerged from our attempt to represent set sort objects in modification theories. However, the generating rules for set sorts can be formulated much more easily than the corresponding modification rules.

16.2 Set Sorts

Let us consider two internal predicates ($\circ\!\!\rightarrow$ and $\bullet\!\!\rightarrow$) with the following informal interpretation:

- $o \circ\!\!\rightarrow P$ means that "the object o has the set of properties P",
- $f \bullet\!\!\rightarrow P$ means that "the fragment f is the cause of the set of properties P".

The desired properties of these predicates have served us as a sample for the development of a system of requirements concerning the way of manifestation of multiplicity of entities (objects) through the properties of internal predicates. First let us consider the following *seemingly trivial* circumstance:

- An object *o possesses a set of properties P*

[1] Major premises of two rules cannot be valid at the same time because of the condition of necessary *disjointness* for the complete modification systems (CMS).

iff
it *possesses each property* from P.

Similar statement holds for the fragment which is a possible cause of the set of properties:

- A fragment f is the *cause of a set of properties P*

  iff
  f is the *cause of each property from P.*

Since we will deal only with *sets of properties* and not with *isolated properties* the above statements should be rewritten as:

- An object o possesses *a set of properties P*

  iff
  it possesses *each singleton (one element subset) of the set of properties P.*

- A fragment f is the *cause of a set of properties P*

  iff
  f is *the cause of each singleton of the set of properties P.*

We consider this example in the logic **St** (see Definition 7.4.9). In the terms of the present chapter the above statements mean that a *positive (universal) atomic condition* w.r.t. the "set of properties" holds for the J-atomic formulas $\mathrm{J}_{+1}(o \circ\!\!\rightarrow P)$ and $\mathrm{J}_{+1}(f \bullet\!\!\rightarrow P)$.

Now we give a more abstract formulation. Let $\mathfrak{s}$ be a set sort (i.e. the entities of the sort s will be considered as sets). Let Φ be a statement on the sets of sort s. We will say that Φ satisfies a *positive (universal) atomic condition w.r.t. the sort* S if:

For any set X of sort s:

- A statement Φ *holds for X* iff it holds *for each singleton of the set X.*

Besides the *positive (universal) atomic condition*, entities of the set sorts may be connected with *negative (existential) atomic condition.* The negative atomic condition may be illustrated with the following statements:

- "object o does not possess the set of properties P " and
- "fragment f is the cause of the absence of the set of properties P ".

These statements can be represented by the J-atomic formulas $\mathrm{J}_{-1}(o \circ\!\!\rightarrow P)$ and $\mathrm{J}_{-1}(f \bullet\!\!\rightarrow P)$, respectively. Obviously:

- An object o *does not possess a set of properties P*

  iff
  there exists a property from P, which o *does not possess.*

- A fragment f *is the cause of the absence of a set of properties P*

  iff
  there exists a property from P, for which f *is the cause of its absence.*

Obviously the above-formulated condititons can be written in the terms of singleton as was done above for the statement "object o possesses the set of properties P" and "fragment f is the cause of the presence of the set of properties P". However, we will immediately describe the general case. Let S be a set sort and let Φ be a statement about the sets of the sort s. We will say that Φ satisfies the *negative (existential) atomic condition w.r.t. the sort* S, if:

For any set X of the sort s:

- A statement Φ holds for X

  iff it *holds for at least one singleton of the set* X.

Besides positive and negative atomic conditions which connect internal predicates with the entities of set sorts, we will suppose that on the objects of set sorts the following are defined:

- Union operation $\cup$
- Inclusion relation $\subseteq$
- Constant $\emptyset$ (empty set)

Operation $\cup$, relation $\subseteq$ and constant $\emptyset$ should satisfy a certain collection of obvious assumptions which will be called in this chapter set axioms.

16.3 Some Properties of the Set Sorts

Let us consier some properties/features of the statements which satisfy the positive atomic condition by the example of the statement "object o possesses the set of properties P". It is to be shown that for this statement the so-called *additivity condition* holds:

Let:

- Object o possess the set of properties P,
- Object o possess the set of properties Q.

Then:

- Object o possesses the set of properties $P \cup Q$.

It is easy to see that *downward inheritance* w.r.t. set-theoretic inclusion takes place here.

Let:

- Object o possess the set of properties P;
- $Q \subseteq P$.

Then:

- Object o possesses the set of properties Q.

I.e. if an object possesses a wider set of properties then it will also possess a narrower one.

Besides the conditions of additivity and downward inheritance we wish to have the following condition of nontriviality (non-emptiness): each object possesses some set of properties. Now we formulate this condition similarly to the above notations of additivity and downward inheritance.

For each object o there exists a set of properties P such that

- Object o possesses the set of properties P.

If downward inheritance takes place then the above-formulated condition of nontriviality is equivalent to the following condition:

An object o possesses the empty set of properties.

Simultaneous fulfilment of the conditions of additivity, downward inheritance and non-triviality means that the collection of the sets of properties which the object o possesses is an *ideal* of the Boolean algebra of all subsets of the universe.[2]

It can be proved that any statement satisfying positive atomic condition w.r.t. some set sort satisfies similar conditions.

Let Φ be a statement about the sets of sort s satisfying positive atomic condition w.r.t. this sort. Then for any sets X and Y of the sort s the following statements hold:

- If Φ holds for X and Y then Φ holds for $X \cup Y$,
- If Φ holds for X and $Y \subseteq X$, then Φ holds for Y,
- Φ is valid for $\emptyset$.

Any statement Φ satisfying the negative atomic condition possesses another collection of useful properties among which there exists *upward inheritance* w.r.t. inclusion:

- If Φ holds for X and $X \subseteq Y$, then Φ holds for Y .

In this section we do not consider properties connected with the negative atomic condition, but these can be found in the further sections of the present chapter.

16.4 Modification Rules and Modification Calculi for Set Sorts

We continue considering the example with objects possessing sets of properties and fragments which can be the causes of these sets of properties. Consider the following candidate rule for searching for possible causes:

---

[2] A *non-empty* collection of sets closed under union and the relation "be a subset" is called an ideal of an algebra of sets. More formally let $\mathfrak{B} = \langle B, \cap, \cup, \rangle$ be a Boolean algebra of sets. An ideal in $\mathfrak{B}$ is called $I \subseteq B$ such that the following conditions hold:

(i) $X \in I$ and $Y \in I$ implies $X \cup Y \in I$;
(ii) $X \in I$ and $Y \subseteq X$ implies $Y \in I$;
(iii) $\emptyset \in I$.

The last condition guarantees that the ideal is non-empty. The set of all subsets of a set is an example ideal. This ideal is called principal.

Let:

- Each object, containing the fragment f, have the set of properties P.

Then:

- The fragment f is a cause of the set of properties P.

Let us consider the premise of this rule:

- Each object containing the fragment f, has the set of properties P.

It can be shown that this statement *satisfies the positive atomic condition* w.r.t. the sort of the set of properties.

Indeed, the premise of a rule which should support set sorts is a more complex construction. What we consider here as a premise is called in this chapter the *kernel of the premise.* We will explain informally what a premise which supports set sorts should look like. This is not yet a modification rule but an intuitive rule on the basis of which a modification rule can be constructed.

Let $\mathrm{K}(f,P)$ denote the *kernel of a premise* of the rule the conclusion of which is the statement "the fragment f is a cause of the set of properties P". Then the rule itself should have the following form:

Let at least one of the following two conditions hold:

- There exists Q such that $P \subseteq Q$ and
  - f is a cause of the set of properties Q or $\mathrm{K}(f,Q)$ holds;
- There are Q and R such that $P = Q \cup R$ and each of the following conditions holds:
  - f is a cause of the set of properties Q or $\mathrm{K}(f,Q)$ holds;
  - f is a cause of the set of properties R or $\mathrm{K}(f,R)$ holds.

Then:

- The fragment f is a cause of the set of properties P.

Now we will formulate this rule even less formally:

The sufficient conditions for the fragment f to be a cause of the set of properties P are the following:

- f will be a cause of an even wider set of properties Q or

- The kernel of the premise would be valid for this wider set and for the fragment f or
- P could be represented in the form of the union of the two sets Q and R such that for each of them:
  - f is a cause of the set of properties or
  - The kernel of the premise holds for f and for the set of properties.

Note that the kernel of the premise satisfies the positive atomic condition w.r.t. the sort of the set of properties. This is an important circumstance. It will be used in the present chapter for the definition of the *perfect* modification calculus. It is for the perfect modification calculi that their *conformability*, and thus, the possibility to define *adequate semantics* for them will be proved. The definition of perfect modification calculus will be given later on.

16.5 Example

We conclude this chapter with the following example. Let us consider a robotic system that supports drug design processes as a robot assistant. The system automatically generates hypotheses to explain the results of investigations. Let the objective of the system be to find a new compound with a certain biological effect.

There are some results of investigations. Namely suppose that the chemical compounds **A**, **B**, **C**, **D** and **E** are investigated for antitumour activity and toxicity. The structural formulas of these compounds may contain a certain fragment **F**. This fragment may be a cause of any of the two target properties: the presence of antitumour activity (**a**) and low toxicity (**l**). However we will consider four sets of properties: $\emptyset$, $\{\mathbf{a}\}$, $\{\mathbf{l}\}$, $\{\mathbf{a},\mathbf{l}\}$.

Thus we have three sorts of objects: C is the sort of chemical compounds , F is the sort of fragments of chemical compounds and P is the sort of sets of (target) properties. Consider two internal predicates: P and C. $\mathrm{P}(f,c,p)$ means "fragment f is contained in the structural formula of compound c, which has the set of properties p". $\mathrm{C}(f,p)$ is understood as "the presence of fragment f in the structural formula is the cause of the appearance of the set of properties p".

We will consider this example in the simplest FPJ-logic **Si**, which has exactly two internal truth values: '?' and '!' (see Definition 7.4.7). Recall that in **Si** we denote $J_?$ and $J_!$ by '?' and '!', respectively. Note that '?' is the value "uncertain" in **Si**, i.e. we use '?' instead of 'τ' in this example.

A model of the description of the initial state can be repesented with Table 16.1.

We will not enumerate all the axioms occuring in the state description the model of which is given in Table 16.1. Only some of them will be given. Note that in our example the union operation $\cup$, the inclusion relation $\subseteq$ and the constant "empty set" $\emptyset$ are defined on the objects of the set sort P. Hence in the state description there

Table 16.1 Internal predicates with set-sort arguments

Predicate P			
F	C	P	I
F	**A**	$\emptyset$	?
F	**A**	{**a**}	!
F	**A**	{**l**}	?
F	**A**	{**a**, **l**}	?
F	**B**	$\emptyset$	?
F	**B**	{**a**}	!
F	**B**	{**l**}	?
F	**B**	{**a**, **l**}	?
F	**C**	$\emptyset$	?
F	**C**	{**a**}	?
F	**C**	{**l**}	!
F	**C**	{**a**, **l**}	?
F	**D**	$\emptyset$	?
F	**D**	{**a**}	?
F	**D**	{**l**}	!
F	**D**	{**a**, **l**}	?
F	**E**	$\emptyset$	?
F	**E**	{**a**}	?
F	**E**	{**l**}	?
F	**E**	{**a**, **l**}	?

Predicate C		
F	P	I
F	$\emptyset$	?
F	{**a**}	?
F	{**l**}	?
F	{**a**, **l**}	?

should occur axioms describing $\cup$, $\subseteq$ and $\emptyset$. Thus the state description will contain the following groups of formulas:

- The group "*description of the operation* $\cup$" contains equalities that completely describe this operation for the objects of the sort P, i.e. those representing the table of this operation. The following formulas are examples of axioms of this group:
  - $\{\mathbf{a}\} \cup \{\mathbf{l}\} = \{\mathbf{a},\mathbf{l}\}$, $\{\mathbf{l}\} \cup \{\mathbf{a}\} = \{\mathbf{a},\mathbf{l}\}$, $\{\mathbf{a}\} \cup \emptyset = \{\mathbf{a}\}$, $\{\mathbf{l}\} \cup \emptyset = \{\mathbf{l}\}$, $\{\mathbf{a}\} \cup \{\mathbf{a},\mathbf{l}\} = \{\mathbf{a},\mathbf{l}\}$.

Altogether this group contains 16 formulas represented in the cells of Table 16.2.

- The group "*description of the external predicate* $\subseteq$" contains formulas that entirely describe this external predicate on the objects of the sort P. The following formulas are examples of axioms of this group:

Table 16.2 Table for the operation ∪

∪	∅	{a}	{l}	{a, l}
∅	∅	{a}	{l}	{a, l}
{a}	{a}	{a}	{a, l}	{a, l}
{l}	{l}	{a, l}	{l}	{a, l}
{a, l}	{a, l}	{a, l}	{a, l}	{a, l}

- $\{\mathbf{a}\} \subseteq \{\mathbf{a},\mathbf{l}\}$, $\{\mathbf{l}\} \subseteq \{\mathbf{a},\mathbf{l}\}$, $\emptyset \subseteq \{\mathbf{a}\}$, $\emptyset \subseteq \{\mathbf{l}\}$, $\{\mathbf{a},\mathbf{l}\} \subseteq \{\mathbf{a},\mathbf{l}\}$, $\{\mathbf{a},\mathbf{l}\} \not\subseteq \{\mathbf{a}\}$, $\{\mathbf{a},\mathbf{l}\} \not\subseteq \{\mathbf{l}\}$, $\{\mathbf{a}\} \not\subseteq \emptyset$

Altogether this group contains 16 formulas represented in the cells of Table 16.3.

Table 16.3 Table for the external predicate ⊆

⊆	∅	{a}	{l}	{a, l}
∅	**t**	**t**	**t**	**t**
{a}	**f**	**t**	**f**	**t**
{l}	**f**	**f**	**t**	**t**
{a, l}	**f**	**f**	**f**	**t**

- The group "*description of the internal predicate* P" contains formulas that entirely describe this internal predicate. The following formulas are examples of axioms of this group:
  - $?\mathrm{P}(\mathbf{F},\mathbf{A},\emptyset)$, $!\mathrm{P}(\mathbf{F},\mathbf{A},\{\mathbf{a}\})$, $?\mathrm{P}(\mathbf{F},\mathbf{A},\{\mathbf{a},\mathbf{l}\})$, $!\mathrm{P}(\mathbf{F},\mathbf{B},\{\mathbf{a}\})$, $?\mathrm{P}(\mathbf{F},\mathbf{C},\{\mathbf{a},\mathbf{l}\})$, $!\mathrm{P}(\mathbf{F},\mathbf{C},\{\mathbf{l}\})$.

Altogether this group contains 20 formulas corresponding to the rows of the table which represents the predicate P (see Table 16.1).

- The group "*description of the internal predicate* C" contains formulas that entirely describe this internal predicate. Here we enumerate all the formulas of this that represents the predicate C (see Table 16.1):
  - $?\mathrm{C}(\mathbf{F},\emptyset)$, $?\mathrm{C}(\mathbf{F},\{\mathbf{a}\})$, $?\mathrm{C}(\mathbf{F},\{\mathbf{l}\})$, $?\mathrm{C}(\mathbf{F},\{\mathbf{a},\mathbf{l}\})$.

Before we formulate the modification rules for the modificational calculus in question, let us note that through the initial external predicates and the constant $\emptyset$ there can be defined the external unary predicate Atom, which has been defined on the objects of sort P. Atom (X) means that X is a singleton.

A set of properties X is called atom if:

- $X \neq \emptyset$;
- for any set of properties Y,
  - $Y \subseteq X$ implies $Y = \emptyset$ or $Y = X$.

I.e. Atom (X) is an abbreviation of the formula

$$X \neq \emptyset \wedge \forall Y \, (Y \subseteq X \rightarrow Y = \emptyset \vee Y = X) \, .$$

The statement "the set Y contains the atom X" will be denoted by $\mathrm{E}(X,Y)$. If a singleton is identified with its only element then $\mathrm{E}(X,Y)$ will denote $X \in Y$. $\mathrm{E}(X,Y)$ is an abbreviation of the formula

$$\mathrm{Atom}\,(X) \wedge X \subseteq Y.$$

$\mathrm{K}(f,p)$ will denote the statement:

- For any atom x that is contained in the set of properties p
  - there exist at least two compounds that contain fragment f and possess the set of properties x.

If a singleton is identified with its only element then this statement can be written as:

- For any $x \in p$ there are at least two compounds that contain fragment f and possess the property x.

$\mathrm{K}(f,p)$ is the abbreviation of the formula

$$\forall x \, (\mathrm{E}\,(x,p) \rightarrow \exists A \exists B \, (A \neq B \wedge !\mathrm{P}\,(f,A,x) \wedge !\mathrm{P}\,(f,B,x))) \, .$$

Taking into account the intuitive meaning of the predicate P, it is easy to show, that for any f the collection of sets of properties which satisfy the condition $\mathrm{K}(f,p)$ is an *ideal of the Boolean algebra of the sets of properties.*

$\mathrm{K}(f,p)$ will be used as the kernel of the premise of +Rule for the predicate C. **!Rule for** C itself will be written as follows:

$$\frac{\begin{array}{l} ?\mathrm{C}\,(f,p)\,, \\ \exists q \, (p \subseteq q \wedge (!\mathrm{C}\,(f,q) \vee \mathrm{K}\,(f,q))) \vee \\ \vee \, \exists q \exists r \, (p = q \cup r \wedge (!\mathrm{C}\,(f,q) \vee \mathrm{K}\,(f,q)) \wedge (!\mathrm{C}\,(f,r) \vee \mathrm{K}\,(f,r))) \end{array}}{!\mathrm{C}\,(f,p)}$$

The intuitive meaning of a similar rule was studied in detail in the previous section.

Note that, for any rule of this form, from the truth of the kernel of the major premise follows the truth of the major premise itself, since each set is a subset of itself.

Let MPC (f,p) denote the major premise of this rule, i.e. the formula

$$\exists q\,(p \subseteq q \wedge (!\mathrm{C}(f,q) \vee \mathrm{K}(f,q))) \vee$$
$$\vee\ \exists q \exists r\,(p = q \cup r \wedge (!\mathrm{C}(f,q) \vee \mathrm{K}(f,q)) \wedge (!\mathrm{C}(f,r) \vee \mathrm{K}(f,r)))\,.$$

Then ?**Rule for** C will be written as

$$\frac{\begin{array}{l} ?\mathrm{C}(f,p), \\ \forall w\,(?\mathrm{C}(f,w) \rightarrow \neg \mathrm{MPC}(f,w)) \end{array}}{?\mathrm{C}(f,p)}$$

This means that the cell of the matrix representing the predicate C will remain undefined iff **!Rule for** C cannot be applied to any of the undefined cells of this matrix.

Now the rules for the predicate P will be defined. First let us define **!Rule for** P. As the major premise for this rule we will use just the formula $!\mathrm{C}(f,p)$.

!Rule for P is the following:

$$\frac{\begin{array}{l} ?\mathrm{P}(f,A,p), \\ \exists q\,(p \subseteq q \wedge (!\mathrm{P}(f,A,q) \vee !\mathrm{C}(f,q))) \vee \\ \vee\ \exists q \exists r\,(p = q \cup r \wedge (!\mathrm{P}(f,A,q) \vee !\mathrm{C}(f,q)) \wedge (!\mathrm{P}(f,A,r) \vee !\mathrm{C}(f,r))) \end{array}}{!\mathrm{P}(f,A,p)}$$

Let $\mathrm{MPC}(f,A,p)$ denote the major premise of this rule, i.e. the formula

$$\exists q\,(p \subseteq q \wedge (!\mathrm{P}(f,A,q) \vee !\mathrm{C}(f,q))) \vee$$
$$\vee\ \exists q \exists r\,(p = q \cup r \wedge (!\mathrm{P}(f,A,q) \vee !\mathrm{C}(f,q)) \wedge (!\mathrm{P}(f,A,r) \vee !\mathrm{C}(f,r)))\,.$$

Then ?**Rule for** P will be formulated in the following way:

$$\frac{\begin{array}{l} ?\mathrm{P}(f,A,p), \\ \forall w\,(?\mathrm{P}(f,A,w) \rightarrow \neg \mathrm{MPC}(f,A,w)) \end{array}}{?\mathrm{P}(f,A,p)}\,.$$

This means that the cell of the matrix representing the predicate P will remain undefined iff **!Rule for** P cannot be applied to any of the undefined cells of this matrix.

Now we will give an example of an inference in the above-given modification calculus in an abbreviated and simplified form. However instead of providing an example of inferences in the modification calculus we will give an example of such informal reasoning which uses informal analogues of the above-introduced rules. This reasoning serves for modification of the matrices representing the predicates C and P.

(i) There exist two compounds (**A** and **B**), that contain the fragment **F** and possess the set of properties $\{\mathbf{a}\}$. Then the kernel of the major premise of the rule **!Rule**

for C is valid and thus the major premise itself is also valid. Therefore **F** is a cause of the set of properties $\{\mathbf{a}\}$.

(ii) There exist two compounds (**C** and **D**), that contain the fragment **F** and possess the set of properties $\{\mathbf{l}\}$. Then the kernel of the major premise of the rule **!Rule for** C is valid and thus the major premise itself is also valid. Therefore, **F** is a cause of the set of properties $\{\mathbf{l}\}$.

Thus in two cells of the table for C, '?' has been replaced by '!' and Table 16.4 has been obtained.

Table 16.4 Illustration of the reasoning process

Predicate C		
F	P	I
F	$\emptyset$	?
F	$\{\mathbf{a}\}$	!
F	$\{\mathbf{l}\}$	!
F	$\{\mathbf{a}, \mathbf{l}\}$	?

(i) Note that **F** is a cause for a set of properties wider than $\emptyset$ (e.g. $\{\mathbf{a}\}$). Then the major premise of **+Rule for C** is also valid for the first row of our table. Therefore we have to replace '?' by '!' in it.

(ii) Note that $\{\mathbf{a}, \mathbf{l}\} = \{\mathbf{a}\} \cup \{\mathbf{l}\}$. The fragment **F** is a cause of each of the sets $\{\mathbf{a}\}$ and $\{\mathbf{l}\}$. Then the major premise of the rule also holds. Therefore we can replace '?' by '!' also in the fourth row of the table for C. Thus Table 16.5 is obtained.

Table 16.5 Illustration of the reasoning process

Predicate C		
F	P	I
F	$\emptyset$	!
F	$\{\mathbf{a}\}$	!
F	$\{\mathbf{l}\}$	!
F	$\{\mathbf{a}, \mathbf{l}\}$	!

Thus the informal analogue of the module $\langle 1,1\rangle$ has been done. That is, the traversal and the completion of the definition of the predicate C has been done for the first of the two internal predicates. Let us now proceed to the module $\langle 1,2\rangle$, in which predicate P must be processed.

(i) Reasoning in the module $\langle 1,2\rangle$ is trivial. The fragment **f** is a cause of each considered set of properties. Then for each row of the table for P the kernel (i.e. the formula $!\mathrm{C}(f,p)$) of the major premise of **!Rule for** P is valid. Then the major premise itself is also valid. Therefore, in all the rows of the table for P where '?' occurs, '?' should be replaced by '!'.

Thus Table 16.6 will be obtained.

Table 16.6 Illustration of the reasoning process

Predicate P				Predicate P (continuation)			
F	C	P	I	F	C	P	I
F	**A**	∅	!	**F**	**C**	{**l**}	!
F	**A**	{**a**}	!	**F**	**C**	{**a**, **l**}	!
F	**A**	{**l**}	!	**F**	**D**	∅	!
F	**A**	{**a**, **l**}	!	**F**	**D**	{**a**}	!
F	**B**	∅	!	**F**	**D**	{**l**}	!
F	**B**	{**a**}	!	**F**	**D**	{**a**, **l**}	!
F	**B**	{**l**}	!	**F**	**E**	∅	!
F	**B**	{**a**, **l**}	!	**F**	**E**	{**a**}	!
F	**C**	∅	!	**F**	**E**	{**l**}	!
F	**C**	{**a**}	!	**F**	**E**	{**a**, **l**}	!

Chapter 17
Set-Admitting Structures

17.1 Atoms

In this section we study the so-called set sorts. As mentioned above, objects of set sorts are regarded as sets. This is related to a certain number of restrictions, some of which will be considered below. Note that we will first investigate those sets of conditions that are technically simple and transparent. Later on our technique will be much more sophisticated and complex.

Definition 17.1.1. Let σ be a signature, $\mathsf{s} \in \mathsf{Sorts}_\sigma$. The sort s is called a *set sort* of signature σ if $\mathrm{Pr}_\sigma(\mathsf{E})$ contains symbol $\subseteq_\mathsf{s}$, and Fu_σ contains symbols $\cup_\mathsf{s}$ and Λ_s such that:

(i) $\mathrm{Ar}_\sigma(\subseteq_\mathsf{s}) = \langle s, s \rangle$,
(ii) $\mathrm{Ar}_\sigma(\cup_\mathsf{s}) = \langle s, s \rangle$, $\mathrm{Sort}_\sigma(\cup_\mathsf{s}) = s$,
(iii) $\mathrm{Ar}_\sigma(\Lambda_\mathsf{s}) = \langle\rangle$, $\mathrm{Sort}_\sigma(\Lambda_\mathsf{s}) = s$.

Below the set-admitting L-structure, where L is an FPJ logic, will be defined. In the regular L-structures the set sorts are properly interpreted, i.e. the objects of the set sorts are regarded as sets. The basic properties of the set-admitting L-structures will be obtained below.

Definition 17.1.2. Let σ be a signature. By definition put

$$\mathsf{SetSorts}_\sigma = \{\mathsf{s} \in \mathsf{Sorts}_\sigma \mid \mathsf{s} \text{ is a set sort of signature } \sigma\}.$$

Definition 17.1.3. Let L be an FPJ logic and σ be a signature. An L-structure $\mathfrak{J}$ of signature σ is called *set-admitting* if for each $\mathsf{s} \in \mathsf{SetSorts}_\sigma$ there exist an $A \subseteq \mathrm{Dom}_\mathfrak{J}(\mathsf{s})$ and an injection

$$\varphi\colon \mathrm{Dom}_\mathfrak{J}(\mathsf{s}) \to 2^A$$

such that

(i) $\varphi(a) = \{a\}$ iff $a \in A$,

O. Anshakov, T. Gergely, *Cognitive Reasoning*, Cognitive Technologies,
DOI 10.1007/978-3-540-68875-4_17,

(ii) $\varphi\left(\Lambda_{\mathsf{s}}^{\mathfrak{J}}\right) = \emptyset$,
(iii) $\varphi\left(b \cup_{\mathsf{s}}^{\mathfrak{J}} c\right) = \varphi(b) \cup \varphi(c)$ for any $b, c \in \mathrm{Dom}_{\mathfrak{J}}(\mathsf{s})$,
(iv) $b \subseteq_{\mathsf{s}}^{\mathfrak{J}} c$ iff $\varphi(b) \subseteq \varphi(c)$ for any $b, c \in \mathrm{Dom}_{\mathfrak{J}}(\mathsf{s})$.

Then A is called *the set of atoms of sort* s *in* $\mathfrak{J}$, and φ is called a *canonic injection of sort* s *in* $\mathfrak{J}$.

Definition 17.1.4. Let L be an FPJ logic and σ be a signature. Let $\mathfrak{J}$ be an L-structure of signature σ and $\mathsf{s} \in \mathsf{SetSorts}_\sigma$. Let by definition

$$\mathrm{At}_{\mathsf{s}}^{\mathfrak{J}} = \Big\{a \in \mathrm{Dom}_{\mathfrak{J}}(\mathsf{s}) \mid a \neq \Lambda_{\mathsf{s}}^{\mathfrak{J}} \text{ and for any } b \in \mathrm{Dom}_{\mathfrak{J}}(\mathsf{s}),$$
$$b \subseteq_{\mathsf{s}}^{\mathfrak{J}} a \text{ implies } b = \Lambda_{\mathsf{s}}^{\mathfrak{J}} \text{ or } b = a\Big\}.$$

Let $\mathrm{Cn}_{\mathsf{s}}^{\mathfrak{J}}$ denote the map from $\mathrm{Dom}_{\mathfrak{J}}(\mathsf{s})$ to the set $2^{\mathrm{At}_{\mathsf{s}}^{\mathfrak{J}}}$ defined as follows: for any $b \in \mathrm{Dom}_{\mathfrak{J}}(\mathsf{s})$,

$$\mathrm{Cn}_{\mathsf{s}}^{\mathfrak{J}}(b) = \Big\{a \in \mathrm{At}_{\mathsf{s}}^{\mathfrak{J}} \mid a \subseteq_{\mathsf{s}}^{\mathfrak{J}} b\Big\}.$$

Lemma 17.1.5. Let L be an FPJ logic and σ be a signature. Let $\mathfrak{J}$ be a set-admitting L-structure of signature σ, Then for each $\mathsf{s} \in \mathsf{SetSorts}_\sigma$ there exist a unique set of atoms and a unique canonical injection, namely $\mathrm{At}_{\mathsf{s}}^{\mathfrak{J}}$ is the set of atoms of sort s, and the canonical injection of sort s equals $\mathrm{Cn}_{\mathsf{s}}^{\mathfrak{J}}$.

Proof. Suppose that there exist an $A \subseteq \mathrm{Dom}_{\mathfrak{J}}(\mathsf{s})$ and an injection

$$\varphi\colon \mathrm{Dom}_{\mathfrak{J}}(\mathsf{s}) \to 2^A$$

such that the conditions (i)–(iv) of Definition 17.1.3 hold. Let us show that $A = \mathrm{At}_{\mathsf{s}}^{\mathfrak{J}}$.

($\subseteq$) Let $a \in A$. Then $a \neq \Lambda_{\mathsf{s}}^{\mathfrak{J}}$. (In the converse case by (i) and (ii) of 17.1.3 we would obtain $\{a\} = \emptyset$, which is a contradiction.) Suppose $b \subseteq_{\mathsf{s}}^{\mathfrak{J}} a$. Then by 17.1.3 (see (i) and (iv)), $\varphi(b) \subseteq \{a\}$, that is possible only if $\varphi(b) = \emptyset$ or $\varphi(b) = \{a\}$, which are equivalent to $b = \Lambda_{\mathsf{s}}^{\mathfrak{J}}$ and $b = a$, respectively (see 17.1.3 (i) and (ii)). Thus it has been proved that the characteristic property of the set $\mathrm{At}_{\mathsf{s}}^{\mathfrak{J}}$ holds for a, therefore $a \in \mathrm{At}_{\mathsf{s}}^{\mathfrak{J}}$.

($\supseteq$) Let $a \in \mathrm{At}_{\mathsf{s}}^{\mathfrak{J}}$. Since φ is a injection then by 17.1.3 (ii)

$$\varphi(a) = \emptyset \quad \text{iff} \quad a = \Lambda_{\mathsf{s}}^{\mathfrak{J}}.$$

But $\Lambda_{\mathsf{s}}^{\mathfrak{J}} \notin \mathrm{At}_{\mathsf{s}}^{\mathfrak{J}}$. Therefore $\varphi(a) \neq \emptyset$. Let us show that $\varphi(a) \subseteq \{a\}$. Let $b \in \varphi(a)$. Then $b \in A$ and $\varphi(b) = \{b\} \subseteq \varphi(a)$. Then by 17.1.3 (iv), $b \subseteq_{\mathsf{s}}^{\mathfrak{J}} a$. By the characteristic property of the set $\mathrm{At}_{\mathsf{s}}^{\mathfrak{J}}$ in this case $b = \Lambda_{\mathsf{s}}^{\mathfrak{J}}$ or $b = a$. But the equality $b = \Lambda_{\mathsf{s}}^{\mathfrak{J}}$ does not hold since $b \in A$. Hence $b = a$. Thus $\varphi(a) \subseteq \{a\}$ and $\varphi(a) \neq \emptyset$. Then $\varphi(a) = \{a\}$ and by Definition 17.1.3 (i), we obtain $a \in A$.

Thus, $A = \mathrm{At}_{\mathsf{s}}^{\mathfrak{J}}$.

Now, let us show that $\varphi(b) = \mathrm{Cn}_{\mathsf{s}}^{\mathfrak{J}}(b)$ for any $b \in \mathrm{Dom}_{\mathfrak{J}}(\mathsf{s})$,

($\subseteq$) Let $a \in \varphi(b)$. Then

$$a \in A = \mathrm{At}_{\mathsf{s}}^{\mathfrak{J}} \quad \text{and} \quad \varphi(a) = \{a\} \subseteq \varphi(b).$$

In this case $a \subseteq_{\mathsf{s}}^{\mathfrak{J}} b$ by 17.1.3 (iv). Therefore, $a \in \mathrm{Cn}_{\mathsf{s}}^{\mathfrak{J}}(b)$ (see 17.1.4)

($\supseteq$) Let $a \in \mathrm{Cn}_{\mathsf{s}}^{\mathfrak{J}}(b)$. Then $a \in \mathrm{At}_{\mathsf{s}}^{\mathfrak{J}} = A$ and $a \subseteq_{\mathsf{s}}^{\mathfrak{J}} b$. In this case, by 17.1.3 (iv): $\varphi(a) = \{a\} \subseteq \varphi(b)$. Then $a \in \varphi(b)$. □

Further on we will work with modification calculi whose models are necessarily finite. In the case of finite models canonical injection is, at the same time, a surjection. Therefore we will speak about canonical bijection, which establishes an isomorphism between our semilattice and the semilattice of all subsets of the set of atoms.

Definition 17.1.6. Let L be an FPJ logic and σ be a signature. Let $\mathfrak{J}$ be an L-structure of signature σ, $\mathsf{s} \in \mathsf{SetSorts}_{\sigma}$, $A \subseteq \mathrm{Dom}_{\mathfrak{J}}(s)$, A is finite. Define the expression

$$\bigcup_{c \in A}{}_{s}^{\mathfrak{J}}\, c$$

by induction on the cardinality of A.

BASIS. $\bigcup_{c \in \emptyset}{}_{s}^{\mathfrak{J}}\, c = \Lambda_{\mathsf{s}}^{\mathfrak{J}}$.

INDUCTION STEP. $\bigcup_{c \in B \cup \{b\}}{}_{s}^{\mathfrak{J}}\, c = \left(\bigcup_{c \in B}{}_{s}^{\mathfrak{J}}\, c\right) \cup_{\mathsf{s}}^{\mathfrak{J}} b$, where $b \notin B$.

Lemma 17.1.7. Let L be an FPJ logic and σ be a signature. Let $\mathfrak{J}$ be a set-admitting L-structure of signature σ, $\mathsf{s} \in \mathsf{SetSorts}_{\sigma}$. Suppose $A \subseteq \mathrm{At}_{\mathsf{s}}^{\mathfrak{J}}$, A is finite. Then

$$\mathrm{Cn}_{\mathsf{s}}^{\mathfrak{J}}\left(\bigcup_{c \in A}{}_{s}^{\mathfrak{J}}\, c\right) = A.$$

Proof. Induction on the cardinality of A.

BASIS. Let the cardinality of A be equal to 0. Then $A = \emptyset$. We see that

$$\begin{aligned} \mathrm{Cn}_{\mathsf{s}}^{\mathfrak{J}}\left(\bigcup_{c \in \emptyset}{}_{s}^{\mathfrak{J}}\, c\right) &= \mathrm{Cn}_{\mathsf{s}}^{\mathfrak{J}}\left(\Lambda_{\mathsf{s}}^{\mathfrak{J}}\right) && \text{(by 17.1.6)} \\ &= \emptyset && \text{(by 17.1.3 (ii))} \end{aligned}$$

INDUCTION STEP. Let the statement be valid if the cardinality of A is equal to n. Assume that cardinality of A is equal to $n+1$. Then $A = B \cup \{b\}$, where $b \notin B$. Hence the cardinality of B is equal to n. Then

$$\begin{aligned} \mathrm{Cn}_{\mathsf{s}}^{\mathfrak{J}}\left(\bigcup_{c \in B \cup \{b\}}{}_{s}^{\mathfrak{J}}\, c\right) &= \mathrm{Cn}_{\mathsf{s}}^{\mathfrak{J}}\left(\left(\bigcup_{c \in B}{}_{s}^{\mathfrak{J}}\, c\right) \cup_{\mathsf{s}}^{\mathfrak{J}} b\right) && \text{(by 17.1.6)} \\ &= \mathrm{Cn}_{\mathsf{s}}^{\mathfrak{J}}\left(\bigcup_{c \in B}{}_{s}^{\mathfrak{J}}\, c\right) \cup \{b\} && \text{(by 17.1.3 (i), (iii))} \\ &= B \cup \{b\} && \text{(by the induction hypothesis)} \end{aligned}$$

□

Lemma 17.1.8. Let L be an FPJ logic and σ be a signature. Let $\mathfrak{J}$ be a set-admitting L-structure of signature σ, $\mathsf{s} \in \mathsf{SetSorts}_\sigma$. Suppose $\mathrm{Dom}_{\mathfrak{J}}(\mathsf{s})$ is finite. Then the canonical injection $\mathrm{Cn}_{\mathsf{s}}^{\mathfrak{J}}$ is a bijection of $\mathrm{Dom}_{\mathfrak{J}}(\mathsf{s})$ onto $2^{\mathrm{At}_{\mathsf{s}}^{\mathfrak{J}}}$.

Proof. We will show that the injection $\mathrm{Cn}_{\mathsf{s}}^{\mathfrak{J}}$ is a surjection. Let $A \subseteq \mathrm{At}_{\mathsf{s}}^{\mathfrak{J}}$. Then A is finite. Let

$$a = \bigcup_{c \in A}{}_{s}^{\mathfrak{J}}\, c.$$

Then by Lemma 17.1.7, $\mathrm{Cn}_{\mathsf{s}}^{\mathfrak{J}}(a) = A$. □

Lemma 17.1.9. Let L be an FPJ logic and σ be a signature. Let $\mathfrak{J}$ be a set-admitting L-structure of signature σ, $\mathsf{s} \in \mathsf{SetSorts}_\sigma$ and Φ be an external L-formula of signature σ. Then the following statements hold:

(i) For any $a \in \mathrm{At}_{\mathsf{s}}^{\mathfrak{J}}$ and any $b \in \mathrm{Dom}_{\mathfrak{J}}(\mathsf{s})$, $a \subseteq_{\mathsf{s}}^{\mathfrak{J}} b$ iff $a \in \mathrm{Cn}_{\mathsf{s}}^{\mathfrak{J}}(b)$;
(ii) $a \subseteq_{\mathsf{s}}^{\mathfrak{J}} \Lambda_{\mathsf{s}}^{\mathfrak{J}}$ does not hold for an arbitrary $a \in \mathrm{At}_{\mathsf{s}}^{\mathfrak{J}}$
(iii) Let $b \subseteq_{\mathsf{s}}^{\mathfrak{J}} c$ and $\mathfrak{J} \models_L \Phi(a)$ hold for any $a \in \mathrm{At}_{\mathsf{s}}^{\mathfrak{J}}$ such that $a \subseteq_{\mathsf{s}}^{\mathfrak{J}} c$; then $\mathfrak{J} \models_L \Phi(a)$ holds also for any $a \in \mathrm{At}_{\mathsf{s}}^{\mathfrak{J}}$ such that $a \subseteq_{\mathsf{s}}^{\mathfrak{J}} b$;
(iv) Let

- $\mathfrak{J} \models_L \Phi(a)$ hold for any $a \in \mathrm{At}_{\mathsf{s}}^{\mathfrak{J}}$ such that $a \subseteq_{\mathsf{s}}^{\mathfrak{J}} b$,
- $\mathfrak{J} \models_L \Phi(a)$ hold for any $a \in \mathrm{At}_{\mathsf{s}}^{\mathfrak{J}}$ such that $a \subseteq_{\mathsf{s}}^{\mathfrak{J}} c$;

then $\mathfrak{J} \models_L \Phi(a)$ holds for any $a \in \mathrm{At}_{\mathsf{s}}^{\mathfrak{J}}$ such that $a \subseteq_{\mathsf{s}}^{\mathfrak{J}} b \cup_{\mathsf{s}}^{\mathfrak{J}} c$.

Proof. (i) directly follows from 17.1.4.

(ii) Let $a \in \mathrm{At}_{\mathsf{s}}^{\mathfrak{J}}$. Then by (i), $a \subseteq_{\mathsf{s}}^{\mathfrak{J}} \Lambda_{\mathsf{s}}^{\mathfrak{J}}$ implies $a \in \emptyset$ (see. 17.1.3 (ii)).

(iii) According to (i) and 17.1.3 (iv), the assumpion of (iii) can be rewritten as follows: Let $\mathfrak{J} \models_L \Phi(a)$ hold for any $a \in \mathrm{Cn}_{\mathsf{s}}^{\mathfrak{J}}(c)$, where $\mathrm{Cn}_{\mathsf{s}}^{\mathfrak{J}}(b) \subseteq \mathrm{Cn}_{\mathsf{s}}^{\mathfrak{J}}(c)$. Then, obviously, $\mathfrak{J}| =_L \Phi(a)$ holds for any $a \in \mathrm{Cn}_{\mathsf{s}}^{\mathfrak{J}}(b)$, i.e. for any $a \in \mathrm{At}_{\mathsf{s}}^{\mathfrak{J}}$ such that $a \subseteq_{\mathsf{s}}^{\mathfrak{J}} b$ (see (i)).

(iv) According to (i) the assumption of (iv) can be rewritten in the following way: Let $\mathfrak{J} \models_L \Phi(a)$ for all $a \in \mathrm{Cn}_{\mathsf{s}}^{\mathfrak{J}}(b)$ and all $a \in \mathrm{Cn}_{\mathsf{s}}^{\mathfrak{J}}(c)$. Then, obviously, $\mathfrak{J} \models_L \Phi(a)$ for all $a \in \mathrm{Cn}_{\mathsf{s}}^{\mathfrak{J}}(b) \cup \mathrm{Cn}_{\mathsf{s}}^{\mathfrak{J}}(c) = \mathrm{Cn}_{\mathsf{s}}^{\mathfrak{J}}\left(b \cup_{s}^{\mathfrak{J}} c\right)$. I.e., $\mathfrak{J} \models_L \Phi(a)$ for all $a \in \mathrm{At}_{\mathsf{s}}^{\mathfrak{J}}$ such that $a \subseteq_{\mathsf{s}}^{\mathfrak{J}} b \cup_{\mathsf{s}}^{\mathfrak{J}} c$ (see (i)). □

Now we formulate the so-called set axioms and prove that any set-admitting L-structure is a model of these axioms. Set axioms provide us with a method of first-order characterisation of the class of set-admitting L-structures. We need this first-order description for correct definition of the set of external **St**-formulas w.r.t. which the generating rules and structure generators will be formulated.

Definition 17.1.10. Let σ be a signature, $\mathsf{s} \in \mathsf{SetSorts}_\sigma$, z and w are different individual variables, $\mathrm{Sort}(z) = \mathrm{Sort}(w) = \mathsf{s}$. By definition put:

$$\mathrm{Atom}_{\mathsf{s}}(z) = (z \neq \Lambda_{\mathsf{s}} \wedge \forall w(w \subseteq_{\mathsf{s}} z \rightarrow w = z \vee w = \Lambda_{\mathsf{s}})).$$

Lemma 17.1.11. Let L be an FPJ logic and σ be a signature. Let $\mathfrak{J}$ be a set-admitting L-structure of signature σ, $\mathsf{s} \in \mathsf{SetSorts}_\sigma$. Then for any $a \in \mathrm{Dom}_\mathfrak{J}(\mathsf{s})$,

$$a \in \mathrm{At}_\mathsf{s}^\mathfrak{J} \quad \text{iff} \quad \mathfrak{J} \models_L \mathrm{Atom}_\mathsf{s}(a).$$

Proof. By direct checking (see Definitions 17.1.4 and 17.1.10). □

17.2 Set Axioms

Definition 17.2.1. Let σ be a signature, $\mathsf{s} \in \mathsf{SetSorts}_\sigma$, x, y, z be different variables of the sort s. Then the following formulas are called *set axioms for* x, y, z:

(i) $x \cup_\mathsf{s} \Lambda_\mathsf{s} = x$,
(ii) $x \cup_\mathsf{s} x = x$,
(iii) $x \cup_\mathsf{s} y = y \cup_\mathsf{s} x$,
(iv) $x \cup_\mathsf{s} (y \cup_\mathsf{s} z) = (x \cup_\mathsf{s} y) \cup_\mathsf{s} z$,
(v) $x \subseteq_\mathsf{s} y \leftrightarrow x \cup_\mathsf{s} y = y$,
(vi) $x \neq \Lambda_\mathsf{s} \to \exists z (\mathrm{Atom}_\mathsf{s}(z) \wedge z \subseteq_\mathsf{s} x)$,
(vii) $\mathrm{Atom}_\mathsf{s}(z) \to (z \subseteq_\mathsf{s} x \cup_\mathsf{s} y \to z \subseteq_\mathsf{s} x \vee z \subseteq_\mathsf{s} y)$,
(viii) $\forall z (\mathrm{Atom}_\mathsf{s}(z) \to (z \subseteq_\mathsf{s} x \leftrightarrow z \subseteq_\mathsf{s} y)) \to x = y$.

$\mathrm{Mult}_\sigma(x,y,z)$ denotes the set of all multiplicity axioms for x, y, z.

It is evident that for each sort $\mathsf{s} \in \mathsf{SetSorts}_\sigma$ we can find three different variables of this sort (since for each sort we have infinitely many variables). Therefore we introduce the notation $\mathrm{Mult}_\sigma(\mathsf{s})$ for the set of multiplicity axioms for the first three variables in some order of the variables of the sort s. Let by definition:

$$\mathrm{MULT}_\sigma = \bigcup_{\mathsf{s} \in \mathsf{SetSorts}_\sigma} \mathrm{Mult}_\sigma(\mathsf{s}).$$

Lemma 17.2.2. Let L be an FPJ logic and σ be a signature. Let $\mathfrak{J}$ be an L-structure of signature σ, $\mathsf{s} \in \mathsf{SetSorts}_\sigma$, x, y, z be different variables of the sort s. Suppose $\mathfrak{J} \models_L \mathrm{MULT}_\sigma$. Then the following formulas are valid in $\mathfrak{J}$:

(i) $x \subseteq_\mathsf{s} x$,
(ii) $x \subseteq_\mathsf{s} y \wedge y \subseteq_\mathsf{s} x \to x = y$,
(iii) $x \subseteq_\mathsf{s} y \wedge y \subseteq_\mathsf{s} z \to x \subseteq_\mathsf{s} z$,
(iv) $\Lambda_\mathsf{s} \subseteq_\mathsf{s} x$,
(v) $x \subseteq_\mathsf{s} x \cup_\mathsf{s} y$,
(vi) $x \subseteq_\mathsf{s} \Lambda_\mathsf{s} \to x = \Lambda_\mathsf{s}$.

Proof. We will prove that (i)–(vi) can be inferred from MULT_σ in L^1.

(i) By Definition 17.2.1 (ii), $x \cup_\mathsf{s} x = x$. Hence by Definition 17.2.1 (v), $x \subseteq_\mathsf{s} x$.

(ii) Let $x \subseteq_\mathsf{s} y$ and $y \subseteq_\mathsf{s} x$. Then by Definition 17.2.1 (v) and (iii), $y = x \cup_\mathsf{s} y = x$.

(iii) Let $x \subseteq_\mathsf{s} y$ and $y \subseteq_\mathsf{s} z$. Then by Definition 17.2.1 (v),

$$x \cup_{\mathsf{s}} y = y, \tag{17.1}$$

$$y \cup_{\mathsf{s}} z = z. \tag{17.2}$$

Then

$$\begin{aligned} x \cup_{\mathsf{s}} z &= x \cup_{\mathsf{s}} (y \cup_{\mathsf{s}} z) && \text{(by (17.2))} \\ &= (x \cup_{\mathsf{s}} y) \cup_{\mathsf{s}} z && \text{(by 17.2.1 (iv))} \\ &= y \cup_{\mathsf{s}} z && \text{(by (17.1))} \\ &= z && \text{(by (17.2))} \end{aligned}$$

Thus, $x \cup_{\mathsf{s}} z = z$. Hence by Definition 17.2.1 (v), $x \subseteq_{\mathsf{s}} z$.

(iv) By Definition 17.2.1 (i), (iii) $\Lambda_{\mathsf{s}} \cup_{\mathsf{s}} x = x$. Then by Definition 17.2.1 (v), $\Lambda_{\mathsf{s}} \subseteq_{\mathsf{s}} x$.

(v) We see that

$$\begin{aligned} x \cup_{\mathsf{s}} (x \cup_{\mathsf{s}} y) &= (x \cup_{\mathsf{s}} x) \cup_{\mathsf{s}} y && \text{(by 17.2.1 (iv))} \\ &= x \cup_{\mathsf{s}} y && \text{(by 17.2.1 (ii))} \end{aligned}$$

Therefore by Definition 17.2.1 (v), $x \subseteq_{\mathsf{s}} x \cup_{\mathsf{s}} y$.

(vi) It is evident that (ii) and (iv) imply (vi). □

We have arrived at the most interesting and important statements of the present section. In the lemma below it will be proved that the class of set-admitting L-structures is equivalent to the class of models of the set MULT_{σ}.

Lemma 17.2.3. Let $\mathfrak{J}$ be a set-admitting L-structure of signature σ. Then $\mathfrak{J} \models_L \mathrm{MULT}_{\sigma}$.

Proof. Let $\mathsf{s} \in \mathsf{SetSorts}_{\sigma}$. Let us show that $\mathfrak{J}$ is a model of the axiom (i) of Definition 17.2.1. The axiom (i) is valid in $\mathfrak{J}$, iff for any $a \in \mathrm{Dom}_{\mathfrak{J}}(\mathsf{s})$, $a \cup_{\mathsf{s}}^{\mathfrak{J}} \Lambda_{\mathsf{s}}^{\mathfrak{J}} = a$. But

$$\begin{aligned} a \cup_{\mathsf{s}}^{\mathfrak{J}} \Lambda_{\mathsf{s}}^{\mathfrak{J}} = a \quad &\text{iff} \quad \mathrm{Cn}_{\mathsf{s}}^{\mathfrak{J}}\left(a \cup_{\mathsf{s}}^{\mathfrak{J}} \Lambda_{\mathsf{s}}^{\mathfrak{J}}\right) = \mathrm{Cn}_{\mathsf{s}}^{\mathfrak{J}}(a) && (\mathrm{Cn}_{\mathsf{s}}^{\mathfrak{J}} \text{ is a bijection}) \\ &\text{iff} \quad \mathrm{Cn}_{\mathsf{s}}^{\mathfrak{J}}(a) \cup \mathrm{Cn}_{\mathsf{s}}^{\mathfrak{J}}\left(\Lambda_{\mathsf{s}}^{\mathfrak{J}}\right) = \mathrm{Cn}_{\mathsf{s}}^{\mathfrak{J}}(a) && \text{(by 17.1.3 (iii))} \\ &\text{iff} \quad \mathrm{Cn}_{\mathsf{s}}^{\mathfrak{J}}(a) \cup \emptyset = \mathrm{Cn}_{\mathsf{s}}^{\mathfrak{J}}(a) && \text{(by 17.1.3 (ii))} \end{aligned}$$

However the last equality in this chain of equivalent equalities is obviously true. The validity of the multiplicity axioms (ii)–(v) (of Definition 17.2.1) in $\mathfrak{J}$ can be checked similarly.

Let us prove that (vi) of Definition 17.2.1 is valid in $\mathfrak{J}$. Suppose $a \neq \Lambda_{\mathsf{s}}^{\mathfrak{J}}$. Then $\mathrm{Cn}_{\mathsf{s}}^{\mathfrak{J}}(a) \neq \emptyset$. Hence there exists $b \in \mathrm{At}_{\mathsf{s}}^{\mathfrak{J}}$ such that $b \in \mathrm{Cn}_{\mathsf{s}}^{\mathfrak{J}}(a)$. Then by Lemma 17.1.11, $\mathfrak{J}| =_L \mathrm{Atom}_{\mathsf{s}}(b)$; by Lemma 17.1.9 (i), $b \subseteq_{\mathsf{s}}^{\mathfrak{J}} a$.

Now let us prove that (vii) of Definition 17.2.1 is valid in $\mathfrak{J}$. If $c \in \mathrm{At}_{\mathrm{s}}^{\mathfrak{J}}$ and $c \subseteq_{s}^{\mathfrak{J}} a \cup_{\mathrm{s}}^{\mathfrak{J}} b$ then

$$c \in \mathrm{Cn}_{\mathrm{s}}^{\mathfrak{J}}\left(a \cup_{\mathrm{s}}^{\mathfrak{J}} b\right) = \mathrm{Cn}_{\mathrm{s}}^{\mathfrak{J}}(a) \cup \mathrm{Cn}_{\mathrm{s}}^{\mathfrak{J}}(b).$$

Therefore $c \in \mathrm{Cn}_{\mathrm{s}}^{\mathfrak{J}}(a)$ or $c \in \mathrm{Cn}_{\mathrm{s}}^{\mathfrak{J}}(b)$, whence $c \subseteq_{\mathrm{s}}^{\mathfrak{J}} a$ or $c \subseteq_{\mathrm{s}}^{\mathfrak{J}} b$.

Let us prove that (viii) of Definition 17.2.1 is also valid in $\mathfrak{J}$. It is evident that validity of (viii) is equivalent to the fact that $\mathrm{Cn}_{\mathrm{s}}^{\mathfrak{J}}$ is an injection. But $\mathrm{Cn}_{\mathrm{s}}^{\mathfrak{J}}$ is really an injection (see Lemma 17.1.5). □

Lemma 17.2.4. Let L be an FPJ logic and σ be a signature. Let $\mathfrak{J}$ be an L-structure of signature σ. Suppose $\mathfrak{J} \vDash_L \mathrm{MULT}_\sigma$. Then $\mathfrak{J}$ is set-admitting.

Proof. Suppose $\mathsf{s} \in \mathsf{SetSorts}_\sigma$. Let us show that $\mathrm{Cn}_{\mathrm{s}}^{\mathfrak{J}}$ is an injection of $\mathrm{Dom}_{\mathfrak{J}}(\mathsf{s})$ to $2^{\mathrm{At}_{\mathrm{s}}^{\mathfrak{J}}}$ such that:

(i) $\mathrm{Cn}_{\mathrm{s}}^{\mathfrak{J}}(a) = \{a\}$ iff $a \in \mathrm{At}_{\mathrm{s}}^{\mathfrak{J}}$,
(ii) $\mathrm{Cn}_{\mathrm{s}}^{\mathfrak{J}}(\Lambda_{\mathrm{s}}^{\mathfrak{J}}) = \emptyset$,
(iii) $\mathrm{Cn}_{\mathrm{s}}^{\mathfrak{J}}(b \cup_{\mathrm{s}}^{\mathfrak{J}} c) = \mathrm{Cn}_{\mathrm{s}}^{\mathfrak{J}}(b) \cup \mathrm{Cn}_{\mathrm{s}}^{\mathfrak{J}}(c)$ for any $b, c \in \mathrm{Dom}_{\mathfrak{J}}(\mathsf{s})$,
(iv) $b \subseteq_{\mathrm{s}}^{\mathfrak{J}} c$ iff $\mathrm{Cn}_{\mathrm{s}}^{\mathfrak{J}}(b) \subseteq \mathrm{Cn}_{\mathrm{s}}^{\mathfrak{J}}(c)$ for any $b, c \in \mathrm{Dom}_{\mathfrak{J}}(\mathsf{s})$.

Let us prove that $\mathrm{Cn}_{\mathrm{s}}^{\mathfrak{J}}$ is an injection. Let $\mathrm{Cn}_{\mathrm{s}}^{\mathfrak{J}}(a) = \mathrm{Cn}_{\mathrm{s}}^{\mathfrak{J}}(b)$. By Definition 17.2.1 (viii), $a = b$.

Now let us show that (i)–(iii) are valid.

(i) ($\Rightarrow$) Let $\mathrm{Cn}_{\mathrm{s}}^{\mathfrak{J}}(a) = \{a\}$. Then $a \in \mathrm{Cn}_{\mathrm{s}}^{\mathfrak{J}}(a)$. Therefore $a \in \mathrm{At}_{\mathrm{s}}^{\mathfrak{J}}$ (see Definition 17.1.4).

($\Leftarrow$) Let $a \in \mathrm{At}_{\mathrm{s}}^{\mathfrak{J}}$. By Lemma 17.2.2 (i), $a \subseteq_{s}^{\mathfrak{J}} a$. Hence by Definition 17.1.4, $a \in \mathrm{Cn}_{\mathrm{s}}^{\mathfrak{J}}(a)$. Therefore $\{a\} \subseteq \mathrm{Cn}_{\mathrm{s}}^{\mathfrak{J}}(a)$. Let us show that $\mathrm{Cn}_{\mathrm{s}}^{\mathfrak{J}}(a) \subseteq \{a\}$. Suppose $b \in \mathrm{Cn}_{\mathrm{s}}^{\mathfrak{J}}(a)$. Then $b \in \mathrm{At}_{\mathrm{s}}^{\mathfrak{J}}$ and $b \subseteq_{\mathrm{s}}^{\mathfrak{J}} a$. Since $a \in \mathrm{At}_{\mathrm{s}}^{\mathfrak{J}}$, $b = a$ or $b = \Lambda_{\mathrm{s}}^{\mathfrak{J}}$. But $b \neq \Lambda_{\mathrm{s}}^{\mathfrak{J}}$ since $b \in \mathrm{At}_{\mathrm{s}}^{\mathfrak{J}}$. Therefore $b = a$, i.e. $b \in \{a\}$. We have proved that $\mathrm{Cn}_{\mathrm{s}}^{\mathfrak{J}}(a) = \{a\}$.

(ii) Assume the converse. Let $a \in \mathrm{Cn}_{\mathrm{s}}^{\mathfrak{J}}(\Lambda_{\mathrm{s}}^{\mathfrak{J}})$. Then $a \in \mathrm{At}_{\mathrm{s}}^{\mathfrak{J}}$ and $a \subseteq_{\mathrm{s}}^{\mathfrak{J}} \Lambda_{\mathrm{s}}^{\mathfrak{J}}$. By Lemma 17.2.2 (iv), $\Lambda_{\mathrm{s}}^{\mathfrak{J}} \subseteq_{\mathrm{s}}^{\mathfrak{J}} a$. Hence by Lemma 17.2.2 (ii) $a = \Lambda_{\mathrm{s}}^{\mathfrak{J}}$. This is a contradiction.

(iii) ($\subseteq$) Let $a \in \mathrm{Cn}_{\mathrm{s}}^{\mathfrak{J}}(b \cup_{\mathrm{s}}^{\mathfrak{J}} c)$. Then $a \in \mathrm{At}_{\mathrm{s}}^{\mathfrak{J}}$ and $a \subseteq_{\mathrm{s}}^{\mathfrak{J}} b \cup_{\mathrm{s}}^{\mathfrak{J}} c$. Hence by Definition 17.2.1 (vii), we get $a \subseteq_{\mathrm{s}}^{\mathfrak{J}} b$ or $a \subseteq_{\mathrm{s}}^{\mathfrak{J}} c$. Therefore $a \in \mathrm{Cn}_{\mathrm{s}}^{\mathfrak{J}}(b)$ or $a \in \mathrm{Cn}_{\mathrm{s}}^{\mathfrak{J}}(c)$. Then $a \in \mathrm{Cn}_{\mathrm{s}}^{\mathfrak{J}}(b) \cup \mathrm{Cn}_{\mathrm{s}}^{\mathfrak{J}}(c)$.

($\supseteq$) Let $a \in \mathrm{Cn}_{\mathrm{s}}^{\mathfrak{J}}(b) \cup \mathrm{Cn}_{\mathrm{s}}^{\mathfrak{J}}(c)$. Then $a \in \mathrm{Cn}_{\mathrm{s}}^{\mathfrak{J}}(b)$ or $a \in \mathrm{Cn}_{\mathrm{s}}^{\mathfrak{J}}(c)$. Hence $a \in \mathrm{At}_{\mathrm{s}}^{\mathfrak{J}}$ and $a \subseteq_{\mathrm{s}}^{\mathfrak{J}} b$ or $a \subseteq_{\mathrm{s}}^{\mathfrak{J}} c$. By Definition 17.2.1 (vii), we get $a \subseteq_{\mathrm{s}}^{\mathfrak{J}} b \cup_{\mathrm{s}}^{\mathfrak{J}} c$. Then $a \in \mathrm{Cn}_{\mathrm{s}}^{\mathfrak{J}}(b \cup_{\mathrm{s}}^{\mathfrak{J}} c)$.

(iv) ($\Rightarrow$) Let $b \subseteq_{\mathrm{s}}^{\mathfrak{J}} c$. We must show that $\mathrm{Cn}_{\mathrm{s}}^{\mathfrak{J}}(b) \subseteq \mathrm{Cn}_{\mathrm{s}}^{\mathfrak{J}}(c)$. Let $a \in \mathrm{Cn}_{\mathrm{s}}^{\mathfrak{J}}(b)$. Then $a \in \mathrm{At}_{\mathrm{s}}^{\mathfrak{J}}$ and $a \subseteq_{\mathrm{s}}^{\mathfrak{J}} b$. By Lemma 17.2.2 (iii), $a \subseteq_{\mathrm{s}}^{\mathfrak{J}} c$. Then $a \in \mathrm{Cn}_{\mathrm{s}}^{\mathfrak{J}}(c)$.

($\Leftarrow$) We see that

$$\begin{aligned}
&\mathrm{Cn}_{\mathsf{s}}^{\mathfrak{J}}(b) \subseteq \mathrm{Cn}_{\mathsf{s}}^{\mathfrak{J}}(c) && \\
&\quad\text{implies}\quad \mathrm{Cn}_{\mathsf{s}}^{\mathfrak{J}}(b) \cup \mathrm{Cn}_{\mathsf{s}}^{\mathfrak{J}}(c) = \mathrm{Cn}_{\mathsf{s}}^{\mathfrak{J}}(c) && \text{(evidently)}\\
&\quad\text{implies}\quad \mathrm{Cn}_{\mathsf{s}}^{\mathfrak{J}}\left(b \cup_{\mathsf{s}}^{\mathfrak{J}} c\right) = \mathrm{Cn}_{\mathsf{s}}^{\mathfrak{J}}(c) && \text{(by (iii))}\\
&\quad\text{implies}\quad b \cup_{\mathsf{s}}^{\mathfrak{J}} c = c && \text{(since } \mathrm{Cn}_{\mathsf{s}}^{\mathfrak{J}} \text{ is an injection)}\\
&\quad\text{implies}\quad b \subseteq_{\mathsf{s}}^{\mathfrak{J}} c && \text{(by 17.2.1 (v))}
\end{aligned}$$

□

Let us now turn our attention to the set-admitting descriptions, i.e. of the syntactic constructions, which require a unique (up to isomorphism) semantic interpretation. It is easy to see that the definitions are similar to those introduced above for set-admitting L-structures.

Definition 17.2.5. Let L be an FPJ logic and σ be a signature. Let D be a σ-vocabulary and Δ be a state description of signature σ over D in L. Δ is called *set-admitting* if for each $\mathsf{s} \in \mathsf{SetSorts}_\sigma$ and constant $\Lambda_\mathsf{s} \in D$ there exist an $A \subseteq \mathrm{Dom}_\sigma(D,\mathsf{s})$ and a injection

$$\varphi\colon \mathrm{Dom}_\sigma(D,\mathsf{s}) \to 2^A$$

such that:

(i) $\varphi(a) = \{a\}$ iff $a \in A$,
(ii) $\varphi(\Lambda_\mathsf{s}) = \emptyset$,
(iii) $(b \cup_\mathsf{s} c = d) \in \Delta$ iff $\varphi(b) \cup \varphi(c) = \varphi(d)$ for any $b, c \in \mathrm{Dom}_\sigma(D,\mathsf{s})$,
(iv) $(b \subseteq_\mathsf{s} c) \in \Delta$ iff $\varphi(b) \subseteq \varphi(c)$.

Here the set A is called the *set of atoms of sort* s *in* Δ, and φ is called *canonical injection of sort* s *for* Δ.

Definition 17.2.6. Let L be an FPJ logic and σ be a signature. Let D be a σ-vocabulary and Δ be a state description of signature σ over D in L, $\mathsf{s} \in \mathsf{SetSorts}_\sigma$. Let by definition

$$\mathrm{At}_\mathsf{s}^\Delta = \{a \in \mathrm{Dom}_\sigma(D,\mathsf{s}) \mid a \neq \Lambda_\mathsf{s} \text{ and for any } b \in \mathrm{Dom}_\sigma(D,\mathsf{s}),\ (b \subseteq_\mathsf{s} a) \in \Delta \text{ implies } b = \Lambda_\mathsf{s} \text{ or } b = a\}.$$

$\mathrm{Cn}_\mathsf{s}^\Delta$ denotes the map of the set $\mathrm{Dom}_\sigma(D,\mathsf{s})$ to the set $2^{\mathrm{At}_\mathsf{s}^\Delta}$ defined as follows: for any $b \in \mathrm{Dom}_\sigma(D,s)$,

$$\mathrm{Cn}_\mathsf{s}^\Delta(b) = \left\{a \in \mathrm{At}_\mathsf{s}^\Delta \mid (a \subseteq_\mathsf{s} b) \in \Delta\right\}.$$

Lemma 17.2.7. Let L be an FPJ logic and σ be a signature. Let D be a σ-vocabulary and Δ be a set-admitting state description of signature σ over D in L. Then for each $\mathsf{s} \in \mathsf{SetSorts}_\sigma$ there exist a unique set of atoms in Δ and a unique canonical injection for Δ. Namely the set of atoms of sort s in Δ is equal to $\mathrm{At}_{\mathsf{s}}^{\Delta}$, and the canonical injection of sort s for Δ is $\mathrm{Cn}_{\mathsf{s}}^{\Delta}$.

Proof. It is similar to the proof of Lemma 17.1.5.

Note that by Lemma 17.1.8, the canonical injection is really a bijection. Moreover, it is an isomorphism of our syntactic semilattice (that is formed from elements of set sort s) to the semilattice of all subsets of $\mathrm{At}_{\mathsf{s}}^{\Delta}$ under the ordinary set-theoretic join and inclusion. □

Lemma 17.2.8. Let L be an FPJ logic and σ be a signature. Let D be a σ-vocabulary and Δ be a set-admitting state description of signature σ over D in L, $\mathfrak{J} = \mathrm{Sint}_L(\Delta, D)$. Then:

(i) $\mathfrak{J}$ is a set-admitting L-structure of the signature σ,
(ii) for each $\mathsf{s} \in \mathsf{SetSorts}_\sigma$, $\mathrm{At}_{\mathsf{s}}^{\mathfrak{J}} = \mathrm{At}_{\mathsf{s}}^{\Delta}$ and $\mathrm{Cn}_{\mathsf{s}}^{\mathfrak{J}} = \mathrm{Cn}_{\mathsf{s}}^{\Delta}$.

Proof. The statement is an obvious consequence of Definition 10.1.16 (ii), 17.1.3, 17.2.5, 17.1.5 and 17.2.7. □

Lemma 17.2.9. Let L be an FPJ logic and σ be a signature. Suppose that:

(i) D is a σ-vocabulary,
(ii) Δ is a set-admitting state description of signature σ over D in L,
(iii) $\mathscr{M} \in \mathrm{MC}_\sigma(L)$ is such that $\mathrm{Univ}_{\mathscr{M}} = D$, $\mathrm{Descr}_{\mathscr{M}} = \Delta$,
(iv) $\vec{\mathfrak{a}}$ is an $\mathscr{M}$-inference from Δ.

Then $\mathrm{CSD}(\vec{\mathfrak{a}})$ is a set-admitting state description of signature σ over D.

Proof. By trivial induction on the length of the $\mathscr{M}$-inference. Those changes that happen to the current state description in the inference extension do not affect the description of the functional symbols and external predicate symbols. Thus the formulas that describe the functional symbols and external predicate symbols in $\mathrm{CSD}(\vec{\mathfrak{a}})$ will be the same as in Δ. □

Corollary 17.2.10. Let L be an FPJ logic and σ be a signature. Let $\mathscr{M} \in \mathrm{MC}_\sigma(L)$, $\vec{\mathfrak{a}}$ be an $\mathscr{M}$-inference from $\mathrm{Descr}_{\mathscr{M}}$ and $\mathfrak{J} = \mathrm{Sint}_L(\mathrm{CSD}(\vec{\mathfrak{a}}), \mathrm{Univ}_{\mathscr{M}})$. Suppose $\mathrm{Descr}_{\mathscr{M}}$ is a set-admitting state description. Then:

(i) $\mathfrak{J}$ is a set-admitting L-structure of the signature σ,
(ii) For each $\mathsf{s} \in \mathsf{SetSorts}_\sigma$, $\mathrm{At}_{\mathsf{s}}^{\mathfrak{J}} = \mathrm{At}_{\mathsf{s}}^{\Delta}$ and $\mathrm{Cn}_{\mathsf{s}}^{\mathfrak{J}} = \mathrm{Cn}_{\mathsf{s}}^{\Delta}$,
(iii) $\mathfrak{J} \models_L \mathrm{MULT}_\sigma$,
(iv) $\mathrm{Cn}_{\mathsf{s}}^{\Delta}$ is a bijection.

Proof. (i) and (ii) are direct consequences of 17.2.8 and Lemma 17.2.9.

(iii) follows from (i), (ii) and 17.2.3. In particular

$$\mathrm{Sint}_L(\mathrm{Descr}_{\mathscr{M}}, \mathrm{Univ}_{\mathscr{M}}) \models_{\mathbf{St}} \mathrm{MULT}_\sigma,$$

since $\mathrm{CSD}(\vec{\mathfrak{a}}) = \mathrm{Descr}_{\mathscr{M}}$ for any $\mathscr{M}(0)$-inference (see Definition 11.3.15).

(iv) follows from (i) and the fact that, for any $\mathsf{s} \in \mathsf{SetSorts}_\sigma$, $\mathrm{Dom}_\sigma(D, \mathsf{s})$ is finite (see Lemma 17.1.8). □

Chapter 18
Set Sorts in Modification Calculi

18.1 Positive and Negative Connection w.r.t. Set Sorts

In the present section we introduce the tools used to represent the sets of properties in the modification calculi. First of all we define the definition method for modification rules. Then we define the transformation of modification rules into generator rules. There will be a wider class of modification calculi defined than the class of atomic modification calculi. Further on a modification calculus from this wider class will be called *regular*. A set of external L-formulas will be established for which the structure generators that correspond to the regular modification rules will be robust. However, the regular modification calculi are not yet the calculi for which adequate semantics can be defined. In the next section we define perfect calculi for which we prove that adequate semantics can be constructed for them.

Definition 18.1.1. Let L be an FPJ logic, σ be a signature, $p \in \mathrm{Pr}_\sigma(\mathsf{I})$. $\vec{x} \in Var^{\diamond} \cap \mathrm{Dom}_\sigma(p)$. Assume that $M = \{\Phi_\varepsilon \mid \varepsilon \in \mathscr{V}(L,\mathsf{I})\}$ is a rule-lifting system for $p(\vec{x})$. Suppose

$$S = \{\mathrm{J}_\tau\, p(\vec{x}),\ \Psi_\varepsilon \Vdash \mathrm{J}_\varepsilon\, p(\vec{x}) \mid \varepsilon \in \mathscr{V}(L,\mathsf{I})\}$$

is a complete modification set (CMS) for p. Let $\varepsilon \in \mathbf{DTS}$,

$$\Theta_\varepsilon = (\mathrm{J}_\varepsilon\, p(\vec{x}) \vee \Phi_\varepsilon),$$

$\mathsf{s} \in \mathsf{SetSorts}_\sigma$. Then we say that:

(i) S and M are *positively* $\mathsf{s}\varepsilon$*-connected* (*+*$\mathsf{s}\varepsilon$*-connected*), if:

- There exists a unique occurence of the sort s in $\mathrm{Ar}_\sigma(p)$;
- $\Psi_\varepsilon = \exists z\left(y \subseteq_\mathsf{s} z \wedge \Theta_\varepsilon{}^{y}_{z}\right) \vee \exists z \exists w\left(y = z \cup_\mathsf{s} w \wedge \Theta_\varepsilon{}^{y}_{z} \wedge \Theta_\varepsilon{}^{y}_{w}\right)$, where:
  - y, z and w are distinct variables of the sort s;
  - y occurs freely in $p(\vec{x})$ and Φ_ε;
  - z and w have no occurences in both $p(\vec{x})$ and Φ_ε;
- $\Psi_\delta = \perp$, if $\delta \in \mathscr{D}^\tau(L)$ and $\delta \neq \varepsilon$;
- $\Psi_\tau = \forall\vec{x}\,(\mathrm{J}_\tau\, p(\vec{x}) \rightarrow \neg\Psi_\varepsilon)$;

O. Anshakov, T. Gergely, *Cognitive Reasoning*, Cognitive Technologies,
DOI 10.1007/978-3-540-68875-4_18, © Springer-Verlag Berlin Heidelberg 2010

(ii) S and M are *negatively* sε-*connected* ($-$sε-*connected*), if:

- There exists a unique occurence of s in $\mathrm{Ar}_\sigma(p)$;
- $\Psi_\varepsilon = \exists z \left(z \subseteq_\mathsf{s} y \wedge \Theta_{\varepsilon}{}_z^y \right)$, where:
  - y and z are distinct variables of the sort s;
  - y occurs freely in $p(\vec{x})$ and Φ_ε;
  - z does not occur in $p(\vec{x})$;
- $\Psi_\delta = \perp$, if $\delta \in \mathscr{D}^\tau(L)$ and $\delta \neq \varepsilon$;
- $\Psi_\tau = \forall \vec{x} \left(\mathrm{J}_\tau\, p(\vec{x}) \rightarrow \neg \Psi_\varepsilon \right)$.

Let us explain the intuitive meaning of +sε- and $-$sε-connections. Suppose s is a sort of property sets. For simplicity we will consider $\mathrm{Ar}_\sigma(p) = \langle t, s \rangle$, where t is the sort of objects that can possess properties. Then in the case of +sε-connection, the rule for ε is as follows:

$$\frac{\begin{array}{l} \mathrm{J}_\tau\, p(x,y), \\ \exists z \left(y \subseteq_\mathsf{s} z \wedge \left(\mathrm{J}_\varepsilon\, p(x,z) \vee \Phi_\varepsilon(x,z) \right) \right) \vee \\ \vee\, \exists z \exists w \left(y = z \cup_\mathsf{s} w \wedge \left(\mathrm{J}_\varepsilon\, p(x,z) \vee \Phi_\varepsilon(x,z) \right) \wedge \left(\mathrm{J}_\varepsilon\, p(x,w) \vee \Phi_\varepsilon(x,w) \right) \right) \end{array}}{\mathrm{J}_\varepsilon\, p(x,y)}$$

$\mathrm{J}_\varepsilon\, p(x,y)$ is interpreted as the fact that the object x has the set of properties y (i.e. x has all the properties of y). According to the major premise of the considered rule, x possesses all the properties from the set y if at least one of the following conditions holds:

(i) x has all the properties of a set that includes y or the condition Φ_ε holds for this wider set;

(ii) y is the union of two property sets such that at least one of the following conditions hold for each of them: (17.1) x possesses all the properties of the given constituent set or (17.2) this constituent set satisfies Φ_ε.

In the case of $-$sε-connection $\mathrm{J}_\varepsilon\, p(x,y)$ may be interpreted as the fact that the object x does not possess the set of properties y, i.e. it does not possess *at least one* property of the set y. The rule for ε in this case may be described as follows:

$$\frac{\begin{array}{l} \mathrm{J}_\tau\, p(x,y), \\ \exists z \left(z \subseteq_\mathsf{s} y \wedge \left(\mathrm{J}_\varepsilon\, p(x,z) \vee \Phi_\varepsilon(x,z) \right) \right) \end{array}}{\mathrm{J}_\varepsilon\, p(x,y)}$$

According to the major premise of the considered rule, x does not contain at least one property from the set y if for a more restricted set at least one of the folowing conditions holds:

(i) x does not possess at least one property from a set narrower than the set y,

(ii) The fact that x does not possess at least one property from a set narrower than the set y is guaranteed by the fulfilment of the condition Φ_ε.

Definition 18.1.2. Let L be an FPJ logic and σ be a signature. Let $R \in \mathrm{MRS}_\sigma(L)$, $p \in \mathrm{Pr}_\sigma(\mathsf{I})$,

$$M = \{\Phi_\varepsilon \mid \varepsilon \in \mathscr{V}(L,\mathsf{I})\}$$

be a RL-system for p such that the set of rules

$$S = \{R[p,\varepsilon] \mid \varepsilon \in \mathscr{V}(L,\mathsf{I})\}$$

is either A- or +sε- or –sε-connected with M.

Then for any $\varepsilon \in \mathscr{V}(L,\mathsf{I})$ the formula Φ_ε is called the *kernel* of the rule $R[p,\varepsilon]$ and is denoted by $\mathrm{Ker}[R,p,\varepsilon]$.

The formula $\mathrm{J}_\varepsilon\, p(\vec{x}) \vee \Phi_\varepsilon$, where $\mathrm{J}_\varepsilon\, p(\vec{x}) = \mathrm{Con}[R,p,\varepsilon]$ (see 10.2.4), is called the *locally sufficient condition* of $R[p,\varepsilon]$ and is denoted by $\mathrm{Suf}[R,p,\varepsilon]$.

Let $\mathscr{M} \in \mathrm{MC}_\sigma(L)$. Then by definition put

$$\mathrm{Ker}[\mathscr{M},p,\varepsilon] = \mathrm{Ker}[\mathrm{MR}_{\mathscr{M}},p,\varepsilon],$$

$$\mathrm{Suf}[\mathscr{M},p,\varepsilon] = \mathrm{Suf}[\mathrm{MR}_{\mathscr{M}},p,\varepsilon].$$

Remark 18.1.3. It is obvious that, under the conditions of Definition 18.1.2, the formula $\mathrm{Suf}[R,p,\varepsilon]$ is local w.r.t. the pure atomic formula $\mathrm{Con}[R,p,\varepsilon]$ (see Definition 15.1.2).

Definition 18.1.4. Let L be an FPJ logic, σ be a signature, $p \in \mathrm{Pr}_\sigma(\mathsf{I})$, $\varepsilon \in \mathscr{D}^\tau(L)$, $\mathsf{s} \in \mathsf{SetSorts}_\sigma$. We say that p is *+sε-multiple in R* if there exists an RL-system M +sε-connected with $R[p]$. Similarly we say that p is *–sε-multiple in R* if there exists an RL-system M –sε-connected with $R[p]$. By definition put

$$+\mathsf{s}\varepsilon[R] = \{p \in \mathrm{Pr}_\sigma(\mathsf{I}) \mid p \text{ is } +\mathsf{s}\varepsilon\text{-multiple in } R\},$$

$$-\mathsf{s}\varepsilon[R] = \{p \in \mathrm{Pr}_\sigma(\mathsf{I}) \mid p \text{ is } -\mathsf{s}\varepsilon\text{-multiple in } R\},$$

$$+[R] = \bigcup_{\substack{\mathsf{s}\in\mathsf{SetSorts}_\sigma\\ \varepsilon\in\mathscr{D}^\tau(L)}} +\mathsf{s}\varepsilon[R],$$

$$-[R] = \bigcup_{\substack{\mathsf{s}\in\mathsf{SetSorts}_\sigma\\ \varepsilon\in\mathscr{D}^\tau(L)}} -\mathsf{s}\varepsilon[R].$$

Let $\mathscr{M} \in \mathrm{MC}_\sigma(L)$, $p \in \mathrm{Pr}_\sigma(\mathsf{I})$, $\varepsilon \in \mathscr{D}^\tau(L)$, $\mathsf{s} \in \mathsf{SetSorts}_\sigma$. We say that p is *+sε-multiple in $\mathscr{M}$* if p is +sε-multiple in $\mathrm{MR}_{\mathscr{M}}$. Similarly we say that p is *–sε-multiple in $\mathscr{M}$* if p is –sε-multiple in $\mathrm{MR}_{\mathscr{M}}$. By definition put

$$+\mathsf{s}\varepsilon[\mathscr{M}] = +\mathsf{s}\varepsilon[\mathrm{MR}_{\mathscr{M}}],$$

$$-\mathsf{s}\varepsilon[\mathscr{M}] = -\mathsf{s}\varepsilon[\mathrm{MR}_{\mathscr{M}}],$$

$$+[\mathscr{M}] = +[\mathrm{MR}_{\mathscr{M}}],$$

$$-[\mathscr{M}] = -[\mathrm{MR}_{\mathscr{M}}].$$

Definition 18.1.5. Let L be an FPJ logic and σ be a signature. Let $R \in \mathrm{MRS}_\sigma(L)$. R is called *regular* if for each $p \in \mathrm{Pr}_\sigma(\mathsf{I})$ one and only one of the following conditions holds:

(i) $p \in A[R]$,
(ii) There exist a unique $\mathsf{s} \in \mathsf{SetSorts}_\sigma$ and a unique $\varepsilon \in \mathscr{D}^\tau(L)$ such that $p \in +\mathsf{s}\varepsilon[\mathscr{M}]$,
(iii) There exist a unique $\mathsf{s} \in \mathsf{SetSorts}_\sigma$ and a unique $\varepsilon \in \mathscr{D}^\tau(L)$ such that $p \in -\mathsf{s}\varepsilon[\mathscr{M}]$.

By definition let:

$$\mathrm{RMRS}_\sigma(L) = \{R \in \mathrm{MRS}_\sigma(L) \mid R \text{ is regular}\}.$$

In order to use the above-defined modification rules for the set sorts the internal predicates whose arity includes such sorts should be *binary*. Moreover, they can have at most two internal truth values, one of which is the undefined value τ.

Below we introduce the abbreviations for the formulas and for the sets of formulas which provide a first-order description (specification) of this requirement. The binarity requirement for the predicates with set sorts is an important part of the requirement system for the modification calculus for which we intend to define adequate semantics.

Definition 18.1.6. Let L be an FPJ logic, σ be a signature, $p \in \mathrm{Pr}_\sigma(\mathsf{I})$, $\vec{x} \in \mathrm{Var}^\diamond \cap \mathrm{Dom}_\sigma(p)$, and $\varepsilon \in \mathscr{D}^\tau(L)$. By definition put

$$\mathrm{Adm}[p,\varepsilon](\vec{x}) = \left(\bigwedge_{\delta \in \mathscr{D}^\tau(L),\ \delta \neq \varepsilon} \neg \mathrm{J}_\delta\, p(\vec{x}) \right).$$

The formula $\mathrm{Adm}[p,\varepsilon](\vec{x})$ means that a predicate represented by p can assume only one of two possible values, namely ε or τ.

Definition 18.1.7. Let L be an FPJ logic, σ be a signature, $R \in \mathrm{MRS}_\sigma(L)$, and $\mathsf{s} \in \mathsf{SetSorts}_\sigma$. The set of formulas

$$\left\{\mathrm{Adm}[p,\varepsilon](\vec{x}) \mid p \in +\mathsf{s}\varepsilon[R] \cup -\mathsf{s}\varepsilon[R],\ \vec{x} \in \mathrm{Var}^\diamond \cap \mathrm{Dom}_\sigma(p),\ \varepsilon \in \mathscr{D}^\tau(L)\right\}$$

is called a *two-valued constraint for R* and is denoted by $\mathrm{TVC}(R)$.

Atomic completeness is the next important requirement for modification structures. Its intuitive meaning is that *the objects of a set sort should be interpreted as sets of atoms.*

Definition 18.1.8. Let L be an FPJ logic, σ be a signature, $\mathsf{s} \in \mathsf{SetSorts}_\sigma$ and Φ be an external L-formula of signature σ. Assume that $y,z \in \mathrm{Var}$, $\mathrm{Sort}(z) = \mathrm{Sort}(y) = \mathsf{s}$, and z has not any occurrence in Φ. The formula

$$\Phi \leftrightarrow \forall z \left(\mathrm{Atom}_\mathsf{s}(z) \wedge z \subseteq_\mathsf{s} y \to \Phi_z^y\right)$$

is said to be *universal (positively) atomic condition* for Φ over x w.r.t. s. This formula is denoted by $+\mathsf{sA}[\Phi](y)$ or simply $+\mathsf{sA}[\Phi]$.

Similarly the formula

$$\Phi \leftrightarrow \exists z\left(\mathrm{Atom}_{\mathsf{s}}(z) \wedge z \subseteq_{\mathsf{s}} y \wedge \Phi_z^y\right)$$

is said to be *existential (negatively) atomic condition* for Φ over y w.r.t. s. This formula is denoted by $-\mathsf{sA}[\Phi](y)$ or simply $-\mathsf{sA}[\Phi]$.

Definition 18.1.9. Let L be an FPJ logic, σ be a signature, $\mathsf{s} \in \mathsf{SetSorts}_\sigma$ and $R \in \mathrm{MRS}_\sigma(L)$, $\varepsilon \in \mathscr{D}^\tau(L)$. By definition put

$$+\mathsf{sA}[R,\varepsilon] = \left\{+\mathsf{sA}[\mathrm{J}_\varepsilon\, p(\vec{x})] \mid p \in +\mathsf{s}\varepsilon[R],\, \vec{x} \in \mathrm{Var}^\diamond \cap \mathrm{Dom}_\sigma(p)\right\},$$

$$-\mathsf{sA}[R,\varepsilon] = \left\{-\mathsf{sA}[\mathrm{J}_\varepsilon\, p(\vec{x})] \mid p \in -\mathsf{s}\varepsilon[R],\, \vec{x} \in \mathrm{Var}^\diamond \cap \mathrm{Dom}_\sigma(p)\right\}.$$

The set of formulas

$$\bigcup_{\varepsilon \in \mathscr{D}^\tau(L),\ \mathsf{s} \in \mathsf{SetSorts}_\sigma} +\mathsf{sA}[R,\varepsilon] \cup -\mathsf{sA}[R,\varepsilon]$$

is called the *atomic completeness constraint* for R and is denoted by $\mathrm{ACC}(R)$.

Now we can put together all our requirements into a unique system, which will be called sufficient set restriction.

Definition 18.1.10. Let L be an FPJ logic, σ be a signature, $\mathsf{s} \in \mathsf{SetSorts}_\sigma$ and $R \in \mathrm{RMRS}_\sigma(L)$. The set of formulas

$$\mathrm{MULT}_\sigma \cup \mathrm{ACC}(R) \cup \mathrm{TVC}(R)$$

is called a *sufficient set constraint* for R and is denoted by $\mathrm{SSC}(R)$.

18.2 Generating Rules for Modification Rule Systems with Set Sorts

We define a function which will construct the corresponding generator rules through the modification rules.

Definition 18.2.1. Let L be an FPJ logic, σ be a signature, $R \in \mathrm{MRS}_\sigma(L)$, $\mathsf{s} \in \mathsf{SetSorts}_\sigma$, $\varepsilon \in \mathscr{D}^\tau(L)$, $p \in \mathrm{Pr}_\sigma(\mathsf{I})$, $\vec{x} \in \mathrm{Var}^\diamond \cap \mathrm{Dom}_\sigma(p)$, $y \in \mathrm{Set}(\vec{x})$, $z \in \mathrm{Var} \setminus \mathrm{Set}(\vec{x})$, $\mathrm{Sort}(z) = \mathrm{Sort}(y) = \mathsf{s}$.

(i) Suppose $p \in +\mathsf{s}\varepsilon[R]$, $\delta \in \mathscr{V}(L,\mathsf{I})$. Then the generating δ-rule

$$\Phi \Vdash \mathrm{J}_\delta\, p(\vec{x}),$$

where

$$\Phi = \begin{cases} \forall z\left(\mathrm{Atom}_{\mathsf{s}}(z) \wedge z \subseteq_{\mathsf{s}} y \rightarrow \mathrm{Suf}[R,p,\varepsilon]\left(\vec{x}_z^y\right)\right) & \text{if } \delta = \varepsilon, \\ \perp & \text{if } \delta \in \mathscr{D}^\tau(L),\ \delta \neq \varepsilon, \\ \exists z\left(\mathrm{Atom}_{\mathsf{s}}(z) \wedge z \subseteq_{\mathsf{s}} y \wedge \neg \mathrm{Suf}[R,p,\varepsilon]\left(\vec{x}_z^y\right)\right) & \text{if } \delta = \tau, \end{cases}$$

is called the +sε-*generating image* (+sεG-image) of the modification rule $R[p,\delta]$ and is denoted by $+\mathsf{s}\varepsilon\mathrm{G}[R,p,\delta]$. The formula Φ (the generating condition of $+\mathsf{s}\varepsilon\mathrm{G}[R,p,\delta]$) is denoted by $\mathrm{GC}[+\mathsf{s}\varepsilon\mathrm{G}(R),p,\delta]$.

(i) Suppose $p \in -\mathsf{s}\varepsilon[R]$, $\delta \in \mathscr{V}(L,\mathsf{I})$. Then the generating δ-rule

$$\Phi \Vdash \mathrm{J}_\delta\, p(\vec{x}),$$

where

$$\Phi = \begin{cases} \exists z\left(\mathrm{Atom}_{\mathsf{s}}(z) \wedge z \subseteq_{\mathsf{s}} y \wedge \mathrm{Suf}[R,p,\varepsilon]\left(\vec{x}_z^y\right)\right) & \text{if } \delta = \varepsilon, \\ \perp & \text{if } \delta \in \mathscr{D}^\tau(L),\ \delta \neq \varepsilon, \\ \forall z\left(\mathrm{Atom}_{\mathsf{s}}(z) \wedge z \subseteq_{\mathsf{s}} y \rightarrow \neg \mathrm{Suf}[R,p,\varepsilon]\left(\vec{x}_z^y\right)\right) & \text{if } \delta = \tau, \end{cases}$$

is called the –sε-*generating image* (+sεG-image) of the modification rule $R[p,\delta]$ and is denoted by $-\mathsf{s}\varepsilon\mathrm{G}[R,p,\delta]$. The formula Φ (the generating condition of $-\mathsf{s}\varepsilon\mathrm{G}[R,p,\delta]$) is denoted by $\mathrm{GC}[-\mathsf{s}\varepsilon\mathrm{G}(R),p,\delta]$.

Definition 18.2.2. Let L be an FPJ logic, σ be a signature, $R \in \mathrm{MRS}_\sigma(L)$, $\mathsf{s} \in \mathsf{SetSorts}_\sigma$, $\varepsilon \in \mathbf{DTS}$, $p \in \mathrm{Pr}_\sigma(\mathsf{I})$.

(i) Suppose $p \in +\mathsf{s}\varepsilon[R]$. Then the set

$$\{+\mathsf{s}\varepsilon\mathrm{G}[R,p,\delta] \mid \delta \in \mathscr{V}(L,\mathsf{I})\}$$

is called the +sε-*generating image* (+sεG-image) of $R[p]$ (written: $+\mathsf{s}\varepsilon\mathrm{G}[R,p]$).

(i) Suppose $p \in -\mathsf{s}\varepsilon[R]$. Then the set

$$\{-\mathsf{s}\varepsilon\mathrm{G}[R,p,\delta] \mid \delta \in \mathscr{V}(L,\mathsf{I})\}$$

is called the $-$sε-*generating image* ($-$sεG-image) of $R[p]$ (written: $-\mathsf{s}\varepsilon\mathrm{G}[R,p]$).

Our task is to show that the function to be defined is really well defined. The next statement will make it possible to prove the correctness of the definition of the function that transforms modification rules into generator rules.

Proposition 18.2.3. Let L be an FPJ logic, σ be a signature, $R \in \mathrm{MRS}_\sigma(L)$, $\mathsf{s} \in \mathsf{SetSorts}_\sigma$, $\varepsilon \in \mathbf{DTS}$, $p \in \mathrm{Pr}_\sigma(\mathsf{I})$.

(i) Suppose $p \in +\mathsf{s}\varepsilon[R]$. Then the set $+\mathsf{s}\varepsilon\mathrm{G}[R,p]$ is a complete generating set (CGS) for p w.r.t. $\mathrm{SSC}(R)$.

(ii) Suppose $p \in -\mathsf{s}\varepsilon[R]$. Then the set $-\mathsf{s}\varepsilon\mathrm{G}[R,p]$ is a complete generating set (CGS) for p w.r.t. $\mathrm{SSC}(R)$.

Proof. (i) We must establish the validity of the conditions (i)–(iii) from Definition 12.2.3. By Definition 18.2.1,

$$\mathrm{GC}[+\mathsf{s}\varepsilon\mathrm{G}(R),p,\tau] \text{ is equivalent to } \neg\mathrm{GC}[+\mathsf{s}\varepsilon\mathrm{G}(R),p,\varepsilon],$$

$\mathrm{GC}[+\mathsf{s}\varepsilon\mathrm{G}(R),p,\delta] = \bot$ if $\delta \neq \varepsilon$ and $\delta \neq \tau$. Hence conditions (i) and (ii) are valid.

Now we must prove that (iii) from Definition 12.2.3 is valid, i.e. for any $\delta \in \mathscr{D}^{\tau}(L)$,

$$\mathrm{SSC}(R) \vdash_L \mathrm{J}_\delta\, p(\vec{x}) \to \mathrm{GC}[+\mathsf{s}\varepsilon\mathrm{G}(R),p,\delta].$$

If $\delta \neq \varepsilon$ then it is evident that

$$\mathrm{TVC}(R) \vdash_L \mathrm{J}_\delta\, p(\vec{x}) \to \mathrm{GC}[+\mathsf{s}\varepsilon\mathrm{G}(R),p,\delta].$$

Let $\delta = \varepsilon$. Suppose $\mathrm{MULT}_\sigma \cup \mathrm{ACC}(R)$ is true, $\mathrm{J}_\varepsilon\, p(\vec{x})$ is also true. Let y be a variable of sort s that occurs in $\vec{x}$. Obviously, in this case the formula

$$\left(\mathrm{Atoms}_{\mathsf{s}}(z) \wedge z \subseteq_{\mathsf{s}} y \to \mathrm{J}_\varepsilon\, p\left(\vec{x}^{y}_{z}\right)\right)$$

is valid. Therefore $\mathrm{GC}[+\mathsf{s}\varepsilon\mathrm{G}(R),p,\varepsilon]$, i.e. the formula

$$\forall z\left(\mathrm{Atoms}_{\mathsf{s}}(z) \wedge z \subseteq_{\mathsf{s}} y \to \mathrm{Suf}[R,p,\varepsilon]\left(\vec{x}^{y}_{z}\right)\right)$$

is also valid. (Recall that by Definition 18.1.2, $\mathrm{Suf}[R,p,\varepsilon]\left(\vec{x}^{y}_{z}\right)$ is the formula $\mathrm{J}_\varepsilon\, p\left(\vec{x}^{y}_{z}\right) \vee \mathrm{Ker}[R,p,\varepsilon]\left(\vec{x}^{y}_{z}\right)$.)

(ii) can be proved similarly. □

Definition 18.2.4. Let L be an FPJ logic, σ be a signature and $R \in \mathrm{RMRS}_\sigma(L)$. Denote by $\mathscr{G}_\sigma$ the set of all generating rule of signature σ. Denote by SG^R the map of the set

$$\{R[p,\delta] \mid p \in \mathrm{Pr}_\sigma(\mathsf{I}), \delta \in \mathscr{V}(L,\mathsf{I})\}$$

to the set $\mathscr{G}_\sigma$ such that for any $p \in \mathrm{Pr}_\sigma(\mathsf{I})$, $\varepsilon \in \mathscr{V}(L,\mathsf{I})$

$$\mathrm{SG}^R(R[p,\delta]) = \begin{cases} \mathrm{AG}[R,p,\delta], & p \in \mathrm{A}[R], \\ +\mathsf{s}\varepsilon\mathrm{G}[R,p,\delta], & p \in +\mathsf{s}\varepsilon[R] \text{ for some } \varepsilon \in \mathscr{D}^{\tau}(L), \\ -\mathsf{s}\varepsilon\mathrm{G}[R,p,\delta], & p \in -\mathsf{s}\varepsilon[R] \text{ for some } \varepsilon \in \mathscr{D}^{\tau}(L). \end{cases}$$

(see Definitions 15.2.2 and 18.1.4). The map SG^R is called the *set generating function for R*.

Since R is regular (see Definition 18.1.5), the map SG^R is well defined.

Proposition 18.2.5. Let L be an FPJ logic, σ be a signature and $R \in \mathrm{RMRS}_\sigma(L)$. Then the map SG^R is a conforming function w.r.t. $\mathrm{SSC}(R)$.

Proof. We must check the conditions (i)–(iii) from Definition 14.1.12. The validity of (i) and (ii) is evident. (iii) is true due to Propositions 15.2.7 and 18.2.3 (see Definition 18.2.4). □

Corollary 18.2.6. Let L be an FPJ logic, σ be a signature and $R \in \mathrm{RMRS}_\sigma(L)$. Then the collection

$$\left\{\mathrm{SG}^R(R[p]) \mid p \in \mathrm{Pr}_\sigma(\mathsf{I})\right\}$$

is a GRS_σ w.r.t. $\mathrm{SSC}(R)$.

Proof. This statement follows directly from Proposition 18.2.5 and 14.1.13. □

It has been shown that the map SG^R, which transforms the modification rules into the generator rules, is correctly defined.

Now we proceed to prove $\mathrm{SSC}(R)$-stability of the obtained generator rule system and for this we need a number of technical lemmas.

Lemma 18.2.7. Let L be an FPJ logic and σ be a signature. Let $R \in \mathrm{MRS}_\sigma(L)$, $p \in \mathrm{Pr}_\sigma(\mathsf{I})$, $\mathfrak{J}$ be an L-structure of signature σ and Φ be an external L-formula of signature σ. Suppose p does not occur in Φ. Assume that $\mathfrak{J} \models_L \Phi$. Then $\mathrm{Succ}(R, \mathfrak{J}, p) \models_L \Phi$.

Proof. The statement can be proved by straightforward induction on external L-formulas. □

Lemma 18.2.8. Let L be an FPJ logic and σ be a signature. Let $R \in \mathrm{RMRS}_\sigma(L)$, $\mathsf{s} \in \mathsf{SetSorts}_\sigma$, $p \in +\mathsf{s}\varepsilon[R] \cup -\mathsf{s}\varepsilon[R]$, $\mathfrak{J}$ be a set-admitting L-structure of signature σ, $c \in \mathrm{At}_{\mathsf{s}}^{\mathfrak{J}}$. Suppose $\vec{a} \in \mathrm{Dom}\left(p^{\mathfrak{J}}\right)$ and $b \in \mathrm{Dom}_{\mathfrak{J}}(\mathsf{s}) \cap \mathrm{Set}(\vec{a})$. Then

$$\mathfrak{J} \models_L \mathrm{GC}\left[\mathrm{SG}^R(R), p, \varepsilon\right]\left(\vec{a}_c^b\right) \quad \text{iff} \quad \mathfrak{J} \models_L \mathrm{Suf}[R, p, \varepsilon]\left(\vec{a}_c^b\right).$$

Proof. Since $c \in \mathrm{At}_{\mathsf{s}}^{\mathfrak{J}}$, $\mathrm{Cn}_{\mathsf{s}}^{\mathfrak{J}}(c) = \{c\}$. Consider the possible cases.

CASE 1. Let $p \in +\mathsf{s}\varepsilon[R]$. Then the condition

$$\mathfrak{J} \models_L \mathrm{GC}\left[\mathrm{SG}^R(R), p, \varepsilon\right]\left(\vec{a}_c^b\right) \tag{18.1}$$

holds if and only if

$$\text{for any } e \in \{c\},\ \mathfrak{J} \models_L \mathrm{Suf}[R, p, \varepsilon]\left(\vec{a}_e^b\right) \tag{18.2}$$

(see Definitions 18.2.4 and 18.2.1). It is evident that (18.2) is equivalent to

$$\mathfrak{J} \models_L \mathrm{Suf}[R, p, \varepsilon]\left(\vec{a}_c^b\right). \tag{18.3}$$

CASE 2. Let $p \in -\mathsf{s}\varepsilon[R]$. Then the condition (18.1) holds if and only if

$$\text{there exists an } e \in \{c\} \text{ such that } \mathfrak{J} \vDash_L \mathrm{Suf}[R, p, \varepsilon]\left(\vec{a}_e^b\right) \tag{18.4}$$

(see Definitions 18.2.4 and 18.2.1). It is evident that (18.4) is equivalent to (18.3). □

Now we are ready to prove the statement about SSC (R)-stability of the generator rule system obtained by the map SG^R. Note that we need the SSC (R)-stability in order to be able to construct a structure generator by a modification calculus.

Proposition 18.2.9. Let L be an FPJ logic, σ be a signature and $R \in \mathrm{RMRS}_\sigma(L)$. Then $\mathrm{SG}^R(R)$ is SSC (R)-robust.

Proof. Let $p \in \mathrm{Pr}_\sigma(\mathsf{l})$ and $\mathfrak{J}$ be an L-structure of signature σ. Suppose $\mathfrak{J} \vDash_{\mathbf{St}}$ SSC (R). Let us prove that

$$\mathrm{Succ}(R, \mathfrak{J}, p) \vDash_L \mathrm{SSC}(R).$$

Note that $\mathfrak{J}$ is set-admitting, since $\mathfrak{J} \vDash_L \mathrm{MULT}_\sigma$ (see Lemma 17.2.4 and Definition 18.1.9). Consider the possible cases.

CASE 1. $p \in \mathrm{A}[R]$. Then p does not occur in any formula of SSC (R). Then by Lemma 18.2.7, $\mathrm{Succ}(R, \mathfrak{J}, p) \vDash_L \mathrm{SSC}(R)$.

CASE 2. $p \in +\mathsf{s}\varepsilon[R]$, where $\varepsilon \in \mathscr{D}^\tau(L)$, $\mathsf{s} \in \mathsf{SetSorts}_\sigma$. Let $\Phi \in \mathrm{SSC}(R)$. If p does not occur in Φ then by Lemma 18.2.7, $\mathrm{Succ}(R, \mathfrak{J}, p) \vDash_L \Phi$. Suppose p occurs in Φ. Consider the possible subcases.

SUBCASE 2.1. $\Phi \in \mathrm{TVC}(R)$. Then Φ has the form

$$\bigwedge_{\delta \in \mathscr{D}^\tau(L),\ \delta \neq \varepsilon} \neg \mathrm{J}_\delta\, p(\vec{x}).$$

Let us show that $\mathrm{Succ}(R, \mathfrak{J}, p) \vDash_L \Phi$. Assume the converse. Then there exist a $\delta \in \mathscr{D}^\tau(L)$ $(\delta \neq \varepsilon)$ and an $\vec{a} \in \mathrm{Dom}\left(p^{\mathrm{Succ}(R,\mathfrak{J},p)}\right)$ such that $p^{\mathrm{Succ}(R,\mathfrak{J},p)}(\vec{a}) = \delta$. Hence by Definition 12.2.8,

$$\mathfrak{J} \vDash_L \mathrm{GC}\left[\mathrm{SG}^R(R), p, \delta\right](\vec{a}).$$

Then by Definition 18.2.4,

$$\mathfrak{J} \vDash_L \mathrm{GC}[+\mathsf{s}\varepsilon\mathrm{G}(R), p, \delta].$$

But this is impossible since by Definition 18.2.1,

$$\mathrm{GC}[+\mathsf{s}\varepsilon\mathrm{G}(R), p, \delta] = \bot.$$

Therewore our supposition is wrong. Hence $\mathrm{Succ}(R, \mathfrak{J}, p) \mid=_L \Phi$.

SUBCASE 2.2. $\Phi \in \mathrm{ACC}(R)$. Then Φ has the form

$$\mathrm{J}_\varepsilon\, p(\vec{x}) \leftrightarrow \forall z \left(\mathrm{Atoms}_{\mathsf{s}}(z) \wedge z \subseteq_{\mathsf{s}} y \rightarrow \mathrm{J}_\varepsilon\, p\left(\vec{x}_z^y\right)\right),$$

where $\vec{x} \in \mathrm{Var}^{\diamond} \cap \mathrm{Dom}_{\sigma}(p)$, $y, z \in \mathrm{Var}$, $\mathrm{Sort}(z) = \mathrm{Sort}(y) = \mathsf{s}$, y occurs in $\mathrm{J}_{\varepsilon}\, p(\vec{x})$ and z does not occur in $\mathrm{J}_{\varepsilon}\, p(\vec{x})$. In this subcase, $\mathrm{Succ}(R, \mathfrak{J}, p) \vDash_{\mathbf{St}} \Phi$ if and only if for any $\vec{a} \in \mathrm{Dom}\left(p^{\mathrm{Succ}(R,\mathfrak{J},p)}\right)$ and $b \in \mathrm{Dom}_{\mathrm{Succ}(R,\mathfrak{J},p)}(\mathsf{s}) \cap \mathrm{Set}(\vec{a})$,

$$p^{\mathrm{Succ}(R,\mathfrak{J},p)}(\vec{a}) = \varepsilon \quad \text{iff} \quad \text{for any } c \in \mathrm{Cn}_{\mathsf{s}}^{\mathrm{Succ}(R,\mathfrak{J},p)}(b),\ p^{\mathrm{Succ}(R,\mathfrak{J},p)}\left(\vec{a}_c^b\right) = \varepsilon. \tag{18.5}$$

Let us prove that (18.5) holds.

(⇒) Let $p^{\mathrm{Succ}(R,\mathfrak{J},p)}(\vec{a}) = \varepsilon$. Then by Definition 12.2.8,

$$\mathfrak{J} \vDash_L \mathrm{GC}\left[\mathrm{SG}^R(R), p, \varepsilon\right](\vec{a}). \tag{18.6}$$

Then by Definitions 18.2.1 and 18.2.4 we get that for any $c \in \mathrm{Cn}_{\mathsf{s}}^{\mathfrak{J}}(b)$,

$$\mathfrak{J} \vDash_L \mathrm{Suf}[R, p, \varepsilon]\left(\vec{a}_c^b\right). \tag{18.7}$$

By Lemma 18.2.8, (18.7) is equivalent to

$$\mathfrak{J} \vDash_L \mathrm{GC}\left[\mathrm{SG}^R(R), p, \varepsilon\right]\left(\vec{a}_c^b\right) \tag{18.8}$$

(see Definition 18.2.1). Hence by Definition 12.2.8 for any $c \in \mathrm{Cn}_{\mathsf{s}}^{\mathrm{Succ}(R,\mathfrak{J},p)}(b)$,

$$p^{\mathrm{Succ}(R,\mathfrak{J},p)}\left(\vec{a}_c^b\right) = \varepsilon,$$

as was to be proved.

(⇐) Let for any $c \in \mathrm{Cn}_{\mathsf{s}}^{\mathrm{Succ}(R,\mathfrak{J},p)}(b)$, $p^{\mathrm{Succ}(R,\mathfrak{J},p)}\left(\vec{a}_c^b\right) = \varepsilon$. Hence by Definition 12.2.8 for any $c \in \mathrm{Cn}_{\mathsf{s}}^{\mathrm{Succ}(R,\mathfrak{J},p)}(b)$, (18.8) holds. Then By Lemma 18.2.8, for any $c \in \mathrm{Cn}_{\mathsf{s}}^{\mathrm{Succ}(R,\mathfrak{J},p)}(b)$, (18.7) is valid. Hence by Definition 12.2.8, (18.6) is true. Then $p^{\mathrm{Succ}(R,\mathfrak{J},p)}(\vec{a}) = \varepsilon$, as was to be proved.

CASE 3. $p \in -\mathsf{s}\varepsilon[R]$, where $\varepsilon \in \mathscr{D}^{\tau}(L)$, $\mathsf{s} \in \mathsf{SetSorts}_{\sigma}$. Let $\Phi \in \mathrm{SSC}(R)$. If p does not occur in Φ then by Lemma 18.2.7, $\mathrm{Succ}(R, \mathfrak{J}, p) \vDash_L \Phi$. Suppose p occurs in Φ. Consider the possible subcases.

SUBCASE 3.1. $\Phi \in \mathrm{TVC}(R)$. The analysis of this subcase is similar to the consideration of SUBCASE 2.1.

SUBCASE 3.2. $\Phi \in \mathrm{ACC}(R)$. Then Φ has the form

$$\mathrm{J}_{\varepsilon}\, p(\vec{x}) \leftrightarrow \exists z\left(\mathrm{Atoms}_{\mathsf{s}}(z) \wedge z \subseteq_{\mathsf{s}} y \wedge \mathrm{J}_{\varepsilon}\, p\left(\vec{x}_z^y\right)\right),$$

where $\vec{x} \in \mathrm{Var}_{()}{}^{\diamond} \cap \mathrm{Dom}_{\sigma}(p)$, $y, z \in \mathrm{Var}$, $\mathrm{Sort}(z) = \mathrm{Sort}(y) = \mathsf{s}$, y occurs in $\mathrm{J}_{\varepsilon}\, p(\vec{x})$, z does not occur in $\mathrm{J}_{\varepsilon}\, p(\vec{x})$. In this subcase, $\mathrm{Succ}(R, \mathfrak{J}, p) \vDash_{\mathbf{St}} \Phi$ if and only if for any $\vec{a} \in \mathrm{Dom}\left(p^{\mathrm{Succ}(R,\mathfrak{J},p)}\right)$ and $b \in \mathrm{Dom}_{\mathrm{Succ}(R,\mathfrak{J},p)}(s) \cap \mathrm{Set}(\vec{a})$,

$$p^{\mathrm{Succ}(R,\mathfrak{J},p)}(\vec{a}) = \varepsilon \quad \text{iff}$$

$$\text{there exists a } c \in \mathrm{Cn}_{\mathrm{s}}^{\mathrm{Succ}(R,\mathfrak{J},p)}(b) \text{ such that } p^{\mathrm{Succ}(R,\mathfrak{J},p)}\left(\vec{a}_c^b\right) = \varepsilon. \quad (18.9)$$

It is evident that (18.9) can be proved similarly to (18.5). □

Corollary 18.2.10. Let L be an FPJ logic, σ be a signature and $R \in \mathrm{RMRS}_\sigma(L)$. Then SG^R is an $\mathrm{SSC}(R)$-robust conforming function.

Proof. By Proposition 18.2.5, SG^R is a conforming function for R w.r.t. $\mathrm{SSC}(R)$. By Proposition 18.2.9 and Definition 14.1.15, SG^R is $\mathrm{SSC}(R)$-robust. □

Notation 18.2.11. Let L be an FPJ logic, σ be a signature and $\mathscr{M} \in \mathrm{RMC}_\sigma(L)$.

(i) Denote by $\mathrm{SG}^{\mathscr{M}}$ the function $\mathrm{SG}^{\mathrm{MR}_{\mathscr{M}}}$.
(ii) Denote by $\mathrm{SSC}(\mathscr{M})$ the set $\mathrm{SSC}(\mathrm{MR}_{\mathscr{M}})$.

Definition 18.2.12. Let L be an FPJ logic and σ be a signature. An $\mathscr{M} \in \mathrm{MC}_\sigma(L)$ is called *regular* if the following conditions are satisfied:

(i) $\mathrm{MR}_{\mathscr{M}} \in \mathrm{RMRS}_\sigma(L)$;
(ii) $\mathrm{Descr}_{\mathscr{M}} \vdash_L \mathrm{SSC}(\mathscr{M})$.

By definition let:

$$\mathrm{RMC}_\sigma(L) = \{\mathscr{M} \in \mathrm{MC}_\sigma(L) \mid \mathscr{M} \text{ is regular}\}$$

Proposition 18.2.13. Let L be an FPJ logic, σ be a signature, $\mathscr{M} \in \mathrm{RMC}_\sigma(L)$. Then $\mathrm{SG}^{\mathscr{M}}(\mathscr{M})$ is an SG_σ w.r.t. $\mathrm{SSC}(\mathscr{M})$.

Proof. By Definition 18.2.12, $\mathrm{MR}_{\mathscr{M}} \in \mathrm{RMRS}_\sigma(L)$. Then, by Proposition 18.2.9, $\mathrm{SG}^{\mathscr{M}}$ is an $\mathrm{SSC}(\mathscr{M})$-robust conforming function. By Definition 18.2.12,

$$\mathrm{Descr}_{\mathscr{M}} \vdash_{\mathbf{St}} \mathrm{SSC}(\mathscr{M}).$$

Therefore $\mathrm{Sint}_L(\mathscr{M}) \models_L \mathrm{SSC}(\mathscr{M})$. Then

$$\mathrm{SG}^{\mathscr{M}}(\mathscr{M}) = \left\langle \mathrm{Sint}_L(\mathscr{M}), \mathrm{SG}^{\mathscr{M}}(\mathrm{MR}_{\mathscr{M}}), \mathrm{Ord}_{\mathscr{M}} \right\rangle$$

is a structure generator of signature σ w.r.t. $\mathrm{SSC}(\mathscr{M})$ in L (see Proposition 14.1.18). □

From here on we are going to prove that $\mathrm{SG}^{\mathscr{M}}$ is $\mathrm{SSC}(\mathscr{M})$-$M$-compatible. Then $\mathrm{SG}^{\mathscr{M}}(\mathscr{M})$ is $\mathrm{SSC}(\mathscr{M})$-conformable and consequently there exists adequate semantics for $\mathrm{SG}^{\mathscr{M}}(\mathscr{M})$ (see Theorem 14.1.11). However, we cannot do this for regular modification calculi. In the next chapter the notions of perfect rule system and of perfect modification calculus will be introduced, and for this perfect calculus we will prove the statements on consistency and conformability.

Chapter 19
Perfect Modification Calculi (PMC)

19.1 Coherent Inferences in Perfect Modification Calculi (General Properties)

The notion of perfect modification calculi may seem unexpected and abstract, but it is introduced because of the needs arising from practical applications. Below we provide a fairly simple artificial example for the rule that allows the analysis of property sets.

Suppose that we have three binary predicate symbols: an external S and two internal C and P. $\mathrm{S}(s,o)$ is regarded as "the fragment s is included into the object o", $\mathrm{C}(s,p)$ is regarded as "the fragment s is the cause of the set of properties p", $\mathrm{P}(o,p)$ is regarded as "the object o possesses the set of properties p".

The predicate $\mathrm{P}(o,p)$ can have only two possible values: +1 that means "yes, it has the set of properties", and τ that means "it is unknown whether it has the set of properties". Similarly $\mathrm{C}(s,p)$ can have only two possible values: +1 that means "yes, it is the cause of the set of properties", and τ that means "it is unknown whether it has any connection to the given set of properties".

Moreover, we wish the statement "the object o possesses the set of properties p" to be valid if and only if this object possesses each of the properties from the set p. Similarly, we want the statement "the fragment s is the cause of the set of properties p" to be true if and only if this fragment is the cause of each of the properties from the set p.

A very simple (and rigid) hypothesis generation rule can be formulated about the cause of the presence of properties. Let in each case when the fragment s is included into the object o, the object o possess the set of properties p. Then the fragment s is the cause of the set of properties p.

If we formulate this rule as a modification rule then we get the following expression:

$$\mathrm{J}_{\tau}\,\mathrm{C}(s,p)\,,\forall o\,(S(s,o) \rightarrow \mathrm{J}_{+1}\,\mathrm{P}(o,p)) \Vdash \mathrm{J}_{+1}\,\mathrm{C}(s,p)\,.$$

But this rule does not make use of the fact that p is a set of properties. This set is interpreted as a whole and the potential advantages of p being a set are not used at

O. Anshakov, T. Gergely, *Cognitive Reasoning*, Cognitive Technologies, DOI 10.1007/978-3-540-68875-4_19, © Springer-Verlag Berlin Heidelberg 2010

all. In order to take the pecularities of p as a set into account we have to consider the formula $\forall o\,(\mathrm{S}(s,o) \rightarrow \mathrm{J}_{+1}\,\mathrm{P}(o,p))$ not as the major premise of a rule, but as *the kernel of the major premise*. Let us denote this formula by $\mathrm{K}(s,p)$, and the rule will be formulated as follows:

$$\frac{\begin{array}{l}\mathrm{J}_{\tau}\,\mathrm{C}(s,p),\\ \exists q\,(p \subseteq q \wedge (\mathrm{J}_{+1}\,\mathrm{C}(s,q) \vee \mathrm{K}(s,q))) \vee \\ \vee\, \exists q_1 \exists q_2\,(p = q_1 \cup q_2 \wedge (\mathrm{J}_{+1}\,\mathrm{C}(s,q_1) \vee \mathrm{K}(s,q_1)) \wedge (\mathrm{J}_{+1}\,\mathrm{C}(s,q_2) \vee \mathrm{K}(s,q_2)))\end{array}}{\mathrm{J}_{+1}\,\mathrm{C}(s,p)}$$

This expression seems awkward. If we had not introduced a shortcut for the kernel of the major premise it would have been illegible. However, despite its awkwardness the rule has a natural interpretation. The conclusion of the rule "the fragment s is the cause of the set of properties p" can be obtained when at least one of the following conditions holds:

(i) "The fragment s is the cause of a greater set of properties q" (or the kernel $\mathrm{K}(s,q)$ is valid for this greater set);
(ii) The set of properties p can be represented as the union of two sets q_1 and q_2 such that the fragment s is the cause of the set of properties q_1 (or $\mathrm{K}(s,q_1)$ is valid) and s is the cause of the set of properties q_2 (or $\mathrm{K}(s,q_2)$ is valid).

$\mathrm{K}(s,q)$ is regarded as the sufficient condition for s to be the cause of set of properties q.

Remember that the predicates P and C satisfy the restriction of positive atomic completeness w.r.t. the sort "set of properties". I.e. as was mentioned above "the object o possesses the set of properties p" iff this object possesses each property of the set p, and also "the fragment s is the cause of the set of properties p" iff this fragment is the cause of each property of the set p.

In order to work correctly with the rules it is necessary that the predicate K also satisfies the same condition of positive atomic completeness. $\mathrm{K}(s,p)$ is regarded as the statement that is *sufficient* for the fragment s to be the cause of the set of properties p. If K does not satisfy the condition of positive atomic completeness at some stage then it cannot be guaranteed that the predicate C would satisfy this condition at the next stage.

Thus it is necessary for $\mathrm{K}(s,p)$ to be valid if and only if $\mathrm{K}(s,e)$ is valid for each property $e \in p$. It can be proved that for the given example this is the case. K really satisfies the condition of positive atomic completeness w.r.t. the sort "set of properties". While proving this fact we will use the case that P satisfies the restriction of positive atomic completeness.

Definition 19.1.1. Let L be an FPJ logic, σ be a signature, $R \in \mathrm{RMRS}_{\sigma}(L)$. R is called *perfect* if for any $\mathsf{s} \in \mathsf{SetSorts}_{\sigma}$ and $\varepsilon \in \mathbf{DTS}$ the following conditions hold:

(i) For all $p \in +\mathsf{s}\varepsilon\,[R]$,

$$\mathrm{SSC}(R) \vdash_L +\mathrm{sA}[\mathrm{Ker}[R,p,\varepsilon]](y);$$

(ii) For all $p \in -\mathrm{s}\varepsilon[R]$,

$$\mathrm{SSC}(R) \vdash_L -\mathrm{sA}[\mathrm{Ker}[R,p,\varepsilon]](y),$$

where y occurs in $\mathrm{Con}[R,p,\varepsilon]$, $Sort(y) = s$ (see Definition 18.1.8).

By definition let:

$$\mathrm{PMRS}_\sigma(L) = \{R \in \mathrm{RMRS}_\sigma(L) \mid R \text{ is perfect}\}.$$

Definition 19.1.2. Let L be an FPJ logic, σ be a signature. An $\mathscr{M} \in \mathrm{RMC}_\sigma(L)$ is called *perfect* if $\mathrm{MR}_{\mathscr{M}} \in \mathrm{PMRS}_\sigma(L)$.

By definition let:

$$\mathrm{PMC}_\sigma(L) = \{\mathscr{M} \in \mathrm{RMC}_\sigma(L) \mid \mathscr{M} \text{ is perfect}\}.$$

In the process of inference, records with formulas are added which result in the change of the knowledge state. The current knowledge state for the pure $\mathscr{M}$-inference $\vec{a}$ is described by the set of formulas $\mathrm{CSD}(\vec{a})$. The $\mathscr{M}$-inferences compatible with the system of restrictions $\mathrm{SSC}(\mathscr{M})$ will be of special interest. I.e. we are interested in the $\mathscr{M}$-inferences, for which the corresponding syntactic structure will be a model of the set $\mathrm{SSC}(\mathscr{M})$. These $\mathscr{M}$-inferences can be considered as an intermediate result. We have already succeeded in extending $\mathscr{M}$-inference so that it satisfies our initial requirements.

Thus we may suppose that these requirements are satisfied, e.g. by the $\mathscr{M}$-inferences which are obtained after having completed the work with the current internal predicate, i.e. by the $\mathscr{M}(s,m,\mathbf{D})$-inferences. Some other $\mathscr{M}$-inferences may also be compatible with $\mathrm{SSC}(\mathscr{M})$. From now on all such $\mathscr{M}$-inferences will be called *coherent*.

Definition 19.1.3. Let L be an FPJ logic, σ be a signature, $\mathscr{M} \in \mathrm{RMC}_\sigma(L)$ and $\vec{\mathfrak{a}}$ be a pure $\mathscr{M}$-inference. $\vec{\mathfrak{a}}$ is called *coherent* if

$$\mathrm{CSD}(\vec{\mathfrak{a}}) \vdash_L \mathrm{SSC}(\mathscr{M}).$$

Here we start a series of technical lemmas which say that coherent inferences preserve the property of positive and negative atomic completeness for both J-atomic formulas and rule kernels. Note that, in the case of kernels, perfectness is required.

Lemma 19.1.4. Let L be an FPJ logic, σ be a signature, $\mathsf{s} \in \mathsf{SetSorts}_\sigma$, $\mathscr{M} \in \mathrm{RMC}_\sigma(L)$ and $\vec{\mathfrak{a}}$ be a coherent $\mathscr{M}$-inference. Let $p \in +\mathrm{s}\varepsilon[R]$, $\vec{a} \in \mathrm{Dom}_{\mathscr{M}}(p)$, $b \in \mathrm{Set}(\vec{a}) \cap \mathrm{Dom}_{\mathscr{M}}(\mathsf{s})$ and $\Delta = \mathrm{Descr}_{\mathscr{M}}$. Then

$$\mathrm{J}_\varepsilon\, p\left(\vec{a}^b_b\right) \in \mathrm{CSD}(\vec{\mathfrak{a}}) \text{ iff for any } e \in \mathrm{Cn}^\Delta_\sigma(b),\ \mathrm{J}_\varepsilon\, p\left(\vec{a}^b_e\right) \in \mathrm{CSD}(\vec{\mathfrak{a}}).$$

Proof. Since $\vec{\mathfrak{a}}$ is coherent,

$$\mathrm{CSD}(\vec{\mathfrak{a}}) \vdash_L \mathrm{SSC}(\mathscr{M})$$

(see Definition 19.1.3). Therefore

$$\mathrm{CSD}(\vec{\mathfrak{a}}) \vdash_L +\mathrm{sA}\left[\mathrm{J}_\varepsilon\, p\left(\vec{a}_b^b\right)\right]$$

(see Definition 18.1.9). I.e.,

$$\mathrm{CSD}(\vec{\mathfrak{a}}) \vdash_L \mathrm{J}_\varepsilon\, p\left(\vec{a}_b^b\right) \leftrightarrow \forall z\left(\mathrm{Atom}_{\mathsf{s}}(z) \wedge z \subseteq_{\mathsf{s}} b \rightarrow \mathrm{J}_\varepsilon\, p\left(\vec{a}_z^b\right)\right)$$

(see Definition 18.1.8). Then, it is easy to show that

$$\mathrm{CSD}(\vec{\mathfrak{a}}) \vdash_L \mathrm{J}_\varepsilon\, p\left(\vec{a}_b^b\right) \text{ iff for any } e \in \mathrm{Cn}_\sigma^\Delta(b),\ \mathrm{CSD}(\vec{\mathfrak{a}}) \vdash_L \mathrm{J}_\varepsilon\, p\left(\vec{a}_e^b\right).$$

Hence by Lemma 15.1.14 we obtain

$$\mathrm{J}_\varepsilon\, p\left(\vec{a}_b^b\right) \in \mathrm{CSD}(\vec{\mathfrak{a}}) \text{ iff for any } e \in \mathrm{Cn}_\sigma^\Delta(b),\ \mathrm{J}_\varepsilon\, p\left(\vec{a}_e^b\right) \in \mathrm{CSD}(\vec{\mathfrak{a}}),$$

as was to be proved. □

Lemma 19.1.5. Let L be an FPJ logic, σ be a signature, $\mathsf{s} \in \mathsf{SetSorts}_\sigma$, $\mathscr{M} \in \mathrm{RMC}_\sigma(L)$ and $\vec{\mathfrak{a}}$ be a coherent $\mathscr{M}$-inference. Let $p \in -\mathsf{s}\varepsilon[R]$, $\vec{a} \in \mathrm{Dom}_{\mathscr{M}}(p)$, $b \in \mathrm{Set}(\vec{a}) \cap \mathrm{Dom}_{\mathscr{M}}(s)$ and $\Delta = \mathrm{Descr}_{\mathscr{M}}$. Then

$$\mathrm{J}_\varepsilon\, p\left(\vec{a}_b^b\right) \in \mathrm{CSD}(\vec{\mathfrak{a}}) \quad \text{iff}$$
$$\text{there exists an } e \in \mathrm{Cn}_\sigma^\Delta(b) \text{ such that } \mathrm{J}_\varepsilon\, p\left(\vec{a}_e^b\right) \in \mathrm{CSD}(\vec{\mathfrak{a}}).$$

Proof. This lemma can be proved similarly to Lemma 19.1.4. □

Lemma 19.1.6. Let L be an FPJ logic, σ be a signature, $\mathsf{s} \in \mathsf{SetSorts}_\sigma$, $\mathscr{M} \in \mathrm{PMC}_\sigma(L)$ and $\vec{\mathfrak{a}}$ be a coherent $\mathscr{M}$-inference. Let $p \in +\mathsf{s}\varepsilon[R]$, $\vec{a} \in \mathrm{Dom}_{\mathscr{M}}(p)$, $b \in \mathrm{Set}(\vec{a}) \cap \mathrm{Dom}_{\mathscr{M}}(s)$ and $\Delta = \mathrm{Descr}_{\mathscr{M}}$. Then

$$\mathrm{CSD}(\vec{\mathfrak{a}}) \vdash_L \mathrm{Ker}[R, p, \varepsilon]\left(\vec{a}_b^b\right)$$

if and only if for any $e \in \mathrm{Cn}_\sigma^\Delta(b)$,

$$\mathrm{CSD}(\vec{\mathfrak{a}}) \vdash_L \mathrm{Ker}[R, p, \varepsilon]\left(\vec{a}_e^b\right).$$

Proof. Since $\vec{\mathfrak{a}}$ is coherent and $\mathscr{M}$ is perfect, we have

$$\mathrm{CSD}(\vec{\mathfrak{a}}) \vdash_L \mathrm{SSC}(\mathscr{M}) \quad \text{and} \quad \mathrm{SSC}(\mathscr{M}) \vdash_L +\mathrm{sA}\left[\mathrm{Ker}[R, p, \varepsilon]\left(\vec{a}_b^b\right)\right]$$

(see Definitions 19.1.1–19.1.3). Then

$$\mathrm{CSD}(\vec{\mathfrak{a}}) \vdash_L +\mathrm{sA}\left[\mathrm{Ker}[R,p,\varepsilon]\left(\vec{a}_b^b\right)\right].$$

We can finish this proof similarly to the proof of Lemma 19.1.4. □

Lemma 19.1.7. Let L be an FPJ logic, σ be a signature, $\mathsf{s} \in \mathsf{SetSorts}_\sigma$, $\mathscr{M} \in \mathrm{PMC}_\sigma(L)$ and $\vec{\mathfrak{a}}$ be a coherent $\mathscr{M}$-inference. Let $p \in -\mathrm{s}\varepsilon[R]$, $\vec{a} \in \mathrm{Dom}_{\mathscr{M}}(p)$, $b \in \mathrm{Set}(\vec{a}) \cap \mathrm{Dom}_{\mathscr{M}}(s)$ and $\Delta = \mathrm{Descr}_{\mathscr{M}}$. Then

$$\mathrm{CSD}(\vec{\mathfrak{a}}) \vdash_L \mathrm{Ker}[R,p,\varepsilon]\left(\vec{a}_b^b\right)$$

if and only if there exists an $e \in \mathrm{Cn}_\sigma^\Delta(b)$ such that

$$\mathrm{CSD}(\vec{\mathfrak{a}}) \vdash_L \mathrm{Ker}[R,p,\varepsilon]\left(\vec{a}_e^b\right).$$

Proof. We can prove this lemma similarly to Lemma 19.1.6. □

We consider several statements that connect internal predicates, whose arity contains set sorts with the operations and relations on the corresponding domains. Particularly, in the next statement we maintain that the set of objects of the sort s, which satisfy the condition set by a J_ε-atomic formula (whose predicate symbol is $+\mathrm{s}\varepsilon$-connected) is an ideal in the semilattice of the s-sorted constants.

Proposition 19.1.8. Let L be an FPJ logic, σ be a signature, $\mathsf{s} \in \mathsf{SetSorts}_\sigma$, $\mathscr{M} \in \mathrm{RMC}_\sigma(L)$ and $\vec{\mathfrak{a}}$ be a coherent $\mathscr{M}$-inference. Suppose $\Delta = \mathrm{Descr}_{\mathscr{M}}$, $p \in +\mathrm{s}\varepsilon[R]$, $\vec{a} \in \mathrm{Dom}_{\mathscr{M}}(p)$, $b \in \mathrm{Set}(\vec{a}) \cap \mathrm{Dom}_{\mathscr{M}}(\mathsf{s})$, and $c,d \in \mathrm{Dom}_{\mathscr{M}}(\mathsf{s})$. Then the following conditions hold:

(i) $\mathrm{J}_\varepsilon\, p\left(\vec{a}_c^b\right) \in \mathrm{CSD}(\vec{\mathfrak{a}})$ and $\mathrm{J}_\varepsilon\, p\left(\vec{a}_d^b\right) \in \mathrm{CSD}(\vec{\mathfrak{a}})$ imply

$$\mathrm{J}_\varepsilon\, p\left(\vec{a}_{c\cup_\mathsf{s} d}^b\right) \in \mathrm{CSD}(\vec{\mathfrak{a}});$$

(ii) $\mathrm{J}_\varepsilon\, p\left(\vec{a}_c^b\right) \in \mathrm{CSD}(\vec{\mathfrak{a}})$ and $\mathrm{Cn}_\sigma^\Delta(d) \subseteq \mathrm{Cn}_\sigma^\Delta(c)$ imply $\mathrm{J}_\varepsilon\, p\left(\vec{a}_d^b\right) \in \mathrm{CSD}(\vec{\mathfrak{a}})$;

(iii) $\mathrm{J}_\varepsilon\, p\left(\vec{a}_{\Lambda_\mathsf{s}}^b\right) \in \mathrm{CSD}(\vec{\mathfrak{a}})$.

Proof. This proposition directly follows from Lemma 19.1.4 and the properties of canonic injection $\mathrm{Cn}_\sigma^\Delta$ (see Definition 17.2.5 and Lemma 17.2.7). □

Proposition 19.1.9. Let L be an FPJ logic, σ be a signature, $\mathsf{s} \in \mathsf{SetSorts}_\sigma$, $\mathscr{M} \in \mathrm{RMC}_\sigma(L)$ and $\vec{\mathfrak{a}}$ be a coherent $\mathscr{M}$-inference. Suppose $\Delta = \mathrm{Descr}_{\mathscr{M}}$, $p \in -\mathrm{s}\varepsilon[R]$, $\vec{a} \in \mathrm{Dom}_{\mathscr{M}}(p)$, $b \in \mathrm{Set}(\vec{a}) \cap \mathrm{Dom}_{\mathscr{M}}(\mathsf{s})$, and $c,d \in \mathrm{Dom}_{\mathscr{M}}(\mathsf{s})$. Then the following conditions hold:

(i) $\mathrm{J}_\varepsilon\, p\left(\vec{a}_c^b\right) \in \mathrm{CSD}(\vec{\mathfrak{a}})$ and $\mathrm{Cn}_\sigma^\Delta(c) \subseteq \mathrm{Cn}_\sigma^\Delta(d)$ imply $\mathrm{J}_\varepsilon\, p\left(\vec{a}_d^b\right) \in \mathrm{CSD}(\vec{\mathfrak{a}})$;

(ii) $\mathrm{J}_\varepsilon\, p\left(\vec{a}_{\Lambda_\mathsf{s}}^b\right) \notin \mathrm{CSD}(\vec{\mathfrak{a}})$;

(iii) if $\mathrm{J}_\varepsilon\, p\left(\vec{a}_c^b\right) \in \mathrm{CSD}(\vec{\mathfrak{a}})$ and $c \neq \Lambda_\mathsf{s}$ and $c \notin \mathrm{At}_\mathsf{s}^\Delta$ then there exists a $g \in \mathrm{Dom}_{\mathscr{M}}(\mathsf{s})$ such that $\mathrm{Cn}_\sigma^\Delta(g) \subset \mathrm{Cn}_\sigma^\Delta(c)$ and $\mathrm{J}_\varepsilon\, p\left(\vec{a}_g^b\right) \in \mathrm{CSD}(\vec{\mathfrak{a}})$.

Proof. This proposition directly follows from Lemma 19.1.5 and the properties of canonic injection $\mathrm{Cn}_\sigma^\Delta$ (see Definition 17.2.5 and Lemma 17.2.7). □

Now we are going to prove a few statements valid only for the perfect calculi. The kernels of the modification rules occur in these statements, which will be similar to the statements proved above for J-atomic formulas.

Proposition 19.1.10. Let L be an FPJ logic, σ be a signature, $\mathsf{s} \in \mathsf{SetSorts}_\sigma$, $\mathscr{M} \in \mathrm{PMC}_\sigma(L)$ and $\vec{\mathfrak{a}}$ be a coherent $\mathscr{M}$-inference. Suppose $\Delta = \mathrm{Descr}_{\mathscr{M}}$, $p \in +\mathsf{s}\varepsilon[R]$, $\vec{a} \in \mathrm{Dom}_{\mathscr{M}}(p)$, $b \in \mathrm{Set}(\vec{a}) \cap \mathrm{Dom}_{\mathscr{M}}(\mathsf{s})$, and $c, d \in \mathrm{Dom}_{\mathscr{M}}(\mathsf{s})$. Then the following conditions hold:

(i) $\mathrm{CSD}(\vec{\mathfrak{a}}) \vdash_L \mathrm{Ker}[R,p,\varepsilon]\left(\vec{a}_c^b\right)$ and $\mathrm{CSD}(\vec{\mathfrak{a}}) \vdash_L \mathrm{Ker}[R,p,\varepsilon]\left(\vec{a}_d^b\right)$ imply $\mathrm{CSD}(\vec{\mathfrak{a}}) \vdash_L \mathrm{Ker}[R,p,\varepsilon]\left(\vec{a}_{c\cup_{\mathsf{s}}d}^b\right)$;

(ii) $\mathrm{CSD}(\vec{\mathfrak{a}}) \vdash_L \mathrm{Ker}[R,p,\varepsilon]\left(\vec{a}_c^b\right)$ and $\mathrm{Cn}_\sigma^\Delta(d) \subseteq \mathrm{Cn}_\sigma^\Delta(c)$ imply $\mathrm{CSD}(\vec{\mathfrak{a}}) \vdash_L \mathrm{Ker}[R,p,\varepsilon]\left(\vec{a}_d^b\right)$;

(iii) $\mathrm{CSD}(\vec{\mathfrak{a}}) \vdash_L \mathrm{Ker}[R,p,\varepsilon]\left(\vec{a}_{\Lambda_{\mathsf{s}}}^b\right)$.

Proof. This proposition follows directly from Lemma 19.1.6 and the properties of canonic injection $\mathrm{Cn}_\sigma^\Delta$ (see Definition 17.2.5 and Lemma 17.2.7). □

Proposition 19.1.11. Let L be an FPJ logic, σ be a signature, $\mathsf{s} \in \mathsf{SetSorts}_\sigma$, $\mathscr{M} \in \mathrm{PMC}_\sigma(L)$ and $\vec{\mathfrak{a}}$ be a coherent $\mathscr{M}$-inference. Suppose $\Delta = \mathrm{Descr}_{\mathscr{M}}$, $p \in -\mathsf{s}\varepsilon[R]$, $\vec{a} \in \mathrm{Dom}_{\mathscr{M}}(p)$, $b \in \mathrm{Set}(\vec{a}) \cap \mathrm{Dom}_{\mathscr{M}}(\mathsf{s})$, and $c, d \in \mathrm{Dom}_{\mathscr{M}}(\mathsf{s})$. Then the following conditions hold:

(i) $\mathrm{CSD}(\vec{\mathfrak{a}}) \vdash_L \mathrm{Ker}[R,p,\varepsilon]\left(\vec{a}_c^b\right)$ and $\mathrm{Cn}_\sigma^\Delta(c) \subseteq \mathrm{Cn}_\sigma^\Delta(d)$ imply $\mathrm{CSD}(\vec{\mathfrak{a}}) \vdash_L \mathrm{Ker}[R,p,\varepsilon]\left(\vec{a}_d^b\right)$;

(ii) $\mathrm{CSD}(\vec{\mathfrak{a}}) \vdash_L \neg\mathrm{Ker}[R,p,\varepsilon]\left(\vec{a}_{\Lambda_{\mathsf{s}}}^b\right)$;

(iii) if $\mathrm{CSD}(\vec{\mathfrak{a}}) \vdash_L \mathrm{Ker}[R,p,\varepsilon]\left(\vec{a}_c^b\right)$ and $c \neq \Lambda_{\mathsf{s}}$ and $c \notin \mathrm{At}_{\mathsf{s}}^\Delta$ then there exists a $g \in \mathrm{Dom}_{\mathscr{M}}(\mathsf{s})$ such that $\mathrm{Cn}_\sigma^\Delta(g) \subset \mathrm{Cn}_\sigma^\Delta(c)$ and $\mathrm{CSD}(\vec{\mathfrak{a}}) \vdash_L \mathrm{Ker}[R,p,\varepsilon]\left(\vec{a}_g^b\right)$.

Proof. This proposition directly follows from Lemma 19.1.7 and the properties of canonic injection $\mathrm{Cn}_\sigma^\Delta$ (see Definition 17.2.5 and Lemma 17.2.7). □

We intend to transfer the previous results on the locally sufficient condition (formula Suf from Definition 18.1.2) of the modification rules. This will not always be possible. But the statements to be proved will be sufficient for further work.

Lemma 19.1.12. Let L be an FPJ logic, σ be a signature, $\mathsf{s} \in \mathsf{SetSorts}_\sigma$, $\mathscr{M} \in \mathrm{PMC}_\sigma(L)$ and $\vec{\mathfrak{a}}$ be a coherent $\mathscr{M}$-inference. Let $p \in +\mathsf{s}\varepsilon[R]$, $\vec{a} \in \mathrm{Dom}_{\mathscr{M}}(p)$, $b \in \mathrm{Set}(\vec{a}) \cap \mathrm{Dom}_{\mathscr{M}}(s)$ and $\Delta = \mathrm{Descr}_{\mathscr{M}}$. Then

$$\mathrm{CSD}(\vec{\mathfrak{a}}) \vdash_L \mathrm{Suf}[R,p,\varepsilon]\left(\vec{a}_b^b\right)$$

implies that,

$$\text{for any } e \in \mathrm{Cn}_{\sigma}^{\Delta}(b),\ \mathrm{CSD}(\vec{\mathfrak{a}}) \vdash_L \ \mathrm{Suf}[R,p,\varepsilon]\left(\vec{a}_e^b\right).$$

Proof. Recall that

$$\mathrm{Suf}[R,p,\varepsilon](\vec{a}) = (\mathrm{J}_{\varepsilon}\, p(\vec{a}) \vee \mathrm{Ker}[R,p,\varepsilon](\vec{a}))$$

(see Definition 18.1.2). Then, using Lemmas 19.1.4 and 19.1.6 by simple case analysis, we can get the conclusion of this lemma. □

Note that, generally speaking, the converse statement of Lemma 19.1.12 is wrong.

Lemma 19.1.13. Let L be an FPJ logic, σ be a signature, $\mathsf{s} \in \mathsf{SetSorts}_{\sigma}$, $\mathcal{M} \in \mathrm{PMC}_{\sigma}(L)$ and $\vec{\mathfrak{a}}$ be a coherent $\mathcal{M}$-inference. Let $p \in -\mathsf{s}\varepsilon[R]$, $\vec{a} \in \mathrm{Dom}_{\mathcal{M}}(p)$, $b \in \mathrm{Set}(\vec{a}) \cap \mathrm{Dom}_{\mathcal{M}}(s)$ and $\Delta = \mathrm{Descr}_{\mathcal{M}}$ and $\Delta = \mathrm{Descr}_{\mathcal{M}}$. Then

$$\mathrm{CSD}(\vec{\mathfrak{a}}) \vdash_L \ \mathrm{Suf}[R,p,\varepsilon]\left(\vec{a}_b^b\right)$$

if and only if

$$\text{there exists an } e \in \mathrm{Cn}_{\sigma}^{\Delta}(b) \text{ such that } \mathrm{CSD}(\vec{\mathfrak{a}}) \vdash_L \ \mathrm{Suf}[R,p,\varepsilon]\left(\vec{a}_e^b\right).$$

Proof. We can prove this lemma by a simple case analysis, where we use Lemmas 19.1.5 and 19.1.7. □

Proposition 19.1.14. Let L be an FPJ logic, σ be a signature, $\mathsf{s} \in \mathsf{SetSorts}_{\sigma}$, $\mathcal{M} \in \mathrm{PMC}_{\sigma}(L)$ and $\vec{\mathfrak{a}}$ be a coherent $\mathcal{M}$-inference. Suppose $\Delta = \mathrm{Descr}_{\mathcal{M}}$, $p \in +\mathsf{s}\varepsilon[R]$, $\vec{a} \in \mathrm{Dom}_{\mathcal{M}}(p)$, $b \in \mathrm{Set}(\vec{a}) \cap \mathrm{Dom}_{\mathcal{M}}(\mathsf{s})$, and $c,d \in \mathrm{Dom}_{\mathcal{M}}(\mathsf{s})$. Then the following conditions hold:

(i) $\mathrm{CSD}(\vec{\mathfrak{a}}) \vdash_L \ \mathrm{Suf}[R,p,\varepsilon]\left(\vec{a}_c^b\right)$ and $\mathrm{Cn}_{\sigma}^{\Delta}(d) \subseteq \mathrm{Cn}_{\sigma}^{\Delta}(c)$ imply
$\mathrm{CSD}(\vec{\mathfrak{a}}) \vdash_L \ \mathrm{Suf}[R,p,\varepsilon]\left(\vec{a}_d^b\right)$;
(ii) $\mathrm{CSD}(\vec{\mathfrak{a}}) \vdash_L \ \mathrm{Suf}[R,p,\varepsilon]\left(\vec{a}_{\Lambda_{\mathsf{s}}}^b\right)$.

Proof. (i) directly follows from Lemma 19.1.12 and the properties of canonic injection $\mathrm{Cn}_{\sigma}^{\Delta}$ (see Definition 17.2.5 and Lemma 17.2.7). (ii) directly follows from Propositions 19.1.8 (iii) and 19.1.10 (iii). □

Proposition 19.1.15. Let L be an FPJ logic, σ be a signature, $\mathsf{s} \in \mathsf{SetSorts}_{\sigma}$, $\mathcal{M} \in \mathrm{PMC}_{\sigma}(L)$ and $\vec{\mathfrak{a}}$ be a coherent $\mathcal{M}$-inference. Suppose $\Delta = \mathrm{Descr}_{\mathcal{M}}$, $p \in -\mathsf{s}\varepsilon[R]$, $\vec{a} \in \mathrm{Dom}_{\mathcal{M}}(p)$, $b \in \mathrm{Set}(\vec{a}) \cap \mathrm{Dom}_{\mathcal{M}}(\mathsf{s})$, and $c,d \in \mathrm{Dom}_{\mathcal{M}}(\mathsf{s})$. Then the following conditions hold:

(i) $\mathrm{CSD}(\vec{\mathfrak{a}}) \vdash_L \ \mathrm{Suf}[R,p,\varepsilon]\left(\vec{a}_c^b\right)$ and $\mathrm{Cn}_{\sigma}^{\Delta}(c) \subseteq \mathrm{Cn}_{\sigma}^{\Delta}(d)$ imply
$\mathrm{CSD}(\vec{\mathfrak{a}}) \vdash_L \ \mathrm{Suf}[R,p,\varepsilon]\left(\vec{a}_d^b\right)$;
(ii) $\mathrm{CSD}(\vec{\mathfrak{a}}) \vdash_L \ \neg\mathrm{Suf}[R,p,\varepsilon]\left(\vec{a}_{\Lambda_{\mathsf{s}}}^b\right)$;
(iii) If $\mathrm{CSD}(\vec{\mathfrak{a}}) \vdash_L \ \mathrm{Suf}[R,p,\varepsilon]\left(\vec{a}_c^b\right)$ and $c \neq \Lambda_{\mathsf{s}}$ and $c \notin \mathrm{At}_{\mathsf{s}}^{\Delta}$ then there exists a $g \in \mathrm{Dom}_{\mathcal{M}}(\mathsf{s})$ such that $\mathrm{Cn}_{\sigma}^{\Delta}(g) \subset \mathrm{Cn}_{\sigma}^{\Delta}(c)$ and $\mathrm{CSD}(\vec{\mathfrak{a}}) \vdash_L \ \mathrm{Suf}[R,p,\varepsilon]\left(\vec{a}_g^b\right)$.

Proof. Using Lemma 19.1.13, Propositions 19.1.9 and 19.1.10 and the properties of canonic injection $\mathrm{Cn}_\sigma^\Delta$ (see Definition 17.2.5 and Lemma 17.2.7), we can easy get the conclusion of this proposition. □

Now we introduce the notion of $\mathscr{M}$-inference ready for a J-atomic formula. This $\mathscr{M}$-inference contains premises of the modification rule with the help of which we can obtain the J-atomic formula in question, but it does not contain this formula itself.

Definition 19.1.16. Let L be an FPJ logic, σ be a signature, $\mathscr{M} \in \mathrm{MC}_\sigma(L)$ and $\vec{\mathfrak{a}}$ be a pure $\mathscr{M}(s,m)$-inference. Suppose $\varepsilon \in \mathscr{V}(L,\mathsf{I})$, $p \in \mathrm{Pr}_\sigma(\mathsf{I})$, $\mathrm{Ord}_{\mathscr{M}}(p) = m$, $\vec{a} \in \mathrm{Dom}_{\mathscr{M}}(p)$, and $b \in \mathrm{Set}(\vec{a}) \cap \mathrm{Dom}_\sigma(D,s)$. $\vec{\mathfrak{a}}$ is called *ready* for $\mathrm{J}_\varepsilon\, p(\vec{a})$ if the following conditions hold:

(i) $\mathrm{J}_\tau\, p(\vec{a}) \in \mathrm{FSet}(\vec{\mathfrak{a}})$,
(ii) $\mathrm{MPr}[R,p,\varepsilon](\vec{a}) \in \mathrm{FSet}(\vec{\mathfrak{a}})$,
(iii) $\vec{\mathfrak{a}}$ contains no record $\mathfrak{k}$ for which $\mathscr{F}(\mathfrak{k}) = \mathrm{J}_\varepsilon\, p(\vec{a})$, $\mathrm{Rank}(\mathfrak{k}) = \langle s,m\rangle$.

In other words an $\mathscr{M}(s,m)$-inference ready for $\mathrm{J}_\varepsilon\, p(\vec{a})$ contains (at stage s and module m) both of the premises of the rule $R[p,\varepsilon]$, but does not contain its conclusion.

Lemma 19.1.17. Let L be an FPJ logic, σ be a signature, $\mathscr{M} \in \mathrm{MC}_\sigma(L)$, $\varepsilon \in \mathscr{V}(L,\mathsf{I})$ and $\vec{\mathfrak{a}}$ be a pure $\mathscr{M}(s,m,d)$-inference, where $s > 0$, $m > 0$. Suppose $p \in \mathrm{Pr}_\sigma(\mathsf{I})$, $\vec{a} \in \mathrm{Dom}_{\mathscr{M}}(p)$, $b \in \mathrm{Set}(\vec{a}) \cap \mathrm{Dom}_\sigma(D,\mathsf{s})$, $\varepsilon \in \mathscr{V}(L,\mathsf{I})$. Assume that $\vec{\mathfrak{a}}$ is ready for $\mathrm{J}_\varepsilon\, p(\vec{a})$. Then $d = \mathbf{M}$.

Proof. Assume the converse. Then for any $p \in \mathrm{Pr}_\sigma(\mathsf{I})$, $\mathrm{Dom}_{\mathscr{M}}(p,\vec{\mathfrak{a}},s) = \mathrm{Dom}_{\mathscr{M}}(p)$. This contradict the fact that $\mathrm{J}_\varepsilon\, p(\vec{a})$ does not occur in $\vec{\mathfrak{a}}$ within a record of rank $\langle s,m\rangle$. □

Lemma 19.1.18. Let L be an FPJ logic, σ be a signature, $\mathscr{M} \in \mathrm{MC}_\sigma(L)$, $\varepsilon \in \mathscr{V}(L,\mathsf{I})$ and $\vec{\mathfrak{a}}$ be a pure $\mathscr{M}(s,m)$-inference, where $s > 0$, $m > 0$. Let $\mathrm{Ord}_{\mathscr{M}}(p) = m$. Suppose that

$$\mathrm{J}_\tau\, p(\vec{a}) \in \mathrm{CSD}\left(\vec{\mathfrak{a}}_{\overleftarrow{s,\,m-1}}\right),$$

and $\vec{\mathfrak{a}}$ contains a record $\mathfrak{k}$ for which $\mathscr{F}(\mathfrak{k}) = \mathrm{J}_\varepsilon\, p(\vec{a})$, $\mathrm{Rank}(\mathfrak{k}) = \langle s,m\rangle$. Then there exists a cut $\left\langle \vec{\mathfrak{b}},\vec{\mathfrak{c}}\right\rangle$ of the $\mathscr{M}(s,m)$-inference $\vec{\mathfrak{a}}$ such that:

(i) $\mathrm{Len}\left(\vec{\mathfrak{a}}_{\overleftarrow{s,\,m-1}}\right) \le \mathrm{Len}\left(\vec{\mathfrak{b}}\right) < \mathrm{Len}(\vec{\mathfrak{a}})$,
(ii) $\vec{\mathfrak{d}} = \mathrm{Resume}\left(\vec{\mathfrak{b}}\right)$ is ready for $\mathrm{J}_\varepsilon\, p(\vec{a})$ pure $\mathscr{M}(s,k,\mathbf{M})$-inference, where

$$m-1 \le k \le m,$$

(iii) $\vec{\mathfrak{d}}_{\overleftarrow{s,\,m-1}} = \vec{\mathfrak{a}}_{\overleftarrow{s,\,m-1}}$.

Proof. (i) Let i be the index of the first record of $\vec{\mathfrak{a}}$ that contains $\mathrm{J}_\varepsilon\, p(\vec{a})$ and has the rank $\langle s,m\rangle$. Let $M = [i..\mathrm{Len}(\vec{\mathfrak{a}})]$. Suppose

$$\text{for any } e \in \mathrm{Cn}_{\sigma}^{\Delta}(b),\ \mathrm{CSD}(\vec{\mathfrak{a}}) \vdash_L \mathrm{Suf}[R,p,\varepsilon]\left(\vec{a}_e^b\right).$$

Proof. Recall that

$$\mathrm{Suf}[R,p,\varepsilon](\vec{a}) = (\mathrm{J}_{\varepsilon}\, p(\vec{a}) \vee \mathrm{Ker}[R,p,\varepsilon](\vec{a}))$$

(see Definition 18.1.2). Then, using Lemmas 19.1.4 and 19.1.6 by simple case analysis, we can get the conclusion of this lemma. □

Note that, generally speaking, the converse statement of Lemma 19.1.12 is wrong.

Lemma 19.1.13. Let L be an FPJ logic, σ be a signature, $\mathsf{s} \in \mathsf{SetSorts}_{\sigma}$, $\mathscr{M} \in \mathrm{PMC}_{\sigma}(L)$ and $\vec{\mathfrak{a}}$ be a coherent $\mathscr{M}$-inference. Let $p \in -\mathsf{s}\varepsilon[R]$, $\vec{a} \in \mathrm{Dom}_{\mathscr{M}}(p)$, $b \in \mathrm{Set}(\vec{a}) \cap \mathrm{Dom}_{\mathscr{M}}(s)$ and $\Delta = \mathrm{Descr}_{\mathscr{M}}$ and $\Delta = \mathrm{Descr}_{\mathscr{M}}$. Then

$$\mathrm{CSD}(\vec{\mathfrak{a}}) \vdash_L \mathrm{Suf}[R,p,\varepsilon]\left(\vec{a}_b^b\right)$$

if and only if

$$\text{there exists an } e \in \mathrm{Cn}_{\sigma}^{\Delta}(b) \text{ such that } \mathrm{CSD}(\vec{\mathfrak{a}}) \vdash_L \mathrm{Suf}[R,p,\varepsilon]\left(\vec{a}_e^b\right).$$

Proof. We can prove this lemma by a simple case analysis, where we use Lemmas 19.1.5 and 19.1.7. □

Proposition 19.1.14. Let L be an FPJ logic, σ be a signature, $\mathsf{s} \in \mathsf{SetSorts}_{\sigma}$, $\mathscr{M} \in \mathrm{PMC}_{\sigma}(L)$ and $\vec{\mathfrak{a}}$ be a coherent $\mathscr{M}$-inference. Suppose $\Delta = \mathrm{Descr}_{\mathscr{M}}$, $p \in +\mathsf{s}\varepsilon[R]$, $\vec{a} \in \mathrm{Dom}_{\mathscr{M}}(p)$, $b \in \mathrm{Set}(\vec{a}) \cap \mathrm{Dom}_{\mathscr{M}}(\mathsf{s})$, and $c, d \in \mathrm{Dom}_{\mathscr{M}}(\mathsf{s})$. Then the following conditions hold:

(i) $\mathrm{CSD}(\vec{\mathfrak{a}}) \vdash_L \mathrm{Suf}[R,p,\varepsilon]\left(\vec{a}_c^b\right)$ and $\mathrm{Cn}_{\sigma}^{\Delta}(d) \subseteq \mathrm{Cn}_{\sigma}^{\Delta}(c)$ imply
$\mathrm{CSD}(\vec{\mathfrak{a}}) \vdash_L \mathrm{Suf}[R,p,\varepsilon]\left(\vec{a}_d^b\right)$;
(ii) $\mathrm{CSD}(\vec{\mathfrak{a}}) \vdash_L \mathrm{Suf}[R,p,\varepsilon]\left(\vec{a}_{\Lambda_\mathsf{s}}^b\right)$.

Proof. (i) directly follows from Lemma 19.1.12 and the properties of canonic injection $\mathrm{Cn}_{\sigma}^{\Delta}$ (see Definition 17.2.5 and Lemma 17.2.7). (ii) directly follows from Propositions 19.1.8 (iii) and 19.1.10 (iii). □

Proposition 19.1.15. Let L be an FPJ logic, σ be a signature, $\mathsf{s} \in \mathsf{SetSorts}_{\sigma}$, $\mathscr{M} \in \mathrm{PMC}_{\sigma}(L)$ and $\vec{\mathfrak{a}}$ be a coherent $\mathscr{M}$-inference. Suppose $\Delta = \mathrm{Descr}_{\mathscr{M}}$, $p \in -\mathsf{s}\varepsilon[R]$, $\vec{a} \in \mathrm{Dom}_{\mathscr{M}}(p)$, $b \in \mathrm{Set}(\vec{a}) \cap \mathrm{Dom}_{\mathscr{M}}(\mathsf{s})$, and $c, d \in \mathrm{Dom}_{\mathscr{M}}(\mathsf{s})$. Then the following conditions hold:

(i) $\mathrm{CSD}(\vec{\mathfrak{a}}) \vdash_L \mathrm{Suf}[R,p,\varepsilon]\left(\vec{a}_c^b\right)$ and $\mathrm{Cn}_{\sigma}^{\Delta}(c) \subseteq \mathrm{Cn}_{\sigma}^{\Delta}(d)$ imply
$\mathrm{CSD}(\vec{\mathfrak{a}}) \vdash_L \mathrm{Suf}[R,p,\varepsilon]\left(\vec{a}_d^b\right)$;
(ii) $\mathrm{CSD}(\vec{\mathfrak{a}}) \vdash_L \neg\mathrm{Suf}[R,p,\varepsilon]\left(\vec{a}_{\Lambda_\mathsf{s}}^b\right)$;
(iii) If $\mathrm{CSD}(\vec{\mathfrak{a}}) \vdash_L \mathrm{Suf}[R,p,\varepsilon]\left(\vec{a}_c^b\right)$ and $c \neq \Lambda_\mathsf{s}$ and $c \notin \mathrm{At}_\mathsf{s}^{\Delta}$ then there exists a $g \in \mathrm{Dom}_{\mathscr{M}}(\mathsf{s})$ such that $\mathrm{Cn}_{\sigma}^{\Delta}(g) \subset \mathrm{Cn}_{\sigma}^{\Delta}(c)$ and $\mathrm{CSD}(\vec{\mathfrak{a}}) \vdash_L \mathrm{Suf}[R,p,\varepsilon]\left(\vec{a}_g^b\right)$.

Proof. Using Lemma 19.1.13, Propositions 19.1.9 and 19.1.10 and the properties of canonic injection $\mathrm{Cn}_{\sigma}^{\Delta}$ (see Definition 17.2.5 and Lemma 17.2.7), we can easy get the conclusion of this proposition. □

Now we introduce the notion of $\mathscr{M}$-inference ready for a J-atomic formula. This $\mathscr{M}$-inference contains premises of the modification rule with the help of which we can obtain the J-atomic formula in question, but it does not contain this formula itself.

Definition 19.1.16. Let L be an FPJ logic, σ be a signature, $\mathscr{M} \in \mathrm{MC}_{\sigma}(L)$ and $\vec{\mathfrak{a}}$ be a pure $\mathscr{M}(s,m)$-inference. Suppose $\varepsilon \in \mathscr{V}(L,\mathsf{I})$, $p \in \mathrm{Pr}_{\sigma}(\mathsf{I})$, $\mathrm{Ord}_{\mathscr{M}}(p) = m$, $\vec{a} \in \mathrm{Dom}_{\mathscr{M}}(p)$, and $b \in \mathrm{Set}(\vec{a}) \cap \mathrm{Dom}_{\sigma}(D,s)$. $\vec{\mathfrak{a}}$ is called *ready* for $\mathrm{J}_{\varepsilon}\, p(\vec{a})$ if the following conditions hold:

(i) $\mathrm{J}_{\tau}\, p(\vec{a}) \in \mathrm{FSet}(\vec{\mathfrak{a}})$,
(ii) $\mathrm{MPr}[R,p,\varepsilon](\vec{a}) \in \mathrm{FSet}(\vec{\mathfrak{a}})$,
(iii) $\vec{\mathfrak{a}}$ contains no record $\mathfrak{k}$ for which $\mathscr{F}(\mathfrak{k}) = \mathrm{J}_{\varepsilon}\, p(\vec{a})$, $\mathrm{Rank}(\mathfrak{k}) = \langle s,m\rangle$.

In other words an $\mathscr{M}(s,m)$-inference ready for $\mathrm{J}_{\varepsilon}\, p(\vec{a})$ contains (at stage s and module m) both of the premises of the rule $R[p,\varepsilon]$, but does not contain its conclusion.

Lemma 19.1.17. Let L be an FPJ logic, σ be a signature, $\mathscr{M} \in \mathrm{MC}_{\sigma}(L)$, $\varepsilon \in \mathscr{V}(L,\mathsf{I})$ and $\vec{\mathfrak{a}}$ be a pure $\mathscr{M}(s,m,d)$-inference, where $s > 0$, $m > 0$. Suppose $p \in \mathrm{Pr}_{\sigma}(\mathsf{I})$, $\vec{a} \in \mathrm{Dom}_{\mathscr{M}}(p)$, $b \in \mathrm{Set}(\vec{a}) \cap \mathrm{Dom}_{\sigma}(D,\mathrm{s})$, $\varepsilon \in \mathscr{V}(L,\mathsf{I})$. Assume that $\vec{\mathfrak{a}}$ is ready for $\mathrm{J}_{\varepsilon}\, p(\vec{a})$. Then $d = \mathbf{M}$.

Proof. Assume the converse. Then for any $p \in \mathrm{Pr}_{\sigma}(\mathsf{I})$, $\mathrm{Dom}_{\mathscr{M}}(p,\vec{\mathfrak{a}},s) = \mathrm{Dom}_{\mathscr{M}}(p)$. This contradict the fact that $\mathrm{J}_{\varepsilon}\, p(\vec{a})$ does not occur in $\vec{\mathfrak{a}}$ within a record of rank $\langle s,m\rangle$. □

Lemma 19.1.18. Let L be an FPJ logic, σ be a signature, $\mathscr{M} \in \mathrm{MC}_{\sigma}(L)$, $\varepsilon \in \mathscr{V}(L,\mathsf{I})$ and $\vec{\mathfrak{a}}$ be a pure $\mathscr{M}(s,m)$-inference, where $s > 0$, $m > 0$. Let $\mathrm{Ord}_{\mathscr{M}}(p) = m$. Suppose that

$$\mathrm{J}_{\tau}\, p(\vec{a}) \in \mathrm{CSD}\left(\vec{\mathfrak{a}}_{\overleftarrow{s,\,m-1}}\right),$$

and $\vec{\mathfrak{a}}$ contains a record $\mathfrak{k}$ for which $\mathscr{F}(\mathfrak{k}) = \mathrm{J}_{\varepsilon}\, p(\vec{a})$, $\mathrm{Rank}(\mathfrak{k}) = \langle s,m\rangle$. Then there exists a cut $\left\langle \vec{\mathfrak{b}},\vec{\mathfrak{c}}\right\rangle$ of the $\mathscr{M}(s,m)$-inference $\vec{\mathfrak{a}}$ such that:

(i) $\mathrm{Len}\left(\vec{\mathfrak{a}}_{\overleftarrow{s,\,m-1}}\right) \le \mathrm{Len}\left(\vec{\mathfrak{b}}\right) < \mathrm{Len}(\vec{\mathfrak{a}})$,
(ii) $\vec{\mathfrak{d}} = \mathrm{Resume}\left(\vec{\mathfrak{b}}\right)$ is ready for $\mathrm{J}_{\varepsilon}\, p(\vec{a})$ pure $\mathscr{M}(s,k,\mathbf{M})$-inference, where

$$m-1 \le k \le m,$$

(iii) $\vec{\mathfrak{d}}_{\overleftarrow{s,\,m-1}} = \vec{\mathfrak{a}}_{\overleftarrow{s,\,m-1}}$.

Proof. (i) Let i be the index of the first record of $\vec{\mathfrak{a}}$ that contains $\mathrm{J}_{\varepsilon}\, p(\vec{a})$ and has the rank $\langle s,m\rangle$. Let $M = [i..\mathrm{Len}(\vec{\mathfrak{a}})]$. Suppose

$$\vec{\mathfrak{b}} = \vec{\mathfrak{a}}\,(\mathrm{IS}\,(\vec{\mathfrak{a}})\setminus M)\,, \quad \vec{\mathfrak{c}} = \vec{\mathfrak{a}}\,(\mathrm{IS}\,(\vec{\mathfrak{a}})\setminus M)\,.$$

Then by Lemma 11.1.11, $\left\langle \vec{\mathfrak{b}}, \vec{\mathfrak{c}} \right\rangle$ is a cut. Obviously,

$$\mathrm{Len}\left(\vec{\mathfrak{b}}\right) = i - 1 < i \leq \mathrm{Len}\,(\vec{\mathfrak{a}})\,. \tag{19.1}$$

Moreover,

$$\mathrm{Len}\left(\vec{\mathfrak{a}}_{\overleftarrow{s,\,m-1}}\right) < i. \tag{19.2}$$

In the opposite case,

$$\mathscr{F}\,(\vec{\mathfrak{a}}\,[i]) = \mathrm{J}_{\varepsilon}\,p\,(\vec{a}) \in \mathrm{FSet}\left(\vec{\mathfrak{a}}_{\overleftarrow{s,\,m-1}}\right),$$

which contradicts the assumption. (19.1) and (19.2) imply (i).

(ii) Above we have supposed that i is the least index of a record of rank $\langle s,m\rangle$ that occurs in $\vec{\mathfrak{a}}$ and contains the formula $\mathrm{J}_{\varepsilon}\,p\,(\vec{a})$. Then Resume $\left(\vec{\mathfrak{b}}\right)$ does not contain any record of rank $\langle s,m\rangle$ that contains the formula $\mathrm{J}_{\varepsilon}\,p\,(\vec{a})$. Then

$$\vec{a} \notin \mathrm{Dom}_{\mathscr{M}}\left(p, \mathrm{Resume}\left(\vec{\mathfrak{b}}\right), s\right).$$

Therefore

$$\mathrm{Dom}_{\mathscr{M}}\left(p, \mathrm{Resume}\left(\vec{\mathfrak{b}}\right), s\right) \neq \mathrm{Dom}_{\mathscr{M}}\,(p)\,.$$

Then by 11.1.22, Resume $\left(\vec{\mathfrak{b}}\right)$ is a pure $\mathscr{M}(s,k,\mathbf{M})$-inference. The fact that $m - 1 \leq k \leq m$ follows from (i). It remains to show that Resume $\left(\vec{\mathfrak{b}}\right)$ is an $\mathscr{M}(s,k,\mathbf{M})$-inference ready for $\mathrm{J}_{\varepsilon}\,p\,(\vec{a})$.

Let us show that $\vec{\mathfrak{a}}\,[i]$ can be added to the inference only by a modification rule; more precisely, just by the rule $R\,[p,\varepsilon]$. The case of the confirmation rule is excluded due to (19.1). Obtaining $\mathrm{J}_{\varepsilon}\,p\,(\vec{a})$ by deductive rules is impossible, since the regime of Resume $\left(\vec{\mathfrak{b}}\right)$ (to which the formula $\vec{\mathfrak{a}}\,[i]$ is added) is $\mathbf{M}$ (see 10.2.31 (Ded)). Thus $\vec{\mathfrak{a}}\,[i]$ is added by a modification rule, therefore in Resume $\left(\vec{\mathfrak{b}}\right)$ the premises of this rule should occur. Hence we conclude that Resume $\left(\vec{\mathfrak{b}}\right)$ is ready for $\mathrm{J}_{\varepsilon}\,p\,(\vec{a})$. (ii) has been proved.

(iii) By (i)

$$\vec{\mathfrak{a}} = \vec{\mathfrak{g}} + \vec{\mathfrak{h}} + \vec{\mathfrak{c}},$$

where

$$\vec{\mathfrak{g}} = \mathrm{Lo}\,(\vec{a}, s, m-1)\,, \tag{19.3}$$

$$\vec{\mathfrak{b}} = \vec{\mathfrak{g}} + \vec{\mathfrak{h}}. \tag{19.4}$$

Then

$$\begin{aligned}\vec{\mathfrak{d}} &= \text{Resume}\left(\vec{\mathfrak{b}}\right) && \text{(by (ii))}\\ &= \text{Resume}\left(\vec{\mathfrak{g}}+\vec{\mathfrak{h}}\right) && \text{(by (19.4))}\\ &= \text{Resume}\left(\vec{\mathfrak{g}},\text{Len}\left(\vec{\mathfrak{b}}\right)\right)+\text{Resume}\left(\vec{\mathfrak{h}},\text{Len}\left(\vec{\mathfrak{b}}\right)\right) && \text{(by 11.1.12,11.1.15)}\end{aligned}$$

Then it is easy to show that

$$\text{Lo}\left(\vec{\mathfrak{d}},s,m-1\right)=\text{Resume}\left(\vec{\mathfrak{g}},\text{Len}\left(\vec{\mathfrak{b}}\right)\right). \tag{19.5}$$

(The operation Resume does not change the rank of the record.) Then

$$\begin{aligned}\vec{\mathfrak{d}}_{\overleftarrow{s,\,m-1}} &= \text{Resume}\left(\text{Resume}\left(\vec{\mathfrak{g}},\text{Len}\left(\vec{\mathfrak{b}}\right)\right)\right) && \text{(by (19.5) and 14.1.4)}\\ &= \text{Resume}\left(\text{Resume}\left(\vec{\mathfrak{g}},\text{Len}\left(\vec{\mathfrak{b}}\right)\right),\text{Len}\left(\vec{\mathfrak{g}}\right)\right) && \text{(by 11.1.12)}\\ &= \text{Resume}\left(\vec{\mathfrak{g}},\text{Len}\left(\vec{\mathfrak{g}}\right)\right) && \text{(by 11.1.14)}\\ &= \vec{\mathfrak{a}}_{\overleftarrow{s,\,m-1}} && \text{(by (19.3) and 14.1.4)}\end{aligned}$$

Thus, (iii) has been proved. □

19.2 Modification Rules Within Coherent Inferences (Positive Case)

Let us now turn to lemmas on the basic properties of the predicate symbols +sε-multiple in modification structures. One of the most important results of the present section will be established.

Lemma 19.2.1. Let L be an FPJ logic, σ be a signature and $\mathscr{M}\in\text{PMC}_{\sigma}(L)$. Suppose $\varepsilon\in\mathscr{D}^{\tau}(L)$, $p\in+\text{s}\varepsilon\,[R]$, $\vec{a}\in\text{Dom}_{\mathscr{M}}(p)$, $b\in\text{Set}(\vec{a})\cap\text{Dom}_{\mathscr{M}}(\text{s})$. Let $\vec{\mathfrak{a}}$ be a coherent $\mathscr{M}(s)$-inference ready for $\text{J}_{\varepsilon}\,p\left(\vec{a}_b^b\right)$, where $s>0$, and $\Delta=\text{Descr}_{\mathscr{M}}$. Then for any $e\in\text{Cn}_{\text{s}}^{\Delta}(b)$,

$$\text{CSD}\left(\vec{\mathfrak{a}}\right)\vdash_L\ \text{Suf}\left[R,\text{P}_m^{\mathscr{M}},\varepsilon\right]\left(\vec{a}_e^b\right).$$

Proof. Let $\mathfrak{J}=\text{Sint}_L\left(\text{CSD}\left(\vec{\mathfrak{a}}\right),\text{Univ}_{\mathscr{M}}\right)$. The fact that $\vec{\mathfrak{a}}$ is a ready for $\text{J}_{\varepsilon}\,p\left(\vec{a}_b^b\right)$ implies $\text{MPr}\,[R,p,\varepsilon]\left(\vec{a}_b^b\right)\in\text{FSet}\vec{a}$. Then by Corollary 11.4.12,

$$\text{CSD}\left(\vec{a}\right)\vdash_L\ \text{MPr}\,[R,p,\varepsilon]\left(\vec{a}_b^b\right),$$

therefore (see Lemma 14.1.1),

$$\mathfrak{J}\vDash_L\ \text{MPr}\,[R,p,\varepsilon]\left(\vec{a}_b^b\right). \tag{19.6}$$

Since $p \in +\mathrm{s}\varepsilon[R]$, $\mathrm{MPr}[R,p,\varepsilon](\vec{a})$ has the form

$$\exists z\left(b \subseteq_{\mathsf{s}} z \wedge \mathrm{Suf}[R,p,\varepsilon]\left(\vec{a}^{b}_{z}\right)\right) \vee \\ \exists z \exists w\left(b = z \cup_{\mathsf{s}} w \wedge \mathrm{Suf}[R,p,\varepsilon]\left(\vec{a}^{b}_{z}\right) \wedge \mathrm{Suf}[R,p,\varepsilon]\left(\vec{a}^{b}_{w}\right)\right). \quad (19.7)$$

Consider the possible cases.

Case 1. Let

$$\mathfrak{J} \vDash_L \exists z\left(b \subseteq_{\mathsf{s}} z \wedge \mathrm{Suf}[R,p,\varepsilon]\left(\vec{a}^{b}_{z}\right)\right)$$

hold. Then

$$b \subseteq^{\mathfrak{J}}_{\mathsf{s}} g \quad \text{and} \quad \mathfrak{J} \vDash_L \mathrm{Suf}[R,p,\varepsilon]\left(\vec{a}^{b}_{g}\right),$$

hold for some $g \in \mathrm{Dom}_{\mathscr{M}}(\mathsf{s})$, that by Definitions 17.2.5 and 17.2.6 and Lemma 14.1.1 is equivalent to

$$\mathrm{Cn}^{\Delta}_{\mathsf{s}}(b) \subseteq_{\mathsf{s}} \mathrm{Cn}^{\Delta}_{\mathsf{s}}(g) \quad \text{and} \quad \mathrm{CSD}(\vec{\mathfrak{a}}) \vdash_L \mathrm{Suf}[R,p,\varepsilon]\left(\vec{a}^{b}_{g}\right), \quad (19.8)$$

respectively. Hence by Proposition 19.1.14 (i) we get

$$\mathrm{CSD}(\vec{\mathfrak{a}}) \vdash_L \mathrm{Suf}[R,p,\varepsilon]\left(\vec{a}^{b}_{b}\right),$$

whence by Lemma 19.1.12 we obtain that for any $e \in \mathrm{Cn}^{\Delta}_{\mathsf{s}}(b)$,

$$\mathrm{CSD}(\vec{\mathfrak{a}}) \vdash_L \mathrm{Suf}\left[R, \mathrm{P}^{\mathscr{M}}_{m}, \varepsilon\right]\left(\vec{a}^{b}_{e}\right), \quad (19.9)$$

as was to be proved.

Case 2. Let

$$\mathfrak{J} \vDash_L \exists z \exists w\left(b = z \cup_{\mathsf{s}} w \wedge \mathrm{Suf}[R,p,\varepsilon]\left(\vec{a}^{b}_{z}\right) \wedge \mathrm{Suf}[R,p,\varepsilon]\left(\vec{a}^{b}_{w}\right)\right). \quad (19.10)$$

Then there are $g,h \in \mathrm{Dom}_{\mathscr{M}}(\mathsf{s})$ such that the following hold:

$$b = g \cup^{\mathfrak{J}}_{\mathsf{s}} h, \quad (19.11)$$

$$\mathfrak{J} \vDash_L \mathrm{Suf}[R,p,\varepsilon]\left(\vec{a}^{b}_{g}\right), \quad (19.12)$$

$$\mathfrak{J} \vDash_L \mathrm{Suf}[R,p,\varepsilon]\left(\vec{a}^{b}_{h}\right). \quad (19.13)$$

By Definitions 17.2.5 and 17.2.6, (19.11) implies

$$\mathrm{Cn}^{\Delta}_{\mathsf{s}}(b) = \mathrm{Cn}^{\Delta}_{\mathsf{s}}(g) \cup \mathrm{Cn}^{\Delta}_{\mathsf{s}}(h). \quad (19.14)$$

From (19.12) and (19.13) by Lemma 14.1.1 we obtain

$$\mathrm{CSD}(\vec{\mathfrak{a}}) \vdash_L \mathrm{Suf}[R,p,\varepsilon]\left(\vec{a}_g^b\right), \tag{19.15}$$

$$\mathrm{CSD}(\vec{\mathfrak{a}}) \vdash_L \mathrm{Suf}[R,p,\varepsilon]\left(\vec{a}_h^b\right), \tag{19.16}$$

respectively. By Lemma 19.1.12 it follows from (19.15) and (19.16) that (19.9) holds for any $e \in \mathrm{Cn}_{\mathsf{s}}^{\Delta}(g)$ and for any $e \in \mathrm{Cn}_{\mathsf{s}}^{\Delta}(h)$, respectively. Then by (19.14), (19.9) holds also for any $e \in \mathrm{Cn}_{\mathsf{s}}^{\Delta}(b)$. This concludes the proof. □

Lemma 19.2.2. Let L be an FPJ logic, σ be a signature and $\mathscr{M} \in \mathrm{PMC}_\sigma(L)$. Suppose:

(a) $\varepsilon \in \mathscr{D}^\tau(L), p \in +\mathsf{s}\varepsilon[R], \vec{a} \in \mathrm{Dom}_{\mathscr{M}}(p), b \in \mathrm{Set}(\vec{a}) \cap \mathrm{Dom}_{\mathscr{M}}(\mathsf{s})$;
(b) $\vec{\mathfrak{a}}$ is an $\mathscr{M}(s,k)$-inference from $\Delta = \mathrm{Descr}_{\mathscr{M}}$ ready for $\mathrm{J}_\varepsilon\, p\left(\vec{a}_b^b\right)$, where $s > 0$, $0 \le m-1 \le k \le m \le Q_\sigma$;
(c) $\vec{\mathfrak{a}}_{\overleftarrow{s,\,m-1}}$ is a coherent $\mathscr{M}(s,m-1,\mathbf{D})$-inference.

Then

(d) $\mathrm{CSD}\left(\vec{\mathfrak{a}}_{\overleftarrow{s,\,m-1}}\right) \vdash_L \mathrm{Suf}[R,p,\varepsilon]\left(\vec{a}_e^b\right)$

holds for any $e \in \mathrm{Cn}_{\mathsf{s}}^{\Delta}(b)$.

Proof. By induction on the length of $\mathrm{Hi}(\vec{\mathfrak{a}},s,m-1)$ (see Definition 11.2.20).

Basis. $\mathrm{Len}(\mathrm{Hi}(\vec{\mathfrak{a}},s,m-1)) = 0$. Then $\vec{\mathfrak{a}}_{\overleftarrow{s,\,m-1}} = \vec{\mathfrak{a}}$. Then $\vec{\mathfrak{a}}_{\overleftarrow{s,\,m-1}}$ is an $\mathscr{M}(s)$-inference, which is simultaneously coherent and ready for the $\mathrm{J}_\varepsilon\, p\left(\vec{a}_b^b\right)$, where $s > 0$. Then by Lemma 19.2.1, for any $e \in \mathrm{Cn}_{\mathsf{s}}^{\Delta}(b)$, (d) holds.

Induction step. Suppose that the lemma holds for any $\mathscr{M}(s,k)$-inference $\vec{\mathfrak{d}}$, whose length of the upper segment of the $(s,\ m-1)$-cut is less than or equal to n. Suppose that $\mathrm{Len}(\mathrm{Hi}(\vec{\mathfrak{a}},s,m-1)) = n+1$.

Let $\mathfrak{J} = \mathrm{Sint}_L(\mathrm{CSD}(\vec{\mathfrak{a}}), D)$. The fact that $\vec{\mathfrak{a}}$ is an $\mathscr{M}$-inference ready for $\mathrm{J}_\varepsilon\, p\left(\vec{a}_b^b\right)$ implies $\mathrm{MPr}[R,p,\varepsilon]\left(\vec{a}_b^b\right) \in \mathrm{FSet}(\vec{\mathfrak{a}})$. But then by Corollary 11.4.12,

$$\mathrm{CSD}(\vec{\mathfrak{a}}) \vdash_L \mathrm{MPr}[R,p,\varepsilon]\left(\vec{a}_b^b\right),$$

therefore (see Lemma 14.1.1),

$$\mathfrak{J} \models_L \mathrm{MPr}[R,p,\varepsilon]\left(\vec{a}_b^b\right). \tag{19.17}$$

Since $p \in +\mathsf{s}\varepsilon[R]$, $\mathrm{MPr}[R,p,\varepsilon]\left(\vec{a}_b^b\right)$ has the form

$$\exists z\left(b \subseteq_{\mathsf{s}} z \wedge \mathrm{Suf}[R,p,\varepsilon]\left(\vec{a}_z^b\right)\right) \vee \\ \exists z \exists w\left(b = z \cup_{\mathsf{s}} w \wedge \mathrm{Suf}[R,p,\varepsilon]\left(\vec{a}_z^b\right) \wedge \mathrm{Suf}[R,p,\varepsilon]\left(\vec{a}_w^b\right)\right). \tag{19.18}$$

Then (19.17) is true iff at least one of the following conditions holds:

Since $p \in +\mathsf{s}\varepsilon[R]$, $\mathrm{MPr}[R,p,\varepsilon](\vec{a})$ has the form

$$\exists z\left(b \subseteq_{\mathsf{s}} z \wedge \mathrm{Suf}[R,p,\varepsilon]\left(\vec{a}_z^b\right)\right) \vee \\ \exists z \exists w\left(b = z \cup_{\mathsf{s}} w \wedge \mathrm{Suf}[R,p,\varepsilon]\left(\vec{a}_z^b\right) \wedge \mathrm{Suf}[R,p,\varepsilon]\left(\vec{a}_w^b\right)\right). \quad (19.7)$$

Consider the possible cases.

CASE 1. Let

$$\mathfrak{J} \vDash_L \exists z\left(b \subseteq_{\mathsf{s}} z \wedge \mathrm{Suf}[R,p,\varepsilon]\left(\vec{a}_z^b\right)\right)$$

hold. Then

$$b \subseteq_{\mathsf{s}}^{\mathfrak{J}} g \quad \text{and} \quad \mathfrak{J} \vDash_L \mathrm{Suf}[R,p,\varepsilon]\left(\vec{a}_g^b\right),$$

hold for some $g \in \mathrm{Dom}_{\mathscr{M}}(\mathsf{s})$, that by Definitions 17.2.5 and 17.2.6 and Lemma 14.1.1 is equivalent to

$$\mathrm{Cn}_{\mathsf{s}}^{\Delta}(b) \subseteq_{\mathsf{s}} \mathrm{Cn}_{\mathsf{s}}^{\Delta}(g) \quad \text{and} \quad \mathrm{CSD}(\vec{\mathfrak{a}}) \vdash_L \mathrm{Suf}[R,p,\varepsilon]\left(\vec{a}_g^b\right), \quad (19.8)$$

respectively. Hence by Proposition 19.1.14 (i) we get

$$\mathrm{CSD}(\vec{\mathfrak{a}}) \vdash_L \mathrm{Suf}[R,p,\varepsilon]\left(\vec{a}_b^b\right),$$

whence by Lemma 19.1.12 we obtain that for any $e \in \mathrm{Cn}_{\mathsf{s}}^{\Delta}(b)$,

$$\mathrm{CSD}(\vec{\mathfrak{a}}) \vdash_L \mathrm{Suf}\left[R,\mathrm{P}_m^{\mathscr{M}},\varepsilon\right]\left(\vec{a}_e^b\right), \quad (19.9)$$

as was to be proved.

CASE 2. Let

$$\mathfrak{J} \vDash_L \exists z \exists w\left(b = z \cup_{\mathsf{s}} w \wedge \mathrm{Suf}[R,p,\varepsilon]\left(\vec{a}_z^b\right) \wedge \mathrm{Suf}[R,p,\varepsilon]\left(\vec{a}_w^b\right)\right). \quad (19.10)$$

Then there are $g, h \in \mathrm{Dom}_{\mathscr{M}}(\mathsf{s})$ such that the following hold:

$$b = g \cup_{\mathsf{s}}^{\mathfrak{J}} h, \quad (19.11)$$

$$\mathfrak{J} \vDash_L \mathrm{Suf}[R,p,\varepsilon]\left(\vec{a}_g^b\right), \quad (19.12)$$

$$\mathfrak{J} \vDash_L \mathrm{Suf}[R,p,\varepsilon]\left(\vec{a}_h^b\right). \quad (19.13)$$

By Definitions 17.2.5 and 17.2.6, (19.11) implies

$$\mathrm{Cn}_{\mathsf{s}}^{\Delta}(b) = \mathrm{Cn}_{\mathsf{s}}^{\Delta}(g) \cup \mathrm{Cn}_{\mathsf{s}}^{\Delta}(h). \quad (19.14)$$

From (19.12) and (19.13) by Lemma 14.1.1 we obtain

$$\mathrm{CSD}(\vec{\mathfrak{a}}) \vdash_L \mathrm{Suf}[R,p,\varepsilon]\left(\vec{a}_g^b\right), \tag{19.15}$$

$$\mathrm{CSD}(\vec{\mathfrak{a}}) \vdash_L \mathrm{Suf}[R,p,\varepsilon]\left(\vec{a}_h^b\right), \tag{19.16}$$

respectively. By Lemma 19.1.12 it follows from (19.15) and (19.16) that (19.9) holds for any $e \in \mathrm{Cn}_{\mathsf{s}}^{\Delta}(g)$ and for any $e \in \mathrm{Cn}_{\mathsf{s}}^{\Delta}(h)$, respectively. Then by (19.14), (19.9) holds also for any $e \in \mathrm{Cn}_{\mathsf{s}}^{\Delta}(b)$. This concludes the proof. □

Lemma 19.2.2. Let L be an FPJ logic, σ be a signature and $\mathscr{M} \in \mathrm{PMC}_{\sigma}(L)$. Suppose:

(a) $\varepsilon \in \mathscr{D}^{\tau}(L), p \in +\mathrm{s}\varepsilon[R], \vec{a} \in \mathrm{Dom}_{\mathscr{M}}(p), b \in \mathrm{Set}(\vec{a}) \cap \mathrm{Dom}_{\mathscr{M}}(\mathsf{s})$;
(b) $\vec{\mathfrak{a}}$ is an $\mathscr{M}(s,k)$-inference from $\Delta = \mathrm{Descr}_{\mathscr{M}}$ ready for $\mathrm{J}_{\varepsilon}\, p\left(\vec{a}_b^b\right)$, where $s > 0$, $0 \leq m-1 \leq k \leq m \leq Q_{\sigma}$;
(c) $\vec{\mathfrak{a}}_{\overleftarrow{s,\,m-1}}$ is a coherent $\mathscr{M}(s,m-1,\mathbf{D})$-inference.

Then

(d) $\mathrm{CSD}\left(\vec{\mathfrak{a}}_{\overleftarrow{s,\,m-1}}\right) \vdash_L \mathrm{Suf}[R,p,\varepsilon]\left(\vec{a}_e^b\right)$

holds for any $e \in \mathrm{Cn}_{\mathsf{s}}^{\Delta}(b)$.

Proof. By induction on the length of $\mathrm{Hi}(\vec{\mathfrak{a}},s,m-1)$ (see Definition 11.2.20).

Basis. $\mathrm{Len}(\mathrm{Hi}(\vec{\mathfrak{a}},s,m-1)) = 0$. Then $\vec{\mathfrak{a}}_{\overleftarrow{s,\,m-1}} = \vec{\mathfrak{a}}$. Then $\vec{\mathfrak{a}}_{\overleftarrow{s,\,m-1}}$ is an $\mathscr{M}(s)$-inference, which is simultaneously coherent and ready for the $\mathrm{J}_{\varepsilon}\, p\left(\vec{a}_b^b\right)$, where $s > 0$. Then by Lemma 19.2.1, for any $e \in \mathrm{Cn}_{\mathsf{s}}^{\Delta}(b)$, (d) holds.

Induction step. Suppose that the lemma holds for any $\mathscr{M}(s,k)$-inference $\vec{\mathfrak{d}}$, whose length of the upper segment of the (s, m –1)-cut is less than or equal to n. Suppose that $\mathrm{Len}(\mathrm{Hi}(\vec{\mathfrak{a}},s,m-1)) = n+1$.

Let $\mathfrak{J} = \mathrm{Sint}_L(\mathrm{CSD}(\vec{\mathfrak{a}}), D)$. The fact that $\vec{\mathfrak{a}}$ is an $\mathscr{M}$-inference ready for $\mathrm{J}_{\varepsilon}\, p\left(\vec{a}_b^b\right)$ implies $\mathrm{MPr}[R,p,\varepsilon]\left(\vec{a}_b^b\right) \in \mathrm{FSet}(\vec{\mathfrak{a}})$. But then by Corollary 11.4.12,

$$\mathrm{CSD}(\vec{\mathfrak{a}}) \vdash_L \mathrm{MPr}[R,p,\varepsilon]\left(\vec{a}_b^b\right),$$

therefore (see Lemma 14.1.1),

$$\mathfrak{J} \vDash_L \mathrm{MPr}[R,p,\varepsilon]\left(\vec{a}_b^b\right). \tag{19.17}$$

Since $p \in +\mathrm{s}\varepsilon[R]$, $\mathrm{MPr}[R,p,\varepsilon]\left(\vec{a}_b^b\right)$ has the form

$$\exists z\left(b \subseteq_{\mathsf{s}} z \wedge \mathrm{Suf}[R,p,\varepsilon]\left(\vec{a}_z^b\right)\right) \vee$$
$$\exists z \exists w\left(b = z \cup_{\mathsf{s}} w \wedge \mathrm{Suf}[R,p,\varepsilon]\left(\vec{a}_z^b\right) \wedge \mathrm{Suf}[R,p,\varepsilon]\left(\vec{a}_w^b\right)\right). \tag{19.18}$$

Then (19.17) is true iff at least one of the following conditions holds:

$$\mathfrak{J} \vDash_L \exists z \left(b \subseteq_{\mathsf{s}} z \wedge \mathrm{Suf}[R, p, \varepsilon] \left(\vec{a}^{\,b}_{z} \right) \right) \tag{19.19}$$

or

$$\mathfrak{J} \vDash_L \exists z \exists w \left(b = z \cup_{\mathsf{s}} w \wedge \mathrm{Suf}[R, p, \varepsilon] \left(\vec{a}^{\,b}_{z} \right) \wedge \mathrm{Suf}[R, p, \varepsilon] \left(\vec{a}^{\,b}_{w} \right) \right). \tag{19.20}$$

Consider the possible cases.

CASE 1. Let (19.19) hold. Then

$$b \subseteq^{\mathfrak{J}}_{\mathsf{s}} g \tag{19.21}$$

and

$$\mathfrak{J} \vDash_L \mathrm{Suf}[R, p, \varepsilon] \left(\vec{a}^{\,b}_{g} \right). \tag{19.22}$$

hold for some $g \in \mathrm{Dom}_{\sigma}(D, \mathsf{s})$. By Definitions 17.2.5 and 17.2.6, (19.5) implies

$$\mathrm{Cn}^{\Delta}_{\mathsf{s}}(b) \subseteq \mathrm{Cn}^{\Delta}_{\mathsf{s}}(g). \tag{19.23}$$

But by Lemma 14.1.1, (19.11) implies

$$\mathrm{CSD}(\vec{\mathfrak{a}}) \vdash_L \mathrm{Suf}[R, p, \varepsilon] \left(\vec{a}^{\,b}_{g} \right). \tag{19.24}$$

First, let us show that $\mathrm{J}_{\varepsilon}\, p\left(\vec{a}^{\,b}_{g}\right) \notin \mathrm{CSD}\left(\vec{\mathfrak{a}}_{\overleftarrow{s,\, m-1}}\right)$. Suppose the converse. Let

$$\mathrm{J}_{\varepsilon}\, p \left(\vec{a}^{\,b}_{g} \right) \in \mathrm{CSD}\left(\vec{\mathfrak{a}}_{\overleftarrow{s,\, m-1}} \right). \tag{19.25}$$

By the assumption $\vec{\mathfrak{a}}_{\overleftarrow{s,\, m-1}}$ is a coherent inference. Then by Proposition 19.1.9 (i) from (19.12) and (19.14) we get $\mathrm{J}_{\varepsilon}\, p\left(\vec{a}^{\,b}_{b}\right) \in \mathrm{CSD}\left(\vec{\mathfrak{a}}_{\overleftarrow{s,\, m-1}}\right)$. Hence, by Corollary 15.1.17, we get $\mathrm{J}_{\varepsilon}\, p(\vec{a}) \in \mathrm{CSD}(\vec{\mathfrak{a}})$ that contradicts the assumption. (According to the assumption $\vec{\mathfrak{a}}$ is an $\mathscr{M}$-inference ready for the $\mathrm{J}_{\varepsilon}\, p(\vec{a})$. Therefore, it should not contain $\mathrm{J}_{\varepsilon}\, p(\vec{a})$ (see Definition 19.1.16). Thus $\mathrm{J}_{\varepsilon}\, p\left(\vec{a}^{\,b}_{g}\right) \notin \mathrm{CSD}\left(\vec{\mathfrak{a}}_{\overleftarrow{s,\, m-1}}\right)$. Hence,

$$\mathrm{J}_{\tau}\, p \left(\vec{a}^{\,b}_{g} \right) \in \mathrm{CSD}\left(\vec{\mathfrak{a}}_{\overleftarrow{s,\, m-1}} \right). \tag{19.26}$$

Now let us consider the possible cases w.r.t. the belonging of the formula $\mathrm{J}_{\varepsilon}\, p\left(\vec{a}^{\,b}_{g}\right)$ to the set $\mathrm{CSD}(\vec{\mathfrak{a}})$.

SUBCASE 1.1. Let $\mathrm{J}_{\varepsilon}\, p\left(\vec{a}^{\,b}_{g}\right) \notin \mathrm{CSD}(\vec{\mathfrak{a}})$. Recall that $\mathrm{Suf}[R, p, \varepsilon]\left(\vec{a}^{\,b}_{g}\right)$ is local $\mathrm{J}_{\varepsilon}\, p\left(\vec{a}^{\,b}_{g}\right)$ (see Remark 18.1.3). Then from (19.11) and (19.15) by Corollary 15.1.12 we get

$$\mathrm{CSD}\left(\vec{\mathfrak{a}}_{\overleftarrow{s,\, m-1}} \right) \vdash_{\mathbf{St}} \mathrm{Suf}[R, p, \varepsilon] \left(\vec{a}^{\,b}_{g} \right). \tag{19.27}$$

By the assumption, $\vec{\mathfrak{a}}_{\overleftarrow{s,\, m-1}}$ is coherent. Then by Lemma 19.1.12 (d) holds for any $e \in \mathrm{Cn}^{\Delta}_{\sigma}(g)$. Then by (19.12), (d) holds for any $e \in \mathrm{Cn}^{\Delta}_{\sigma}(b)$, as was to be proved.

SUBCASE 1.2. Let
$$\mathrm{J}_{\varepsilon}\, p\left(\vec{a}_{g}^{b}\right) \in \mathrm{CSD}(\vec{\mathfrak{a}}). \tag{19.28}$$

Then from (19.15) and (19.28) by Lemma 19.1.18 we obtain that there exists a cut $\left\langle \vec{\mathfrak{b}}, \vec{\mathfrak{c}} \right\rangle$ of $\mathscr{M}(s,m)$-inference $\vec{\mathfrak{a}}$ such that

$$\mathrm{Len}\left(\vec{\mathfrak{a}}_{\overleftarrow{s,\,m-1}}\right) \leq \mathrm{Len}\left(\vec{b}\right) < \mathrm{Len}(\vec{\mathfrak{a}}), \tag{19.29}$$

$\vec{\mathfrak{d}} = \mathrm{Resume}\left(\vec{\mathfrak{b}}\right)$ is an $\mathscr{M}(s,k,\mathbf{M})$-inference from $\mathrm{Descr}_{\mathscr{M}}$ ready for $\mathrm{J}_{\varepsilon}\, p\left(\vec{a}_{g}^{b}\right)$, where $m-1 \leq k \leq m$,
$$\vec{\mathfrak{d}}_{\overleftarrow{s,\,m-1}} = \vec{\mathfrak{a}}_{\overleftarrow{s,\,m-1}}. \tag{19.30}$$

By (19.29), $\mathrm{Len}\left(\vec{\mathfrak{d}}\right) = \mathrm{Len}\left(\vec{\mathfrak{b}}\right) \leq n$. Then we can apply the inductive assumption to $\vec{\mathfrak{d}}$. Then (d) holds for any $e \in \mathrm{Cn}_{\mathsf{s}}^{\Delta}(g)$. By (19.12) (d) also holds for any $e \in \mathrm{Cn}_{\mathsf{s}}^{\Delta}(b)$, as was to be proved.

CASE 2. Let (19.20) be valid. Then

$$b = g \cup_{\mathsf{s}}^{\mathfrak{J}} h \tag{19.31}$$

holds for some $g,h \in \mathrm{Dom}_{\sigma}(D,\mathrm{s})$, and (19.11) also holds and

$$\mathfrak{J} \models_{L} \mathrm{Suf}[R,p,\varepsilon]\left(\vec{a}_{h}^{b}\right). \tag{19.32}$$

From (19.31) by Definitions 17.2.5 and 17.2.6 we get

$$\mathrm{Cn}_{\mathsf{s}}^{\Delta}(b) = \mathrm{Cn}_{\mathsf{s}}^{\Delta}(g) \cup \mathrm{Cn}_{\mathsf{s}}^{\Delta}(h). \tag{19.33}$$

Emphasise that (19.11) holds like in CASE 1. Thus, argumenting like in that case we obtain two subcases: $\mathrm{J}_{\varepsilon}\, p\left(\vec{a}_{g}^{b}\right) \notin \mathrm{CSD}(\vec{\mathfrak{a}})$ and $\mathrm{J}_{\varepsilon}\, p\left(\vec{a}_{g}^{b}\right) \in \mathrm{CSD}(\vec{\mathfrak{a}})$. By using Corollary 15.1.12 for the first subcase and the inductive assumption for the second subcase we can prove that (d) holds for any $e \in \mathrm{Cn}_{\sigma}^{\Delta}(g)$ just like in CASE 1.

Similarly, by using (19.32) instead of (19.11), two subcases can be considered: $\mathrm{J}_{\varepsilon}\, p\left(\vec{a}_{h}^{b}\right) \notin \mathrm{CSD}(\vec{\mathfrak{a}})$ and $\mathrm{J}_{\varepsilon}\, p\left(\vec{a}_{h}^{b}\right) \in \mathrm{CSD}(\vec{\mathfrak{a}})$. As a result we obtain that (d) holds for any $e \in \mathrm{Cn}_{\sigma}^{\Delta}(h)$. Taking into account (19.2) we conclude that (d) holds for any $e \in \mathrm{Cn}_{\sigma}^{\Delta}(b)$, which was to be proved.

All cases have been analysed. □

Lemma 19.2.3. Let L be an FPJ logic, σ be a signature, $\mathsf{s} \in \mathsf{SetSorts}_{\sigma}$ and $\mathscr{M} \in \mathrm{PMC}_{\sigma}(L)$. Let $\vec{\mathfrak{a}}$ be an $\mathscr{M}(s,m,\mathbf{D})$-inference from $\Delta = \mathrm{Descr}_{\mathscr{M}}$, where $s>0, m>0$. Suppose $\varepsilon \in \mathscr{D}^{\tau}(L)$, $\mathrm{P}_{m}^{\mathscr{M}} \in R[+,s,\varepsilon]$, $\vec{a} \in \mathrm{Dom}_{\mathscr{M}}\left(\mathrm{P}_{m}^{\mathscr{M}}\right)$, $b \in \mathrm{Set}(\vec{a}) \cap \mathrm{Dom}_{\mathscr{M}}(\mathsf{s})$. Let:

(a) $\vec{\mathfrak{a}}_{\overleftarrow{s,\,m-1}}$ be a coherent $\mathscr{M}(s,m-1,\mathbf{D})$-inference from $\mathrm{Descr}_{\mathscr{M}}$,
(b) $\mathrm{J}_{\varepsilon}\, \mathrm{P}_{m}^{\mathscr{M}}\left(\vec{a}_{b}^{b}\right) \in \mathrm{CSD}(\vec{\mathfrak{a}})$.

Then for any $e \in \mathrm{Cn}_{\mathsf{s}}^{\Delta}(b)$ it holds that

(c) $\mathrm{CSD}\left(\vec{\mathfrak{a}}_{\overleftarrow{s,\,m-1}}\right) \vdash_L \mathrm{Suf}\left[R, \mathrm{P}_m^{\mathscr{M}}, \varepsilon\right]\left(\vec{a}_e^b\right)$.

Proof. Consider the possible cases.

CASE 1. $\mathrm{J}_\varepsilon \mathrm{P}_m^{\mathscr{M}}\left(\vec{a}_b^b\right) \in \mathrm{CSD}\left(\vec{\mathfrak{a}}_{\overleftarrow{s,\,m-1}}\right)$. Then, since $\vec{\mathfrak{a}}_{\overleftarrow{s,\,m-1}}$ is coherent by Lemma 19.1.4, for any $e \in \mathrm{Cn}_{\sigma}^{\Delta}(b)$

$$\mathrm{J}_\varepsilon \mathrm{P}_m^{\mathscr{M}}\left(\vec{a}_e^b\right) \in \mathrm{CSD}\left(\vec{\mathfrak{a}}_{\overleftarrow{s,\,m-1}}\right).$$

Thus (c) holds for any $e \in \mathrm{Cn}_{\sigma}^{\Delta}(b)$ (see Definition 18.1.2).

CASE 2. $\mathrm{J}_\varepsilon \mathrm{P}_m^{\mathscr{M}}(\vec{a}) \notin \mathrm{CSD}\left(\vec{\mathfrak{a}}_{\overleftarrow{s,\,m-1}}\right)$. Then

$$\mathrm{J}_\tau \mathrm{P}_m^{\mathscr{M}}(\vec{a}) \in \mathrm{CSD}\left(\vec{\mathfrak{a}}_{\overleftarrow{s,\,m-1}}\right). \tag{19.34}$$

By Lemma 19.1.18 it follows from (17.1) and (b) that there exists an initial segment $\vec{\mathfrak{b}}$ of the $\mathscr{M}(s,m)$-inference $\vec{\mathfrak{a}}$, such that $\vec{\mathfrak{d}} = \mathrm{Resume}\left(\vec{\mathfrak{b}}\right)$ is an $\mathscr{M}(s,k,\mathbf{M})$-inference from $\mathrm{Descr}_{\mathscr{M}}$ ready for $\mathrm{J}_\varepsilon \mathrm{P}_m^{\mathscr{M}}\left(\vec{a}_b^b\right)$, where $m-1 \le k \le m$. Then by Lemma 19.2.2, (c) holds for any $e \in \mathrm{Cn}_{\sigma}^{\Delta}(b)$. □

Lemma 19.2.4. Let L be an FPJ logic, σ be a signature and $\mathscr{M} \in \mathrm{PMC}_\sigma(L)$. Suppose:

(a) $\varepsilon \in \mathscr{D}^\tau(L)$, $p \in +\mathsf{s}\varepsilon[R]$, $\vec{a} \in \mathrm{Dom}_{\mathscr{M}}(p)$, $b \in \mathrm{Set}(\vec{a}) \cap \mathrm{Dom}_{\mathscr{M}}(\mathsf{s})$;
(b) $\vec{\mathfrak{a}}$ is an $\mathscr{M}(s,k)$-inference from $\Delta = \mathrm{Descr}_{\mathscr{M}}$ ready for $\mathrm{J}_\tau p\left(\vec{a}_b^b\right)$, where $s > 0$, $0 \le m-1 \le k \le m \le Q_\sigma$;
(c) $\vec{\mathfrak{a}}_{\overleftarrow{s,\,m-1}}$ is a coherent $\mathscr{M}(s,m-1,\mathbf{D})$-inference from $\mathrm{Descr}_{\mathscr{M}}$.

Then there exists $e \in \mathrm{Cn}_{\mathsf{s}}^{\Delta}(b)$ such that

(d) $\mathrm{CSD}\left(\vec{\mathfrak{a}}_{\overleftarrow{s,\,m-1}}\right) \nvdash_L \mathrm{Suf}[R,p,\varepsilon]\left(\vec{a}_e^b\right)$.

Proof. Let us suppose the converse. Let

$$\mathrm{CSD}\left(\vec{\mathfrak{a}}_{\overleftarrow{s,\,m-1}}\right) \vdash_{\mathbf{St}} \mathrm{Suf}[R,p,\varepsilon]\left(\vec{a}_e^b\right)$$

hold for any $e \in \mathrm{Cn}_{\mathsf{s}}^{\Delta}(b)$. Then by Lemma 14.1.1,

$$\mathfrak{J} \vDash_L \mathrm{Suf}[R,p,\varepsilon]\left(\vec{a}_e^b\right)$$

holds for any $e \in \mathrm{Cn}_{\mathsf{s}}^{\Delta}(b)$, where

$$\mathfrak{J} = \mathrm{Sint}_L\left(\mathrm{CSD}\left(\vec{\mathfrak{a}}_{\overleftarrow{s,\,m-1}}\right), D\right).$$

By Definition 18.1.2, $\mathrm{Suf}[R,p,\varepsilon]\left(\vec{a}_e^b\right)$ is the formula $\mathrm{J}_\varepsilon\, p\left(\vec{a}_e^b\right) \vee \mathrm{Ker}[R,p,\varepsilon]\left(\vec{a}_e^b\right)$, which is closed. Thus

$$\mathfrak{J} \vDash_L \mathrm{J}_\varepsilon\, p\left(\vec{a}_e^b\right)$$

or

$$\mathfrak{J} \vDash_L \mathrm{Ker}[R,p,\varepsilon]\left(\vec{a}_e^b\right),$$

for any $e \in \mathrm{Cn}_{\mathsf{s}}^{\Delta}(b)$. Let

$$G = \left\{e \in \mathrm{Cn}_{\mathsf{s}}^{\Delta}(b) \mid \mathfrak{J} \vDash_{\mathbf{St}} \mathrm{J}_\varepsilon\, p\left(\vec{a}_e^b\right)\right\}, \quad H = \mathrm{Cn}_{\mathsf{s}}^{\Delta}(b) \setminus G. \tag{19.35}$$

Suppose

$$g = \left(\mathrm{Cn}_{\mathsf{s}}^{\Delta}\right)^{-1}(G), \quad h = \left(\mathrm{Cn}_{\mathsf{s}}^{\Delta}\right)^{-1}(H). \tag{19.36}$$

Then for any $e \in \mathrm{Cn}_{\mathsf{s}}^{\Delta}(g)$

$$\mathrm{J}_\varepsilon\, p\left(\vec{a}_e^b\right) \in \mathrm{CSD}\left(\vec{\mathfrak{a}}_{\overleftarrow{s,\,m-1}}\right),$$

but for any $e \in \mathrm{Cn}_{\mathsf{s}}^{\Delta}(h)$

$$\mathrm{CSD}\left(\vec{\mathfrak{a}}_{\overleftarrow{s,\,m-1}}\right) \vdash_{\mathbf{St}} \mathrm{Ker}[R,p,\varepsilon]\left(\vec{a}_e^b\right).$$

Then by Lemmas 19.1.4 and 19.1.6, we get

$$\mathrm{J}_\varepsilon\, p\left(\vec{a}_g^b\right) \in \mathrm{CSD}\left(\vec{\mathfrak{a}}_{\overleftarrow{s,\,m-1}}\right), \tag{19.37}$$

$$\mathrm{CSD}\left(\vec{\mathfrak{a}}_{\overleftarrow{s,\,m-1}}\right) \vdash_L \mathrm{Ker}[R,p,\varepsilon]\left(\vec{a}_h^b\right). \tag{19.38}$$

From (19.37) (taking into account Lemma 15.1.14) and from (19.38) by Corollary 15.1.13 we obtain

$$\mathrm{CSD}(\vec{\mathfrak{a}}) \vdash_L \mathrm{J}_\varepsilon\, p\left(\vec{a}_g^b\right), \tag{19.39}$$

$$\mathrm{CSD}(\vec{\mathfrak{a}}) \vdash_L \mathrm{Ker}[R,p,\varepsilon]\left(\vec{a}_h^b\right). \tag{19.40}$$

Moreover, (19.35) and (19.36), clearly, imply

$$\mathrm{CSD}(\vec{\mathfrak{a}}) \vdash_L b = g \cup_{\mathsf{s}} h. \tag{19.41}$$

Results (19.39), (19.40) and (19.41) imply

$$\mathrm{CSD}(\vec{\mathfrak{a}}) \vdash_L$$
$$\exists z \exists w \left(b = z \cup_{\mathsf{s}} w \wedge \mathrm{Suf}[R,p,\varepsilon]\left(\vec{a}_z^b\right) \wedge \mathrm{Suf}[R,p,\varepsilon]\left(\vec{a}_w^b\right)\right). \tag{19.42}$$

However by the assumption, $\vec{\mathfrak{a}}$ is an $\mathscr{M}$-inference from $\mathrm{Descr}_{\mathscr{M}}$ ready for $\mathrm{J}_\tau\, p(\vec{a})$. Then

$$\mathrm{CSD}(\vec{\mathfrak{a}}) \vdash_L \ \mathrm{MPr}[R,p,\varepsilon]\left(\vec{a}_b^b\right).$$

But then (see Definition 18.1.1)

$$\mathrm{CSD}(\vec{\mathfrak{a}}) \vdash_L$$
$$\neg\exists z \exists w \left(b = z \cup_{\mathsf{s}} w \wedge \mathrm{Suf}[R,p,\varepsilon]\left(\vec{a}_z^b\right) \wedge \mathrm{Suf}[R,p,\varepsilon]\left(\vec{a}_w^b\right)\right). \quad (19.43)$$

Results (19.42) and (19.43) imply that $\mathrm{CSD}(\vec{\mathfrak{a}})$ is inconsistent, which is impossible by Lemma 14.1.1.

Thus our supposition does not hold. Therefore, there exists $e \in \mathrm{Cn}_{\mathsf{s}}^{\Delta}(b)$ such that

$$\mathrm{CSD}\left(\vec{\mathfrak{a}}_{\overleftarrow{s,\,m-1}}\right) \nvdash_L \ \mathrm{Suf}[R,p,\varepsilon]\left(\vec{a}_e^b\right).$$

□

Lemma 19.2.5. Let L be an FPJ logic, σ be a signature, $\mathscr{M} \in \mathrm{PMC}_\sigma(L)$ and $\vec{\mathfrak{a}}$ be an $\mathscr{M}(s,m,\mathbf{D})$-inference from $\Delta = \mathrm{Descr}_{\mathscr{M}}$, where $s > 0$, $m > 0$. Suppose $\varepsilon \in \mathscr{D}^\tau(L)$, $\mathrm{P}_m^{\mathscr{M}} \in +\mathsf{s}\varepsilon[R]$, $\vec{a} \in \mathrm{Dom}_{\mathscr{M}}(p)$, $b \in \mathrm{Set}(\vec{a}) \cap \mathrm{Dom}_{\mathscr{M}}(\mathsf{s})$. Let:

(a) $\vec{\mathfrak{a}}_{\overleftarrow{s,\,m-1}}$ be a coherent $M(s,m-1,\mathbf{D})$-inference from $\Delta = \mathrm{Descr}_{\mathscr{M}}$,

(b) For any $e \in \mathrm{Cn}_{\mathsf{s}}^{\Delta}(b)$, $\mathrm{CSD}\left(\vec{\mathfrak{a}}_{\overleftarrow{s,\,m-1}}\right) \vdash_L \ \mathrm{Suf}\left[R,\mathrm{P}_m^{\mathscr{M}},\varepsilon\right]\left(\vec{a}_e^b\right)$.

Then

(c) $\mathrm{J}_\varepsilon\, \mathrm{P}_m^{\mathscr{M}}\left(\vec{a}_b^b\right) \in \mathrm{CSD}(\vec{\mathfrak{a}})$.

Proof. Let us suppose the converse: $\mathrm{J}_\varepsilon\, \mathrm{P}_m^{\mathscr{M}}\left(\vec{a}_b^b\right) \notin \mathrm{CSD}(\vec{\mathfrak{a}})$. Then $\mathrm{J}_\tau\, \mathrm{P}_m^{\mathscr{M}}\left(\vec{a}_b^b\right) \in \mathrm{CSD}(\vec{\mathfrak{a}})$ and in the inference $\vec{\mathfrak{a}}$ there exists a record of rank $\langle s,m\rangle$with the formula $\mathrm{J}_\tau\, \mathrm{P}_m^{\mathscr{M}}\left(\vec{a}_b^b\right)$. Then by Lemma 19.1.18, there exists an initial segment $\vec{\mathfrak{b}}$ of $\vec{\mathfrak{a}}$ such that $\vec{\mathfrak{d}} = \mathrm{Resume}\left(\vec{\mathfrak{b}}\right)$ is an $\mathscr{M}(s,m)$-inference from $\mathrm{Descr}_{\mathscr{M}}$ ready for $\mathrm{J}_\tau\, \mathrm{P}_m^{\mathscr{M}}\left(\vec{a}_b^b\right)$, where $m-1 \leq k \leq m$, and also $\vec{\mathfrak{d}}_{\overleftarrow{s,\,m-1}} = \vec{\mathfrak{a}}_{\overleftarrow{s,\,m-1}}$. Then by Lemma 19.2.4, there exists an $e \in \mathrm{Cn}_{\mathsf{s}}^{\Delta}(b)$ such that $\mathrm{CSD}\left(\vec{\mathfrak{a}}_{\overleftarrow{s,\,m-1}}\right) \nvdash_L \ \mathrm{Suf}\left[R,\mathrm{P}_m^{\mathscr{M}},\varepsilon\right]\left(\vec{a}_e^b\right)$, which contradicts the assumption. □

Below is the second key lemma of this section (the first one is Lemma 15.2.11, which deals with the predicate symbols being atomic in modification structures). The lemma is about the predicate symbols $+\mathsf{s}\varepsilon$-multiple in modification sructures. It will be necessary to prove the statement about the $\mathrm{SSC}(\mathscr{M})$-comfomability of perfect modification calculi. According to Proposition 14.1.21 an adequate class of models for the comformable calculi can be defined.

Lemma 19.2.6. Let L be an FPJ logic and σ be a signature. Let $\mathscr{M} \in \mathrm{PMC}_\sigma(L)$ and $\vec{\mathfrak{a}}$ be an $\mathscr{M}(s,m,\mathbf{D})$-inference from $\Delta = \mathrm{Descr}_{\mathscr{M}}$, where $s > 0$, $m > 0$. Suppose

$\varepsilon \in \mathscr{D}^{\tau}(L)$, $\mathrm{P}_m^{\mathscr{M}} \in +\mathrm{s}\varepsilon[R]$, $\vec{a} \in \mathrm{Dom}_{\mathscr{M}}(p)$, $b \in \mathrm{Set}(\vec{a}) \cap \mathrm{Dom}_{\mathscr{M}}(\mathrm{s})$. Let $\vec{\mathfrak{a}}_{\overleftarrow{s,\, m-1}}$ be a coherent $\mathscr{M}(s, m-1, \mathbf{D})$-inference from $\mathrm{Descr}_{\mathscr{M}}$. Then $\mathrm{J}_{\varepsilon}\, \mathrm{P}_m^{\mathscr{M}}\left(\vec{a}_b^b\right) \in \mathrm{CSD}(\vec{\mathfrak{a}})$ holds if and only if

$$\mathrm{CSD}\left(\vec{\mathfrak{a}}_{\overleftarrow{s,\, m-1}}\right) \vdash_{\mathrm{St}} \mathrm{Suf}\left[R, \mathrm{P}_m^{\mathscr{M}}, \varepsilon\right]\left(\vec{a}_e^b\right).$$

holds for any $e \in \mathrm{Cn}_{\mathrm{s}}^{\Delta}(b)$

Proof. This lemma follows directly from Lemmas 19.2.3 and 19.2.5. □

The lemma proved above is the last statement about the predicate symbols $+\mathrm{s}\varepsilon$-multiple in modification sructures. Now we are going to investigate the properties of $-\mathrm{s}\varepsilon$-multiple predicate symbols. We intend to establish properties similar to the ones obtained above.

19.3 Modification Rules Within Coherent Inferences (Negative Case)

Lemma 19.3.1. Let L be an FPJ logic and σ be a signature. Let $\mathscr{M} \in \mathrm{PMC}_{\sigma}(L)$ and $\vec{\mathfrak{a}}$ be a pure $\mathscr{M}(s, m, \mathbf{D})$-inference, where $s > 0$, $m > 0$. Suppose $\varepsilon \in \mathscr{D}^{\tau}(L)$, $p \in -\mathrm{s}\varepsilon[R]$, $\vec{a} \in \mathrm{Dom}_{\mathscr{M}}(p)$, $b \in \mathrm{Set}(\vec{a}) \cap \mathrm{Dom}_{\mathscr{M}}(\mathrm{s})$. Let:

(a) $\vec{\mathfrak{a}}_{\overleftarrow{s,\, m-1}}$ be a coherent $\mathscr{M}(s, m-1, \mathbf{D})$-inference from $\Delta = \mathrm{Descr}_{\mathscr{M}}$,
(b) $e \in \mathrm{At}_{\mathrm{s}}^{\Delta}$;
(c) $\mathrm{J}_{\varepsilon}\, p\left(\vec{a}_e^b\right) \in \mathrm{CSD}(\vec{\mathfrak{a}})$.

Then

(d) $\mathrm{CSD}\left(\vec{\mathfrak{a}}_{\overleftarrow{s,\, m-1}}\right) \vdash_L \mathrm{Suf}[R, p, \varepsilon]\left(\vec{a}_e^b\right)$.

Proof. Consider the possible cases.

CASE 1. $\mathrm{J}_{\varepsilon}\, p\left(\vec{a}_e^b\right) \in \mathrm{CSD}\left(\vec{\mathfrak{a}}_{\overleftarrow{s,\, m-1}}\right)$. Obviously, in this case (d) holds (see Lemma 15.1.14 and Definition 18.1.2).

CASE 2. $\mathrm{J}_{\varepsilon}\, p\left(\vec{a}_e^b\right) \notin \mathrm{CSD}\left(\vec{\mathfrak{a}}_{\overleftarrow{s,\, m-1}}\right)$. Then

$$\mathrm{J}_{\tau}\, p\left(\vec{a}_e^b\right) \in \mathrm{CSD}\left(\vec{\mathfrak{a}}_{\overleftarrow{s,\, m-1}}\right). \tag{19.44}$$

We obtain from (19.44) and (c) by Lemma 19.1.18 that there exists an initial segment $\vec{\mathfrak{b}}$ of the $\mathscr{M}$-inference $\vec{\mathfrak{a}}$ such that $\vec{\mathfrak{d}} = \mathrm{Resume}\left(\vec{\mathfrak{b}}\right)$ is an $\mathscr{M}(s, k, \mathbf{M})$-inference from $\mathrm{Descr}_{\mathscr{M}}$ ready for $\mathrm{J}_{\varepsilon}\, p\left(\vec{a}_{\Lambda_{\mathrm{s}}}^b\right)$, where $m-1 \leq k \leq m$; moreover, $\vec{\mathfrak{d}}_{\overleftarrow{s,\, m-1}} = \vec{\mathfrak{a}}_{\overleftarrow{s,\, m-1}}$.

Then $\vec{\mathfrak{d}}$ contains the premise $\mathrm{MPr}[R, p, \varepsilon]\left(\vec{a}_e^b\right)$. Therefore,

$$\mathrm{CSD}\left(\vec{\mathfrak{d}}\right) \vdash_L \exists z \left(z \subseteq_{\mathrm{s}} e \wedge \mathrm{Suf}\left[R,p,\varepsilon\right]\left(\vec{a}_z^b\right)\right) \tag{19.45}$$

(see Definition 18.1.1). Since $\mathrm{Cn}_{\mathrm{s}}^{\Delta}(e) = \{e\}$ it is easy to show that (19.45) implies

$$\mathrm{CSD}\left(\vec{\mathfrak{d}}\right) \vdash_L \mathrm{Suf}\left[R,p,\varepsilon\right]\left(\vec{a}_e^b\right) \tag{19.46}$$

(see, for example, the proof of Lemma 18.2.8). Since $\vec{\mathfrak{d}}$ is an $\mathscr{M}$-inference ready for $\mathrm{J}_{\varepsilon}\, p\left(\vec{a}_e^b\right)$, and $\mathrm{Suf}\left[R,p,\varepsilon\right]\left(\vec{a}_e^b\right)$ is local w.r.t. $p\left(\vec{a}_e^b\right)$, by Corollary 15.1.12, (19.46) implies

$$\mathrm{CSD}\left(\vec{\mathfrak{a}}_{\overleftarrow{s,\,m-1}}\right) \vdash_{\mathbf{St}} \mathrm{Suf}\left[R,p,\varepsilon\right]\left(\vec{a}_e^b\right),$$

as was to be proved. □

Lemma 19.3.2. Let L be an FPJ logic and σ be a signature. Let $\mathscr{M} \in \mathrm{PMC}_{\sigma}(L)$ and $\vec{\mathfrak{a}}$ be a pure $\mathscr{M}(s,m,\mathbf{D})$-inference, where $s > 0$, $m > 0$. Suppose $\varepsilon \in \mathscr{D}^{\tau}(L)$, $p \in -\mathrm{s}\varepsilon[R]$, $\vec{a} \in \mathrm{Dom}_{\mathscr{M}}(p)$, $b \in \mathrm{Set}(\vec{a}) \cap \mathrm{Dom}_{\mathscr{M}}(\mathrm{s})$. Assume that

$$\mathrm{J}_{\varepsilon}\, p\left(\vec{a}_{\Lambda_{\mathrm{s}}}^b\right) \notin \mathrm{CSD}\left(\vec{\mathfrak{a}}\right).$$

Let $\vec{\mathfrak{a}}_{\overleftarrow{s,\,m-1}}$ be a pure coherent $\mathscr{M}(s,m-1,\mathbf{D})$-inference. Then

$$\mathrm{CSD}\left(\vec{\mathfrak{a}}\right) \nvdash_L \mathrm{Suf}\left[R,p,\varepsilon\right]\left(\vec{a}_{\Lambda_{\mathrm{s}}}^b\right).$$

Proof. Assume the converse. Let

$$\mathrm{CSD}\left(\vec{\mathfrak{a}}\right) \vdash_L \mathrm{Suf}\left[R,p,\varepsilon\right]\left(\vec{a}_{\Lambda_{\mathrm{s}}}^b\right).$$

By the assumption, $\mathrm{J}_{\varepsilon}\, p\left(\vec{a}_{\Lambda_{\mathrm{s}}}^b\right) \notin \mathrm{CSD}\left(\vec{\mathfrak{a}}\right)$. Then, since $\vec{\mathfrak{a}}_{\overleftarrow{s,\,m-1}}$ is coherent,

$$\mathrm{J}_{\varepsilon}\, p\left(\vec{a}_{\Lambda_{\mathrm{s}}}^b\right) \notin \mathrm{CSD}\left(\vec{\mathfrak{a}}\right).$$

Then

$$\mathrm{J}_{\tau}\, p\left(\vec{a}_{\Lambda_{\mathrm{s}}}^b\right) \notin \mathrm{CSD}\left(\vec{\mathfrak{a}}\right) \quad \text{and} \quad \mathrm{J}_{\tau}\, p\left(\vec{a}_{\Lambda_{\mathrm{s}}}^b\right) \notin \mathrm{CSD}\left(\vec{\mathfrak{a}}_{\overleftarrow{s,\,m-1}}\right).$$

Therefore, for any $\varepsilon \in \mathscr{V}(L,\mathsf{I})$,

$$\mathrm{J}_{\varepsilon}\, p\left(\vec{a}_{\Lambda_{\mathrm{s}}}^b\right) \in \mathrm{FSet}\left(\vec{\mathfrak{a}}\right) \quad \text{implies} \quad \mathrm{J}_{\varepsilon}\, p\left(\vec{a}\right) \in \mathrm{FSet}\left(\vec{\mathfrak{a}}_{\overleftarrow{s,\,m-1}}\right).$$

Then by Lemma 15.1.10 and Remark 18.1.3,

$$\mathrm{CSD}\left(\vec{\mathfrak{a}}_{\overleftarrow{s,\,m-1}}\right) \vdash_L \mathrm{Suf}\left[R,p,\varepsilon\right]\left(\vec{a}_{\Lambda_{\mathrm{s}}}^b\right),$$

Which contradicts Proposition 19.1.15 (ii). Our supposition is wrong. □

Lemma 19.3.3. Let L be an FPJ logic and σ be a signature. Let $\mathscr{M} \in \mathrm{PMC}_\sigma(L)$ and $\vec{\mathfrak{a}}$ be a pure $\mathscr{M}(s,m,\mathbf{D})$-inference, where $s > 0$, $m > 0$. Suppose $\varepsilon \in \mathscr{D}^\tau(L)$, $p \in -\mathrm{s}\varepsilon[R]$, $\vec{a} \in \mathrm{Dom}_{\mathscr{M}}(p)$, $b \in \mathrm{Set}(\vec{a}) \cap \mathrm{Dom}_{\mathscr{M}}(\mathrm{s})$. Let $\vec{\mathfrak{a}}_{\overleftarrow{s,\,m-1}}$ be a coherent $\mathscr{M}(s,m-1,\mathbf{D})$-inference. Then

$$\mathrm{J}_\varepsilon\, p\left(\vec{a}^{b}_{\Lambda_\mathrm{s}}\right) \notin \mathrm{CSD}(\vec{\mathfrak{a}}).$$

Proof. Since $\vec{\mathfrak{a}}_{\overleftarrow{s,\,m-1}}$ is a coherent $\mathscr{M}(s,m-1,\mathbf{D})$-inference, by Proposition 19.1.9 (ii), we obtain

$$\mathrm{J}_\varepsilon\, p\left(\vec{a}^{b}_{\Lambda_\mathrm{s}}\right) \notin \mathrm{CSD}\left(\vec{\mathfrak{a}}_{\overleftarrow{s,\,m-1}}\right).$$

Then

$$\mathrm{J}_\tau\, p\left(\vec{a}^{b}_{\Lambda_\mathrm{s}}\right) \in \mathrm{CSD}\left(\vec{\mathfrak{a}}_{\overleftarrow{s,\,m-1}}\right). \tag{19.47}$$

Now let us prove that $\mathrm{J}_\varepsilon\, p\left(\vec{a}^{b}_{\Lambda_\mathrm{s}}\right) \notin \mathrm{CSD}(\vec{\mathfrak{a}})$. Suppose the converse. Let

$$\mathrm{J}_\varepsilon\, p\left(\vec{a}^{b}_{\Lambda_\mathrm{s}}\right) \in \mathrm{CSD}(\vec{\mathfrak{a}}). \tag{19.48}$$

Then from (17.1) and (17.2) by Lemma 19.1.18 we obtain that there exists an initial segment $\vec{\mathfrak{b}}$ of $\vec{\mathfrak{a}}$ such that $\vec{\mathfrak{d}} = \mathrm{Resume}\left(\vec{\mathfrak{b}}\right)$ is a pure $\mathscr{M}(s,k,\mathbf{M})$-inference ready for $\mathrm{J}_\varepsilon\, p\left(\vec{a}^{b}_{\Lambda_\mathrm{s}}\right)$, where $m-1 \le k \le m$, and $\vec{\mathfrak{d}}_{\overleftarrow{s,\,m-1}} = \vec{\mathfrak{a}}_{\overleftarrow{s,\,m-1}}$. Then $\vec{\mathfrak{d}}$ does not contain the formula $\mathrm{J}_\varepsilon\, p\left(\vec{a}^{b}_{\Lambda_\mathrm{s}}\right)$, but it contains the major premise of the corresponding rule. Then

$$\mathrm{CSD}\left(\vec{\mathfrak{d}}\right) \vdash_L \exists z\left(z \subseteq_\mathrm{s} \Lambda_\mathrm{s} \wedge \mathrm{Suf}[R,p,\varepsilon]\left(\vec{a}^{b}_{z}\right)\right) \tag{19.49}$$

(see Definition 18.1.1). Let $\mathfrak{K} = \mathrm{Sint}_L\left(\mathrm{CSD}\left(\vec{\mathfrak{d}}\right), \mathrm{Univ}_{\mathscr{M}}\right)$. Then (19.49) implies

$$\mathfrak{K} \vDash_L \exists z\left(z \subseteq_\mathrm{s} \Lambda_\mathrm{s} \wedge \mathrm{Suf}[R,p,\varepsilon]\left(\vec{a}^{b}_{z}\right)\right). \tag{19.50}$$

By Lemma 17.2.9, $\mathfrak{K}$ is a set-admitting L-structure. Then by Lemma 17.2.3, $\mathfrak{K} \vDash_L \mathrm{MULT}_\sigma$. Hence by Lemma 17.2.2 (vi),

$$\mathfrak{K} \vDash_L \forall z(z \subseteq_\mathrm{s} \Lambda_\mathrm{s} \to z = \Lambda_\mathrm{s}). \tag{19.51}$$

(19.50) and (19.51) imply

$$\mathfrak{K} \vDash_L \mathrm{Suf}[R,p,\varepsilon]\left(\vec{a}^{b}_{\Lambda_\mathrm{s}}\right).$$

Then by Lemma 14.1.1,

$$\mathrm{CSD}\left(\vec{\mathfrak{d}}\right) \vdash_L \mathrm{Suf}[R,p,\varepsilon]\left(\vec{a}^{b}_{\Lambda_\mathrm{s}}\right). \tag{19.52}$$

On the other hand, $\vec{\mathfrak{d}}$ does not contain the formula $\mathrm{J}_\varepsilon\, p\left(\vec{a}^{b}_{\Lambda_s}\right)$. Therefore by Lemma 15.1.14,

$$\mathrm{J}_\varepsilon\, p\left(\vec{a}^{b}_{\Lambda_s}\right) \notin \mathrm{CSD}\left(\vec{\mathfrak{d}}\right).$$

Then by Lemma 19.3.2,

$$\mathrm{CSD}\left(\vec{\mathfrak{d}}\right) \nvdash_L \mathrm{Suf}\left[R, p, \varepsilon\right]\left(\vec{a}^{b}_{\Lambda_s}\right). \tag{19.53}$$

But (19.52) contradicts (19.53). □

Lemma 19.3.4. Let L be an FPJ logic and σ be a signature. Let $\mathcal{M} \in \mathrm{PMC}_\sigma(L)$ and $\vec{\mathfrak{a}}$ be a pure $\mathcal{M}(s, m, \mathbf{D})$-inference, where $s > 0$, $m > 0$. Suppose $\varepsilon \in \mathscr{D}^\tau(L)$, $p \in -\mathrm{s}\varepsilon[R]$, $\vec{a} \in \mathrm{Dom}_{\mathcal{M}}(p)$, $b \in \mathrm{Set}(\vec{a}) \cap \mathrm{Dom}_{\mathcal{M}}(\mathrm{s})$. Let $\vec{\mathfrak{a}}_{\overleftarrow{s,\,m-1}}$ be a coherent $\mathcal{M}(s, m-1, \mathbf{D})$-inference. Then

$$\mathrm{CSD}\left(\vec{\mathfrak{a}}\right) \nvdash_L \mathrm{Suf}\left[R, p, \varepsilon\right]\left(\vec{a}^{b}_{\Lambda_s}\right).$$

Proof. By Lemma 19.3.3 $\mathrm{J}_\varepsilon\, p\left(\vec{a}^{b}_{\Lambda_s}\right) \notin \mathrm{CSD}(\vec{a})$. Then by Lemma 19.3.2,

$$\mathrm{CSD}\left(\vec{\mathfrak{a}}\right) \nvdash_L \mathrm{Suf}\left[R, p, \varepsilon\right]\left(\vec{a}^{b}_{\Lambda_s}\right),$$

as was to be proved. □

Lemma 19.3.5. Let L be an FPJ logic and σ be a signature. Let $\mathcal{M} \in \mathrm{PMC}_\sigma(L)$ and $\vec{\mathfrak{a}}$ be a pure $\mathcal{M}(s, m, \mathbf{D})$-inference, where $s > 0$, $m > 0$. Suppose $\varepsilon \in \mathscr{D}^\tau(L)$, $\mathrm{P}^{\mathcal{M}}_m \in -\mathrm{s}\varepsilon[R]$, $\vec{a} \in \mathrm{Dom}_{\mathcal{M}}\left(\mathrm{P}^{\mathcal{M}}_m\right)$, $b \in \mathrm{Set}(\vec{a}) \cap \mathrm{Dom}_{\mathcal{M}}(\mathrm{s})$, and $\Delta = \mathrm{Descr}_{\mathcal{M}}$. Let:

(a) $\vec{\mathfrak{a}}_{\overleftarrow{s,\,m-1}}$ be a coherent $\mathcal{M}(s, m-1, \mathbf{D})$-inference,
(b) $\mathrm{J}_\varepsilon \mathrm{P}^{\mathcal{M}}_m\left(\vec{a}^{b}_{b}\right) \in \mathrm{CSD}\left(\vec{\mathfrak{a}}\right)$.

Then there exists an $e \in \mathrm{Cn}^{\Delta}_{\mathrm{s}}(b)$ such that

(c) $\mathrm{CSD}\left(\vec{\mathfrak{a}}_{\overleftarrow{s,\,m-1}}\right) \vdash_L \mathrm{Suf}\left[R, \mathrm{P}^{\mathcal{M}}_m, \varepsilon\right]\left(\vec{a}^{b}_{e}\right)$.

Proof. We show by induction on the cardinality of $\mathrm{Cn}^{\Delta}_{\mathrm{s}}(b)$ that (b) implies (c).

BASIS. Let $\mathrm{Cn}^{\Delta}_{\mathrm{s}}(b) = \emptyset$. Then $b = \Lambda_{\mathrm{s}}$. Then (b) is false by Lemma 19.3.3. Therefore the statement holds since the premise is false.

INDUCTION STEP. Let, for any $g \in \mathrm{Dom}_{\mathcal{M}}(\mathrm{s})$ such that the cardinality of $\mathrm{Cn}^{\Delta}_{\mathrm{s}}(g)$ is less than or equal to n,

$$\mathrm{J}_\varepsilon \mathrm{P}^{\mathcal{M}}_m\left(\vec{a}^{b}_{g}\right) \in \mathrm{CSD}\left(\vec{\mathfrak{a}}\right) \tag{19.54}$$

imply the fact that there exists an $e \in \mathrm{Cn}^{\Delta}_{\mathrm{s}}(g)$ such that

$$\mathrm{CSD}\left(\vec{\mathfrak{a}}_{\overleftarrow{s,\,m-1}}\right) \vdash_{\mathbf{St}} \mathrm{Suf}\left[R, \mathrm{P}^{\mathcal{M}}_m, \varepsilon\right]\left(\vec{a}^{b}_{e}\right)$$

(i.e. (c) holds). Suppose that the cardinality of $\mathrm{Cn}_{\mathsf{s}}^{\Delta}(b)$ is equal to $n+1$ and (b) holds. Consider the possible cases.

CASE 1. $\mathrm{J}_{\varepsilon}\mathrm{P}_{m}^{\mathscr{M}}(\vec{a}_{b}^{b}) \in \mathrm{CSD}\left(\vec{\mathfrak{a}}_{\overleftarrow{s,m-1}}\right)$. Then by Lemma 19.1.5, there exists an $e \in \mathrm{Cn}_{\mathsf{s}}^{\Delta}(b)$ such that

$$\mathrm{J}_{\varepsilon}\mathrm{P}_{m}^{\mathscr{M}}\left(\vec{a}_{e}^{b}\right) \in \mathrm{CSD}\left(\vec{\mathfrak{a}}_{\overleftarrow{s,m-1}}\right).$$

Then by Definition 18.1.2 and Lemma 15.1.14, we get that (c) is valid.

CASE 2. $\mathrm{J}_{\varepsilon}\mathrm{P}_{m}^{\mathscr{M}}(\vec{a}) \notin \mathrm{CSD}\left(\vec{\mathfrak{a}}_{\overleftarrow{s,m-1}}\right)$. Then

$$\mathrm{J}_{\tau}\mathrm{P}_{m}^{\mathscr{M}}(\vec{a}) \in \mathrm{CSD}\left(\vec{\mathfrak{a}}_{\overleftarrow{s,m-1}}\right). \tag{19.55}$$

From (19.55) and (b) it follows by Lemma 19.1.18 that there is an initial segment $\vec{\mathfrak{b}}$ of the $\mathscr{M}(s,m)$-inference $\vec{\mathfrak{a}}$ such that $\vec{\mathfrak{d}} = \mathrm{Resume}\left(\vec{\mathfrak{b}}\right)$ is a pure $\mathscr{M}(s,k,\mathbf{M})$-inference ready for $\mathrm{J}_{\varepsilon}\mathrm{P}_{m}^{\mathscr{M}}(\vec{a})$, where $m-1 \leq k \leq m$, moreover $\vec{\mathfrak{d}}_{\overleftarrow{s,m-1}} = \vec{\mathfrak{a}}_{\overleftarrow{s,m-1}}$. Then $\mathrm{MPr}\left[R, \mathrm{P}_{m}^{\mathscr{M}}, \varepsilon\right]$ occurs in $\vec{\mathfrak{d}}$. Therefore,

$$\mathrm{CSD}\left(\vec{\mathfrak{d}}\right) \vdash_{L} \exists z\left(z \subseteq_{\mathsf{s}} b \wedge \mathrm{Suf}[R,p,\varepsilon]\left(\vec{a}_{z}^{b}\right)\right) \tag{19.56}$$

(see Definition 18.1.1). Let $\mathfrak{K} = \mathrm{Sint}_{L}\left(\mathrm{CSD}\left(\vec{\mathfrak{d}}\right), \mathrm{Univ}_{\mathscr{M}}\right)$. Then (19.56) implies

$$\mathfrak{K} \vDash_{L} \exists z\left(z \subseteq_{\mathsf{s}} b \wedge \mathrm{Suf}[R,p,\varepsilon]\left(\vec{a}_{z}^{b}\right)\right).$$

Hence

$$\mathrm{Cn}_{\mathsf{s}}^{\Delta}(g) \subseteq_{\mathsf{s}} \mathrm{Cn}_{\mathsf{s}}^{\Delta}(b) \quad \text{and} \quad \mathfrak{K} \vDash_{L} \mathrm{Suf}[R,p,\varepsilon]\left(\vec{a}_{g}^{b}\right)$$

hold for some $g \in \mathrm{Dom}_{\sigma}(D,\mathsf{s})$. Then by Lemma 14.1.1 we get

$$\mathrm{CSD}\left(\vec{\mathfrak{d}}\right) \vdash_{L} \mathrm{Suf}[R,p,\varepsilon]\left(\vec{a}_{g}^{b}\right). \tag{19.57}$$

Since $\vec{\mathfrak{d}}$ is an $\mathrm{M}(s,k)$-inference ready for $\mathrm{J}_{\varepsilon}\mathrm{P}_{m}^{\mathscr{M}}(\vec{a})$ we can apply Corollary 15.1.12 and from (19.57) we can obtain

$$\mathrm{CSD}\left(\vec{\mathfrak{a}}_{\overleftarrow{s,m-1}}\right) \vdash_{\mathbf{St}} \mathrm{Suf}[R,p,\varepsilon]\left(\vec{a}_{g}^{b}\right). \tag{19.58}$$

Consider the possible subcases.

SUBCASE 2.1. $g = \Lambda_{\mathsf{s}}$. By Lemma 19.3.4, in this subcase the condition (19.58) could not hold. Therefore this subcase is impossible.

SUBCASE 2.2. $g = b$. Then (19.57) can be turned into

$$\mathrm{CSD}\left(\vec{\mathfrak{d}}\right) \vdash_{L} \mathrm{Suf}[R,p,\varepsilon]\left(\vec{a}_{b}^{b}\right). \tag{19.59}$$

Since $\vec{\mathfrak{d}}$ is an M(s, k)-inference ready for $\mathrm{J}_{\varepsilon}\,\mathrm{P}_{m}^{\mathscr{M}}(\vec{a})$ we can apply Corollary 15.1.12 and from (19.59) we can obtain

$$\mathrm{CSD}\left(\vec{\mathfrak{a}}_{\overleftarrow{s,\,m-1}}\right) \vdash_{\mathbf{St}} \mathrm{Suf}[R,p,\varepsilon]\left(\vec{a}_{b}^{b}\right).$$

Then by Lemma 19.1.13, there exists an $e \in \mathrm{Cn}_{\mathrm{s}}^{\Delta}(b)$ such that (c) holds.

SUBCASE 2.3. $g \neq b$ and $g \neq \Lambda_{\mathrm{s}}$. Then the cardinalty of g is less than or equal to n. From (19.57) by Definition 18.1.2 we get

$$\mathrm{CSD}(\vec{\mathfrak{d}}) \vdash_{L} \mathrm{J}_{\varepsilon}\,\mathrm{P}_{m}^{\mathscr{M}}\left(\vec{a}_{g}^{b}\right) \tag{19.60}$$

or

$$\mathrm{CSD}(\vec{\mathfrak{d}}) \vdash_{L} \mathrm{Ker}[R,p,\varepsilon]\left(\vec{a}_{g}^{b}\right). \tag{19.61}$$

If (19.60) is valid then by Lemma 15.1.14 we get

$$\mathrm{J}_{\varepsilon}\,\mathrm{P}_{m}^{\mathscr{M}}\left(\vec{a}_{g}^{b}\right) \in \mathrm{CSD}(\vec{\mathfrak{d}}).$$

Hence (see Lemma 15.1.15)

$$\mathrm{J}_{\varepsilon}\,\mathrm{P}_{m}^{\mathscr{M}}\left(\vec{a}_{g}^{b}\right) \in \mathrm{CSD}(\vec{\mathfrak{a}}),$$

i.e. (19.54) is valid. By the inductive assumption, there exists an $e \in \mathrm{Cn}_{\mathrm{s}}^{\Delta}(g) \subseteq \mathrm{Cn}_{\mathrm{s}}^{\Delta}(b)$ such that (c) holds.

If (19.60) is not valid then by Lemma 15.1.14 we get

$$\mathrm{J}_{\varepsilon}\,\mathrm{P}_{m}^{\mathscr{M}}\left(\vec{a}_{g}^{b}\right) \notin \mathrm{FSet}(\vec{\mathfrak{d}}). \tag{19.62}$$

If (19.60) is not valid then (19.61) holds. Recall that

$$\vec{\mathfrak{d}}_{\overleftarrow{s,\,m-1}} = \vec{\mathfrak{a}}_{\overleftarrow{s,\,m-1}}. \tag{19.63}$$

By Corollary 15.1.12, it follows from (19.61–19.63) that

$$\mathrm{CSD}\left(\vec{\mathfrak{a}}_{\overleftarrow{s,\,m-1}}\right) \vdash_{\mathbf{St}} \mathrm{Ker}[R,p,\varepsilon]\left(\vec{a}_{g}^{b}\right).$$

Then by Lemma 19.1.7 there exists an $e \in \mathrm{Cn}_{\mathrm{s}}^{\Delta}(g) \subseteq \mathrm{Cn}_{\mathrm{s}}^{\Delta}(b)$ such that

$$\mathrm{CSD}\left(\vec{\mathfrak{a}}_{\overleftarrow{s,\,m-1}}\right) \vdash_{\mathbf{St}} \mathrm{Ker}[R,p,\varepsilon]\left(\vec{a}_{e}^{b}\right).$$

Hence, by Definition 18.1.2, (c) holds. □

Lemma 19.3.6. Let L be an FPJ logic and σ be a signature. Let $\mathscr{M} \in \mathrm{PMC}_{\sigma}(L)$ and $\vec{\mathfrak{a}}$ be a pure $\mathscr{M}(s,m,\mathbf{D})$-inference, where $s > 0$, $m > 0$. Suppose $\varepsilon \in \mathscr{D}^{\tau}(L)$, $\mathrm{P}_{m}^{\mathscr{M}} \in -\mathrm{s}\varepsilon[R]$, $\vec{a} \in \mathrm{Dom}_{\mathscr{M}}\left(\mathrm{P}_{m}^{\mathscr{M}}\right)$, $b \in \mathrm{Set}(\vec{a}) \cap \mathrm{Dom}_{\mathscr{M}}(\mathrm{s})$, and $\Delta = \mathrm{Descr}_{\mathscr{M}}$. Let:

(a) $\vec{\mathfrak{a}}_{\overleftarrow{s,m-1}}$ be a coherent $M(s,m-1,\mathbf{D})$-inference,

(b) There exists an $e \in \mathrm{Cn}_{\mathsf{s}}^{\Delta}(b)$ such that $\mathrm{CSD}\left(\vec{\mathfrak{a}}_{\overleftarrow{s,m-1}}\right) \vdash_L \mathrm{Suf}\left[R, \mathrm{P}_m^{\mathscr{M}}, \varepsilon\right]\left(\vec{a}_e^b\right)$.

Then

(c) $\mathrm{J}_\varepsilon \mathrm{P}_m^{\mathscr{M}}\left(\vec{a}_b^b\right) \in \mathrm{CSD}(\vec{\mathfrak{a}})$.

Proof. Let

$$\mathrm{CSD}\left(\vec{\mathfrak{a}}_{\overleftarrow{s,m-1}}\right) \vdash_{\mathbf{St}} \mathrm{Suf}\left[R, \mathrm{P}_m^{\mathscr{M}}, \varepsilon\right]\left(\vec{a}_e^b\right). \tag{19.64}$$

Let us show that $\mathrm{J}_\varepsilon \mathrm{P}_m^{\mathscr{M}}\left(\vec{a}_b^b\right) \in \mathrm{CSD}(\vec{\mathfrak{a}})$. Assume the converse. Then

$$\mathrm{J}_\tau \mathrm{P}_m^{\mathscr{M}}(\vec{a}) \in \mathrm{CSD}(\vec{\mathfrak{a}}) \tag{19.65}$$

and there is a record of rank $\langle s,m\rangle$ in $\vec{\mathfrak{a}}$ that contains the formula $\mathrm{J}_\tau \mathrm{P}_m^{\mathscr{M}}(\vec{a})$. By Lemma 19.1.18, there exists a cut $\left\langle \vec{\mathfrak{b}}, \vec{\mathfrak{c}}\right\rangle$ of the $\mathscr{M}(s,m,\mathbf{D})$-inference $\vec{\mathfrak{a}}$ such that

(i) $\mathrm{Len}\left(\vec{\mathfrak{a}}_{\overleftarrow{s,m-1}}\right) \le \mathrm{Len}\left(\vec{\mathfrak{b}}\right) < \mathrm{Len}(\vec{\mathfrak{a}})$,

(ii) $\vec{\mathfrak{d}} = \mathrm{Resume}\left(\vec{\mathfrak{b}}\right)$ is ready for $\mathrm{J}_\tau \mathrm{P}_m^{\mathscr{M}}(\vec{a})$ $\mathscr{M}(s,k,\mathbf{M})$-inference from $\mathrm{Descr}_{\mathscr{M}}$, where $m-1 \le k \le m$,

(iii) $\vec{\mathfrak{d}}_{\overleftarrow{s,m-1}} = \vec{a}_{\overleftarrow{s,m-1}}$.

Then $\mathrm{MPr}\left[R, \mathrm{P}_m^{\mathscr{M}}, \tau\right] \in \mathrm{FSet}\left(\vec{\mathfrak{d}}\right)$, whence by Corollary 11.4.12

$$\mathrm{CSD}\left(\vec{\mathfrak{d}}\right) \vdash_L \mathrm{MPr}\left[R, \mathrm{P}_m^{\mathscr{M}}, \tau\right],$$

i.e.

$$\mathrm{CSD}\left(\vec{\mathfrak{d}}\right) \vdash_L \forall \vec{x}\left(\mathrm{J}_\tau \mathrm{P}_m^{\mathscr{M}}(\vec{x}) \to \neg\, \mathrm{MPr}\left[R, \mathrm{P}_m^{\mathscr{M}}, \varepsilon\right]\right). \tag{19.66}$$

By Lemma 15.1.15, (19.65) implies that $\mathrm{J}_\tau \mathrm{P}_m^{\mathscr{M}}(\vec{a}) \in \mathrm{CSD}\left(\vec{\mathfrak{d}}\right)$. Hence by Lemma 15.1.14,

$$\mathrm{CSD}\left(\vec{\mathfrak{d}}\right) \vdash_L \mathrm{J}_\tau \mathrm{P}_m^{\mathscr{M}}(\vec{a}). \tag{19.67}$$

It follows fom (19.66) and (19.67) that

$$\mathrm{CSD}\left(\vec{\mathfrak{d}}\right) \vdash_L \neg\, \mathrm{MPr}\left[R, \mathrm{P}_m^{\mathscr{M}}, \varepsilon\right](\vec{a}).$$

Then by Definitions 18.1.1 and 18.1.2,

$$\mathrm{CSD}\left(\vec{\mathfrak{d}}\right) \vdash_L \neg \exists z\left(z \subseteq_{\mathsf{s}} b \wedge \mathrm{Suf}\left[R, \mathrm{P}_m^{\mathscr{M}}, \varepsilon\right]\left(\vec{a}_z^b\right)\right). \tag{19.68}$$

Let us show that

$$\mathrm{CSD}\left(\vec{\mathfrak{d}}\right) \vdash_L \mathrm{Suf}\left[R, \mathrm{P}_m^{\mathscr{M}}, \varepsilon\right]\left(\vec{a}_e^b\right). \tag{19.69}$$

Assume the converse. Then by the use of Lemma 14.1.1 it is easy to prove that

$$\mathrm{CSD}\left(\vec{\mathfrak{d}}\right) \vdash_L \neg\mathrm{Suf}\left[R, \mathrm{P}_m^{\mathcal{M}}, \varepsilon\right]\left(\vec{a}_e^b\right). \tag{19.70}$$

Recall that the formula $\neg\mathrm{Suf}\left[R, \mathrm{P}_m^{\mathcal{M}}, \varepsilon\right]\left(\vec{a}_e^b\right)$ is local w.r.t. $\mathrm{P}_m^{\mathcal{M}}\left(\vec{a}_e^b\right)$. Note that (19.67) implies $\mathrm{J}_\varepsilon \mathrm{P}_m^{\mathcal{M}}(\vec{a}) \notin \mathrm{CSD}\left(\vec{\mathfrak{d}}\right)$. Then by 15.1.12, (19.70) implies

$$\mathrm{CSD}\left(\vec{\mathfrak{d}}_{\overleftarrow{s,\, m-1}}\right) \vdash_{\mathbf{St}} \neg\mathrm{Suf}\left[R, \mathrm{P}_m^{\mathcal{M}}, \varepsilon\right]\left(\vec{a}_e^b\right). \tag{19.71}$$

But by (iii), (19.71) is equivalent to

$$\mathrm{CSD}\left(\vec{a}_{\overleftarrow{s,\, m-1}}\right) \vdash_{\mathbf{St}} \neg\mathrm{Suf}\left[R, \mathrm{P}_m^{\mathcal{M}}, \varepsilon\right]\left(\vec{a}_e^b\right). \tag{19.72}$$

However, (19.64) contradicts (19.72). Our supposition is wrong. Therefore,

$$\mathrm{J}_\varepsilon \mathrm{P}_m^{\mathcal{M}}(\vec{a}) \in \mathrm{CSD}(\vec{a}),$$

as was to be proved. □

Now we turn to the third key statement of this section. [We recall that the first is about the predicate symbols atomic in modification calcult (Lemma 15.2.11)], while the second is about the properties of the predicate symbols $+\mathrm{s}\varepsilon$-multiple in modification structures (Lemma 19.2.6). The forthcoming statement is about the properties of the predicate symbols $-\mathrm{s}\varepsilon$-multiple in modification structures. It will play an important role in the proof of the statement about $\mathrm{SSC}(\mathcal{M})$-conformability of the perfect modification calculi. For the conformable calculi in accordance with Proposition 14.1.21 we can define an adequate class of models.

Lemma 19.3.7. Let L be an FPJ logic and σ be a signature. Let $\mathcal{M} \in \mathrm{PMC}_\sigma(L)$ and $\vec{\mathfrak{a}}$ be a pure $\mathcal{M}(s,m,\mathbf{D})$-inference, where $s > 0$, $m > 0$. Suppose $\varepsilon \in \mathcal{D}^\tau(L)$, $\mathrm{P}_m^{\mathcal{M}} \in -\mathrm{s}\varepsilon[R]$, $\vec{a} \in \mathrm{Dom}_{\mathcal{M}}\left(\mathrm{P}_m^{\mathcal{M}}\right)$, $b \in \mathrm{Set}(\vec{a}) \cap \mathrm{Dom}_{\mathcal{M}}(\mathsf{s})$. Let $\vec{\mathfrak{a}}_{\overleftarrow{s,\, m-1}}$ be a coherent $\mathcal{M}(s, m-1, \mathbf{D})$-inference from $\Delta = \mathrm{Descr}_{\mathcal{M}}$. Then $\mathrm{J}_\varepsilon \mathrm{P}_m^{\mathcal{M}}\left(\vec{a}_b^b\right) \in \mathrm{CSD}(\vec{\mathfrak{a}})$ if and only if there exists an $e \in \mathrm{Cn}_{\mathsf{s}}^\Delta(b)$ such that

$$\mathrm{CSD}\left(\vec{\mathfrak{a}}_{\overleftarrow{s,\, m-1}}\right) \vdash_{\mathbf{St}} \mathrm{Suf}\left[R, \mathrm{P}_m^{\mathcal{M}}, \varepsilon\right]\left(\vec{a}_e^b\right).$$

Proof. Use Lemmas 19.3.5 and 19.3.6. □

In the next lemma we integrate the results of Lemmas 15.2.11, 19.2.6 and 19.3.7.

19.4 Conformability (Set Case)

Lemma 19.4.1. Let L be an FPJ logic and σ be a signature. Let $\mathcal{M} \in \mathrm{PMC}_\sigma(L)$ and $\vec{\mathfrak{a}}$ be a pure $\mathcal{M}(s,m,\mathbf{D})$-inference, where $s > 0$, $m > 0$. Suppose $\Delta = \mathrm{Descr}_{\mathcal{M}}$, $\varepsilon \in \mathcal{D}^\tau(L)$, $\vec{a} \in \mathrm{Dom}_{\mathcal{M}}\left(\mathrm{P}_m^{\mathcal{M}}\right)$, $b \in \mathrm{Set}(\vec{a}) \cap \mathrm{Dom}_{\mathcal{M}}(\mathsf{s})$. Let $\vec{\mathfrak{a}}_{\overleftarrow{s,\, m-1}}$ be a pure coherent $\mathcal{M}(s, m-1, \mathbf{D})$-inference. Then

$$\mathrm{J}_\varepsilon \mathrm{P}_m^{\mathscr{M}}(\vec{a}) \in \mathrm{CSD}(\vec{\mathfrak{a}})$$

if and only if the following conditions hold:

(i) if $\mathrm{P}_m^{\mathscr{M}} \in R[A]$ then
$\mathrm{J}_\varepsilon \mathrm{P}_m^{\mathscr{M}}(\vec{a}) \in \mathrm{CSD}\left(\vec{\mathfrak{a}}_{\overleftarrow{s,\,m-1}}\right)$ or
$\mathrm{J}_\tau \mathrm{P}_m^{\mathscr{M}}(\vec{a}) \in \mathrm{CSD}\left(\vec{\mathfrak{a}}_{\overleftarrow{s,\,m-1}}\right)$ and $\mathrm{CSD}\left(\vec{\mathfrak{a}}_{\overleftarrow{s,\,m-1}}\right) \vdash_L \mathrm{Ker}\left[R, \mathrm{P}_m^{\mathscr{M}}, \varepsilon\right](\vec{a})$;

(ii) if $\mathrm{P}_m^{\mathscr{M}} \in +\mathrm{s}\varepsilon[R]$ then for any $e \in \mathrm{Cn}_{\mathrm{s}}^{\Delta}(b)$,

$$\mathrm{CSD}\left(\vec{\mathfrak{a}}_{\overleftarrow{s,\,m-1}}\right) \vdash_L \mathrm{Suf}\left[R, \mathrm{P}_m^{\mathscr{M}}, \varepsilon\right]\left(\vec{a}_e^b\right);$$

(iii) if $\mathrm{P}_m^{\mathscr{M}} \in -\mathrm{s}\varepsilon[R]$ then there exists an $e \in \mathrm{Cn}_{\mathrm{s}}^{\Delta}(b)$ such that

$$\mathrm{CSD}\left(\vec{\mathfrak{a}}_{\overleftarrow{s,\,m-1}}\right) \vdash_L \mathrm{Suf}\left[R, \mathrm{P}_m^{\mathscr{M}}, \varepsilon\right]\left(\vec{a}_e^b\right).$$

Proof. (i) By Lemma 15.2.11. (ii) By Lemma 19.2.6. (iii) By Lemma 19.3.7. □

At the end of this section we investigate the $\mathscr{M}$-compatibility of the function $\mathrm{SG}^{\cdot\mathscr{M}}$ and the $\mathscr{M}$-conformability of the perfect modification calculi. The Corollary 19.4.3 below is the generalisation of Corollary 15.2.13, since each atomic modification calculus is obviously perfect. Note that Corollary 19.4.3 is *the main result of the present section.* It implies by Proposition 14.1.21 that for the perfect calculi we can determine an appropriate class of models for which the analogue of the theorems on correctness and completeness holds.

Lemma 19.4.2. Let $\mathscr{M} \in \mathrm{PMC}_\sigma(L)$. Then $\mathrm{SG}^{\cdot\mathscr{M}}(\mathscr{M})$ is an $\mathscr{M}$-compatible SG_σ.

Proof. By Definition 19.1.2, $\mathscr{M} \in \mathrm{RMC}_\sigma(L)$. By Propositions 18.2.13, the ordered triple $\mathrm{SG}^{\cdot\mathscr{M}}(\mathscr{M})$ is an SG_σ. We must establish the validity of the conditions (i), (ii) and (iii) from Definition 14.1.10. By Definition 14.1.17,

$$\mathrm{SG}^{\cdot\mathscr{M}}(\mathscr{M}) = \left\langle \mathrm{Sint}_L(\mathscr{M}), \mathrm{SG}^{\cdot\mathscr{M}}(\mathrm{MR}_{\mathscr{M}}), \mathrm{Ord}_{\mathscr{M}} \right\rangle.$$

Then (i) and (ii) from Definition 14.1.10 are valid. By Lemma 19.4.1 and Definitions 15.2.5 and 18.2.1 we get that for any $\varepsilon \in \mathscr{D}^\tau(L)$, $\vec{a} \in \mathrm{Dom}_{\mathscr{M}}\left(\mathrm{P}_m^{\mathscr{M}}\right)$,

$$\mathrm{J}_\varepsilon \mathrm{P}_m^{\mathscr{M}}(\vec{a}) \in \mathrm{CSD}(\vec{\mathfrak{a}}) \quad \text{iff} \quad \mathrm{CSD}\left(\vec{a}_{\overleftarrow{s,\,m-1}}\right) \mid\!-_L \mathrm{GC}\left[\mathrm{SG}^{\cdot\mathscr{M}}(\mathrm{MR}_{\mathscr{M}}), \mathrm{P}_m^{\mathscr{M}}, \varepsilon\right](\vec{a}).$$

Then also

$$\mathrm{J}_\tau \mathrm{P}_m^{\mathscr{M}}(\vec{a}) \in \mathrm{CSD}(\vec{\mathfrak{a}}) \quad \text{iff} \quad \mathrm{CSD}\left(\vec{\mathfrak{a}}_{\overleftarrow{s,\,m-1}}\right) \vdash_{\mathbf{St}} \mathrm{GC}\left[\mathrm{SG}^{\cdot\mathscr{M}}(\mathrm{MR}_{\mathscr{M}}), \mathrm{P}_m^{\mathscr{M}}, \tau\right](\vec{a}),$$

since there exists a unique $\delta \in \mathscr{V}(L, \mathsf{I})$ such that $\mathrm{J}_\delta \mathrm{P}_m^{\mathscr{M}}(\vec{a}) \in \mathrm{CSD}(\vec{\mathfrak{a}})$, and there exists a unique $\delta \in \mathscr{V}(L, \mathsf{I})$ such that

$$\mathrm{CSD}\left(\vec{\mathfrak{a}}_{\overleftarrow{s,\,m-1}}\right) \vdash_L \mathrm{GC}\left[\mathrm{SG}^{\mathscr{M}}\left(\mathrm{MR}_{\mathscr{M}}\right), \mathrm{P}_m^{\mathscr{M}}, \delta\right](\vec{a}).$$

Therefore (iii) is also valid. □

Corollary 19.4.3. Let L be an FPJ logic, σ be a signature and $\mathscr{M} \in \mathrm{PMC}_\sigma(L)$. Then the following conditions hold:

(i) $\mathrm{SG}^{\mathscr{M}}$ is an $\mathrm{SSC}(\mathscr{M})$-$\mathscr{M}$-compatible conforming function;
(ii) $\mathscr{M}$ is $\mathrm{SSC}(\mathscr{M})$-conformable.

Proof. (i) By Corollary 18.2.10, $\mathrm{SG}^{\mathscr{M}}$ is an $\mathrm{SSC}(\mathscr{M})$-robust conforming function. By Lemma 19.4.2,

$$\mathrm{SG}^{\mathscr{M}}(\mathscr{M}) \in \mathrm{SG}_\sigma(L, \mathscr{M}).$$

Then by Definition 14.1.19, $\mathrm{SG}^{\mathscr{M}}$ is $\mathrm{SSC}(\mathscr{M})$-$\mathscr{M}$-compatible.
(ii) follows from (i) (see Definition 14.1.20). □

Part V
JSM Theories

Chapter 20
Introductory Explanation

In the present part we will consider a set of examples of modification theories related to one of the methods of logical data analysis and knowledge acqusition, the so-called JSM method of automatic hypothesis generation. This method was proposed by V. K. Finn at the beginning of the 1980s.

The abbreviation JSM originates from the initials of John Stuart Mill, a British logician and philosopher of the 19th century who formulated methods of inductive reasoning [Mill, 1843]. The JSM method uses formalised plausible reasoning, and particularly inductive reasoning, in the spirit of Mill's canons for data analysis. V. K. Finn defines JSM method as a result of the synthesis of three cognitive procedures: induction, analogy and abduction. For detailed description of JSM method see [Finn, 1999, 1991]. The JSM method is currently being advanced by a group of Russian researchers.

The authors of the present work collaborated with V. K. Finn and his group. Undoubtedly, the JSM method (in both conceptual and technical aspects) is one of the main sources of the approach proposed here. Moreover, for the time being the JSM method is almost the only example of cognitive technology which is (basically) included in the cognitive frames discussed in this book, and which has been brought to practical realisation and application, and successfully used for data analysis, e.g. in medical diagnosis, pharmacology and sociology (see, e.g., [Blinova et al, 2001, 2003; Finn and Mikheyenkova, 1993a,b; Mikheyenkova et al, 1996; Zabezhailo et al, 1995]).

However, in this chapter we explain not the JSM method itself but modification theories associated with various versions of the JSM method. Formal description of this method was usually done with the use of a language of such theories which we call iterative (cf. [Anshakov et al, 1989, 1991, 1995]). But here we consider modification theories since they demonstrate the dynamics of cognitive reasoning better than the iterative theories. (The fact that for each modification theory a corresponding iterative theory can be constructed and vice versa was shown in Chap. 14; see Theorem 14.2.35).

We emphasise that the theories presented below are associated with the JSM method and do not provide an exact translation of the selected versions of the JSM

O. Anshakov, T. Gergely, *Cognitive Reasoning*, Cognitive Technologies,
DOI 10.1007/978-3-540-68875-4_20, © Springer-Verlag Berlin Heidelberg 2010

method into the language of modification theories. These theories are considered here only to demonstrate the potential of the approach developed in this book and not to introduce the JSM method. Although the correspondence between the variants of the JSM method and the modification theories considered below is not complete, it is strong enough to address these theories as JSM theories. However, the terms and notations used below for the description of JSM theories can differ from the corresponding terms and notations used in works dealing with the JSM method.

We describe only those pecularities of the JSM method which are necessary for understanding the intuitive meaning of the formal constructions introduced below.

The JSM method accomplishes the following two tasks:

- Acquisition of a characteristic relation between structured objects and sets of target properties
- Prediction of the presence or absence of given target properties for the objects with known structure and unknown presence or absence of these target properties

Traditionally the JSM method interprets the characteristic relation between the structure of an object and target properties as a cause–effect relation. I.e., either a fragment of the strucure is considered as a possible cause of the presence or absence of target properties (the direct JSM method), or, conversly, the family of target properties is considered as a possible cause of the structural peculiarities of an object (the inverse JSM method).

The JSM method presumes that both the presence and absence of properties should have their own cause, considered as a structural pecularity (fragment).

Note that below we define not concrete modification theories, but classes of such theories. These classes are defined by requirements to:

- The signature (which should include a definite collection of symbols of given arities and sorts)
- The system of modification rules
- The system of modification rules
- The non-logical axioms of the theory

We do not set any explicit requirements on state description, which should be necessarily present in any modification theory. However, if we wish to work effectively in the framework of a theory we require that it should be consistent at least in the zeroth stage ((0,0)-consistent). In order to meet this requirement the non-logical axioms of the theory should be valid in the syntactical model of state description.

In the present chapter we will consider the logical–mathematical basis of the JSM method in our interpretation. We will give an interpretation of this basis from a more general point of view in order to (1) insert it into the context of cognitive theories considered in the previous chapters, and (2) provide a uniform description for the rules of various versions of the JSM method.

Chapter 21
Simple JSM Theories

21.1 Basic JSM theories

Pure J logics are the logical basis for the JSM method. Both the modification and the iterative theories built over the **St** or **It** logics, respectively, can be suggested for the mathematical representation of various versions of the JSM method

Now we consider the formalisation of the JSM method by the use of modification theories.

Definition 21.1.1. A signature σ is called a *basic JSM signature* if it meets the following conditions:

(i) $\mathsf{Sorts}_\sigma = \{\mathsf{O}, \mathsf{S}, \mathsf{P}\}$ where O is called *the sort of objects*, S is called *the sort of structural pecularities* (or subobjects, or fragments), P is called the sort of target properties;

(ii) Fu_σ contains a binary functional symbol $\sqcap$, which denotes the so-called *similarity operation*;

(iii) $\mathrm{Ar}_\sigma(\sqcap) = \langle \mathsf{S}, \mathsf{S} \rangle$, $\mathrm{Sort}_\sigma(\sqcap) = \mathsf{S}$, i.e. the operation $\sqcap$ is defined on the set of pairs of fragments and the result of this operation is also a fragment (structural pecularity);

(iv) Fu_σ contains the 0-ary functional symbol, Λ, which denotes *the empty fragment*; $\mathrm{Ar}_\sigma(\Lambda) = \langle \rangle$, $\mathrm{Sort}_\sigma(\Lambda) = \mathsf{S}$;

(v) Fu_σ contains the unary functional symbol Struct, which denotes the so-called *structure function*; $\mathrm{Ar}_\sigma(\mathrm{Struct}) = \langle \mathsf{O} \rangle$, $\mathrm{Sort}_\sigma(\mathrm{Struct}) = \mathsf{S}$, i.e. the operation Struct is defined on the set of objects; the result of this operation is an object structure, i.e. a fragment that contains all components of the object (more precisely the result of this operation is its argument considered not as a whole but as a complex structure).

Now we give an intuitive explanation of the above-mentioned operations and predicates. Let us suppose that s and s' are structure fragments and o is an object. Then:

(i) $s \sqcap s'$ is the greatest common part (intersection) of fragments s and s';

DOI 10.1007/978-3-540-68875-4_21,

(ii) Operations Λ and Struct have been described in detail above;

(iii) $s \lessdot o = \begin{cases} \mathbf{t} & \text{if } s \text{ is included in } o, \\ \mathbf{f} & \text{otherwise.} \end{cases}$

Definition 21.1.2. Let σ be a basic JSM signature. A modification theory T of signature σ is called a *basic JSM theory*, if the theory contains the following axioms:

(S1) $s \sqcap s = s$,

(S2) $s \sqcap s' = s' \sqcap s$,

(S3) $(s \sqcap s') \sqcap s'' = s \sqcap (s' \sqcap s'')$,

(S4) $s \sqcap \Lambda = \Lambda$,

where s, s', s'' are variables of sort S;

(SA) $s \lessdot o \leftrightarrow s = s \sqcap \mathrm{Struct}(o)$,

where s is a variable of sort S, o is a variable of sort O.

Axioms (S1)–(S4) are called *similarity axioms*. Axiom (SA) is called the *structure axiom*. Axioms (S1)–(S3) say that the set of structural pecularities (of fragments) forms a semilattice w.r.t. the operation $\sqcap$. Then external predicate $\sqsubseteq$ of arity $\langle \mathsf{S}, \mathsf{S} \rangle$ can be defined as follows: let $s \sqsubseteq s'$ if $s = s \sqcap s'$. The relation corresponding to $\sqsubseteq$ is a weak partial order relation, for which $s \sqcap s' = \inf\{s, s'\}$. Below we will identify any external predicate with the relation corresponding to it.

The axiom (S4) implies that the empty fragment Λ is the least element of the set of fragments w.r.t. the partial order $\sqsubseteq$.

If the strucure of objects are described in the form of sets (as is often the case in the applications of JSM method) then the similarity operation is the usual intersection operation of sets ($\cap$); relation $\sqsubseteq$ is the usual set-theoretical inclusion ($\subseteq$); and the empty fragment Λ is simply the empty set ($\emptyset$).

The axiom (SA) states that a fragment is contained in the object if and only if it is included into its structure. By using the notation $\sqsubseteq$ this axiom can be rewritten as $s \lessdot o \leftrightarrow s \sqsubseteq \mathrm{Struct}(o)$.

21.2 Simple JSM Theories

Definition 21.2.1. A basic JSM signature σ is called a *simple JSM signature*, if the following conditions hold:

(i) Pr_σ contains the binary predicate symbols $\circ\!\!\rightarrow$ and $\bullet\!\!\rightarrow$ such that $\mathrm{Ar}_\sigma(\circ\!\!\rightarrow) = \langle \mathsf{O}, \mathsf{P} \rangle$, $\mathrm{Ar}_\sigma(\bullet\!\!\rightarrow) = \langle \mathsf{S}, \mathsf{P} \rangle$, $\mathrm{Sort}_\sigma(\circ\!\!\rightarrow) = \mathrm{Sort}_\sigma(\bullet\!\!\rightarrow) = \mathsf{I}$, i.e. these are the internal predicate symbols; $\circ\!\!\rightarrow$ denotes the predicate "*...possesses the property ...*" and $\bullet\!\!\rightarrow$ denotes the predicate "*... is a possible cause of the property ...*";

(ii) Pr_σ contains binary predicate symbols M^+ and M^- such that $\mathrm{Ar}_\sigma(\mathrm{M}^+) = \mathrm{Ar}_\sigma(\mathrm{M}^-) = \langle \mathsf{S}, \mathsf{P} \rangle$, $\mathrm{Sort}_\sigma(\mathrm{M}^+) = \mathrm{Sort}_\sigma(\mathrm{M}^-) = \mathsf{E}$, i.e. these are external predicate symbols; M^+ denotes the predicate "*... is a candidate for the possible cause of the presence of property ...*", M^- denotes the predicate "*... is a candidate for the possible cause of the absence of the property ...*".

(iii) Pr_σ contains binary predicate symbols Π^+ and Π^- such that $\mathrm{Ar}_\sigma(\Pi^+) = \mathrm{Ar}_\sigma(\Pi^-) = \langle \mathsf{O}, \mathsf{P} \rangle$, $\mathrm{Sort}_\sigma(\Pi^+) = \mathrm{Sort}_\sigma(\Pi^-) = \mathsf{E}$, i.e. these are external predicate symbols; Π^+ denotes the predicate "*... is a candidate for the presence of the property ...*", Π^- denotes the predicate "*... is a candidate for the absence of the property ...*".

Let us see the intuitive meaning of the operations and predicates of the simple JSM signature. Suppose that s is a structural fragment, o is an object and p is a target property. Then:

(i) $$o \circ\!\!\rightarrow p = \begin{cases} +1 \text{ if } o \text{ possesses } p, \\ -1 \text{ if } o \text{ does not possess } p, \\ 0 \text{ if } o \text{ simultanuosly possesses and does not possess } p, \\ \tau \text{ if we do not know whether } o \text{ possesses } p; \end{cases}$$

(ii) $$s \bullet\!\!\rightarrow p = \begin{cases} +1 \text{ if } s \text{ is a cause of the presence of } p, \\ -1 \text{ if } s \text{ is a cause of the absence of } p, \\ 0 \text{ if } s \text{ can be a cause both of the presence and the absence of } p, \\ \tau \text{ if we know nothing about the cause–effect relation between} \\ \quad s \text{ and } p. \end{cases}$$

(iii) The predicates M^+ and M^- are used in the rules for obtaining hypotheses about the possible causes. The predicates Π^+ and Π^- play a similar role in the rules for predicting the presence or absence of the target properties of objects.

The presence of essential reason for accepting some statement is the main component of the intuitive meaning of these predicates. $\mathrm{M}^+(s,p)$ means that we have essential reasons (enough arguments "for" or enough confirming examples) for accepting the hypothesis that s is a possible cause of the presence of property p. Similarly, $\mathrm{M}^-(s,p)$ means that we have essential reasons for accepting the hypothesis that s is a possible cause of the absence of property p.

However, essential reasons are not necessarily sufficient. Therefore, we have described above the intuitive meaning of the predicates M^+ and M^- with the following sentence fragment: "...is a candidate for possible causes of presence/absence of the property ...". It is just a candidate and nothing more.

(iv) The intuitive meaning of the predicates Π^+ and Π^- can be also explained in the terms of essential reasons. E.g. $\Pi^+(o,p)$ means that there are essential reasons for accepting the hypothesis that the object o possesses the property p (the object o is a candidate for the presence of the property p). The predicate Π^- can be explained in the same way.

Definition 21.2.2. Let σ be a simple JSM signature. A basic JSM theory of signature σ is called *simple*, if the traversal sequence for internal predicates of this theory is $\langle \bullet\!\!\rightarrow, \circ\!\!\rightarrow \rangle$ and it contains the following modification rules:

$$\text{(CR+)} \qquad \frac{J_\tau (s \bullet\!\!\rightarrow p), \quad M^+(s,p) \wedge \neg M^-(s,p)}{J_{+1}(s \bullet\!\!\rightarrow p)},$$

$$\text{(CR−)} \qquad \frac{J_\tau (s \bullet\!\!\rightarrow p), \quad M^-(s,p) \wedge \neg M^+(s,p)}{J_{-1}(s \bullet\!\!\rightarrow p)},$$

$$\text{(CR0)} \qquad \frac{J_\tau (s \bullet\!\!\rightarrow p), \quad M^+(s,p) \wedge M^-(s,p)}{J_{0}(s \bullet\!\!\rightarrow p)},$$

$$\text{(CR}\tau\text{)} \qquad \frac{J_\tau (s \bullet\!\!\rightarrow p), \quad \neg M^+(s,p) \wedge \neg M^-(s,p)}{J_{\tau}(s \bullet\!\!\rightarrow p)},$$

where s is a variable of sort S, p is a variable of sort P. Rules of the group CR are called causal (or cause–effect) rules.

$$\text{(PR+)} \qquad \frac{J_\tau (o \circ\!\!\rightarrow p), \quad \Pi^+(o,p) \wedge \neg \Pi^-(o,p)}{J_{+1}(o \circ\!\!\rightarrow p)},$$

$$\text{(PR−)} \qquad \frac{J_\tau (o \circ\!\!\rightarrow p), \quad \Pi^-(o,p) \wedge \neg \Pi^+(o,p)}{J_{-1}(o \circ\!\!\rightarrow p)},$$

$$\text{(PR0)} \qquad \frac{J_\tau (o \circ\!\!\rightarrow p), \quad \Pi^+(o,p) \wedge \Pi^-(o,p)}{J_{0}(o \circ\!\!\rightarrow p)},$$

$$\text{(PR}\tau\text{)} \qquad \frac{J_\tau (o \circ\!\!\rightarrow p), \quad \neg \Pi^+(o,p) \wedge \neg \Pi^-(o,p)}{J_{+1}(o \circ\!\!\rightarrow p)},$$

where o is a variable of sort O and p is a variable of sort P. Rules of the group PR are called prediction rules.

Proposition 21.2.3. The class of simple JSM theories is well defined.

Proof. It is enough to verify that the introduced set of modification rules forms a system of modification rules in the sense of Definition 10.2.3. Hence, it is enough to prove that the set of causal rules as well as the set of prediction rules are disjoint complete modification sets in the sense of Definition 10.2.2. Therefore, it is enough to prove the disjointness of the major premises of the rules from each above mentioned set, which can be done by a simple direct check. □

21.3 Causal and Prediction Rules

Causal rules[1] permit the formulation of hypotheses about the possible causes of properties. The intuitive meaning of the rule (CR+) can be described as follows:

Let:

- There be no information on the cause–effect relationship between s and p
- s be a candidate of the possible causes of the presence of the property p
- s be not a candidate of the possible causes of the absence of the property p

Then:

- s is the (possible) cause of the presence of the property p.

The intuitive meaning of the rule (CR−) can be described similarly. The rule (CR0) can be informally represented as follows:

Let:

- There be no information on the cause–effect relationship between s and p
- s be a candidate of the possible causes of the presence of the property p
- s be a candidate of the possible causes of the absence of the property p

Then:

- s is a (possible) cause of both the presence and the absence of the property p.

The rule (CRτ) can be represented as follows:

Let:

- There be no information on the cause–effect relationship between s and p
- s be not a candidate of the possible causes of the presence of the property p
- s be not a candidate of the possible causes of the absence of the property p

Then:

[1] In the original publications on the JSM method, the causal rules are called rules of the first type. These represent inductive reasoning. The prediction rules are called rules of the second type. These rules represent a variant of reasoning by analogy.

- As before we have no information on the cause–effect relationship between *s* and *p*.

Predictive rules permit the formulation of hypotheses about the presence or absence of properties. The intuitive meaning of the rule (PR+) can be formulated as follows:

Let:

- There be no information on whether *o* possesses the property (the set of properties) *p*
- *o* be a candidate of the presence of *p*
- *o* be not a candidate of the absence of *p*

Then *o* possesses the property *p*.

The intuitive meaning of the other prediction rules can be described similarly.

21.4 Defining Axioms for the Simple JSM Theories

Various versions of simple JSM theories can variously define the interrelation between the predicates M^+, M^-, Π^+ and Π^- (which look *complicated*) and the *elementary*, basic predicates $\circ\!\!\rightarrow$, $\bullet\!\!\rightarrow$ and $<$. It would be of interest to represent the complicated predicates through the basic ones by using defining axioms. It is simpler to do this for the predicates Π^+ and Π^-. According to the traditions of the JSM method we can propose the following defining axioms for these predicates:

$$(\mathrm{D}\Pi^+) \qquad \Pi^+(o,p) \leftrightarrow \exists s\,(s < o \wedge \mathrm{J}_{+1}\,(s \bullet\!\!\rightarrow p)),$$

$$(\mathrm{D}\Pi^-) \qquad \Pi^-(o,p) \leftrightarrow \exists s\,(s < o \wedge \mathrm{J}_{-1}\,(s \bullet\!\!\rightarrow p)),$$

where s is a variable of sort S.

Informally these axioms can be described as follows:

- Object *o* is a candidate for the presence of the property (set of properties) *p* iff *o* contains a structural fragment *s*, which is a possible cause of the presence of *p*;
- Object *o* is a candidate for the absence of the property (set of properties) *p* iff *o* contains a structural fragment *s*, which is a possible cause of the absence of *p*.

The situation is more difficult for the predicates M^+ and M^-. If we follow the Method of Agreement introduced by J. S. Mill (1867), then $\mathrm{M}^+(s,p)$ should be understood as:

- *s is a non-empty similarity of at least two different objects* possessing the property (set of properties) p.

If we generalise the Method of Agreement by introducing an arbitrary threshold for decision-making not necessarily equal to two, then $\mathrm{M}^{+}(s,p)$ can be understood as:

- *s is a non-empty similarity of sufficient amount of different objects* possessing the property (set of properties) p.

This "sufficient amount" may be calculated, e.g., by using statistical criterias.

Further on for the sake of simplicity we will follow the Method of Agreement of J. S. Mill.

How should we understand the sentence "s is a non-empty similarity of at least two different objects possessing the property (set of properties) p"? Before answering this question let us introduce by definition some conditions imposed on the pair $\langle s,p\rangle$.

Definition 21.4.1. We say that a fragment s and a property (set of properties) p satisfy the *positive Mill's condition*, if s is contained in two different objects possessing the property (set of properties) p. Symbolically, the positive Mill's condition for a pair $\langle s,p\rangle$ will be denoted by $\mathrm{Mill}^{+}(s,p)$ and will be considered as an abbreviation of the following formula:

$$\exists o_1 \exists o_2 (o_1 \neq o_2 \wedge s \lessdot o_1 \wedge s \lessdot o_2 \wedge \mathrm{J}_{+1}(o_1 \circ\!\!\rightarrow p) \wedge \mathrm{J}_{+1}(o_2 \circ\!\!\rightarrow p)).$$

We say that a fragment s and a property (set of properties) p satisfy the *negative Mill's condition*, if s is contained in two different objects which do not possess the property (set of properties) p. In symbolic form, the negative Mill's condition for the pair $\langle s,p\rangle$ will be denoted by $\mathrm{Mill}^{-}(s,p)$ and will considered as an abbreviation of the formula:

$$\exists o_1 \exists o_2 (o_1 \neq o_2 \wedge s \lessdot o_1 \wedge s \lessdot o_2 \wedge \mathrm{J}_{+1}(o_1 \circ\!\!\rightarrow p) \wedge \mathrm{J}_{+1}(o_2 \circ\!\!\rightarrow p)).$$

Now let us see the informal understanding of the sentence: "*s is a non-empty similarity of at least two objects possessing a property (set of properties) p*". The simplest interpretation of this sentence in terms of elementary predicates ($\circ\!\!\rightarrow$ and $\lessdot$) is:

- A non-empty *s is contained in two different objects* possessing a property (set of properties).

Using the above-defined Mill's condition we can introduce the following defining axioms for M^{+} and M^{-} for this simplest interpretation:

(DM^{+}S) $\mathrm{M}^{+}(s,p) \leftrightarrow s \neq \Lambda \wedge \mathrm{Mill}^{+}(s,p)$,

(DM^{-}S) $\mathrm{M}^{-}(s,p) \leftrightarrow s \neq \Lambda \wedge \mathrm{Mill}^{-}(s,p)$.

The letter 'S' means "simpliest" in the notations of these axioms.

It is clear that if we require that the number of examples confirming the fact about the presence (absence) of property would not be two but some other quantity, then it would be necessary to modify the defining axioms by substituting the Mill's condition with another one.

However, the works on JSM method use the operation $\sqcap$ instead of the relation $\lessdot$ for representing the predicates M^+ and M^-. These representations are formulated differently for the case when the elements of sort TPr are considered as (atomic) properties and for the case when it is considered as a set of properties.

First let us consider a simpler case. We will assume that the elements of sort TPr are properties and not sets of properties.

Then $M^+(s, p)$ means that:

- A non-empty s is equal to the similarity (intersection) of at least two objects possessing the property p.

In spite of the simplicity of this naive description of predicate M^+ and in spite of its resemblance to the previous naive description we cannot directly formulate a defining axiom corresponding to this description without falling outside of the limits of the first-order logic.

The problem of representing the predicates M^+ and M^- by elementary predicates and the operation $\sqcap$ by using first-order formulas was investigated in detail by Vinogradov [2000]. He showed that if we restrict ourselves to the *finite* (**St**- or **It**-) structures then it is possible to propose a first-order formula expressing the predicate M^+ by the predicates $\lessdot$, $\circ\!\!\rightarrow$ and the operation $\sqcap$. Moreover, the same formula is appropriate for any finite structure. Vinogradov gave an example of such an infinite structure within which the predicate M^+ cannot be expressed by the predicates $\lessdot$, $\circ\!\!\rightarrow$ and the operation $\sqcap$ with a first-order formula.

Models of modification theories are constructed by the use of finite **St**-structures. But infinite **It**-structures can serve as models of iterative theories. Here we consider modification theories only. Thus according to the results of Vinogradov [2000] the predicate M^+ can be expressed by the predicates $\lessdot$, $\circ\!\!\rightarrow$ and operation $\sqcap$ with a first-order formula. This formula will not be very shocking although it looks slightly artificial. The results of D. Vinogradov will be explained below.

21.5 Simple JSM Theories with Exclusion of Counterexamples

The so-called simple JSM method with exclusion of counterexamples is often used in practical applications of the JSM method. In this method, the hypotheses evaluated by the value 0 (contradictory hypotheses) are not generated. Moreover, this method is highly rigid and usually yields very precise (but sometimes very incomplete) predictions. This method proved good in the tasks of prediction of biological activity of chemical compounds in medical chemistry.

Definition 21.5.1. Let σ be a simple JSM signature. A basic JSM theory of signature σ is called *a simple JSM theory with exclusion of counterexamples* if it meets the following conditions:

- The traversal sequence for the set of internal predicates of this theory is $\langle \bullet\!\!\rightarrow \circ\!\!\rightarrow \rangle$,
- It contains the following axioms:

(C+) $\mathrm{M}^{+}(s,p) \rightarrow \mathrm{C}^{+}(s,p)$,

(C−) $\mathrm{M}^{-}(s,p) \rightarrow \mathrm{C}^{-}(s,p)$.

- It contains the following modification rules:

$$(\mathrm{CR}+\mathrm{C})\quad \frac{\mathrm{J}_{\tau}(s \bullet\!\!\rightarrow p), \quad \mathrm{M}^{+}(s,p) \wedge \neg \mathrm{C}^{-}(s,p)}{\mathrm{J}_{+1}(s \bullet\!\!\rightarrow p)},$$

$$(\mathrm{CR}-\mathrm{C})\quad \frac{\mathrm{J}_{\tau}(s \bullet\!\!\rightarrow p), \quad \mathrm{M}^{-}(s,p) \wedge \neg \mathrm{C}^{+}(s,p)}{\mathrm{J}_{-1}(s \bullet\!\!\rightarrow p)},$$

$$(\mathrm{CR0C})\quad \frac{\mathrm{J}_{\tau}(s \bullet\!\!\rightarrow p), \quad \bot}{\mathrm{J}_{0}(s \bullet\!\!\rightarrow p)},$$

$$(\mathrm{CR}\tau\mathrm{C})\quad \frac{\mathrm{J}_{\tau}(s \bullet\!\!\rightarrow p), \quad \neg(\mathrm{M}^{+}(s,p) \wedge \neg \mathrm{C}^{-}(s,p)) \wedge \neg(\mathrm{M}^{-}(s,p) \wedge \neg \mathrm{C}^{+}(s,p))}{\mathrm{J}_{\tau}(s \bullet\!\!\rightarrow p)},$$

where s is a variable of sort S and p is a variable of sort TPr,

$\mathrm{C}^{+}(s,p)$ is the following formula

$$\exists o\,(s \lessdot o \wedge \mathrm{J}_{+}(o \circ\!\!\rightarrow p)),$$

$\mathrm{C}^{-}(s,p)$ is the formula

$$\exists o\,(s \lessdot o \wedge \mathrm{J}_{-}(o \circ\!\!\rightarrow p)),$$

where o is a variable of sort O and $\bot$ is the identically false formula,

$$(\mathrm{PR}+\mathrm{C})\quad \frac{\mathrm{J}_{\tau}(o \circ\!\!\rightarrow p), \quad \Pi^{+}(o,p) \wedge \neg \Pi^{-}(o,p)}{\mathrm{J}_{+1}(o \circ\!\!\rightarrow p)},$$

$$(\mathrm{PR}-\mathrm{C})\quad \frac{\mathrm{J}_{\tau}(o \circ\!\!\rightarrow p), \quad \Pi^{-}(o,p) \wedge \neg \Pi^{+}(o,p)}{\mathrm{J}_{-1}(o \circ\!\!\rightarrow p)},$$

$$(\mathrm{PR0C})\quad \frac{\mathrm{J}_{\tau}(o \circ\!\!\rightarrow p), \quad \bot}{\mathrm{J}_{0}(o \circ\!\!\rightarrow p)},$$

$$(\mathrm{PR}\tau\mathrm{C})\quad \frac{\mathrm{J}_{\tau}(o \circ\!\!\rightarrow p), \quad \neg(\Pi^{+}(o,p) \wedge \neg \Pi^{-}(o,p)) \wedge \neg(\Pi^{-}(o,p) \wedge \neg \Pi^{+}(o,p))}{\mathrm{J}_{\tau}(o \circ\!\!\rightarrow p)},$$

where o is a variable of sort O, p is a variable of sort P and $\bot$ is an identically false formula.

Since the major premise of the rules (CR0C) and (PR0C) is an identically false formula $\bot$, the conclusions of the form $\mathrm{J}_{0}(s \bullet\!\!\rightarrow p)$ and $\mathrm{J}_{0}(o \circ\!\!\rightarrow p)$ cannot be obtained.

Proposition 21.5.2. The class of JSM theories with exclusion of counterexamples is well defined.

Proof. The proof is similar to the proof of Proposition 21.2.3. The axioms (C+) and (C–) can be used to prove the disjointness of the sets of rules. These axioms can be proved if we accept the defining axioms ($\mathrm{DM}^+\mathrm{C}$) and ($\mathrm{DM}^-\mathrm{C}$). □

Proposition 21.5.3. In a simple JSM theory with exclution of counterexamples the major premises of the rules ($\mathrm{CR}\tau\mathrm{C}$) and ($\mathrm{PR}\tau\mathrm{C}$) are equivalent to the formulas

$$\left(\neg \mathrm{M}^+(s,p) \wedge \neg \mathrm{M}^-(s,p)\right) \vee \left(\mathrm{C}^-(s,p) \wedge \mathrm{C}^+(s,p)\right)$$

and

$$\left(\neg \Pi^+(o,p) \wedge \neg \Pi^-(o,p)\right) \vee \left(\Pi^-(o,p) \wedge \Pi^+(o,p)\right),$$

respectively.

Proof. By direct checking. □

Chapter 22
Advanced JSM Theories

22.1 Generalised JSM Theories

The generalised JSM method was proposed as a method that provides finer data analysis than that provided by the simple JSM method. The main supposition of the generalised JSM method is: *each possible cause may be associated with a set of its own inhibitors, which interfere with the appearence of effects even if the cause exists.* Therefore, we can conclude that an object possesses a certain property (according to the generalised JSM method) only if this object contains the cause of this property and contains none of the inhibitors of this cause.

In this section we describe the generalised JSM theories, which are based on the idea of the generalised JSM method. However, the proposed here technique strongly differs from that proposed in [Finn, 1989].

Definition 22.1.1. A basic JSM signature is called a *generalised JSM signature* if it meets the following conditions:

- σ contains all the external predicate symbols which are contained in the simple JSM signature with the same arities and sorts.
- Pr_σ contains the internal predicate symbols $\circ\!\!\rightarrow$, $\bullet\!\!\rightarrow_+$ and $\bullet\!\!\rightarrow_-$ such that $\mathrm{Ar}_\sigma(\circ\!\!\rightarrow) = \langle \mathsf{O}, \mathsf{P}\rangle$, $\mathrm{Ar}_\sigma(\bullet\!\!\rightarrow_+) = \mathrm{Ar}_\sigma(\bullet\!\!\rightarrow_-) = \langle \mathsf{S}, \mathsf{P}\rangle$.
  The intuitive meaning of the predicates corresponding to these symbols is the same as of the predicates that correspond to the symbols $\circ\!\!\rightarrow$ and $\bullet\!\!\rightarrow$ from the simple JSM signature. $\bullet\!\!\rightarrow_+$ and $\bullet\!\!\rightarrow_-$ are like to be the halves of the predicate, $\bullet\!\!\rightarrow$ corresponding to the cause of presence and to the cause of absence, respectively. Below we will give a more detailed description of the intuitive meaning of the predicates denoted by these symbols.
- Pr_σ contains ternary internal predicate symbols Inh^+ and Inh^- , $\mathrm{Ar}_\sigma\left(\mathrm{Inh}^+\right) = \mathrm{Ar}_\sigma\left(\mathrm{Inh}^-\right) = \langle \mathsf{S}, \mathsf{S}, \mathsf{P}\rangle$; Inh^+ denotes the predicate "*...in combination with... inhibits the presence of the property...*", and Inh^- denotes the predicate "*... in combination with... inhibits the absence of the property...*".
- Pr_σ contains ternary external predicate symbols H^+ and H^-; H^+ denotes the predicate "*...is a candidate for inhibiting in combination with the presence of*

O. Anshakov, T. Gergely, *Cognitive Reasoning*, Cognitive Technologies,
DOI 10.1007/978-3-540-68875-4_22,

the property..."; H^- denotes the predicate "*...is a candidate for inhibiting in combination with the absence of the property...*"

Now we discuss the intuitive meaning of the operations and predicates denoted by the symbols of the generalised JSM signature. First, we note that all operations have the same meaning as in the case of the simple JSM theory. The predicates $\lessdot$ and $\circ\!\!\rightarrow$ also have the same meaning as in the case of the simple JSM theory. Suppose that s and t are structural fragments, o is an object and p is the target property or the set of target properties. Then:

$$s \bullet\!\!\rightarrow_+ p = \begin{cases} +1 & \text{if we certanly know} \\ & \text{that } s \text{ forces a presence of } p, \\ \tau & \text{if we do not know certainly} \\ & \text{whether } s \text{ forces a presence of } p, \end{cases}$$

$$s \bullet\!\!\rightarrow_- p = \begin{cases} +1 & \text{if we certanly know} \\ & \text{that } s \text{ forces an absence of } p, \\ \tau & \text{if we do not know certainly} \\ & \text{whether } s \text{ forces an absence of } p, \end{cases}$$

$$\mathrm{Inh}^+(t,s,p) = \begin{cases} +1 & \text{if it is known} \\ & \text{that } t \text{ with } s \text{ inhibits a presence of } p, \\ \tau & \text{if it is not known} \\ & \text{whether } t \text{ with } s \text{ inhibits a presence of } p, \end{cases}$$

$$\mathrm{Inh}^-(t,s,p) = \begin{cases} +1 & \text{if it is known} \\ & \text{that } t \text{ with } s \text{ inhibits an absence of } p, \\ \tau & \text{if it is not known} \\ & \text{whether } t \text{ with } s \text{ inhibits an absence of } p. \end{cases}$$

It can be seen from the above description that new internal predicates cannot obtain values different from +1 and τ.

The predicates H^+ and H^- are used in the rules for obtaining hypotheses about the inhibitors of the possible causes. They play the same role in these rules as the predicates M^+ and M^- in the rules for obtaining hypotheses about the possible causes and the predicates Π^+ and Π^- in the rules for predicting the presence or absence of the target properties in the considered objects.

The presence of essential reasons for obtaining some statement is the main component of the intuitive meaning of all these predicates. $H^+(t,s,p)$ means that we have essential reasons (sufficient number of arguments "for", sufficient number of confirming examples) for accepting the hypothesis that t is an inhibitor for the possible cause s of the presence of the property p. Similarly, $\mathrm{H}^{-}(t,s,p)$ can be understood as essential reasons for accepting the hypothesis that t is an inhibitor for the possible cause s of the absence of property p.

Definition 22.1.2. Let σ be a signature. A basic JSM theory of signature σ is called a *generalised JSM theory* if the traversal sequence for the set of internal predicates is $\langle \bullet\!\!\rightarrow_{+}, \bullet\!\!\rightarrow_{-}, \mathrm{Inh}^{+}, \mathrm{Inh}^{-}, \circ\!\!\rightarrow \rangle$ and it contains the following modification rules:

$$(\mathrm{CR}++) \qquad \frac{\mathrm{J}_{\tau}(s \bullet\!\!\rightarrow_{+} p), \quad \mathrm{M}^{+}(s,p)}{\mathrm{J}_{+1}(s \bullet\!\!\rightarrow_{+} p)},$$

$$(\mathrm{CR}+-) \qquad \frac{\mathrm{J}_{\tau}(s \bullet\!\!\rightarrow_{+} p), \quad \bot}{\mathrm{J}_{-1}(s \bullet\!\!\rightarrow_{+} p)},$$

$$(\mathrm{CR}+0) \qquad \frac{\mathrm{J}_{\tau}(s \bullet\!\!\rightarrow_{+} p), \quad \bot}{\mathrm{J}_{0}(s \bullet\!\!\rightarrow_{+} p)},$$

$$(\mathrm{CR}+\tau) \qquad \frac{\mathrm{J}_{\tau}(s \bullet\!\!\rightarrow_{+} p), \quad \neg\mathrm{M}^{+}(s,p)}{\mathrm{J}_{\tau}(s \bullet\!\!\rightarrow_{+} p)}.$$

The rules processing $\bullet\!\!\rightarrow_{+}$ are completed and now we describe the rules processing $\bullet\!\!\rightarrow_{-}$.

$$(\mathrm{CR}-+) \qquad \frac{\mathrm{J}_{\tau}(s \bullet\!\!\rightarrow_{-} p), \quad \mathrm{M}^{-}(s,p)}{\mathrm{J}_{+1}(s \bullet\!\!\rightarrow_{-} p)},$$

$$(\mathrm{CR}--) \qquad \frac{\mathrm{J}_{\tau}(s \bullet\!\!\rightarrow_{-} p), \quad \bot}{\mathrm{J}_{-1}(s \bullet\!\!\rightarrow_{-} p)},$$

$$(\mathrm{CR}-0) \qquad \frac{\mathrm{J}_{\tau}(s \bullet\!\!\rightarrow_{-} p), \quad \bot}{\mathrm{J}_{0}(s \bullet\!\!\rightarrow_{-} p)},$$

$$(\mathrm{CR}-\tau) \qquad \frac{\mathrm{J}_{\tau}(s \bullet\!\!\rightarrow_{-} p), \quad \neg\mathrm{M}^{-}(s,p)}{\mathrm{J}_{\tau}(s \bullet\!\!\rightarrow_{-} p)},$$

where s is a variable of sort S and p is a variable of sort P. Note that by using these rules the predicates $\bullet\!\!\rightarrow_{+}$ and $\bullet\!\!\rightarrow_{-}$ can obtain only the values +1 and τ. The rules of the group CR are called causal (cause–effect) rules.

$$(\mathrm{IR}++) \qquad \frac{\mathrm{J}_{\tau}\left(\mathrm{Inh}^{+}(t,s,p)\right), \quad \mathrm{H}^{+}(t,s,p)}{\mathrm{J}_{+1}\left(\mathrm{Inh}^{+}(t,s,p)\right)},$$

$$(\mathrm{IR}+-) \qquad \frac{\mathrm{J}_{\tau}\left(\mathrm{Inh}^{+}(t,s,p)\right), \quad \bot}{\mathrm{J}_{-1}\left(\mathrm{Inh}^{+}(t,s,p)\right)},$$

$$(\mathrm{IR}+0)\qquad \frac{\mathrm{J}_{\tau}\left(\mathrm{Inh}^{+}(t,s,p)\right),\quad \perp}{\mathrm{J}_{-1}\left(\mathrm{Inh}^{+}(t,s,p)\right)},$$

$$(\mathrm{IR}+\tau)\qquad \frac{\mathrm{J}_{\tau}\left(\mathrm{Inh}^{+}(t,s,p)\right),\quad \neg\mathrm{H}^{+}(t,s,p)}{\mathrm{J}_{\tau}\left(\mathrm{Inh}^{+}(t,s,p)\right)}.$$

The rules processing Inh^{+} are completed and now we start to describe the rules processing Inh^{-}.

$$(\mathrm{IR}-+)\qquad \frac{\mathrm{J}_{\tau}\left(\mathrm{Inh}^{-}(t,s,p)\right),\quad \mathrm{Inh}^{-}(t,s,p)}{\mathrm{J}_{+1}\left(\mathrm{Inh}^{-}(t,s,p)\right)},$$

$$(\mathrm{IR}--)\qquad \frac{\mathrm{J}_{\tau}\left(\mathrm{Inh}^{-}(t,s,p)\right),\quad \perp}{\mathrm{J}_{-1}\left(\mathrm{Inh}^{-}(t,s,p)\right)},$$

$$(\mathrm{IR}-0)\qquad \frac{\mathrm{J}_{\tau}\left(\mathrm{Inh}^{-}(t,s,p)\right),\quad \perp}{\mathrm{J}_{-1}\left(\mathrm{Inh}^{-}(t,s,p)\right)},$$

$$(\mathrm{IR}-\tau)\qquad \frac{\mathrm{J}_{\tau}\left(\mathrm{Inh}^{-}(t,s,p)\right),\quad \neg\mathrm{H}^{-}(t,s,p)}{\mathrm{J}_{\tau}\left(\mathrm{Inh}^{-}(t,s,p)\right)},$$

where t and s are variables of sort S and p is a variable of sort P. Note that by the use of these rules predicates Inh^{+} and Inh^{-} can obtain only the values +1 and τ. The rules of the group IR will be called *inhibitor* rules .

$$(\mathrm{PR}+)\qquad \frac{\mathrm{J}_{\tau}(o \circ\!\!\rightarrow p),\quad \Pi^{+}(o,p)\wedge\neg\Pi^{-}(o,p)}{\mathrm{J}_{+1}(o \circ\!\!\rightarrow p)},$$

$$(\mathrm{PR}-)\qquad \frac{\mathrm{J}_{\tau}(o \circ\!\!\rightarrow p),\quad \Pi^{-}(o,p)\wedge\neg\Pi^{+}(o,p)}{\mathrm{J}_{-1}(o \circ\!\!\rightarrow p)},$$

$$(\mathrm{PR}0)\qquad \frac{\mathrm{J}_{\tau}(o \circ\!\!\rightarrow p),\quad \Pi^{+}(o,p)\wedge\Pi^{-}(o,p)}{\mathrm{J}_{0}(o \circ\!\!\rightarrow p)},$$

$$(\mathrm{PR}\tau)\qquad \frac{\mathrm{J}_{\tau}(o \circ\!\!\rightarrow p),\quad \Pi^{+}(o,p)\wedge\Pi^{-}(o,p)}{\mathrm{J}_{\tau}(o \circ\!\!\rightarrow p)},$$

where o is a variable of sort O and p is a variable of sort P. The rules of the group PR will be called prediction rules. These rules are formulated exactly as the prediction rules of the simple JSM theory. However, the defining axioms for the predicates Π^{+} and Π^{-} will be completely different.

Proposition 22.1.3. The class of generalised JSM theories is well defined.

Proof. The proof is similar to the proof of Proposition 21.2.3. The proof of the fact that the sets of rules are disjoint is trivial. □

22.2 Defining Axioms for the Generalised JSM Theories

We can raise the problem of the expressibility of the predicates M^+, M^-, Π^+ and Π^- and of the new predicates H^+ and H^- (all these predicates look complicated) through the elementary, basic predicates $\circ\!\!\rightarrow$, $\bullet\!\!\rightarrow_+$, $\bullet\!\!\rightarrow_-$ and $\lessdot$ for the generalised JSM theories as well as for the simple ones. The most interesting representation of complicated predicates would be through basic ones by the use of the defining axioms. As in the case of the simple JSM theories, it is easiest to do this for the predicates Π^+ and Π^-. For the generalised JSM method the following defining axioms are proposed.

$$(\mathrm{D}\Pi^+\mathrm{G}) \quad \Pi^+(o,p) \leftrightarrow \exists s\left(s \lessdot o \wedge \mathrm{J}_{+1}(s \bullet\!\!\rightarrow_+ p) \wedge \forall t\left(\mathrm{J}_{+1}\,\mathrm{Inh}^+(t,\ s,\ p) \rightarrow \neg(t \lessdot o)\right)\right),$$

$$(\mathrm{D}\Pi^-\mathrm{G}) \quad \Pi^-(o,p) \leftrightarrow \exists s\left(s \lessdot o \wedge \mathrm{J}_{+1}(s \bullet\!\!\rightarrow_- p) \wedge \forall t\left(\mathrm{J}_{+1}\,\mathrm{Inh}^-(t,\ s,\ p) \rightarrow \neg(t \lessdot o)\right)\right),$$

where s and t are variables of sort S.

Informally these axioms can be described in the following way:

- Object o is a candidate for the presence of the property (set of properties) p iff o contains a structural fragment s, which is a possible cause of the presence of p and does not contain any inhibitor for this cause;
- Object o is a candidate for the absence of the property (set of properties) p iff o contains a structural fragment s, which is a possible cause of the absence of p and does not contain any inhibitor for this cause.

The intuitive meaning and the formalisation of the predicates M^+ and M^- can be done for the generalised JSM theory in just the same way as for the simple JSM theory. I.e. $\mathrm{M}^+(s,p)$ is understood as:

- *s is a non-empty similarity of at least two different objects* possessing the property (set of properties) p.

The simplest interpretation of this sentence in the terms of elementary predicates $\circ\!\!\rightarrow$ and $\lessdot$ is as follows:

- A non-empty *s is contained in two* different objects possessing the property (set of properties) p.

We can introduce defining axioms for M^+ and M^- for this simplest interpretation by using Mill's condition 21.4.1 in just the same way as for simple JSM theories.

If we wish to represent the predicates M^+ and M^- through operation Π then we can use the results of Vinogradov for this, which will be described in this part later on (cf. [Vinogradov, 2000]).

Now we turn to the description of the predicates H^+ and H^-. According to the description of the informal meaning of the predicate H^- from Definition 22.1.1 the expression $\mathrm{H}^+(t,s,p)$ means "t is a candidate for inhibiting in combination with s the presence of the property p". In the case of searching for inhibitors, similarly to the case of searching for possible causes, we will implicitly use the method of agreement of J. S. Mill. Therefore, we will informally understand $\mathrm{H}^{+}(t,s,p)$ as:

- t is a non-empty, non-coinciding with s similarity of at least two objects that contain s and do not possess the property p.

We can now ask the question of how to understand the sentence "it is similarity of not fewer than two objects" as well as for the description of the predicates M^{+} and M^{-}. Once more in the simplest case t should be contained in at least two different objects and in more complicated cases t should be the intersection of at least two different objects.

Now we define the generalised Mill's condition, which will be about the combination of two structural peculiarities (i.e. about the combination of two factors that promote or inhibit the manifestation of a property).

Definition 22.2.1. We will say that a fragment t in combination with a fragment s and a property (set of properties) p satisfy the generalised positive Mill's condition if t together with s are contained in two different objects possessing the property (set of properties) p. In symbols, the generalised positive Mill's condition for the triple $\langle t,s,p\rangle$ will be denoted by $\mathrm{GMill}^{+}(t,s,p)$ and it will be considered as the abbreviation for the formula:

$$\exists o_1 \exists o_2 (o_1 \neq o_2 \wedge \mathrm{J}_{+1}(o_1 \multimap p) \wedge \mathrm{J}_{+1}(o_2 \multimap p) \wedge t \lessdot o_1 \wedge t \lessdot o_2 \wedge s \lessdot o_1 \wedge s \lessdot o_2).$$

We will say that a fragment t in combination with a fragment s and a property (set of properties) p satisfy the *generalised negative Mill's condition* if t together with s are contained in two different objects which do not possess the property (set of properties) p. Symbolically, the generalised negative Mill's condition for the triple $\langle t,s,p\rangle$ will be denoted by $\mathrm{GMill}^{-}(t,s,p)$ and it will be considered as the abbreviation for the formula:

$$\exists o_1 \exists o_2 (o_1 \neq o_2 \wedge \mathrm{J}_{-1}(o_1 \multimap p) \wedge \mathrm{J}_{-1}(o_2 \multimap p) \wedge t \lessdot o_1 \wedge t \lessdot o_2 \wedge s \lessdot o_1 \wedge s \lessdot o_2).$$

By using the generalised Mill's conditions we can introduce defining axioms for H^{+} and H^{-} for this simplest interpretation:

(DH^{+}G) $\quad \mathrm{H}^{+}(t,s,p) \leftrightarrow t \neq \Lambda \wedge t \neq s \wedge \mathrm{GMill}^{-}(t,s,p)$,

(DH^{-}G) $\quad \mathrm{H}^{-}(t,s,p) \leftrightarrow t \neq \Lambda \wedge t \neq s \wedge \mathrm{GMill}^{+}(t,s,p)$.

The letter "S" means "simpliest" in the notations of these axioms (as well as in the notations of axioms (D M^{+} S) and (D M^{+} S)).

Note that the results of Vinogradov [2000] — on the representation of the statement about similarity of objects in the form of first-order formula — permit the description of the predicates $\mathrm{H}^{+}(t,s,p)$ and $\mathrm{H}^{-}(t,s,p)$ in the form of first-order formula even in the case when the sentence "is a similarity of not fewer than two objects" is understood as "is equal to the intersection of not fewer than two objects". These results of D. Vinogradov will be discussed later on in this part.

Note that according to the modification rules of the generalised JSM theories one and the same fragment can be considered as a possible cause of presence as well as of absence of a property. This may mean that this fragment is a part of some fragment, which is a cause of presence of a property, and also it is a part of another fragment, which is a cause of absence of the same property. Then the inhibitors of the given fragment, i.e. of the negative cause (cause of absence of the property), are actually the missing parts of the same fragment. These inhibitors are necessary for obtaining positive cause (cause of presence of the property). We have a symmetric situation for the inhibitors of positive cause.

22.3 Non-symmetric JSM Theories

The non-symmetric JSM method was proposed by Vinogradov [2001]. This method resembles the generalised JSM method. It also uses the notion of inhibitor of possible cause. It can be considered as a certain simplification of the generalised JSM method. The non-symmetric JSM method has the following pecularities:

- It permits to avoid the modification of the basic set of internal predicate symbols, i.e. it permits not to introduce two symbols for the predicate "...is a possible cause...",
- It eliminates the necessity of using rule iteration (the procedure of hypothesis generation by the use of the non-symmetric JSM method terminates in one step).

With the help of the non-symmetric JSM method only the causes of *presence* of properties and their inhibitors are sought. Therefore it is not necessary to introduce an additional symbol for the predicate "... is a cause of absence of the property...". The predicates "...in combination with...inhibit the absence of the property..." and "...is a candidate for *inhibiting* in combination with the absence of the property...".

The non-symmetric JSM theories to be defined below will rather precisely reproduce the non-symmetric JSM method.

Since we have already defined the generalised JSM signature the signature for the non-symmetric JSM theories can be defined as a reduction of the generalised case, and a corresponding JSM theory can be defined as a modification of the generalised JSM theory.

Definition 22.3.1. A signature σ is called a *non-symmetric* JSM signature, if it is obtained from the generalised JSM signature by elimination of the symbols $\bullet\!\!\longrightarrow_{-}$, Inh^{-}, M^{-}, H^{-} and Π^{-}.

The intuitive interpretation of the remaining symbols in the non-symmetric JSM theories will be the same as in the case of generalised JSM theories.

In a non-symmetric JSM signature the symbols dual to the symbols $\bullet\!\!\longrightarrow_{+}$, Inh^{+}, M^{+}, H^{+} and Π^{+} are missed. Therefore, in the description of the JSM method the subscript or the superscript "+" could be omitted, e.g. we could simply write $\bullet\!\!\longrightarrow$.

However, we will avoid such omission in order to have a possibility to define the non-symmetric JSM theories by the generalised ones.

Definition 22.3.2. Let σ be a non-symmetric JSM signature. A basic JSM theory of signature σ is called a *non-symmetric basic JSM theory*, if the traversal sequence for the set of internal predicates of this theory is $\langle \bullet\!\!\rightarrow_{+}, \mathrm{Inh}^{+}, \circ\!\!\rightarrow \rangle$ and it contains the following modification rules:

- The rules (CR++), (CR+−), (CR+0), (CR+τ) from the definition of generalised JSM theory (these rules process $\bullet\!\!\rightarrow_{+}$)
- The rules (IR++), (IR+-), (IR+0), (IR+τ) from the definition of generalised JSM theory (these rules process Inh^{+})
- The following rules for processing *possess*:

$$(\mathrm{PR+NS})\qquad \frac{\mathrm{J}_{\tau}(o \circ\!\!\rightarrow p),\quad \Pi^{+}(o,p)}{\mathrm{J}_{+1}(o \circ\!\!\rightarrow p)},$$

$$(\mathrm{PR-NS})\qquad \frac{\mathrm{J}_{\tau}(o \circ\!\!\rightarrow p),\quad \neg\Pi^{+}(o,p)}{\mathrm{J}_{-1}(o \circ\!\!\rightarrow p)},$$

$$(\mathrm{PR0NS})\qquad \frac{\mathrm{J}_{\tau}(o \circ\!\!\rightarrow p),\quad \bot}{\mathrm{J}_{0}(o \circ\!\!\rightarrow p)},$$

$$(\mathrm{PR\tau NS})\qquad \frac{\mathrm{J}_{\tau}(o \circ\!\!\rightarrow p),\quad \bot}{\mathrm{J}_{\tau}(o \circ\!\!\rightarrow p)},$$

where o is a variable of sort O and p is a variable of sort P. The rules from the group PR will be called, as above, *prediction rules*.

Proposition 22.3.3. The class of non-symmetric JSM theories is well defined.

Proof. It is similar to the proof of Proposition 21.2.3. The fact that the sets of the causal and the inhibitor rules are disjoint follows from Proposition 22.1.3. The fact that the set of the predictive rules is disjoint is obvious. □

We take the axiom (D Π^{+} G) for generalised JSM theories (from Sect. 22.2) as the defining axiom for Π^{+}.

The structure of predictive rules of a non-symmetric JSM theory is such that the reasoning process will necessarily terminate in the first stage. In the first stage for each object it will be determined whether it possesses a certain property or does not. In this case the saturation condition will hold and any new hypotheses about the internal predicates cannot be obtained. In fact in the non-symmetric JSM method the hypothesis of a closed world is accepted and negation is interpreted as failure. Such assumptions are not typical for the JSM method. In the JSM method failure is usually interpreted as preservation of uncertainty.

However, the non-symmetric JSM method can be easily modified, retaining the possibility of repeated iteration of applying rules. In the *modified non-symmetric JSM method* we will only be interested in prediction of presence of properties and possible causes of presence of properties with their inhibitors.

Definition 22.3.4. Let σ be a non-symmetric JSM signature. A basic JSM theory of signature σ is called *modified non-symmetric* if it is obtained from the non-symmetric JSM theory by the use of the following substitutions:

- The rule (PR-NS) is substituted by the rule

$$(\mathrm{PR}-\mathrm{MNS})\quad \frac{\mathrm{J}_{\tau}(o \circ\!\!\rightarrow p),\quad \perp}{\mathrm{J}_{-1}(o \circ\!\!\rightarrow p)},$$

- The rule (PRτNS) is substituted by the rule

$$(\mathrm{PR}\tau\mathrm{MNS})\quad \frac{\mathrm{J}_{\tau}(o \circ\!\!\rightarrow p),\quad \neg\Pi^{+}(o,p)}{\mathrm{J}_{\tau}(o \circ\!\!\rightarrow p)}.$$

If we could not justify that o possesses a property p in a modified non-symmetric JSM theory then we assume that the question about whether o possesses the property p remains open (uncertainty is reserved). In this case the possibility to iterate rules remains.

Proposition 22.3.5. The class of modified non-symmetric JSM theories is well defined.

Proof. It is obvious. □

Chapter 23
Similarity Representation

23.1 Basic Concepts

Definition 23.1.1. Let σ be a basic JSM signature and $\mathfrak{J}$ be an L-structure of signature σ, where L is a PJ logic. Then $\mathfrak{J}$ is called a *basic JSM-structure* (*BJSM-structure*), if it is a model of the axioms (S1)–(S4) and (SA) of the basic JSM theory.

Notation 23.1.2. Let us introduce the symbol $\sqsubset$ and we will use the expression $s_1 \sqsubset s_2$ as an abbreviation for $s_1 \sqsubseteq s_2 \& s_1 \neq s_2$.

Notation 23.1.3. Let us define by induction the abbreviation of the form

$$\prod_{i=1}^{k} t_i \qquad (k \geqslant 1).$$

BASIS. $\prod_{i=1}^{1} t_i = t_1$.

INDUCTION STEP. $\prod_{i=1}^{k+1} t_i = \left(\prod_{i=1}^{k} t_i \right) \sqcap t_{k+1}$.

Here $t_1, \ldots, t_k, t_{k+1}$ are terms of sort S.

Notation 23.1.4. Let $\Phi(x)$ be an external formula of some PJ logic, which we consider as a formula of variable x . (In these cases instead of Φ , as usual, we write $\Phi(x)$, and instead of Φ_t^x we write $\Phi(t)$.) Let $t_1, \ldots, t_k, t_{k+1}$ be terms of the same sort as the variable x. We define by induction the abbreviation of the form

$$\bigwedge_{i=1}^{k} \Phi(t_i) \qquad (k \geqslant 1).$$

BASIS. $\bigwedge_{i=1}^{1} \Phi(t_i) = \Phi(t_1)$.

INDUCTION STEP. $\bigwedge_{i=1}^{k+1} \Phi(t_i) = \left(\bigwedge_{i=1}^{k} \Phi(t_i) \right) \wedge \Phi(t_{k+1})$.

O. Anshakov, T. Gergely, *Cognitive Reasoning*, Cognitive Technologies,
DOI 10.1007/978-3-540-68875-4_23, © Springer-Verlag Berlin Heidelberg 2010

The abbreviation

$$\bigvee_{i=1}^{k} \Phi(t_i) \qquad (k \geqslant 1)$$

is defined similarly.

Notation 23.1.5. Let L be a PJ logic and $\Phi(x)$ be an external L-formula of signature σ. Let $\mathfrak{J}$ be an L-structure of signature σ and a be an element of one of the universes of the L-structure $\mathfrak{J}$, which has the same sort as the variable x. In this case we will use the expression

$$\mathfrak{J} \models_L \Phi(a)$$

as an abbreviation of the fact that, for any valuation v in the L-structure $\mathfrak{J}$,

$$\Phi^{\mathfrak{J},\, v[x \to a]} = \mathbf{t}$$

holds. If it is clear from the context what L-structure we speak about then instead of $\mathfrak{J} \models_L \Phi(a)$ we simply write $\Phi(a)$.

Agreement 23.1.6. We will decrease the preciseness of the presentation in order to make it simpler. Thus we will omit the superscripts, which indicate that we consider not a (functional or predicate) symbol, but its interpretation in a model. I.e. we will write, e.g., $a \sqsubset b$ and $a \sqsubseteq b$ instead of $a \sqsubset^{\mathfrak{J}} b$ and $a \sqsubseteq^{\mathfrak{J}} b$, respectively. Earlier we introduced a similar convention for the propositional logic by using the same symbols for the operations of formula algebra and for logic algebra.

Definition 23.1.7. Let $\mathfrak{J}$ be a basic JSM-structure of signature σ and $\Phi(x)$ be an external formula of signature σ. We say that a structural fragment $s \in \mathrm{Dom}_{\mathfrak{J}}(S)$ is a *similarity of the objects* from $\mathfrak{J}$ constrained by Φ, if for some $k \geq 1$ there are objects $o_1, \ldots, o_k \in \mathrm{Dom}_{\mathfrak{J}}(O)$ such that:

- $s = \bigsqcap_{i=1}^{k} \mathrm{Struct}(o_i)$,
- $\mathfrak{J} \models_L \bigwedge_{i=1}^{k} \Phi(o_i)$.

The case that s is a similarity of objects from $\mathfrak{J}$ constrained by Φ will be denoted by $\mathbf{Sim}^{\mathfrak{J}}(s, \Phi)$. If it is clear from the context what BJSM-structure we speak about then we will simply write $\mathbf{Sim}(s, \Phi)$.

Definition 23.1.8. Let $\mathfrak{J}$ be a basic JSM-structure of signature σ, $\Phi(x)$ be an external formula of signature σ and $\mathrm{Sort}_\sigma(x) = \mathsf{O}$. We will say that a structural peculiarity $s \in \mathrm{Dom}_{\mathfrak{J}}(\mathsf{S})$ *satisfies the distinguishability condition w.r.t.* Φ in $\mathfrak{J}$, if:

(i) There is an object $o \in \mathrm{Dom}_{\mathfrak{J}}(\mathsf{O})$ such that $s < o$ and $\Phi(o)$;
(ii) There does not exist a structural fragment s' such that:

  (i) $s \sqsubset s'$,
  (ii) $s < o'$ and $\Phi(o')$ implies $s' < o'$ for any object $o' \in \mathrm{Dom}_{\mathfrak{J}}(O)$.

Intuitively, a fragment s satisfies this condition iff s can be *distinguished* from an s' greater than s by some object o such that s is a fragment of s but s' is not.

We will denote the fact that a structural peculiarity s satisfies the distinguishability condition w.r.t. Φ in $\mathfrak{J}$ by $\mathbf{Dis}^{\mathfrak{J}}(s,\Phi)$. If it will be clear from the context what L-structure we speak about then we will simply write $\mathbf{Dis}(s,\Phi)$.

Definition 23.1.9. Let L be a PJ logic, σ be a basic JSM signature, $\Phi(o)$ be an external L-formula of the signature σ and $\mathrm{Sort}_\sigma(o) = \mathsf{O}$. The following L-formula of signature σ denoted by $\mathrm{Dis}(s,\Phi)$ is called *the distinguishability formula for s and Φ*:

$$\exists o\,(s \lessdot o \wedge \Phi(o)) \wedge \neg\exists s'\left(s \sqsubset s' \wedge \forall o'\left(s \lessdot o' \wedge \Phi\left(o'\right) \to s' \lessdot o'\right)\right).$$

23.2 Distinguishability Condition and Similarity Representation

Proposition 23.2.1. Let L be a PJ logic, σ be a basic JSM signature, $\Phi(o)$ be an external L-formula of signature σ and $\mathrm{Sort}_\sigma(o) = \mathsf{O}$. Let $\mathfrak{J}$ be a BJSM-structure. Then

$$\mathbf{Dis}^{\mathfrak{J}}(s,\Phi) \qquad \text{iff} \qquad \mathfrak{J} \vDash_L \mathrm{Dis}(s,\Phi)$$

for any $s \in \mathrm{Dom}_{\mathfrak{J}}(\mathsf{S})$.

In this statement we state that the distinguishability condition (w.r.t. Φ) $\mathbf{Dis}(s,\Phi)$ is expressible by the use of the first-order formula $\mathrm{Dis}(s,\Phi)$.

Proof. It is obvious. □

Lemma 23.2.2. Let L be a PJ logic, σ be a basic JSM signature, $\Phi(o)$ be an external L-formula of signature σ and $\mathrm{Sort}_\sigma(o) = \mathsf{O}$. Let $\mathfrak{J}$ be a BJSM-structure of signature σ. Then

$$\mathbf{Sim}^{\mathfrak{J}}(s,\Phi) \qquad \text{implies} \qquad \mathbf{Dis}^{\mathfrak{J}}(s,\Phi)$$

for any $s \in \mathrm{Dom}_{\mathfrak{J}}(\mathsf{S})$.

Proof. Let $s \in \mathrm{Dom}_{\mathfrak{J}}(\mathsf{S})$ and $\mathbf{Sim}^{\mathfrak{J}}$ be true. Then by Definition 23.1.7, for some $k \geq 1$ there exist objects $o_1, \ldots, o_k \in \mathrm{Dom}_{\mathfrak{J}}(\mathsf{O})$ such that

$$s = \bigsqcap_{i=1}^{k} \mathrm{Struct}(o_i), \tag{23.1}$$

$$\mathfrak{J} \vDash_L \bigwedge_{i=1}^{k} \Phi(o_i). \tag{23.2}$$

Obviously, (23.1) implies

$$s \sqsubseteq \mathrm{Struct}(o_i) \tag{23.3}$$

for any i, $(1 \leq i \leq k)$ where $k \geq 1$. By the axiom (SA), (23.3) implies

$$s \lessdot o_i \tag{23.4}$$

for each i $(1 \le i \le k)$, $k \ge 1$. □

We show that $s \in \mathrm{Dom}_{\mathfrak{J}}(\mathsf{S})$ satisfies the distinguishability condition. Assume the converse. Then at least one of the conditions (i), (ii) of Definition 23.1.8 does not hold. Let us consider the possible cases.

CASE 1. Let condition (i) not hold. Then there is no object $o \in \mathrm{Dom}_{\mathfrak{J}}(\mathsf{O})$ such that $s \lessdot o$ and $\Phi(o)$ would be true simultaneously. This contradicts the fact that (23.4) holds for each i $(1 \le i \le k)$, $k \ge 1$ and the condition (23.2).

CASE 2. Let condition (ii) not hold. Then there exists an $s' \in \mathrm{Dom}_{\mathfrak{J}}(\mathsf{S})$ such that:

(a) $s \sqsubset s'$,
(b) $s \lessdot o'$ and $\Phi(o')$ imply $s' \lessdot o'$ for any object $o' \in \mathrm{Dom}_{\mathfrak{J}}(\mathsf{O})$.

Then (see (23.2), (23.4) and (b)),

$$s' \lessdot o_i$$

for each i $(1 \le i \le k)$. Therefore

$$s' \sqsubseteq \bigsqcap_{i=1}^{k} \mathrm{Struct}(o_i) = s,$$

which contradicts (a).

Theorem 23.2.3 (D. Vinogradov). Let L be a PJ logic, σ be a basic JSM signature, $\Phi(o)$ be an external L-formula of signature σ and $\mathrm{Sort}_\sigma(o) = \mathsf{O}$. Let $\mathfrak{J}$ be a BJSM-structure of signature σ. Then

$$\mathbf{Sim}^{\mathfrak{J}}(s, \Phi) \quad \text{iff} \quad \mathbf{Dis}^{\mathfrak{J}}(s, \Phi)$$

for any $s \in \mathrm{Dom}_{\mathfrak{J}}(\mathsf{S})$.

Proof. ($\Rightarrow$) By the use of Lemma 23.2.2.

($\Leftarrow$) Let $s \in \mathrm{Dom}_{\mathfrak{J}}(\mathsf{S})$ satisfy the distinguishability condition. Then by (i) from 23.1.8 there exists an $s \in \mathrm{Dom}_{\mathfrak{J}}(\mathsf{O})$ such that $s \lessdot o$ and $\Phi(o)$. Therefore

$$\{o \in \mathrm{Dom}_{\mathfrak{J}}(\mathsf{O}) \,|\, s \lessdot o \wedge \Phi(o)\} \neq \varnothing.$$

Let

$$\{o \in \mathrm{Dom}_{\mathfrak{J}}(\mathsf{O}) \,|\, s \lessdot o \wedge \Phi(o)\} = \{o_1, \ldots, o_k\}. \tag{23.5}$$

where $k \ge 1$. (Since $\mathrm{Dom}_{\mathfrak{J}}(\mathsf{O})$ is a finite set, the number of objects from $\mathrm{Dom}_{\mathfrak{J}}(\mathsf{O})$, which contain the structural peculiarity s, and for which Φ holds, is finite.) Then by the axiom (SA), $s \sqsubseteq \mathrm{Struct}(o_i)$ for any i $(1 \le i \le k)$, i.e. s is the lower bound of the set $\{o_1, \ldots, o_k\}$ w.r.t. the partial order $\sqsubseteq$.

But $\bigsqcap_{i=1}^{k} \mathrm{Struct}(o_i)$ is the greatest lower bound of the set $\{o_1, \ldots, o_k\}$, w.r.t. the partial order $\sqsubseteq$. Therefore

$$s \sqsubseteq \bigsqcap_{i=1}^{k} \text{Struct}(o_i). \tag{23.6}$$

Let us show that

$$s = \bigsqcap_{i=1}^{k} \text{Struct}(o_i). \tag{23.7}$$

Suppose that the equality (23.7) does not hold. Let

$$s' = \bigsqcap_{i=1}^{k} \text{Struct}(o_i). \tag{23.8}$$

We show that in this case the following conditions hold:

(a) $s \sqsubset s'$,
(b) $s \lessdot o'$ and $\Phi(o')$ imply $s' \lessdot o'$ for any object $o' \in \text{Dom}_{\mathfrak{J}}(\mathsf{O})$.

Let us prove (a). Because of (23.6) and (23.8), $s \sqsubseteq s'$ holds. (23.8) and our assumption of the falsity of (23.7) imply $s \neq s'$. Therefore, $s \sqsubset s'$.

Let us prove (b). Let $s \lessdot o'$. Then due to (23.5), $o' = o_i$ for some i $(1 \leq i \leq k)$, and because of (23.8), $s' \sqsubseteq \text{Struct}(o')$, then by the axiom (SA) $s' \lessdot o'$ holds.

Thus, assuming that the equality (23.7) is false we proved that there exists an $s' \in \text{Dom}_{\mathfrak{J}}(\mathsf{S})$ for which the conditions (a) and (b) hold. But this contradicts the assumption that s satisfies the distinguishability condition (see Definition 23.1.8 (ii)). Therefore our supposition is false. The equality (23.7) holds, thus s is a similarity for the objects of that BJSM-structure $\mathfrak{J}$, constrained by Φ (see Definition 23.1.7, i.e. $\mathbf{Sim}^{\mathfrak{J}}(s, \Phi)$ holds. □

Corollary 23.2.4. Let L be a PJ logic, σ be a basic JSM signature, $\Phi(o)$ be an external L-formula of signature σ and $\text{Sort}_\sigma(o) = \mathsf{O}$. Let $\mathfrak{J}$ be a BJSM-structure of signature σ. Then

$$\mathbf{Sim}^{\mathfrak{J}}(s, \Phi) \quad \text{iff} \quad \mathfrak{J} \vDash_L \text{Dis}(s, \Phi)$$

for any $s \in \text{Dom}_{\mathfrak{J}}(\mathsf{S})$.

This statement simply says that in an arbitrary finite BJSM-structure the condition $\mathbf{Sim}^{\mathfrak{J}}(s, \Phi)$ can be expressed by the use of a first-order formula $\text{Dis}(s, \Phi)$.

Proof. By using 23.2.1 and 23.2.3. □

Definition 23.2.5. Let L be a PJ logic, σ be a basic, generalised or non-symmetric JSM signature, o be a variable of the sort O and p be a variable of the sort P. Then the formula $\text{Dis}(s, \text{J}_{+1}(o \multimap p))$ is called the *positive distinguishability formula for s and p*, and it is denoted by $\text{Dis}^+(s, p)$.

The formula $\text{Dis}(s, \text{J}_{-1}(o \multimap p))$ is called the *negative distinguishability formula for s and p*, and it is denoted by $\text{Dis}^-(s, p)$.

Axiom Schemata 23.2.6. Obviously, $\text{Dis}^+(s, p)$ is true iff s is a similarity (intersection) of a non-empty set of objects possessing the property p. Similarly, $\text{Dis}^-(s, p)$

is true iff s is a similarity (intersection) of a non-empty set of objects which do not possess the property p. However, the validity of the formulas $\mathrm{Dis}^{+}(s,p)$ and $\mathrm{Dis}^{-}(s,p)$ does not guarantee that s is a similarity (intersection) of *at least two objects* which do possess or do not possess the property p, as required in the predicates M^{+} and M^{-}. In order to guarantee this we have to add the Mill's condition to the distinguishability formulas (see Definition 21.4.1). Thus we can propose the following defining axioms for the predicates M^{+} and M^{-}:

$$(\mathrm{DM}^{+}\mathrm{D}) \quad \mathrm{M}^{+}(s,p) \leftrightarrow s \neq \Lambda \wedge \mathrm{Mill}^{+}(s,p) \wedge \mathrm{Dis}^{+}(s,p),$$

$$(\mathrm{DM}^{-}\mathrm{D}) \quad \mathrm{M}^{-}(s,p) \leftrightarrow s \neq \Lambda \wedge \mathrm{Mill}^{-}(s,p) \wedge \mathrm{Dis}^{-}(s,p).$$

The letter "D" means in this notation that the distinguishability formulas are used in the description.

By the use of the distinguishability formulas we can reformulate the defining axioms for the predicates H^{+} and H^{-}, by which the hypotheses on the inhibitors of possible causes are formulated. We recall the intuitive meaning of these predicates. $\mathrm{H}^{+}(t,s,p)$ means that t is a non-empty and differing from s similarity of at least two objects that contain s and do not possess property p. Then the defining axiom for H^{+} which uses the distinguishability formula can be described as follows:

$$(\mathrm{DH}^{+}\mathrm{D}) \quad \mathrm{H}^{+}(t,s,p) \leftrightarrow t \neq \Lambda \wedge t \neq s \wedge \mathrm{GMill}^{-}(t,s,p) \wedge \mathrm{Dis}^{-}(t,p).$$

$\mathrm{H}^{-}(t,s,p)$ means, that t is a non-empty and differing from s similarity of at least two objects that contain s and possess the property p. Then we can propose the following defining axiom for H^{-}, which uses the distinguishability formula:

$$(\mathrm{DH}^{+}\mathrm{D}) \quad \mathrm{H}^{+}(t,s,p) \leftrightarrow t \neq \Lambda \wedge t \neq s \wedge \mathrm{GMill}^{+}(t,s,p) \wedge \mathrm{Dis}^{+}(t,p).$$

Chapter 24
JSM Theories for Complex Structures

24.1 JSM Theories with Set Sorts

Suppose that we have a situation where the properties themselves are sets, i.e. the elements of the sort P are sets. If we attempt to represent this situation in some versions of the JSM theories then this yields significant changes and complications in the selected theories. The advantage may be a potential decrease of the computational complexity of the corresponding algorithms caused by the interdependency of the properties (sets of properties).

Suppose that the sort P is a set sort. Intuitively this means that the elements of the sort P are considered not as individual properties but as property sets. Then we obtain the following modifications in the signatures and in the sets of axioms for JSM theories.

Definition 24.1.1. A signature σ is called a *basic JSM signature with property sets* if it meets the following conditions:

(i) σ is a basic JSM signature (see Definition 21.1.1), in which the sort P is considered *as the sort of target property sets*;

(ii) Fu_σ contains the binary functional symbol $\cup$, which denotes the so-called *union operation*; $\mathrm{Ar}_\sigma(\cup) = \langle \mathsf{P}, \mathsf{P} \rangle$, $\mathrm{Sort}_\sigma(\cup) = \mathsf{P}$, i.e. the operation $\cup$ is defined on the set of pairs of property sets, the result of this operation also being a property set;

(iii) Fu_σ contains the 0-ary functional symbol $\curlywedge$, which denotes the *empty set* of properties; $\mathrm{Ar}_\sigma(\curlywedge) = \langle\rangle$, $\mathrm{Sort}_\sigma(\curlywedge) = \mathsf{P}$;

(iv) Pr_σ contains the external binary predicate symbol $\subseteqq$ such that $\mathrm{Ar}_\sigma(\subseteqq) = \langle \mathsf{P}, \mathsf{P} \rangle$, $\subseteqq$ denotes the predicate "...is a subset...";

(v) Pr_σ contains the external unary predicate symbol Atom such that $\mathrm{Ar}_\sigma(\mathrm{Atom}) =$ [illegible] denotes the predicate "...is a one-element set".

Definition 24.1.2. Let σ be a basic JSM signature with property sets. A modification theory T of signature σ is called a *basic JSM theory with property sets*, if it meets the following conditions:

- The theory T is a basic JSM theory (in the sense of Definition 21.1.2);
- The theory T contains the following axioms:

(P1) $p \cup p = p$

(P2) $p \cup p' = p' \cup p$

(P3) $(p \cup p') \cup p'' = p \cup (p \cup p'')$

(P4) $p \cup \curlywedge = p$

(P5) $p \subseteqq p' \leftrightarrow p \cup p' = p'$

(P6) $\mathrm{Atom}(p) \leftrightarrow (p \neq \curlywedge \& \forall p' (p' \subseteqq p \rightarrow p' = p \vee p = \curlywedge))$

(P7) $p \neq \curlywedge \rightarrow \exists p' (\mathrm{Atom}(p') \& p' \subseteqq p)$

(P8) $\mathrm{Atom}(p) \rightarrow (p \subseteqq p' \cup p'' \rightarrow p \subseteqq p' \vee p \subseteqq p'')$

(P9) $\forall p (\mathrm{Atom}(p) \rightarrow (p \subseteqq p' \leftrightarrow p \subseteqq p'')) \rightarrow p' = p''$

where p, p' and p'' are variables of sort P.

The axioms (P1)–(P3) state that the set of property sets forms a semilattice w.r.t. the operation $\cup$.

The axiom (P5) is a defining axiom for the predicate $\subseteqq$. It is easy to show that this predicate is a partial order relation for which $p \cup p' = \sup\{p, p'\}$.

The axiom (P4) implies that the empty set $\curlywedge$ is the least element of the set of property sets with respect to the partial order $\subseteqq$.

The axiom (P6) is a defining axiom for the predicate Atom.

The axioms (P7)–(P9) describe the properties of atoms (one-element property sets). (P7) states that any non-empty set of properties contains atoms. (P9) is the axiom of extensionality; it states that if a set of properties contains those and only those atoms that another set contains then these sets coincide. (P8) describes the relationship between the predicate Atom and the operation $\cup$; this axiom states that if an atom is contained in the union of property sets then it is contained in at least one of these sets.

The axioms (P1)–(P9) seem obvious if we consider that the entities of the sort P are sets. However, this is not necessarily the case. They *should behave as sets* in those cases which are necessary for the construction of a JSM theory. The axioms (P1)–(P9) describe the necessary minimum of requirements for the entities of sort P in order that their behaviour should be similar to that of sets.

We emphasise that a basic JSM theory with property sets, as follows from its definition, is a basic JSM theory and conse

24.2 Simple JSM Theories with Property Sets

Definition 24.2.1. A basic JSM signature σ with property sets will be called a *simple JSM signature with property sets* if the following conditions hold:

(i) Pr_σ contains the internal predicate symbols $\circ\!\!\rightarrow_+$, $\circ\!\!\rightarrow_-$, $\bullet\!\!\rightarrow_+$ and $\bullet\!\!\rightarrow_-$ such that:

$$\mathrm{Ar}_\sigma(\circ\!\!\rightarrow_+) = \mathrm{Ar}_\sigma(\circ\!\!\rightarrow_-) = \langle \mathsf{O}, \mathsf{P} \rangle,$$

$$\mathrm{Ar}_\sigma(\bullet\!\!\rightarrow_+) = \mathrm{Ar}_\sigma(\bullet\!\!\rightarrow_-) = \langle \mathsf{S}, \mathsf{P} \rangle,$$

where $\circ\!\!\rightarrow_+$ denotes the predicate "...possesses the property set...", and $\circ\!\!\rightarrow_-$ denotes the predicate "...does not possess the property set...", $\bullet\!\!\rightarrow_+$ denotes the predicate "... is a possible cause of the presence of the property set ...", $\bullet\!\!\rightarrow_-$ denotes the predicate "... is a possible cause of the absence of the property set ...";

(ii) Pr_σ contains the binary external predicate symbols M^+, M^-, Π^+ and Π^-, which have the same arity and the same intuitive meaning as the corresponding symbols have in the simple JSM signature (without property sets). But now we have to speak about candidates for being a possible cause of presence/absence of *property sets* and about the candidates for the presence/absence of *property sets*.

Now we discuss the intuitive meaning of the operations and predicates of a simple JSM signature with property sets. Suppose that s is a structural fragment, o is an object and p is the set of the target properties. Then:

(i) $$o \circ\!\!\rightarrow_+ p = \begin{cases} +1 \text{ if we know that } o \text{ possesses } p, \\ \tau \text{ if we do not know whether } o \text{ possesses } p, \end{cases}$$

(ii) $$o \circ\!\!\rightarrow_- p = \begin{cases} +1 \text{ if we know that } o \text{ does not possess } p, \\ \tau \text{ if we do not know whether } o \text{ possesses } p, \end{cases}$$

(iii) The intuitive meaning of the predicates $\bullet\!\!\rightarrow_+$ and $\bullet\!\!\rightarrow_-$ can be described in just the same way as was done after the definition of generalised JSM signature (see Definition 22.1.1),

(iv) The intuitive meaning of the predicates $\mathrm{M}^+, \mathrm{M}^-, \Pi^+$ and Π^- can be described in just the same way as was done in the definition of simple JSM signature (see Definition 21.2.1).

Notation 24.2.2. We now introduce abbreviations in order to simplify the formulation of the rules for the simple JSM theories with property sets, which will be defined below. Interpretation of the entities of the sort P as property sets yields to rather complicated modification rules. In this case the predicates M^+, M^-, Π^+ and Π^- are included in the kernel of the rules. The major premise of the modification rules for the set sorts can be frightening. First we introduce notations for the extended predicates. Let:

(i) $\mathrm{XM}^+(s,p)$ be the abbreviation for the formula $\mathrm{J}_{+1}(s \bullet\!\!\rightarrow_+ p) \vee \mathrm{M}^+(s,p)$,
(ii) $\mathrm{XM}^-(s,p)$ be the abbreviation for the formula $\mathrm{J}_{+1}(s \bullet\!\!\rightarrow_- p) \vee \mathrm{M}^-(s,p)$,
(iii) $\mathrm{X\Pi}^+(o,p)$ be the abbreviation for the formula $\mathrm{J}_{+1}(s \bullet\!\!\rightarrow_+ p) \vee \Pi^+(s,p)$,
(iv) $\mathrm{X\Pi}^-(o,p)$ be the abbreviation for the formula $\mathrm{J}_{+1}(s \bullet\!\!\rightarrow_- p) \vee \Pi^-(s,p)$;
XM^+ is called the *extended predicate of* M^+. We call the other introduced abbreviations analogously.
Now we introduce notations for the major premises of some rules. We will use the abbreviation of the form CP^+ indexCP^+ and CP^- which will be called *causal "+"-premises and causal "−"-premises*, respectively. Similarly the formulas PP^+ and PP^- will be called *predictive "+"-premises and predictive "−"-premises*, respectively. Now we give the abbreviations:
(v) $\mathrm{CP}^+(s,p)$ is the abbreviation of the formula

$$\exists p' \left(p \subseteqq p' \wedge \mathrm{XM}^+(s,p')\right) \vee \exists p' \exists p'' \left(p = p' \cup p'' \wedge \mathrm{XM}^+(s,p') \wedge \mathrm{XM}^+(s,p'')\right).$$

(vi) $\mathrm{CP}^-(s,p)$ is the abbreviation of the formula $\exists p' \left(p' \subseteqq p \wedge \mathrm{XM}^-(s,p')\right)$.
(vii) $\mathrm{PP}^+(o,p)$ is the abbreviation of the formula

$$\exists p' \left(p \subseteqq p' \wedge \mathrm{X\Pi}^+(o,p')\right) \vee \exists p' \exists p'' \left(p = p' \cup p'' \wedge \mathrm{X\Pi}^+(o,p') \wedge \mathrm{X\Pi}^+(o,p'')\right).$$

(viii) $\mathrm{PP}^-(s,p)$ is the abbreviation of the formula $\exists p' \left(p' \subseteqq p \wedge \mathrm{X\Pi}^-(s,p')\right)$.

Now we discuss the intuitive meaning of the introduced abbreviations.

(i′) $\mathrm{XM}^+(s,p')$ means that s may be either a (already justified) possible cause of presence of p, or a candidate possible cause of presence of p.
(ii′) $\mathrm{XM}^-(s,p')$ means that s may be either a (already justified) possible cause of absence of p, or a candidate possible cause of absence of p.
(iii′) $\mathrm{X\Pi}^+(s,p')$ means that o certainly may possess a property set p and may also be a candidate for the presence of the property set p.
(iv′) $\mathrm{X\Pi}^-(s,p')$ means that o may certainly not possess a property set p and may also be a candidate for the absence of the property set p.
(v′) $\mathrm{CP}^+(s,p)$ means that at least one of the following two conditions holds:

- s is a possible cause of presence, or a candidate possible cause of presence of a property set p' larger than, or equal to, p.
- p can be represented as the union of two property sets p' and p'' such that s is a possible cause of presence, or a candidate possible cause of presence, of each of these property sets.

(vi′) $\mathrm{CP}^-(s,p)$ means that s is a possible cause of absence, or a candidate possible cause of absence of the property set p' larger than, or equal to, p.
(vii′) $\mathrm{PP}^+(s,p)$ means that at least one of the following two conditions holds:

- o possesses or o is a candidate for possessing the property set p' larger than, or equal to, p,
- p can be represented as the union of two property sets p' and p'' such that o possesses or o is a candidate for possessing each of these property sets.

(viii′) $\mathrm{PP}^{+}(s,p)$ means that o does not possess or o is a candidate for not possessing a set of properties p' smaller than, or equal to, p.

We suppose that o possesses a property set p iff o possesses each property of the set p. Therefore, o does not possess a property set p iff o does not possess at least one of the properties of the set p. We understand similarly the connection between the property sets and their possible causes.

By the use of these assumptions it is easy to justify the following statements:

- If o possesses a larger property set then it possesses a smaller one too,
- If o does not possess a smaller property set, then it does not possess a larger set either,
- If o possesses each of the property sets p' and p'', then it possesses the union of these two property sets.

Similar statements can also be justified for the possible causes of presence/absence of properties.

Similar statements allow us to say that the formulation of the major premises of the causal and predictive rules completely corresponds to their intuitive meaning, i.e. this formulation is correct from the point of view of the informal interpretation.

Definition 24.2.3. Let σ be a simple JSM signature with property sets. A basic JSM theory with property sets of signature σ is called a *simple JSM theory with property sets*, if the traversal sequence for the set of internal predicates of this theory is $\langle \bullet\!\!\rightarrow_{+}, \bullet\!\!\rightarrow_{-}, \circ\!\!\rightarrow_{+}, \circ\!\!\rightarrow_{-} \rangle$ and this theory contains the following axioms and modification rules.

Axioms:

(TV1) $\neg \mathrm{J}_{-1}(s \bullet\!\!\rightarrow_{+} p) \wedge \neg \mathrm{J}_{0}(s \bullet\!\!\rightarrow_{+} p)$,

(TV2) $\neg \mathrm{J}_{-1}(s \bullet\!\!\rightarrow_{-} p) \wedge \neg \mathrm{J}_{0}(s \bullet\!\!\rightarrow_{-} p)$,

(TV3) $\neg \mathrm{J}_{-1}(s \circ\!\!\rightarrow_{+} p) \wedge \neg \mathrm{J}_{0}(s \circ\!\!\rightarrow_{+} p)$,

(TV4) $\neg \mathrm{J}_{-1}(s \circ\!\!\rightarrow_{-} p) \wedge \neg \mathrm{J}_{0}(s \circ\!\!\rightarrow_{-} p)$,

where o is a variable of sort O, s is a variable of sort S and p is a variable of sort P.

Modification rules

(CR + +S) $$\frac{\mathrm{J}_{\tau}(s \bullet\!\!\rightarrow_{+} p), \quad \mathrm{CP}^{+}(s,p)}{\mathrm{J}_{+1}(s \bullet\!\!\rightarrow_{+} p)},$$

(CR + −S) $$\frac{\mathrm{J}_{\tau}(s \bullet\!\!\rightarrow_{+} p), \quad \bot}{\mathrm{J}_{-1}(s \bullet\!\!\rightarrow_{+} p)},$$

(CR + 0S) $$\frac{\mathrm{J}_{\tau}(s \bullet\!\!\rightarrow_{+} p), \quad \bot}{\mathrm{J}_{0}(s \bullet\!\!\rightarrow_{+} p)},$$

$$(\mathrm{CR}+\tau\mathrm{S})\quad \frac{\mathrm{J}_{\tau}(s \bullet\!\!\rightarrow_{+} p),\quad \forall p'''\,(\mathrm{J}_{\tau}(s \bullet\!\!\rightarrow_{+} p''') \rightarrow \neg\mathrm{CP}^{+}(s,p'''))}{\mathrm{J}_{\tau}(s \bullet\!\!\rightarrow_{+} p)},$$

where s is a variable of sort S and p' and p''' are variables of sort P.

The rules for processing $\bullet\!\!\rightarrow_{+}$ are completed and now we turn to continue with the rules for processing $\bullet\!\!\rightarrow_{-}$.

$$(\mathrm{CR}-+\mathrm{S})\quad \frac{\mathrm{J}_{\tau}(s \bullet\!\!\rightarrow_{-} p),\quad \mathrm{CP}^{-}(s,p)}{\mathrm{J}_{+1}(s \bullet\!\!\rightarrow_{-} p)},$$

$$(\mathrm{CR}--\mathrm{S})\quad \frac{\mathrm{J}_{\tau}(s \bullet\!\!\rightarrow_{-} p),\quad \bot}{\mathrm{J}_{-1}(s \bullet\!\!\rightarrow_{-} p)},$$

$$(\mathrm{CR}-0\mathrm{S})\quad \frac{\mathrm{J}_{\tau}(s \bullet\!\!\rightarrow_{-} p),\quad \bot}{\mathrm{J}_{0}(s \bullet\!\!\rightarrow_{-} p)},$$

$$(\mathrm{CR}-\tau\mathrm{S})\quad \frac{\mathrm{J}_{\tau}(s \bullet\!\!\rightarrow_{-} p),\quad \forall p'''\,(\mathrm{J}_{\tau}(s \bullet\!\!\rightarrow_{-} p''') \rightarrow \neg\mathrm{CP}^{-}(s,p'''))}{\mathrm{J}_{\tau}(s \bullet\!\!\rightarrow_{-} p)},$$

where s is a variable of sort S and p and p''' are variables of sort P. Note that by the use of these rules the predicates $\bullet\!\!\rightarrow_{+}$ and $\bullet\!\!\rightarrow_{-}$ can obtain only the values +1 and τ. The rules of the group CR, as usual, will be called *causal* (or cause–effect) rules .

$$(\mathrm{PR}++\mathrm{S})\quad \frac{\mathrm{J}_{\tau}(s \circ\!\!\rightarrow_{+} p),\quad \mathrm{PP}^{+}(s,p)}{\mathrm{J}_{+1}(s \circ\!\!\rightarrow_{+} p)},$$

$$(\mathrm{PR}+-\mathrm{S})\quad \frac{\mathrm{J}_{\tau}(s \circ\!\!\rightarrow_{+} p),\quad \bot}{\mathrm{J}_{-1}(s \circ\!\!\rightarrow_{+} p)},$$

$$(\mathrm{PR}+0\mathrm{S})\quad \frac{\mathrm{J}_{\tau}(s \circ\!\!\rightarrow_{+} p),\quad \bot}{\mathrm{J}_{0}(s \circ\!\!\rightarrow_{+} p)},$$

$$(\mathrm{PR}+\tau\mathrm{S})\quad \frac{\mathrm{J}_{\tau}(s \circ\!\!\rightarrow_{+} p),\quad \forall p'''\,(\mathrm{J}_{\tau}(s \circ\!\!\rightarrow_{+} p''') \rightarrow \neg\mathrm{PP}^{+}(s,p'''))}{\mathrm{J}_{\tau}(s \circ\!\!\rightarrow_{+} p)},$$

where o is a variable of sort O and p and p''' are variables of sort P.

The rules for processing $\circ\!\!\rightarrow_{+}$ are completed and now we turn to continue with the rules for processing $\circ\!\!\rightarrow_{-}$.

$$(\mathrm{PR}-+\mathrm{S})\quad \frac{\mathrm{J}_{\tau}(s \circ\!\!\rightarrow_{-} p),\quad \mathrm{PP}^{-}(s,p)}{\mathrm{J}_{+1}(s \circ\!\!\rightarrow_{-} p)},$$

$$(\mathrm{PR}--\mathrm{S})\quad \frac{\mathrm{J}_{\tau}(s \circ\!\!\rightarrow_{-} p),\quad \bot}{\mathrm{J}_{-1}(s \circ\!\!\rightarrow_{-} p)},$$

$$(\mathrm{PR}-0\mathrm{S})\quad \frac{\mathrm{J}_\tau(s \circ\!\!\rightarrow_- p), \quad \bot}{\mathrm{J}_0(s \circ\!\!\rightarrow_- p)},$$

$$(\mathrm{PR}-\tau\mathrm{S})\quad \frac{\mathrm{J}_\tau(s \circ\!\!\rightarrow_- p), \quad \forall p'''\left(\mathrm{J}_\tau(s \circ\!\!\rightarrow_- p''') \rightarrow \neg \mathrm{PP}^-(s, p''')\right)}{\mathrm{J}_\tau(s \circ\!\!\rightarrow_- p)},$$

where o is a variable of sort O and p and p''' are variables of sort P.

Note that by the use of these rules the predicates $\circ\!\!\rightarrow_+$ and $\circ\!\!\rightarrow_-$ can obtain only the values +1 and τ. The rules of the group PR, as usual, will be called *predictive* rules.

Proposition 24.2.4. (i) A simple JSM theory with property sets is well defined. (ii) The modification rule system of a simple JSM theory with property sets is regular.

Proof. (i) By direct checking of the fact that the system of modification rules satisfies the conditions of Definition 14.2.13 (see also 10.2.6).

(ii) By direct checking of the fact that the system of modification rules satisfies the conditions of Definition 18.1.5 (see also 18.1.1 and 18.1.4). □

It can be shown that the systems of causal and predictive "+"-rules (rules for the predicates $\bullet\!\!\rightarrow_+$ and $\circ\!\!\rightarrow_+$) are positively $P+1$-connected and the systems of causal and predictive "-"-rules (rules for the predicates $\bullet\!\!\rightarrow_-$ and $\circ\!\!\rightarrow_-$) are negatively $P-1$-connected to the appropriate rule-lifting systems (see 18.1.1). This implies by Definiton 18.1.4 that the predicates $\bullet\!\!\rightarrow_+$ and $\circ\!\!\rightarrow_+$ are $+P+1$-multiple and the predicates $\bullet\!\!\rightarrow_-$ and $\circ\!\!\rightarrow_-$ are $-P+1$-multiple.

Hence we obtain by Definition 18.1.5 that the proposed above system of modification rules for the simple JSM theory with property sets is regular.

24.3 Defining Axioms for the Simple JSM Theories with Property Sets

In this section we formulate the defining axioms for the predicates M^+, M^-, Π^+ and Π^-, i.e. we express them by the predicates $\circ\!\!\rightarrow_+$, $\circ\!\!\rightarrow_-$, $\bullet\!\!\rightarrow_+$, $\bullet\!\!\rightarrow_-$ and $<$.

These predicates form the kernels of the major premises of the modification rules in the sense of Definition 18.1.2. We have to propose such axioms which permit the kernels of the modification rules to satisfy the universal (positive) or the existential (negative) atomic conditions from Definition 18.1.8. This can be achieved, for example, by the use of the following defining axioms:

$$(\mathrm{DM}^+\,\mathrm{Set})\quad \mathrm{M}^+(s,p) \leftrightarrow s \neq \Lambda \wedge \forall p'\left(\mathrm{Atom}(p') \wedge p' \subseteqq p \rightarrow \mathrm{Mill}^+(s,p')\right),$$

$$(\mathrm{DM}^-\,\mathrm{Set})\quad \mathrm{M}^-(s,p) \leftrightarrow s \neq \Lambda \wedge \forall p'\left(\mathrm{Atom}(p') \wedge p' \subseteqq p \rightarrow \mathrm{Mill}^-(s,p')\right),$$

$$(\mathrm{D}\Pi^+\,\mathrm{Set})\quad \Pi^+(o,p) \leftrightarrow \forall p'\,(\mathrm{Atom}(p') \wedge p' \subseteqq p \rightarrow \exists s\,(s < o \wedge \mathrm{J}_{+1}(s \bullet\!\!\rightarrow_+ p'))),$$

(DΠ^- Set) $\Pi^-(o,p) \leftrightarrow \forall p' (\mathrm{Atom}(p') \wedge p' \subseteqq p \rightarrow \exists s (s < o \wedge \mathrm{J}_{+1}(s \bullet\!\!\rightarrow_- p')))$.

These axioms assign the following meaning to the predicates M^+, M^-, Π^+ and Π^-:

(i) $\mathrm{M}^+(s,p)$ means that s is non-empty and for each property from the property set p, s is contained in at least two objects possessing this property.
(ii) $\mathrm{M}^-(s,p)$ means that s is non-empty and there exists a property from the property set p such that s is contained at least in two objects possessing this property.
(iii) $\Pi^+(o,p)$ means that for each property from the property set p the object o contains a possible cause of the presence of this property.
(iv) $\Pi^-(o,p)$ means that there exists a property from the set of properties p such that the object o contains a possible cause of the absence of this property.

Proposition 24.3.1. (i) The formulas $\mathrm{M}^+(s,p)$ and $\Pi^+(o,p)$ satisfy the universal (positive) atomic condition for the variable p w.r.t. the sort P. (ii) The formulas $\mathrm{M}^-(s,p)$ and $\Pi^-(o,p)$ satisfy the existential (negative) atomic condition for the variable p w.r.t. the sort P.

Proof. By direct checking of the correspondence to the conditions of Definition 18.1.8. □

Corollary 24.3.2. The modification rule system of a simple JSM theory with property sets is perfect.

Proof. The statement directly follows from Statements 24.2.4 (ii) and 24.3.1 (see Definition 19.1.1) □

Proposition 24.3.3. A (0, 0)-consistent simple JSM theory with property sets is regular.

Proof. A simple JSM theory with property sets similarly to any modification theory should include some state description. The definition of a simple JSM theory with property sets does not impose any explicit constraints on this description. However, if the considered theory is (0, 0)-consistent, then the syntactical structure that corresponds to the state description is a model of the axioms of this theory including the axioms (P1)–(P9) and (TV1)–(TV4).

Let σ be the signature of a simple JSM theory with property sets and R be the set of the modification rules of this theory. By Proposition 24.2.4 (ii), R is regular. Then the set of axioms (P1)–(P9) is MULT_σ, the set of axioms (TV1)–(TV4) is $\mathrm{TVC}(R)$, and Proposition 24.3.1 proved above corresponds to the set of formulas $\mathrm{ACC}(R)$.

Then the union of the sets MULT_σ, $\mathrm{TVC}(R)$ and $\mathrm{ACC}(R)$ is valid in the syntactical structure that corresponds to the state description. By Definition 18.1.9 this means that in this syntactical structure the set of formulas $\mathrm{SSC}(R)$ (*sufficient set constraint* for R) is true.

But in this case the set $\mathrm{SSC}(R)$ will be derivable in the calculus $\mathbf{St}^1$ from the state description of the considered theory. This means that the condition (ii) of the definition of a regular modification calculus (see 18.2.12) is satisfied for the modification calculus of the considered simple JSM theory with property sets. The condition (i) of Definition 18.2.12 holds by Proposition 24.2.4 (ii).

Thus the modification calculus of the considered theory is regular, i.e. this theory itself is also regular. □

Corollary 24.3.4. A (0, 0)-consistent simple JSM theory with property sets is perfect.

Proof. The statement directly follows from Proposition 24.3.3 and 24.3.2 (see. Definition 19.1.2) □

The fact that (0, 0)-consistent simple JSM theories with property sets are perfect permits the definition of a natural semantics for these theories, where sequences of **St**-structures are the models.

Generalised JSM theories with property sets can be defined analogously to simple ones.

The generalised JSM signature with property sets is an expansion of the simple JSM signature with property sets by adding new predicate symbols: Inh^+, Inh^-, H^+ and H^-. The arity, the sort and the intuitive meaning of these symbols are the same as in the case of the generalised JSM signature (without property sets).

A generalised JSM theory with property sets contains all the axioms and modification rules of the simple JSM theory with property sets.

The modification rules for Inh^+ and Inh^- and the defining axioms for H^+ and H^- can be defined similarly to the modification rules for $\bullet\!\!\rightarrow_+$ and $\bullet\!\!\rightarrow_-$ and to the defining axioms for M^+ and M^-, respectively.

The analogues of Propositions 24.2.4, 24.3.1, 24.3.3 and of Corollaries 24.3.2 and 24.3.4 hold for the generalised JSM theories with property sets.

Part VI
Looking Back and Ahead

Chapter 25
Introductory Overview

Understanding cognition and particularly cognitive reasoning and developing artificial cognizing agents are challenging and long-term research problems. In this book we wish to take a step towards a better understanding of cognitive reasoning processes by developing a scientifically well-founded general approach that provides methods and tools for modelling, designing and generating information processing responsible for the cognitive processes of artificial cognizing agents. Our aim is that the required general approach should support the study at three levels of abstraction: conceptual, formal, and realisational.

While developing the general approach at the conceptual level of abstraction the following main postulates were taken into account:

(i) Cognitive processes are able to extract new information and knowledge from the data and facts obtained from the environment and/or from the corresponding subject domain.
(ii) The cognizing agent is the simplest system able to realise cognitive reasoning processes.
(iii) The cognitive reasoning processes can be considered as discrete, internal rule-based manipulation of the represented information.
(iv) A cognizing agent possesses knowledge and experiences about itself and about the situations of the environment it has met.
(v) A cognitive process is a discrete process that includes alternating perception and reasoning phases. A perception phase terminates when the new data and facts are sufficient for starting a new reasoning phase, during which the agent extracts regularities and formulates hypotheses that reduce the domain of uncertainty. The alternating process permits the appearance of relatively stable knowledge states of the agents.

Developing the conceptual CR framework we followed the traditions of the cognitivist approach, i.e. we used a corresponding system of concepts concerned with symbolic representations and rule manipulation and modification.

Continuing the development of the general approach at the formal level of abstraction the following requirements were formulated:

O. Anshakov, T. Gergely, *Cognitive Reasoning*, Cognitive Technologies,
DOI 10.1007/978-3-540-68875-4_25, © Springer-Verlag Berlin Heidelberg 2010

(i) A challenging requirement was handling the dual character of the term cognitive reasoning by appropriate tools. Namely, the proposed approach has been armed with tools that could handle both the reasoning processes and the content of these processes, and also the interplay between static contents and dynamic processes.
(ii) Moreover, we intended to handle the dynamics of the reasoning processes not only in a descriptive way but rather in a constructive one. Concerning the dynamic nature of cognitive reasoning processes we met two aspects: informational and representational. Reasoning, from the former aspect, can produce substantively new information, whereas the representational aspect emphasises the possibility to modify the history of a reasoning process.

Further important requirements were implied by the specific nature of reasoning, namely:

(i) Reasoning processes are indeterminate in the sense that they remain reasoning processes even if they seem incorrect (whereas a wrong or incorrect proof is not a proof).
(ii) Reasoning processes are referential, as they presuppose semantic character connected with meaning, i.e. with the internal representation of objects and their structures. (As for proofs, they are of inferential nature, articulating statements inferentially, according only to their shape and regardless of the reference.)

The constructive character of the proposed approach is manifested at the realisational level. The approach should provide tools and methods to build a program, i.e. a computational realisation of each constituent of the cognitive architecture. That is, at this level the proposed approach completes the frameworks provided at the conceptual and formal levels with computational tools. Thus a computational realisation will realise the models provided at the previous levels of abstraction.

The realisational level of the proposed approach will be presented in the next chapter. An object-oriented architecture will be given, which (i) adequately represents the main characteristics of the cognitive processes, (ii) satisfies the formal logic foundation and (iii) permits the realisation of efficient computation tools for each constituent of the cognitive architecture responsible for the cognitive reasoning processes.

In Chap. 27 we resume our cognitive reasoning framework. In Sect. 27.1 we sum up the methods and tools of the conceptual CR framework provided by the developed approach at the conceptual level. Here the functional architecture realising the cognitive reasoning processes of a cognizing agent will be presented and discussed.

Sect. 27.2 is devoted to a summary of the formal CR framework. Namely, the structure and interconnection of the formal methods and tools will be presented in a coherent form. Moreover, we will provide open problems which may play an important role in the further development of the corresponding formal apparatus.

The proposed CR framework is open for further development. So we can consider it as an ongoing project, toward which the present development is only an initial and perhaps relatively modest contribution. In Chap. 28 we give some open problems, the solution of which could move further this general project. These problems will concern the mathematical theory of the formal CR framework.

The proposed approach of investigating and modelling cognitive reasoning is an integrative one because its abstraction levels reflect the three directions of investigation simultaneously, namely the philosophical–methodological, the logical–mathematical and the engineering–computational ones. The philosophical–methodological direction provides the conceptual background for cognitive reasoning theories. The key concept in most of these theories is that of heuristics, the guide in extracting new information. This explains whether the development of heuristics is one of the focuses in the methodological investigations. In the proposed approach the investigation of heuristics plays a significant role. Namely, at both levels — conceptual and formal — special attention was devoted to the characterisation and formalisation of heuristics as modification rules. We emphasise the modification theories which provide an original method of handling the cognitive processes in the line of the logical–mathematical approach. It is similar at the realisational level of abstraction, which reflects the main expectations of the engineering–computational approach.

Note that this reflection remains sometimes implicit. Therefore, the implicit philosophical–methodological outcomes of the proposed approach will be made explicit in Chap. 29.

Chapter 26
Towards the Realisation

In this chapter we discuss the level of realisation of the proposed CR formalism. We introduce the general scheme (the object model) of an application program for data analysis by means of the formalised reasonings provided by the formal CR framework.

In this program the reasoning rules will be applied. These will be the modification rules (that often are more convenient for computer implementation) or the generation rules corresponding to modification rules. For the development of a module responsible for application of rules, it is possible to use the programming languages oriented to operate with rules (productions), for example CLIPS or Prolog.

However, a reasoning strategy can also be realised by means of the usual procedural programming languages. Application of a modification rule can be interpreted as a conditional statement: satisfaction of a condition is checked; depending on the satisfaction of this condition a decision will be made about the change of the truth value of some predicate.

Data and discovered regularities in them will be represented in a unified form as is in the mathematical theory. Below, the construction representing data or regularities is called a predicate. Visually, the predicate can be represented in the form of a table containing columns with arguments and a distinguished column with truth values. The structure of the arguments is not important for predicate representation; therefore it will be stored separately.

26.1 Object Model Description

The object model of the system realising dynamic reasoning with modification of the truth values is presented in Fig. 26.1. Of course, this scheme includes only the main objects and collections of the system. It is formed by following the principle of "embedding instead of inheritance".

In this model the tool of collection, which is a dynamic data type, is widely used, as in modern programming languages. Usually it is standard or is included in one

O. Anshakov, T. Gergely, *Cognitive Reasoning*, Cognitive Technologies,
DOI 10.1007/978-3-540-68875-4_26, © Springer-Verlag Berlin Heidelberg 2010

of the standard libraries of the programming language. Collections are also called associative arrays, dictionaries or hash tables.

A collection can be understood as a list of elements of any data type; access to an element of a collection can be carried out both by its number (index) and by its name (unique identifier or key). In a class of collections there should be methods permitting the addition and deletion of elements. It is also possible to change elements of the collections because there is an access to them. Some programming languages and/or standard libraries have the extended functionality of collections, for example they contain sorting procedures.

Usually the languages supporting collections contain a special loop statement "**For each** *Item* **In** *Collection*" which permits to traverse all the collections and to carry out the actions of the loop body with all the elements of the collection discussed.

In the model the basic objects are represented by means of double-border boxes; collections of objects are distinguished by bold fonts and by shadowed boxes. Names and signatures of the methods are shown in italics. Boxes of collections are connected to boxes of their items by dashed lines.

The following classes are presented in the model:

(i) *System* (the system itself);
(ii) *Reasoner* (the class implementing functionality of the reasoning subsystem; this class contains the method Execute, which organises the loop of application of the groups of rules);
(iii) *Rule_Group* (the class representing a group of alternative rules processing one predicate; application of this group of rules is implemented by the method Apply);
(iv) *Rule* (the class supporting application of a rule understood as a conditional statement; satisfying the condition is realised by the method Condition);
(v) *Predicate* (the class, in the items of which the information on the predicates represented in a table form is stored; the header of the table is stored in the properties Arity and Sort; the rows of the table are stored in the collection Instances);
(vi) *Instance* (the class, instances of which store the information on rows of the tables representing predicates; the property Arguments stores the information on arguments; the property Value stores the information about the value of a predicate; of course, only the names of the entities (or the references to entities) which are arguments, are stored in the collection Arguments; the structures of these entities is stored separately);
(vii) *Sort* (the class intended for storage of the information about entities of one sort; for the storage of this information the collection Entities is used).
(viii) *Entitiy* (a class intended for storage of the information on the structure of the entity; the structure of the entity is saved in the property Value).

Certainly our object model contains only the main classes of the system. In a concrete implementation the number of classes should inevitably increase. For example, there should be the classes supporting various operations over the structures of the

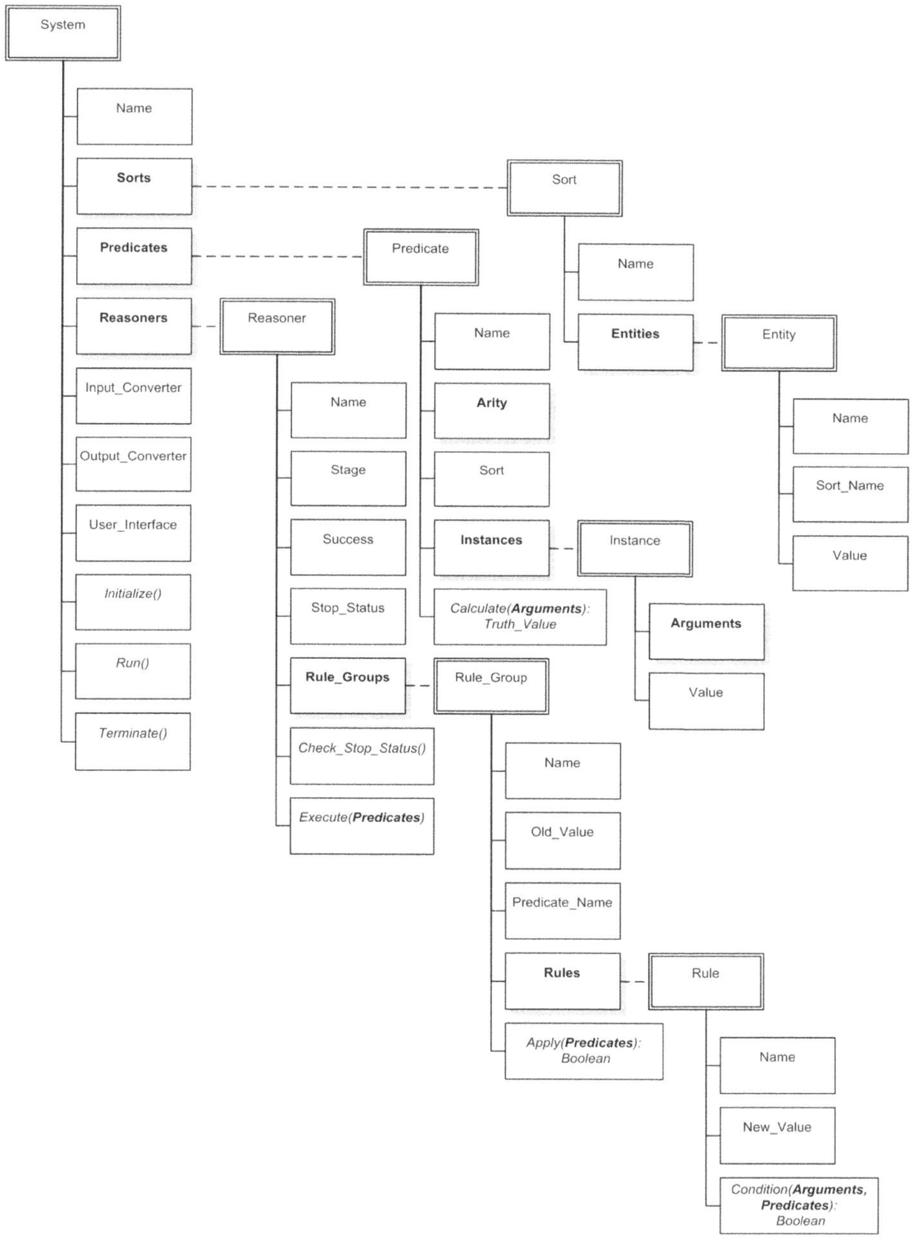

Fig. 26.1 Object model of the data analysis system

entities. There should also be the possibility of customisation for specific tasks of a concrete subject domain.

Now we will consider the classes enumerated above in more details.

Class System

Properties:

- **Name:** the string containing a name (identifier) of the system.
- **Sorts:** the collection of sorts of entities; each item of this collection contains a collection of all the entities of the given sort.
- **Predicates:** the collection of the predicates; each item of this collection contains the table representing it.

The union of the collections Sorts and Predicates can be considered as a model of the external world. The structure of the objects of the external world is represented by entities contained in the items of the collection Sorts. Many-valued relations between the entities are represented in the collection Predicates.

- **Reasoners:** the collection of reasoners; each item of this collection represents an active object that simulates reasoning about a subject domain by the use of a model of the external world.
- **Input_Converter:** the object, the methods of which permit the conversion of the data (represented in various formats) about the external world into internal data representation formats of the system. Generally speaking, different sorts can have different data representation formats.
- **Output_Converter:** the object which solves the task inverse to the task of the object Input_Converter. Methods of the object Output_Converter permit the conversion of data from the internal formats into formats suitable for visualisation, printing, saving on external carriers, and for data exchange.
- **User_Interface:** the object implementing the user interface, including functions of the customisation of parameters of the system, formation of the schedule of the computer experiment, selection of data sources, definition of the ways of saving the results of the computer experiment etc.

Methods:

- ***Initialize ():*** this method carries out the usual operations which run at the start of an application, in particular it restores the saved configuration of the application, checks the possibility of executing the application etc.
- ***Run ():*** the application's operation. As in any modern application this method traces the events and starts the procedures for their processing.
- ***Terminate ():*** end of application's operation, saving of the non-saved data and of the current configuration, i.e. of the customisation and the options of the application.

Class Reasoner

Properties:

- **Name:** a string containing the name of the reasoner.

- **Stage:** an integer, the number of a stage of the reasoning process.
- **Success:** a Boolean property interpreted as an indication of the fact that reasoning resulted in new information.
- **Stop_Status:** a Boolean property interpreted as an indication of the achievement of a certain type of saturation; if this property is equal to True the reasoner should stop the operation.

The properties Stage, Success and maybe Stop_Status too should be defined as private properties accessible only for the methods of the object Reasoner.

- **Rule_Groups:** a collection of groups of rules; consecutive application of the groups of rules forms the basic algorithm of the reasoner's operation represented by the procedure *Execute*.

Methods:

- ***Check_Stop_Status ()*:** check and change the value of the property Stop_Status; namely, call of this procedures at the required moment permits support of the correct value of the property Stop_Status that is necessary for correct termination of the reasoner's activity. Values of the properties Stage and Success can be used for the computation of the value of the property Stop_Status by using the discussed procedure.
- ***Execute (Predicates):*** procedure, which receives a collection of predicates as input. This procedure consists of consecutive application of groups of rules from the collection Rule_Groups of the instance of the considered class. Each group of rules is responsible for one predicate from the input data of the procedure *Execute*.

Class Rule_Group

Properties:

- **Name:** a string containing a name of the group of rules.
- **Old_Value:** old truth value. Any rule from the given group of rules replaces this old truth value by a new one, being its own for each rule of the group. Generally speaking, when developing a system that completely corresponds to the proposed theory it is not necessary to include this property because the old truth value, which is the subject to be replaced, is always equals to τ. This property is nevertheless included for further development of the theory, when other internal truth values will be doubted and become subject to replacement.
- **Predicate_Name:** name of a predicate. All rules from the group of rules operate with a shared predicate.
- **Rules:** a collection of the rules forming a group of rules. We emphasise that, unlike theoretical consideration, this collection includes neither the τ-rules, which keep the old truth value, nor the rules with an identically false premise.

Methods:

- ***Apply (Predicates)*:** application of a group of rules. Application of a group of rules requires the analysis of various predicates and not only of that predicate associated with the given group. Therefore this method accepts the collection of predicates as input data. The method Apply returns a value of Boolean type. The value of the function Apply is equal to True if the application of the group of rules changes the truth value of the predicate associated with this group of rules (for at least one sequence of arguments). Otherwise, the value of the function Apply is equal to False.

Class Rule

Properties:

- **Name:** a string containing the name of the rule.
- **New_Value:** new truth value. As a result of rule application, the old truth value (shared for all the rules of the group) will be replaced by a new one, which is its own for each rule of the group.

Methods:

- ***Condition (Arguments, Predicates)*:** the function defining whether the condition of the rule is met. A condition means the major premise of a modification or generation rule. The given object model does not require the necessary expressibility of this function by a first-order formula. The computation algorithm of the function Condition can use various operations over the structures of entities. It can use statistical reasons or heuristics concerning the subject domain. In the simplest case, this algorithm operates with the predicates (the second argument of the function discussed) and detects whether a certain condition (expressed by a first-order formula containing denotations for these predicates) is satisfied. Input data of the function Condition are the following: Arguments, a sequence (of lists) of argument names; Predicates, a collection of predicates. Function Condition returns a value of Boolean type

Class Predicate

Properties:

- **Name:** a string containing the name of the predicate.
- **Arity:** a collection (sequence) of sort names of the predicate arguments.
- **Sort:** name of the sorts of the predicate values.

We assume that the predicate has a table representation. The properties Arity and Sort together form the row of the column headings of the table, which contains the names of the sorts of the arguments and of the result.

- **Instances:** a collection of rows of the table representing a predicate. Keys of this collection are formed from the sequence of argument names. To each list of argument names a unique key corresponds. A function is defined, which calculates this key by the use of the sequence of argument names. (In the discussed object model this function is not reflected because it is auxiliary.)

Methods:

- ***Calculate(Arguments)*:** the function calculating the value of the predicate for a given list of names of arguments. This calculation in the simplest case is done by a key formed from the sequence of names of arguments; the coresponding instance is found and the property Value of this instance is restored. The function Calculate returns a value of type Truth_Value. Truth_Value can be defined as an enumerated data type, which contains truth values of all sorts: external (Boolean), internal or any other, if the theory development makes this possible.

Class Instance

Properties:

- **Arguments:** a list of argument names.
- **Value:** value of the predicate (the table row which is represented by the class discussed) for the given list of arguments.

Class Sort

Properties:

- **Name:** the string containing the name of the sort.
- **Entities:** a collection of entities.

Class Entity

Properties:

- **Name:** The string containing the name of an entity.
- **Sort_Name:** the string containing the name of the sort of the entity.
- **Value:** value of the entity. In this property the structure of the entity can be stored. On the structures of entities, operations and relations can be defined, which can be used in the computation of the value of the method Condition for the object Rule.

Certainly, the proposed object model is simplified and reduced for the sake of simplicity. In this model many circumstances have not been reflected. Some of these have already been mentioned. While developing a concrete system this model inevitably should be detailed and in places probably changed.

One of the circumstances which can demand modification of this model is the situation when not all entities of a sort are known. Moreover, all entities of this sort will never be known because of the large cardinality of the set of such entities. Then in the object Sort instead of the collection Entities we should use a method which generates entities by an algorithm. If entities of this sort are used in a predicate then in this predicate, instead of the collection Instances, we should define a method that forms a sequence of entities–arguments and computes the result.

Now we consider algorithms of some of the above-mentioned methods in more detail. For the description of algorithms we will use two methods: the classical flow-chart and pseudo-code similar to Visual Basic. Flow-charts are shown in Figs. 26.2 and 26.3. The pseudo-code is discussed below.

Methods of class System

Here we will not consider algorithms of the methods *Initialize ()*, *Run ()* and *Terminate ()*. It is obvious that each programmer will realise these methods in his/her own manner.

Methods of class Reasoner

```
Sub Reasoner.Execute(Predicates)
    Stop_Status = False
    Stage = 0
    Do
        Success = False
        For Each rg In Rule_Groups
            If rg.Apply(Predicates) Then Success = True
        Next
        Stage = Stage + 1
        Check_Stop_Status()
    Loop Until Stop_Status
End Sub
```

Possible variants of the implementation of the method Check_Stop_Status ()

```
Sub Reasoner.Check_Stop_Status()
    Stop_Status = Not Success
End Sub

Sub Reasoner.Check_Stop_Status()
    Stop_Status = (Stage > Max_Stage)
End Sub

Sub Reasoner.Check_Stop_Status ()
    Stop_Status = (Not Success Or Stage > Max_Stage)
End Sub
```

Max_Stage is a parameter which is stored in the global options of the system.

There can be more complex versions of the procedure Check_Stop_Status, for example, Stop_Status is set to True if the size of the uncertainty area of one or more or even all predicates does not exceed a certain limit. This limit can be established by statistical argumentation.

Methods of the class Rule_Group

```
Function Rule_Group.Apply(Predicates) As Boolean
    Dim S As Boolean
    S = False
    For Each ins In Predicates(Predicate_Name).Instances
        For Each r In Rules
            If r.Condition(ins.Arguments, Predicates) Then
               ins.Value = r.New_Value
               S = True
            End If
        Next
    Next
    Apply = S
End Function
```

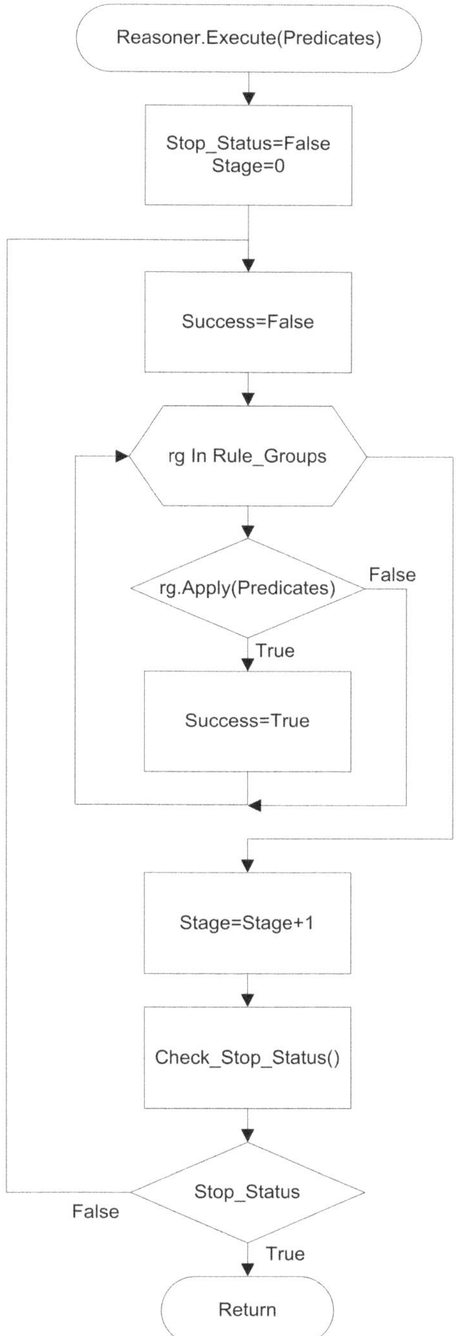

Fig. 26.2 Algorithm of the method *Execute* of the class Reasoner

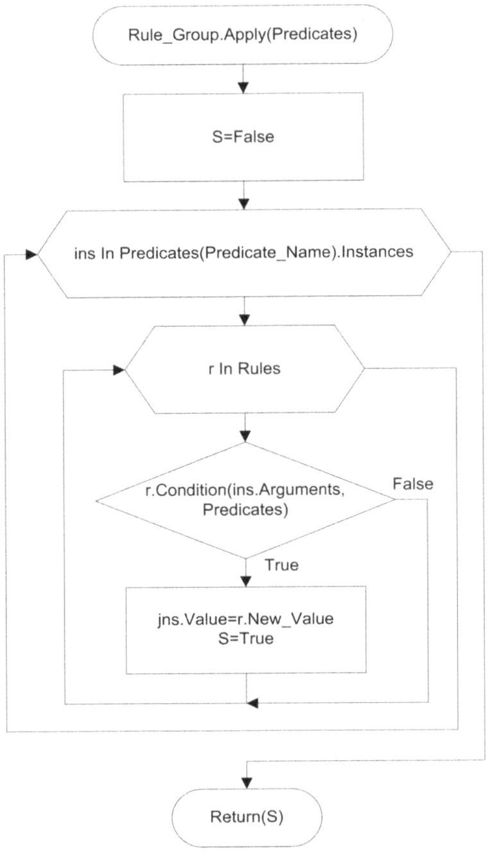

Fig. 26.3 Algorithm of the method *Apply* of the class Rule_Group

Methods of the class Rule

The method *Condition* can be realised in different ways for different problems. The specificity of the reasoner manifests just in the rules used by it. But the rules operate by the use of the method *Condition*. Generally speaking, in this case it is more convenient to use inheritance and to define classes of the concrete rules as descendants of the class Rule. Then the method *Condition* will be declared as virtual or dynamic. Each descendant will have its own realisation of this method.

Methods of the class Predicate

Regarding a possible realisation of the method *Calculate* see the description of class Predicate.

The classes Instance, Sort and Entity have no methods in our model.

26.2 Object Model Application

The object model discussed above can be considered as the *bridge* between theoretical (mathematical) results and concrete program implementation of the essential part (kernel) of the cognizing agent. It can also be considered as an advising system automating data analysis and regularity extraction.

Development of a concrete implementation, within the limits of the given object model, includes the usual stages: construction of a subject domain model, designing and implementation of the application. We will discuss the problems connected with the construction of a subject domain model in more detail.

1. First of all, it is necessary to define the sorts of the entities which will be represented in our model. A sort can be interpreted as a class of objects of one and the same nature (chemical compounds, diseases, patients, animals etc.). Some sorts can represent the properties of the objects to be investigated (the structure of chemical compounds and their biological activity, symptoms of diseases, morphological features of animals etc.).
   Entities of some sorts can have complex structure. Knowledge about a subject domain is used for the formalised representation of this structure and for the definition of operations and relations using structural representation (description) of objects.
2. Having defined the sorts and the functions connected with the structure of objects of certain sorts the relations between the elements of various sorts can be described. These relations in the object model described above are represented by predicates. Some of these predicates can be interpreted as data representation, e.g. results of experiments or predictions, and the others as the description of the regularities in a subject domain, i.e. the knowledge or information.
3. The next step is the definition of a set of plausible rules and of the sequence of their application (strategy) for extraction of regularities in the data about a subject domain. This means the definition of a modification (or generation) rule system of a modification calculus, i.e. the definition of a sequence of groups of rules, where each group of rules works with one predicate.
   The modification (or generation) rule system is interpreted by means of the object model considered above. First of all this interpretation is expressed by the description of the algorithm of the method *Condition* for each rule.

By this the model construction is completed and we can begin the system design and its realisation.

System realisation, beyond its practical advantage, could serve as a stimulus for further development of the theory. Inevitably applied problems might appear and these are out of the scope of the current object model. A change of the model can demand a theoretical foundation and the theory development, in turn, should lead to the following stage of change of the model.

The object model discussed in the present section already assumes the possibility of further development of the theory. It supposes, for example, rules which can change various truth values. The proposed mathematical theory does not handle this. This model does not impose restrictions on the computation algorithm for the value of the major premise of a rule. (The mathematical theory imposes rigid enough restrictions on systems of modification and generation rules.) The considered model also supposes extensions by the use of functions processing the structure of objects.

So, consideration of the not yet realised object model of the kernel of a cognizing agent permits to some questions of theoretical character to be raised which presuppose further development of the proposed mathematical theory. These (and further) open mathematical problems will be discussed in Chap. 28.

Chapter 27
CR Framework

27.1 Conceptual CR Framework

At the conceptual level of abstraction the proposed approach provides the conceptual CR framework. In order to characterise and model the cognitive reasoning processes this framework defines the main actor of cognition, which is the *cognizing agent*. The structure together with all the constituents and functioning of cognizing agents is represented in the form of a cognitive architecture. This framework has modelled cognitive reasoning with respect to this architecture that interacts with its environment in order to obtain data, facts and information, and it processes these in order to extract new information and regularities. The extracted information is used to augment the agent's knowledge about the environment and/or about the corresponding subject domain.

According to the proposed cognitive architecture the structure of a cognizing agent includes the following subsystems:

- A sensing subsystem
- An information processing subsystem
- An effecting subsystem
- The long-term memory which stores history of the activity of the cognizing agent

The information processing subsystem is responsible for the realisation of the processes related to cognitive reasoning, consisting of the following units:

- The interiorisation unit
- The reasoning unit
- The control unit

The *interiorisation unit* is used for the accumulation of information arriving from the environment and for information conversion into an internal representation (an internal format) of the cognizing agent. The *reasoning unit* is the kernel of the considered subsystem. It contains the *fact base*, the *knowledge base* and the *reasoner* (the mechanism of reasoning, that is, the reasoning engine). The fact base and the

O. Anshakov, T. Gergely, *Cognitive Reasoning*, Cognitive Technologies,
DOI 10.1007/978-3-540-68875-4_27,

knowledge base together form the short-term (working) memory. The reasoner introduces changes in the working memory during its operation (in both the fact base and the knowledge base). Changes in the knowledge base can be interpreted as acquisition of new knowledge, while changes in the fact base can be interpreted as generating predictive hypotheses.

The basic restrictions used in the proposed model of the cognizing agent are as follows:

- For knowledge extraction, a cognizing agent uses *reasoning in a broad sense*, which may include deductive inference, plausible argumentation and computing procedures;
- Activity of a cognizing agent is *discrete*; it consists of alternating phases of perception of the environment and reasoning directed at the formation of new knowledge;
- In the course of functioning of the cognizing agent a *history* of its cognizing activity is formed.

The information processing activity of a cognizing agent can be represented in the form of several embedded loops, from which two embedded loops can be distinguished. The internal loop consists of iterative reasoning algorithms until the completion condition is met. The external loop represents repetition of perception and reasoning phases and proceeds from the beginning to the end of the activity of the cognizing agent.

The proposed conceptual theory represents both the internal and the external loops. These loops will determine the cognitive reasoning processes of a cognizing agent. For the representation of these processes that are taking place in the "brain" of the cognizing agent, i.e. in the reasoning unit, the CR framework provides two classes of theories: (i) the open cognitive or quasi-axiomatic theories, and (ii) the modification theories.

The open theories are intended to represent the interaction between the cognizing agent and its environment. They are open for (i) obtaining new facts for analysis, (ii) adopting and internalising knowledge from other similar theories and (iii) modification of the reasoning rules. An *open cognitive theory* represents the activity of a cognizing agent as a history. Thus an open cognitive theory can also serve as an essential component of *self-reflection* of an agent.

The open cognitive theories model the external loops of the operation of a cognizing agent. These theories are related to the second class, i.e. to the modification theories. They represent the internal loops of the activity of the cognizing agent, which are connected with reasoning. These loops are connected with a simple enough procedure of *traversal* of some distinguished part of the working memory including a part of the fact base and a part of the knowledge base. This traversal is accompanied by the analysis of the facts, extraction of knowledge and modification of the fact and knowledge bases.

These theories are named modification theories because, during the construction of a process in such theories that model the reasoning process, not only can a new statement be added, but the history of this process itself can be modified. Modification theories use a special technique of inference for modelling the processes. An inference in the modification theory is called a modification inference. For the construction of a *modification inference* modification rules are used. *Modification rules* modify: (a) the truth values of the statements entering an inference and (b) the modification inference itself. These rules correspond to the plausible reasoning rules.

The modification rules change not only the inference but also the truth value of some statements; they replace the truth value "uncertain" by one of the given truth values, for example by "true" or "false". Thus, modification theories permit the representation of dynamics of cognitive processes, understood as movement from ignorance (uncertainty) to knowledge (definiteness).

Modification rules in the theory are organised into a system which is the analogue of the reasoning algorithm of the cognizing agent. This system controls the construction of a modification inference that is divided into stages, and each stage is an analogue of one iteration of the *traversal* of the fact base and of the knowledge base.

Modification inferences are non-monotonous, which is connected to the deactivation of some formulas entering an inference. That is, non-monotonity is connected to the modification of inferences, whereas those statements for which the truth values have been changed are declared to be inactive. Those statements which depend on the deactivated ones are declared inactive too. The inactive statements cannot be used as premises in further inference extension any more, but these features of a modification inference do not eliminate the possibility of obtaining contradictions in the use of modification rules. However, a modification inference possesses a rather unusual property of *temporal contradictions*. The contradictions which have appeared in an inference can later disappear because of a modification of the inference, yet new contradictions can appear again etc. Thus contradictions in modification theories are handled.

The semantics of modification theories have been defined as a sequence of structures. Each structure reflects on the one hand a knowledge state about a considered fragment of a subject domain whereas, on the other hand, it is an image of the working memory of the cognizing agent at some stage of reasoning. Application of modification rules leads to modification of the structures, i.e. it appears as *modification of the semantics*.

Thus we have seen how the modification theories represent the structure and functions of the reasoning unit of the cognitive architecture.

27.2 Formal CR Framework

The proposed integrated approach considers the cognitive reasoning processes as a *motion from ignorance to knowledge*. In the course of this motion the uncertainty domain is permanently narrowing. At the formal level of abstraction the proposed integrated approach suggests a formal CR framework that provides an original non-monotone derivation technique obtained by adding specific rules to the standard deductive technique to model this motion. This new technique plays the role of the plausible constituent of cognitive reasoning, which supports data analysis and the discovery of principally new relationships and regularities by the use of non-deductive reasoning technique. This discovery has its price, such as the incorrectness of non-deductive reasoning, the possibility of obtaining incorrect conclusions even from valid premises and the possibility of obtaining contradictions.

In order to develop this formal CR framework first the logic foundation of the formal approach was developed, which allowed us to construct a logically well-founded CR framework. The proposed foundation supports two alternative approaches: (i) the simulative one that permits the simulation of cognitive processes by using standard deductive derivations and (ii) the direct one which provides original calculi appropriate to represent and realise cognitive reasoning processes.

The logic foundation is based on a special family of many-valued logics. These logics are the pure J-logics (PJ-logics) which have two sorts of truth values: external and internal. The two sorts of truth values are to represent *the two levels of cognitive reasoning processes*: the empirical (internal) and the theoretical (external) ones. The internal level corresponds to observations and experiments of the cognizing agent who interprets the results of these activities. Therefore the corresponding internal knowledge represents the individual or subjective knowledge. The external knowledge can be considered as the conventional or objective knowledge. At this level there is knowledge about the general regularities and laws of a subject domain which are consolidated in a certain theory, checked several times and generally accepted at the present time.

The external knowledge level, which is but the theoretical (logical or deductive) one, can be considered as the level of *conventional or objective knowledge*.

(ii) The set of J-operators is the *tool for encapsulation* of internal knowledge into external knowledge. By using these operators we can apply empirical knowledge to theoretical constructions.

(iii) The absence of logical connectives different from the classical (external) ones in PJ-logics is the most controversial requirement. This may be cancelled in the further development of the proposed approach.

Two types of PJ-logics were introduced: the finite-valued PJ-logics and a class of infinite-valued ones. Any finite-valued PJ-logic is appropriate to describe and represent *knowledge states* at a certain moment of time (with some presuppositions about the representation of truth values of observational and empirical statements). However, this is a *static* description. Standard deductive techniques do not permit the representation of the dynamics of the cognitive processes in a finite-valued PJ-logic. Any infinite-valued PJ-logic from the above-mentioned class can represent the dynamics of the cognitive processes. Thus the proposed logic foundation provides two possible solutions for the representation of the dynamics of the cognitive processes:

(i) Equip a finite-valued PJ-logic with non-standard inference technique (with *modification rules*) which allows us to generate sequences of knowledge states;

(ii) Use a finite-valued PJ-logic as a basis for the construction of *its iterative version,* which is an infinite-valued PJ-logic from the above class.

The iterative version of a finite-valued PJ-logics permits the description of the history of cognitive processes that includes several repetitions (iterations) of certain cognitive procedures. A cognitive process includes reasoning with the application of plausible inference rules (induction, analogy, abduction etc.). In the iterative logics, to these rules the so-called cognitive axioms correspond. Cognitive axioms and iterative logics allow us to repesent cognitive reasoning dynamics by the use of ordinary logical inferences. In this case we have a *deductive simulation* of plausible reasoning. If we restrict ourselves to finite structures, then a cognitive process, considered as motion from ignorance to knowledge, will be completed necessarily. The sign of completion is the so-called stabilisation or saturation, which has a formal representation in the language of iterative logics in the form of stabilisation axiom schema.

The direct representation of the dynamics of the cognitive processes is based on a finite-valued PJ-logic supplied with a non-standard inference technique, the so-called *modification inference.* As we know, applying traditional inference rules we can only add formulas to an inference but cannot do anything with formulas which are already in the inference. In the proposed formalism, by adding formulas to an inference by non-deductive rules, we may cause modifications in those constituents which are already in the inference. In our formalism this is why we call non-deductive rules as modification ones.

Of course, any formalism imposes certain restrictions and puts the cognitive processes into a certain frame. For example, our view of dynamics of cognitive processes is *discrete* and is rather rigidly regulated. This cognitive process is split into an unlimited number of stages. Each stage is further split into modules (the number of which is finite and the same for all stages). Each module is divided into two phases: the *plausible or rapid one,* where plausible reasoning appears, and the *deductive or torpid one,* where only deductive reasoning appears. A completed module corresponds to some *knowledge state*, which can be expressed by its formal *description*. Thus the structure representing dynamics has three levels of embedding: stage, module and phase.

In spite of the rigid structure of the representation of reasoning our approach permits plausible rules to be applied and even formal logical contradictions to be obtained. An interesting effect of the proposed formalism is the temporal character of the obtained contradictions. A cognitive process may be free of contradiction at the initial stage. Contradiction may appear at one of the internal stages and then may disappear. At the consequent stages of this cognitive process a contradiction may also appear, but for a different reason.

The proposed formalism has a dual (semantic–syntactic) nature. Formally it can be considered as a method of constructing calculus of special form, the so-called *modification calculus*. The definition of a modification calculus consists of:

- A system of *modification rules* (any such rule is a formal analogue of a plausible reasoning rule)

- An *initial state* of knowledge represented by its formal description

The (knowledge) state description is an important constituent of each calculus. It is a collection of axioms which *entirely and uniquely describes the semantic structure,* the latter of which is considered a model of our initial knowledge about the subject domain.

By the use of such calculi, syntactic objects, i.e. inferences, may be generated. The modification inference is an adequate representation of the cognitive process, whose dynamics meets the above restrictions (discrete, split into stages etc.). In Fig. 27.1 we include the notion of modification inference in the scope of modification calculi.

Note that the completion of the rapid phase of each module of a cognitive reasoning process results in a new collection of axioms. These axioms describe a new semantic structure as a model of the knowledge state about the subject domain. The sequence of such structures can be considered as a model of the modification calculus itself.

The modification calculi are used to analyse facts obtained as a result of interiorisation of experiments and observations. The aim of the analysis is the discovery of regularities. This analysis of facts is carried out by the use of the modification rules.

We obtain a *modification theory* if we add a set of non-logical axioms to a modification calculus. These axioms postulate some requirements for the initial knowledge state represented in a modification calculus. For example, these axioms can require some algebraic structure for a model of the subject domain (e.g. to be a lattice, a Boolean algebra, a group etc.). This theory, in general, is temporarily open, which means that a cognizing subject as the carrier of the considered cognitive processes is open to its environment, from where facts may be received at any time.

For each knowledge state it is possible to construct a corresponding modification theory. Therefore we have defined an *open cognitive theory* as a finite sequence of modification theories, each of which represents the current content of the fact base and knowledge base of the cognizing agent obtained:

- Either directly before starting
- Or directly after the completion

of the reasoning process.

Thus an open cognitive theory represents the history of activity of a cognizing agent from the beginning of its activity to a current moment of time. It is obvious that such a theory can always be continued.

The formal components of the proposed formal CR framework are shown in Fig. 27.1.

Note that reasoning processes in a subject domain are strongly related to the language by the use of which the experimental data, the results of observations, the individual facts and general assertions are described and represented. In the CR framework this language is supposed but without providing any restrictions on it. However, when the environment is very complex, special techniques and methods are required in order to represent this complexity. In order to handle such situations the CR framework was augmented:

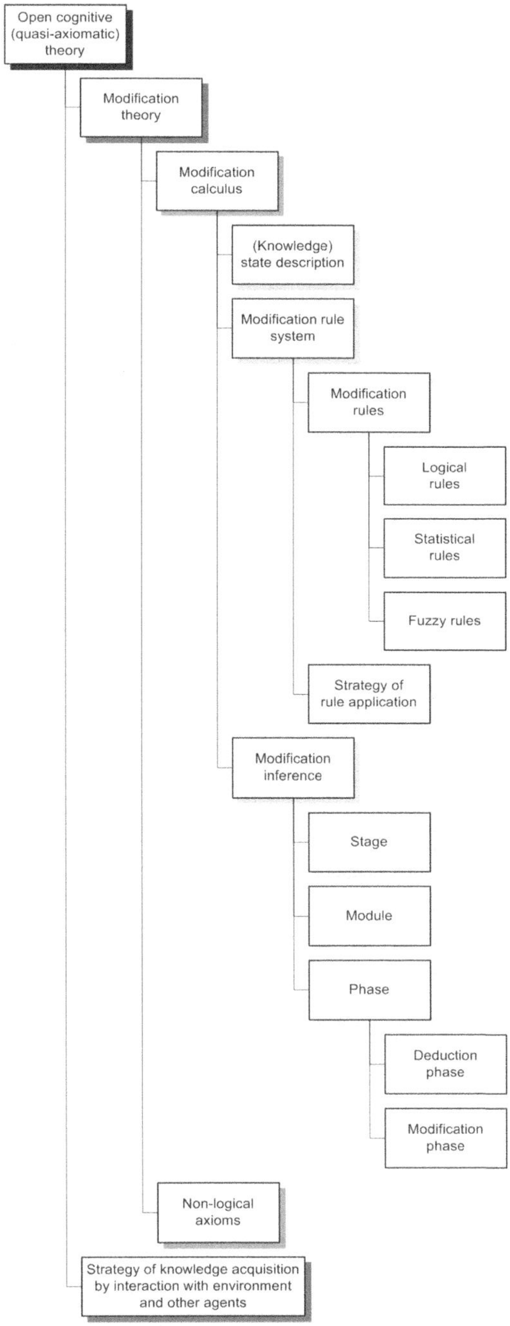

Fig. 27.1 Components of the proposed formal framework

- With a more powerful language that can handle complex structures with interacting components. (This extension concerns both syntactic and semantic aspects. In this language the objects should be interpreted as formal structures such as sets, strings, graphs etc.)
- With a calculus that can handle the complex language units in order to find hidden relationships concerning the features of complex objects and their structures.

The proposed extension is realised with the tools that permit handling of so-called set sorts, i.e. sorts, the objects of which can be interpreted as sets. This permits the consideration of formal structures of complex objects in the form of sets. The set sort objects should behave like sets and be arguments of internal predicates. The extension concerns the languages, their syntax and semantics, the inference technique as well as the modification calculi.

Special tools were introduced to represent the sets of properties in the modification calculi. First of all the definition method for modification rules was defined. Then the transformation of modification rules into generation rules was given. Different classes of modification calculi were investigated.

While we are confident that such a formal CR framework is possible we are far from sure that we have handled all of these aspects in the optimum way. Thus we prefer to regard the proposed logical framework as an ongoing project, toward which the present development is only an initial and perhaps relatively modest contribution. If we are right, this general project, despite its recent neglect, should be utterly central to the cognizing agents and cognitive reasoning agenda. Thus our primary goal was not to resolve fully all of the issues discussed, but rather to bring them to the attention of others who may see further or deeper than we have managed to do so far so the proposed formal CR framework is open for further development

Therefore we give now some open problems, the solution of which could move forward this general project.

Chapter 28
Open Problems

Here we discuss some open problems connected with the further development of the mathematical theory of the formal CR framework developed in the present book. We present the problems in thematically grouped form.

Theme 1. Further development of modification calculi for the purpose of strengthening their non-monotonicity

Modification calculi really possess *non-monotonicity* properties, but this non-monotonicity is limited. Actually we can only change an uncertain truth value on one of the defined truth values. A definite truth values, once justified ("true" or "false"), cannot be further subjected to doubt.

Some problems:

- Problem 1.1: to define modification calculi with broader capabilities for changing the truth values of the internal predicates
- Problem 1.2: to investigate proof-theoretic and model-theoretic properties of these modification calculi
- Problem 1.3: to investigate possibilities of practical application of such calculi for building cognizing agents

Theme 2. The development of logics with two or more sorts of truth values

The idea of using logics with truth values of two sorts (external and internal) appears to have been rather fruitful for both the development of non-monotonic calculi and in practical applications. In our opinion, the development of logics with truth values of two sorts is determined first of all by practical needs concerning formalisation of plausible reasonings, and not only to generalise mathematical results.

Some problems:

- Problem 2.1: to define and study logics with internal truth values of more than one sort allowing the consideration of various kinds of uncertainty,
- Problem 2.2: to define and study continuum-valued logics, the internal truth values of which could be considered as degrees of validity and falsity and calculated

O. Anshakov, T. Gergely, *Cognitive Reasoning*, Cognitive Technologies,
DOI 10.1007/978-3-540-68875-4_28, © Springer-Verlag Berlin Heidelberg 2010

on the basis of statistical reasons or fuzzy rules, so that probability and possibility are introduced in the logics explicitly;

- Problem 2.3: to design and investigate logics with internal truth values represented by tuples, the elements of which are interpreted as aspects of argumentation and in the elementary case can be represented as pairs:

$$\langle \text{weight of arguments for, weight of arguments against} \rangle$$

These logics could be used in those situations when in the reasoning rules it is necessary to account independently for the pros and cons and, in a more general case, for the different aspects of the argumentations;

- Problem 2.4: to define and investigate logics with operations on internal truth values. In our opinion, such operations should be defined not arbitrarily, but with regard to the practical needs of problem solving.

Theme 3. Development of calculi with generalised quantifiers

Another way to increase the expressive possibilities of logics considered in the present book is introducing generalised quantifiers. In order to have a language capable of describing rules that use statistical and other quantitative aspects, we need to introduce generalised quantifiers, such as quantifiers of measure and capacity or more complex statistical quantifiers (see e.g. [Hajek and Havranek, 1978]).

One problem:

- Problem 3.1: to define and investigate modification calculi and theories with the generalised quantifiers

Theme 4. Development of modification theories formalizing concept of similarity

Similarity plays a significant role in the generation of hypotheses about the presence of regularities (laws) interpreted as cause–effect relations. We have already studied formalisation of similarity as an operation which results in the general part of object structures (see Chap. 23). However, this is not the only possibility. It would be interesting to develop various approaches to understanding similarity both as a relation (not necessarily binary) and as a function, the value of which can be interpreted as *degree (or measure) of similarity*. The degree of similarity can be related to the degree of difference, with the latter defined as distance between objects, This, in its turn, will permit the study of possibilities of using various definitions of distance (Euclidean, Hamming, Jacquard etc.) to define degree of similarity.

Some problems:

- Problem 4.1: to define and investigate modification theories with various definitions of similarity, in particular those *with a function of similarity measure*
- Problem 4.2: to investigate the possibility of application of reasoning in such modification theories for classification and clusterisation

Theme 5. Working out algorithms for the search for structures connected with the presence or absence of a target property

Search for structural fragments related to the presence or absence of a target property of objects containing these fragments is one of the basic components of the JSM method. However, detection of such relations may also be needed in the operation of reasoners based on rules differing from those of the JSM method's.

In the JSM method the structural fragment assossiated with a property is usually interpreted as a possible cause of the presence or absence of this property. Unfortunately, the algorithms applied by the JSM method in the search for the possible causes cannot be aimed precisely at such search. Apart from causes they find many extraneous fragments which are actually not causes.

One problem:

- Problem 5.1: to develop an algorithm for accurate search for possible causes and their inhibitors which should be able (i) to find the cause containing all necessary components but nothing superfluous and (ii) to solve practical problems of average complexity in appropriate time limits

At this point we could go on enumerating open problems but we feel these should serve as a sufficient illustration.

Chapter 29
Philosophical–Methodological Implications of the Proposed CR Framework

As mentioned, the proposed approach integrates all three types of methodology of investigation: the philosophical–methodological, the logical–mathematical and the engineering–computational ones. At the formal level of abstraction the approach basically provides methods and tools in the logical–mathematical spirit. However, at the same time the formal CR framework has philosophical–methodological implications which differ from the issues described at the conceptual level of abstraction. Here we discuss some of these.

29.1 Epistemology

The proposed integrated approach deals with processes able to generate new information and to improve knowledge. Moreover, it also deals with the organisation and representation of information and knowledge. Thus knowledge plays a significant role in both information processing and cognitive reasoning .Therefore the proposed approach has a direct connection to the branch of philosophy concerned with the origins, structure, methods and validity of knowledge, i.e. epistemology. This connection makes explicit the necessity of a theory of knowledge that explains how knowledge grows, which requires both describing the structure of knowledge and the inferential procedures by which knowledge can be increased. That is, an epistemology is needed that can be adequate for the proposed integrated approach. So this epistemology should be a precise one and it should consider cognizing agents. However, the problem of creating a precise epistemology for intelligent systems, particularly for cognizing agents and for its systematic study is still unsolved (however, many steps have been taken in this direction, see e.g. [Sowa, 2000; Thagard, 1993]). This epistemology should be an epistemology of a computer-aided cognizing agent which is the opposite of K. Popper's epistemology [Popper, 1967]. The latter rejects both the cognizing agent and dealing with heuristic procedures (particularly with induction). This is so because its central subject is how information discovery influences knowledge development in a general setting, i.e. cognition is

O. Anshakov, T. Gergely, *Cognitive Reasoning*, Cognitive Technologies,
DOI 10.1007/978-3-540-68875-4_29, © Springer-Verlag Berlin Heidelberg 2010

interpreted in a social or cultural sense to describe the emerging development of knowledge and concepts within a group. That is, here epistemology without the cognizing agent is the central subject of study (see e.g. [Popper, 1974])

According to the proposed approach the required epistemology should be interrelated with (i) the formal methodology of dealing with information and knowledge, (ii) the methods of knowledge discovery, (iii) the control of the knowledge generated by the proposed special methods of falsification and justification and (iv) the methods of growing the knowledge base. This epistemology could be related to both a special type of experimental computational epistemology and to evolutionary epistemology. It could deal with the processes of knowledge discovery realised by cognitive reasoning procedures and with the growth of the knowledge base resulting from them. Evolutionary epistemology studies the process of changing knowledge. This process can be considered as a transition from ignorance to knowledge and from approximate solution of problems to posing new problems. The basic formula of evolutionary epistemology is the following:

$$\text{initial problem} \longrightarrow \text{tentative theory} \longrightarrow \text{error elimination} \longrightarrow \text{new problem}$$

This formula can be called Popper's principle of knowledge development. According to it the central idea of evolutionary epistemology is the evolution of problem: formulation of problem, construction of a tentative theory (with respect to this problem), the elimination of mistakes and the emergence of a new problem that changes the initial one. This evolution of problems, which allows for a competition of the theories that correspond to them, does not require an examination of the psychological aspects of the creativity of a human being. Thus, heuristic, as a tool of cognition of the individual or of classes of individuals (for example, some scientific school), does not interest the evolutionary epistemologist. Therefore, evolutionary epistemology (particularly Popper's approach) is an *epistemology without cognizing agent*. Thus the activity of any cognizing agent's heuristics will be out of study of Popper's approach. Yet both controlled reasoning and knowledge representation are objects of epistemological studies. Thus, modern computer science, where intelligent systems play an important role, needs an epistemology with a cognizing agent.

In order to consider cognizing agents both the understanding of logical inference and the understanding of theory should be extended. This means that the idea of logical inference must be extended to the structurally determined notion of reasoning, while deductive (closed) theory should be extended to the notion of empirical (open) theory. These changes are also connected with the requirement of the computer realisation of cognitive reasoning procedures and of the knowledge representation.

Thus the required epistemology is related to a special type of computational epistemology that is related to evolutionary epistemology, too.

The proposed integrated approach for cognitive reasoning processes provides a formal basis in the form of two classes of theories: modification and open (quasi-axiomatic) ones, which at the same time suggests a formal epistemology that corresponds to this cognitive reasoning approach. Its main characteristics are as follows:

(i) For modelling the cognitive reasoning processes a special non-Aristotelian theory of truth is developed, which synthesises three theories of truth: correspondence, coherence and pragmatic theories. The correspondence theory is applied to the fact base in its non-classical version for many-valued logic. The coherence theory is used for the evaluation and acceptance of hypotheses according to the plausible inference rules. The pragmatic theory is used for the acceptance of those hypotheses that were found practically useful.

(ii) In the proposed approach the following types of truth values are used. *First*, all the truth values are divided into two classes: the class of external truth values and the class of internal truth values. There are two external truth values: the classical "true" and "false". There are at least two internal truth values, of which one is called "*uncertain*", which means the complete absence of information in the question of interest. The other(s) may express the *degree and nature* of certainty and how well one is kept informed. Note that this division into two classes is similar to the classical philosophical tradition of dividing truths into analytical and factual ones. *Second*, the factual truth values are subdivided into the values of facts given at the start of the cognitive process and the values of hypotheses generated during this process. *Third*, the values of hypotheses are generated operationally by means of plausible inference rules so that the process of generation of truth values is constructive and iterative. The degree and nature of certainty expressed in the values of hypotheses are connected with the quantity and significance of arguments which are either for or against the statement characterised and evaluated by internal truth values.

(iii) The cognitive reasoning processes synthesised by the CR framework are applied to the facts and knowledge represented in the form of open (quasi-axiomatic) cognitive theories. An open cognitive theory is a finite sequence of modification theories each of which (1) represents the current content of the fact base and knowledge base of the cognizing agent and (2) partially characterises with its set of non-logical axioms the external world under investigation. Thus an open cognitive theory represents the history of activity of a cognizing agent from the beginning of its activity to a current moment of time. It is obvious that such a theory can always be continued. The openness of a knowledge base is a necessary condition for epistemology with a cognizing agent.

(iv) A cognitive process includes reasoning with the application of plausible inference rules (induction, analogy, abduction etc.). In the iterative logics provided by the formal CR framework, the so-called cognitive axioms correspond to these rules. Cognitive axioms and iterative logics allow us to repesent cognitive reasonig dynamics by the use of ordinary logical inferences. In this case we have a *deductive simulation* of plausible reasoning. This provides a certain type of self-reflection.

(v) The CR framework formalises the process of guessing by exactly describing heuristics that generates hypotheses on the basis of data analysis and prediction. Thus the CR framework provides a formalisation of cognitive reasoning, which means that it implements a weak psychologism in the sense that it represents the logical basis of mental cognitive processes connected with heuristics for hypothesis generation at a generic level.

The basic formula of knowledge development in the proposed case of formal epistemology with cognizing agent may have the following form:

initial problem (represented in a modification theory) → plausible modification theory → generation of another modification theory by iterative maximisation of the truth of the theory → new problem

This schema expresses the development of knowledge with respect to the problem-solving activity of the cognizing agent. In this process the openness of the fact base and knowledge base is essential.

A philosophical discussion of the objectivity of knowledge is more substantial if the connection between characterisations of ontology and epistemology are formulated. Now we turn to consider the ontology of the proposed cognitive reasoning approach.

29.2 Ontology

A cognizing agent interacts with its environment and/or with the corresponding subject domain. Usually its actions and reactions agree with its model about the environment, i.e. about the external world. The effectiveness of these actions depends on the adequacy of this model for the environment, i.e. on the adequacy of its knowledge for the world it interacts with. The adequacy depends on how completely the knowledge represents and reflects the peculiarities of the world relevant to the actual situation. These peculiarities are related to the type of world in question. We distinguish two main types: the *deterministic* and the *non-deterministic* ones. In the deterministic case the most general features and relations of the constituents (objects, entities or events) can be represented by a deterministic formal description. In the non-deterministic case this representation requires some non-deterministic formalism such as probabilistic or statistics, fuzzy or possibility-theoretic ones. Of course there are situations where these two types may appear in a mixed form. E.g. a situation may have both random changes of events and deterministic relations among events affecting the final states of this world.

Knowing the type of the actual situation of the external world is important for reacting in an adequate way to the influences of the environment. The adequacy also depends on the model of the world that the cognizing agent formed. The more adequate and flexible the world model of the agent, the more correct its response to the influence of the environment is and the more adapted it will be to the environment. As for the world model itself, it is obvious that the most important role in it belongs to knowledge represented in one or another form. However, this knowledge should be adequate for the actual type of the environment.

Thus the objectivity of knowledge depends on how adequately it represents the type of world under consideration. The adequacy depends on the ontological basis of the cognizing agent, i.e. how it represents the objects, events and other entities that are assumed to exist in the potentially important environments and the relationships that hold among them. The knowledge about a situation of a world, i.e. about the corresponding subject domain, can be represented in a formal language. If this language is descriptive then the constituents of the situation, e.g. the entities, will be represented by definitions which associate the names of entities (e.g. classes, relations, functions or other objects) with descriptions of their meaning, and formal axioms that constrain the interpretation and well-formed use of these terms. Formally, an ontology is represented by a set of statements of a formal theory; in our case it will belong to a modification theory. The ontology influences the efficiency of the cognitive processes and the growth of knowledge. In order to ensure efficiency the adequacy principle should be applied, which means that the cognizing agent should form the cognitive reasoning processes by using modification rules and knowledge adequate for the type of the actual world. That is, if the agent interacts with an environment which is a non-deterministic world, supposing it to be of stochastic nature, then the agent should use statistic rules and the world model should also contain stochastic relationships in order to be able to extract the required regularities of this environment and to ensure objectivity.

The relationships among entities or the characterisation of the entities may often be of cause–effect type, which can be represented by a cause–effect relation. One can consider the arguments of a cause–effect relation as events. Each event can bind some objects (event participants). In this case, we are able to denote event participants by terms, so we can represent an event by a formula of predicate logic. In this model of the cause–effect relation, the cause precedes (in time) the effect. This fact is an argument for the use of temporal logic for formal descriptions in the context of the given model. Note that many researchers regard the cause–effect relationships just within a single space of events. Therefore there exists a common space for the arguments of the cause–effect relation. That is, the cause and the effect are considered as homogenous entities.

Another view of causes and effects is to take them as parameters (values), so one can consider causal relation as functional dependence, where a cause is an argument and an effect is a value (result) of some function (see e.g. [Pearl, 2000]). This approach allows us to take causes and effects from different universes; however, it emphasises the quantitative (non-structural) nature of cause and effect.

A third view is connected with structural causal relationships. It is a relationship between the structure (construction) and function (behaviour) of objects of the environment. This case is of great theoretical and practical importance and unfairly neglected in the literature.

In addition to including the entities and the relevant properties of a subject domain, developing a suitable ontology may be accomplished in various ways. Indeed, the design of a suitable descriptive and identifying system of representation for the entities of a particular subject domain might involve a great deal of creativity. The goal of ontology is a system of representation such that there is a one-to-one correspondence between descriptions and entities: every entity has a unique descriptive or specifying representation in the system.

Ignoring the difference between an entity and its unique representation, we can use the neutral term "object". As an object we may consider either an entity as uniquely represented in the ontology, or else the representation itself of this entity. We can also introduce the term "subobject", which means either some not necessarily proper part of an entity or else a fragment of the total representation of the entity.

Thus, the entities of the subject domain are expressed in terms of an ontology of "objects" and their "subobjects". An example of objects might be chemical compounds, and an example of subobjects might be certain parts of chemical compounds, such as covalent bonds, benzene rings, OH-radicals and the like. The language of representation might be some more or less standard form of representation of chemical compounds, such as three-dimensional chemical diagrams (graphs), or the like.

There is one feature of any ontology of a subject domain, however, that is absolutely crucial if the objects or their subobjects should be analysed in comparison to each other. This necessary feature is that the ontology must admit the definition of similarity and a method for its checking. This is also important in regularity extraction, since in a world of deterministic type, similarity of objects implies the existence of a common cause of an effect and the repetition of the latter.

It is important to remember that we have at least two constituents of ontology definition for representation: first, express what there is, what exists, what the objects are made of and, secondly, say what the most general features and relations of these objects are.

Concerning a cognizing agent we may require the ontological commitment principle, which means that its observable actions are consistent with the definitions in the ontology represented in its knowledge base.

29.3 Methodology

Together with its CR framework the proposed integrated approach provides an efficient foundation for:

(i) The solution of problems that require the extraction of new information and regularities in order to find some solutions, at least hypothetical ones;
(ii) The formalisation of the knowledge of those subject domains that integrate theoretical knowledge with experimentation and measurements, like in medicine, economics, geology, education, sociology etc. One is faced with the ongoing task of bringing a more or less established body of knowledge to bear on problems in unique, although not wholly unfamiliar, situations;[1]
(iii) The reasoning methodology for the subject domains mentioned in (ii).

Now let us discuss these areas of application in more detail.

[1] Situations may arise that are unfamiliar because they involve events that are truly inconsistent with, or are not covered by, established knowledge.

(i) The application of the proposed approach for the solution of various problems requires the satisfaction of some conditions. First, the description of the entities (e.g. objects) of the problem domain should be such that the modification rules and the corresponding computation procedures can be applied to them, e.g. for the comparison of the entities by the use of a similarity relation or operation. Second, the descriptions should be informationally complete in the sense that they should hide the constituents of the regularity to be extracted. In order to meet these conditions the description requires an adequate descriptive language that permits the description and representation of the data, facts and knowledge. This means that the selection of an appropriate descriptive language is an important requirement for the successful application of the CR framework.
Some of the problem situations that can be solved by the proposed framework are:

  (i) Discovery of causal regularities in deterministic and stochastic environments. E.g.:
    (i) Discovery of the cause of a property of an object in relation to its structure
    (ii) Discovery of the cause of the behaviour of a system in relation to the system's architecture
    (iii) Discovery of the cause of the appearance of an event in relation to other events

  Let us consider some examples from medical chemistry for the case 1a: Such an example is the discovery of the chemical compounds that are responsible for certain biological effects. The investigation of the biological effect requires efficient cognitive methods to discover what is responsible for the required or undesirable effect. Ensuring the required biological effect is the focus of drug design. Avoidance of undesirable effects is important for both drug design and environment sciences, e.g. for risk analysis of food and drinking water. In both cases the main problem is to find the main cause, e.g. a chemical substructure that is responsible for the positive (desirable) or negative (undesirable) biological effects. Efficient cognitive procedures are required to discover the corresponding cause–effect relationship. The discovery process involves experimental results, facts, theories and knowledge related to the experiences of the specialists involved in the process.
  (ii) Diagnostic problem situations. E.g.:
    (i) Fault diagnosis with *frequent* fault occurrence: consider a system about which we obtain a large number of status data acquired from on-board sensors and maybe from maintenance personnel by voice observations. Faults occur quite frequently, providing plenty of opportunities for learning from experience. A diagnosis is viewed as a fault assignment to the various components of the system that is consistent with (or that explains) the observations. In this case a

large amount of apriori knowledge may be available in the form of technical documents specifying the structure and functions of the system as well as previous case histories.

(ii) Fault diagnosis with *rare* fault occurrence: consider a system and observations about it with unexpected results. E.g. in the case of emergency management in the *nuclear power plant* area, where emergencies rarely occur and the problem situation is not repeated, however the decision is critical. For the solution of this problem, simulation of possible emergency situations and the collection of relevant information from different sources are important.

(iii) Medical diagnosis: consider a patient and a set of symptoms. A diagnosis is viewed as a disease assignment to the patient status that is consistent with (or that explains) the observed symptoms.

(iv) Medical chemistry: consider a chemical compound and the fact that it is biologically active. A "diagnosis" is viewed as a cause assignment to the various components of the compound that is consistent with (or that explains) the observations concerning certain biological activity. Thus the aim is to find those components that are responsible for the observed biological effect.[2]

(iii) Prediction problem situations, where the agent should extract some regularities that can be used to make future decisions accurately. When an agent solves a prediction problem it may bring together all available past and current data, facts and knowledge as a basis for the development of reasonable expectations about the future. Now let us consider some examples:

(i) Prediction of biological effect of a chemical compound on the basis of extracted structure–biological effect relation

(ii) Prediction of expected effects, given the goals of an agent and its model of the environment. During the planning of its activity it predicts the expected effects on the environment which will be obtained by using the world model for the generation of the activity plan. The agent's expectations can be compared with its later observations.

(ii) The formalisation of the subject domains which integrate empirical and theoretical knowledge. For this aim the proposed formalism of open theories may provide an adequate platform. Moreover this platform supports a very important requirement such as the coherent multidisciplinary application of different

[2] When we are interested in the biological effect of a chemical compound then we have to investigate the interaction of the compound (e.g. drug) molecules with biological structures, such as lipoprotein receptors, enzymes, biomembranes, nucleic acids or small molecules. This interaction triggers a series of steps that ultimately results in a macroscopic, physiological change, which constitutes the biological effect. Only by discovering the interaction between the chemical molecule and a macromolecular structure can we understand the biological activity at the cellular level. The biological activity at the level of organs and whole organisms is immensely more complex than at the level of individual cells, and therefore requires the understanding of many more parameters.

disciplines, theories, methods and approaches with the knowledge and methods of the relevant disciplines to gain a comprehensive understanding of the problems of interest and to solve them more successfully and more efficiently. Multidisciplinary application of different disciplines means the synthesis of diverse fields of knowledge and the complementary combination of their methods for the successful solution of problems. We note that, while rigid knowledge boundaries are barriers to multidisciplinary thinking, appropriately described, represented and structured disciplinary knowledge and methods themselves are essential compasses for generating multidisciplinary space. Knowledge of medical science may be a significant example of knowledge which requires essential tools for unravelling the complexity and organising the constituents into an appropriate multidisciplinary space related to the problems to be solved (cf. [Paton and McNamara, 2005].

(iii) Reasoning, and cognitive reasoning methodology in particular, is very important for extracting new information and regularities from experimental data and facts while considering the above-mentioned complex knowledge. The proposed approach may be used for taking serious steps towards the formalisation of the reasoning methodology of such disciplines as medicine, economics, geology, education, sociology etc. Thus the proposed CR framework with its modification theory may play an efficient role in the development of the cognitive conception of medical theories from a philosophical account into a full-fledged logical–mathematical constructive theory of medical reasoning (see e.g. [Patel et al, 2004; Thagard, 2005]).

We hope that the above-mentioned potential applications will give the reader a wider insight into the many challenges connected with the evolution of the information society. We wonder whether our present ways of dealing with challenges are adequate for the future. One of these challenges is handling distributed knowledge, where data and information repositories, the Internet and other digitised information play a crucial role for professionals. Another challenge is how reasoning can be organised in this distributed information space.

References

Alisheda A (1997) Seeking Explanations: Abduction in Logic, Philosophy of Science and Artificial Intelligence. Dissertation Stanford University, Institute for Logic, Language, and Computation (ILLC)

Alisheda A (2004) Logics in scientific discovery. Found Sci (9):339–363

Anshakov OM, Finn VK, Skvortsov DP (1989) On axiomatization of many-valued logics associated with formalization of plausible reasonings. Studia Logica 47(4):423–447

Anshakov OM, Finn VK, Skvortsov DP (1991) On logical constructions of JSM-method of automatic hypotheses generation. Doklady Math 44(2)

Anshakov OM, Finn VK, Skvortsov DP (1993) On deductive simulation of some variants of JSM-method of automatic hypothesis generation. Semiotika i Informatika 33:164–233 (In Russian)

Anshakov OM, Finn VK, Skvortsov DP (1995) On logical means of the JSM-method of automatic hypotheses generation. In: 10th IEEE International Symposium on Intelligent Control, Monterey, CA, pp 86–88

Antoniou G (1997) Nonmonotonic Reasoning. MIT Press

Bacon F (2000) The New Organon. Cambridge Univ. Press, Cambridge, UK

van Benthem J (1996) Exploring Logical Dynamics. CSLI, Stanford

van Benthem J (2008) Logic and reasoning: do the facts matter? Studia Logica 88:67–84

Blinova VG, Dobrynin DA, Finn VK, Kuznetsov SO, Pankratova ES (2001) Predictive toxicology by means of the JSM-method. In: 5th Conference on Data Mining and Knowledge Discovery: Proc. of the Workshop on Predictive Toxicology Challenge, Freiburg, Germany

Blinova VG, Dobrynin DA, Finn VK, Kuznetsov SO, Pankratova ES (2003) Toxicology analysis by means of JSM-method. Bioinformatics 19(10):1201–1207

Bouchon-Meunier B, Dubois D, Godo L, Prade H (1999) Fuzzy sets and possibility theory in approximate and plausible reasoning. In: Bezdek JC, Dubois D, Prade H (eds) Fuzzy Sets in Approximate Reasoning and Information Systems (The Handbooks of Fuzzy Sets), Springer, pp 15–190

Burks AW (1946) Peirce's theory of abduction. Philos Sci 13:301–306

Chater N, Tenenbaum JB, Yuille A (2006) Probabilistic models of cognition: conceptual foundations. Trends Cogn Sci 10(7):287–291

Cheng PW, Holyoak KJ (1985) Pragmatic reasoning schemas. Cogn Psychol 17:391–416

O. Anshakov, T. Gergely, *Cognitive Reasoning*, Cognitive Technologies,
DOI 10.1007/978-3-540-68875-4,

Clancey WJ (1997) Situated cognition on human knowledge and computer representations. Cambridge University Press, Cambridge

Cosmides L (1989) The logic of social exchange: has natural selection shaped how humans reason? studies with the wason selection task. Cognition 31:187–276

Finn VK (1989) On generalized JSM method of automatic hypothesis generation. Semiotika i Informatika 29:93–123 (In Russian)

Finn VK (1991) Plausible inferences and plausible reasoning. J Sov Math 56(1):2201–2248

Finn VK (1999) The synthesis of cognitive procedures and problem of induction. Nauchno-Technicheskaya Informacia, Ser 2 (1–2):8–45 (In Russian)

Finn VK, Mikheyenkova MA (1993a) Logical frameworks for intellectual systems of public opinion poll analysis. In: East–West Conference on Artificial Intelligence, Moscow, pp 319–323

Finn VK, Mikheyenkova MA (1993b) On the application of JSM-method of automatic hypotheses generation in sociological investigations. Artificial Intelligence News, special issue pp 91–98

Flach PA, Kakas AC (eds) (2000) Abductive and Inductive Reasoning: Essays on Their Relation and Integration, Kluwer Academic, Dordrecht

Flach PA, Kakas AC, Magnani L, Ray O (eds) (2006) Workshop on Abduction and Induction in AI and Scientific Modelling, University of Trento, (Available at www.doc.ic.ac.uk/~or/AIAI06-/AIAI06.pdf), Italy

Gabbay DM, Woods J (2005) A Practical Logic of Cognitive Systems: The Reach of Abduction, Insight and Trial, vol 2. Elsevier

Girard JY (1987) Linear logic. Theor Comput Sci 50:1–10

Hajek P, Havranek T (1978) Mechanizing Hypothesis Formation (Mathematical Foundations for General Theory). Springer, Berlin–Heidelberg–NY

Henkin LA (1949) The completeness of the first-order functional calculus. J Symb Logic 14:159–166

Hume DA (1739) Treatise of Human Nature. John Noon

Jaynes ET (2003) Probability Theory: the Logic of Science. Cambridge Univ. Press

Johnson-Laird PN, Byrne RMJ (1991) Deduction. Erlbaum, Hillsdale, NJ

Joinet JB (2001) Proofs, Reasoning and the Metamorphosis of Logic. (available at http://www-philo.univ-paris1.fr/Joinet/Rio2001LastVersion.pdf)

Langley P, Laird JE, Rogers S (2008) Cognitive architectures: Research issues and challenges. Cognitive Systems Research

Markman AB, Gentner D (2001) Thinking. Ann Rev Psychol

Mendelson E (1997) Introduction to Mathematical Logic, 4th edn. Chapman and Hall

Mikheyenkova MA, Danilova EN, Finn VK, Ivashko VG, Yadov BA (1996) Application of JSM-method of automatic hypotheses generation to some problems of sociology. In: Workshop on Applied Semiotics, European Conference on Artificial Intelligence-96, Budapest, pp 22–25

Mill JS (1843) A System of Logic Ratiocinative and Inductive, Being a Connected View of the Principles of Evidence and the Methods of Scientific Investigation, 1st edn. Parker, Son and Bowin, London

Nourani CF (1991) Planning and plausible reasoning in artificial intelligence. In: Mayoh B (ed) Proceedings of the Scandinavian Conference on Artificial Intelligence - '91 (SCAI '91), IOS Press, pp 150–157

Paoli F (2002) Substructural Logics: a primer. Kluwer, Dordrecht

Patel VL, Arocha JF, Zhang J (2004) Thinking and reasoning in medicine. In: Holyoak K (ed) Cambridge Handbook of Thinking and Reasoning, Cambridge University Press, Cambridge, UK

Paton R, McNamara LA (eds) (2005) Multidisciplinary Approaches to Theory in Medicine, Elsevier, Amsterdam

Pearl J (1997) Probabilistic Reasoning in Intelligent Systems: Networks of Plausible Inference, 2nd edn. Morgan Kaufmann, San Francisco

Peirce CS (1903) Lectures on Pragmatism. (Published in Turisi, P. A. (ed.) Pragmatism as a Principle and Method of Right Thinking: The 1903 Harvard "Lectures on Pragmatism", State University of New York Press, Albany, NY, 1997)

Peirce CS (1995) Abduction and induction. In: Buchler J (ed) Philosophical Writings of Peirce, Dover, NY, pp 150–156

Polya G (1954) Mathematics and Plausible Reasoning. Princeton Univ. Press

Popper KR (1967) Epistemology without a knowing subject. In: First International Congress on Logic, Methodology, and the Philosophy of Science

Popper KR (1974) Objective Knowledge. An Evolutionary Approach. Clarendon, Oxford

Restall G (2000) An Introduction to Substructural Logics. Routledge

Rips LJ (1994) The Psychology of Proof: Deductive Reasoning in Human Thinking. MIT Press, Cambridge, MA

Rosser JB, Turquette AR (1951) Many-valued Logics. Studies in Logic and the Foundations of Mathematics, North-Holland, Amsterdam

Sowa JF (2000) Knowledge Representation: Logical, Philosophical, and Computational Foundations. Brooks Cole, Pacific Grove, CA

Suchman LA (1987) Plans and Situated Actions: The Problem of Human–Machine Communication. Cambridge Univ. Press, Cambridge, UK

Sun R (2004) Desiderata for cognitive architectures. Philos Psychol 17(3):341–373

Sun R (2007a) The challenges of building computational cognitive architectures. In: Duch W, Mandziuk J (eds) Challenges in Computational Intelligence, Springer, Berlin

Sun R (2007b) Introduction to computational cognitive modeling. In: Sun R (ed) The Cambridge Handbook of Computational Cognitive Modeling, Cambridge Univ. Press, NY

Thagard P (1993) Computational Philosophy of Science. MIT Press, Cambridge, MA

Thagard P (2005) What is a medical theory? In: Paton R, McNamara L (eds) Multidisciplinary Approaches to Theory in Medicine, Elsevier, pp 48–63

Troestra AS (1992) Lectures on Linear Logic. Center for the Study of Language and Information, Stanford

Vernon D, Metta G, Sandini G (2007) A survey of artificial cognitive systems: implications for the autonomous development of mental capabilities in computational agents. In: IEEE Trans. Evol. Comput. Special Issue on Autonomous Mental Development, 11, pp 151–180

Vinogradov D (2000) Formalizing plausible arguments in predicate logic. Autom Doc Math Linguist 34(6):6–10

Vinogradov D (2001) Correct logical programs for plausible reasoning. Autom Doc Math Linguist 35(3)

Wason PC, Johnson-Laird PN (1972) Psychology of Reasoning Structure and Content. Routledge, London

Zabezhailo MI, Finn VK, Blinova VG, Fabricantova EF, Ivashko VG, Leibov AE, Mel'nicov NI, Pankratova ES (1995) Reasoning models for decision, making: Applications of JSM-method for intelligent control systems. In: 10th IEEE International Symposium on Intelligent Control, Proceedings of the 1995 ISIC Workshop, Monterey, CA, pp 99–108

Glossary

General Terms[1]

Cognition* is the activity that provides extraction of new information and relationships by the analysis of the initial data, generation of hypotheses, hypotheses checking and acceptance or rejection of hypotheses; the possible sources of information to be extracted are experimentation, systematic observation and theorising the experimental and observational results and knowledge.

Cognitive architecture is the structural and functional model of cognizing agents at the conceptual level.

Cognitive process is a discrete process that includes alternating perception and reasoning phases and is able to extract new information and knowledge from data and facts obtained from the environment and/or from the corresponding subject domain.

Cognitive reasoning* is a common form of reasoning that takes place in the case of insufficient and incomplete information in situations with uncertainties and aims to extract new information or knowledge from raw data received from the external world by observation or experimentation.

Cognizing agent is the main actor that can realise cognitive reasoning and has no other motivation than the formation of an adequate model of the environment in order to be able to *adapt* to the changing conditions of the external environment.

Constructivity is a formal representation which is realisable at the computational level.

CR framework is a coherent collection of appropriate methods and tools necessary for representing and realising cognitive reasoning processes and knowledge.

---

[1] The terms of dual character are marked with $*$. These terms stand for both the intellectual processes and for the content of those processes.

Data signals received from the environment and represented in some form appropriate for analysis and processing.

Data analysis is the use of formal and/or mathematical methods of processing data in order to find general characteristics over the data.

Deductive simulation is a simulation of the dynamics of the (non-monotonic) reasoning by traditional logic (deductive) inference of a special kind.

Fact is a simple statement representing data obtained as a result of observations and/or experiments or as a result of a logical inference which uses heuristics.

Fact base is a set of facts that can be received, used, processed and generated by a cognizing agent.

Formal language is a triple that consists of a non-empty set of formulas, a class of models and the validity relation that connect formulas to their interpretation in the models.

Interiorisation is transformation of data and external knowledge into facts and internal knowledge of a cognizing agent.

Knowledge base is a set of formal representations of information and regularities (which are constituents of knowledge) that can be received, used, processed and generated by a cognizing agent.

Reasoner is the mechanism of reasoning in the architecture of a cognizing agent.

Reasoning* discrete, internal rule-based manipulation of the represented facts, information and knowledge.

Plausible reasoning* is a reasoning the rules of which are such that the validity of their premises does not guarantee the validity of their conclusions.

Reliable reasoning* is a reasoning the rules of which are such that the validity of their premises guarantees the validity of their conclusions.

Logical Terms

Arity is a *string* of *sorts* of arguments of a function (either an operation or a predicate).

External truth value is one of two possible values: "true"' or "false".

FPJ logic (finite PJ logic) is a logic that has finitely many internal truth values and has such a set of J-operators that contains the characteristic functions of all one-element subsets of the set of internal truth values and only these.

Internal truth value is a truth value that is different from "true" and "false"; here internal truth values characterise the nature and/or the degree of certainty, e.g. "certainly true", "certainly false", "uncertain"; we use the internal truth values for assigning statements that describe results of observation, experimentation and plausible reasoning.

Iterative logic is an infinite-valued PJ logic such that for any internal truth values of the source FPJ logic (with the exception of one called "uncertain") there exists an infinite class of internal truth values used for representing the dynamic and stepwise nature of cognitive reasoning; iterative versions of FPJ logics are used for deductive simulation of cognitive reasoning.

J-logic is a logic that contains the so-called J-operators that are used for encapsulation of internal knowledge into external knowledge allowing cognizing agents to apply empirical knowledge to theoretical constructions; we consider a J-logic as a logic with two sorts of truth values: external and internal (internal truth values corresponds to the observational and experimental level of knowledge; external truth values corresponds to the theoretical level of knowledge).

J-operator in this book a J-operator is a map from the set of internal truth values to the set of external truth values; this operator can be considered as a tool for encapsulation of internal knowledge into external knowledge.

Knowledge is an arbitrary representation of general regularities and rules of activity concerning one or more subject domains.

Knowledge state is knowledge that is obtained by a cognizing agent after performing several stages of its cognitive activity.

***L*-structure,** where L is a logic, is an analogue of ordinary mathematical structure (algebraic system, model), which can contain external predicates as well as internal predicates taking values from the set of internal truth values of L.

Logic is a formal language together with a derivation system or calculus.

PJ-logic (pure J-logic) is a J-logic that has no logical connectives defined on and taking values in the set of internal truth values.

Predicate is a function that takes values from the set of truth values, which may be either external (that takes values from the set of external truth values) or internal (that takes values from the set of internal truth values).

Sort is a mathematical analogue for data type.

Sort of an individual is a mathematical analogue for the type of a variable or of a constant.

Sort of an operation (a function) is a sort of values returned by this operation (function).

Sort of a predicate is a sort of values returned by this predicate; there exist two alternative sorts of predicates: E (external) and I (internal); the sort E is a direct analogue of the data type Boolean.

State logic is an FPJ logic with four internal truth values: $+1$, -1, 0 and τ, and with four J-operators (characteristic functions of the one-element subsets of the set of internal truth values).

Truth value are values that may be assigned to the statements (formulas) of a formal language by analogy; any of the values that a semantic theory may accord to a statement.

Special Terms

Certainification* is a process of modification of internal predicates, where the returned value "uncertain" is replaced with a definite truth value (an internal truth value that represents the nature or degree of certainty).

Complete modification set (CMS) is a set of modification rules which is devoted to decreasing the uncertainty domain of exactly one internal predicate.

Cut is a representation of a string by a concatenation of two segments: lower (left) and higher (right).

Modification calculus is a system of *modification rules* with an *initial state* of knowledge represented by its formal description.

Modification inference is a formal analogue of cognitive reasoning including application of reliable and plausible reasoning rules.

Modification rule is a formal analogue of a plausible reasoning rule which may cause modifications in those constituents which are already in the inference; the minor premise of a modification rule indicates the actual uncertainty; the major premise describes the conditions sufficient for replacing the above uncertainty with some kind of certainty.

Modification rule system is a *sequence* of complete modification sets (CMS).

Modification theory is a set of non-logical axioms together with a modification calculus.

Open cognitive theory is a history of the activity of the cognizing agent, beginning with an initial knowledge state and ending at a current time moment. So it is a finite sequence of modification theories, each of which represents the current content of the fact base and knowledge base of the cognizing agent.

Generation rules resemble the modification rules, but they can have only one premise.

Generation rule system is similar to a *modification rule system* but it is defined w.r.t. some set of external formulas that express certain additional restrictions to the model of the subject domain.

JSM is the abbreviation of John Stuart Mill, whose method of agreement underlies the JSM method of hypothesis generation.

JSM method is a method of automatic generation of hypotheses about relations between the structure of some objects of the subject domain and their properties significant for practical use; we distinguish its three variants: simple (where the causes of the presence of properties and causes of the absence of properties are considered independently), generalised (where the causes of the presence of properties as well as causes of the absence of properties can have their own inhibitors) and non-symmetric (where the causes of the presence of properties can have their own inhibitors and the causes of the absence of properties are not considered).

JSM theory is a modification theory that can be either simple if it represents the simple JSM method, or generalised if it represents the generalised JSM method, or non-symmetric if it represents the non-symmetric JSM method.

String is a finite sequence of objects of arbitrary nature.

Structure generator is a semantic construction which contains an initial L-structure (where L is a state logic) and a generating rule system.

Set sort is such a sort, the objects of which can be interpreted as sets; on the domain of such a sort a binary operation, which plays the role of union operation, a binary relation that stands for the inclusion relation and a constant that stands for the empty set are defined.

State description is a set of axioms that describes a *knowledge state*.

Index

Symbols

N

O

P

R

S